MODELING AND CONTROL OF ENGINES AND DRIVELINES

Automotive Series

Series Editor: Thomas Kurfess

MODELING AND CONTROL OF ENGINES AND DRIVELINES

Lars Eriksson and Lars Nielsen

Linköping University, Sweden

This edition first published 2014
© 2014 John Wiley and Sons Ltd

Registered office
John Wiley & Sons Ltd, The Atrium, Southern Gate, Chichester, West Sussex, PO19 8SQ, United Kingdom

For details of our global editorial offices, for customer services and for information about how to apply for permission to reuse the copyright material in this book please see our website at www.wiley.com.

Library of Congress Cataloging-in-Publication Data

Eriksson, Lars, 1970–
    Modeling and control of engines and drivelines / Lars Eriksson and Lars Nielsen.
        1 online resource.
    Includes bibliographical references and index.
    Description based on print version record and CIP data provided by publisher; resource not viewed.   ISBN 978-1-118-53619-3 (ePub) – ISBN 978-1-118-53620-9 (Adobe PDF) – ISBN 978-1-118-47999-5 (cloth) 1. Motor vehicles–Power trains – Simulation methods. 2. Automobiles – Motors – Simulation methods. 3. Motor vehicles–Power trains – Control systems – Design and construction. 4. Automobiles – Motors – Control systems – Design and construction. 5. Automobiles – Electronic equipment. I. Nielsen, Lars, 1955–II. Title.
    TL260
    629.25001′1 – dc23
                    2013035431

A catalogue record for this book is available from the British Library.

ISBN: 978-1-118-47999-5

Typeset in 10/12pt Times by Laserwords Private Limited, Chennai, India

1   2014

To Bodil, Ingrid, and our families.

Contents

Part V DIAGNOSIS AND DEPENDABILITY

Preface

This book provides a complete and up-to-date treatment of modeling and control of engines and drivelines. Models for engine and driveline components have been thoroughly studied, and there are appropriate and validated models that can be used as building blocks in simulation or for design of control and diagnosis systems. Where other books have a perspective of mechanics and fluid dynamics, this book instead has a clear perspective of systems engineering and control systems development. This is a perspective that is currently at the core of overall design of vehicle properties, and here our close collaboration with the automotive industry has given a good picture of the knowledge and skills that practicing engineers need when developing and analyzing control systems for powertrains.

We have three main goals with this book. The first is to provide a thorough understanding of component models, both for teaching and for long-term reference for engineers. Thus, it has been important for us to provide measurements from real processes early in the presentation and treatment of different systems, and then explain the underlying physics, describe the modeling considerations, and validate the resulting models using experimental data. All in all it shows how models are approximations of reality and tailored for engineering. The models are timeless; but as a second important goal for the book we show how they are used in current, and important, control and diagnosis systems design. Examples and case studies are thus used to illustrate control system designs for achieving the desired performance, as well as trade-offs between conflicting goals in these complex systems. The components or system designs are of course never used in isolation, so the third important goal is to provide a complete setting for systems integration and evaluation. This means that the book contains descriptions of complete vehicle models in longitudinal motion together with actual requirements for emission and fuel consumption analysis in driving cycles and simulation.

As mentioned above, our intended audience is both students, learning the subject, and practicing engineers benefiting from reference literature. The material has been developed for both Electrical and Mechanical Engineering students in a course at masters level at Linköping University since 1998. It has also been used for national and international courses, as well as tailored courses for industry. It has, for example, been used in a course in the national Swedish Green Car program. Internationally, examples are at the Powertrain Engineering Program at IFP School in Paris, France, at UPV Valencia in Spain, and at Tianjin University in China. Besides these audiences, there is also an intention to provide a reference for engineers who work within the automotive industry and need to develop and integrate components. Validated models here are an important means of communication between engineers both within an organization and between component suppliers, system manufacturers, and car manufacturers.

The text is written for masters level students or early graduate students. Prerequisites are general engineering courses, like mathematics, mechanics, physics, and a basic course in automatic control or signals and systems. It is helpful, but not necessary, to have a background in thermodynamics. For those interested in using the book as teaching or study material, Section 1.3, Organization of the Book, gives an overview of the subjects. In teaching it is natural to integrate experimental work with computer exercises to follow the chain from data collection, through modeling, to control design and verification. This can be complemented with problem solving sessions, and for the teacher, the active student, or those who want to practice, there is more material available on the homepages,

```
wiley.com/go/powertrain and
www.fs.isy.liu.se/Software
```

where, for example, the complete engine model in Figure 8.27 (LiU-Diesel) can be downloaded. We have prepared the examples and illustrations in the book using mainly Matlab/Simulink, since it is dominant in the automotive industry. However, the focus in the book is on tool-independent properties, like measurement data and equations, which enable a reader to implement the models in any suitable software or modeling environment.

Acknowledgments

Our interest and enthusiasm for the field of automotive modeling and control has led to this book, but it would not have become what it is without the contribution of many others. The material has its foundation in the research on engine and driveline control at the Vehicular Systems group and it has, to a large extent, been performed in close collaboration with the automotive industry. It all started with our own engine lab in 1994 and the first course in 1998. The material then evolved in symbiosis with our many collaborations inside and outside the university, so there is a large number of persons that have contributed to the final result and this list is too long to provide here.

In our group at the university there has been a collaborative effort to provide courses of high relevance and quality for our students, and many of our PhD students have contributed to discussions concerning the subject area, and how it can be approached while learning. Hence, this book is also a result of the joint research and discussions with our PhD students, and all our previous and current PhD students are greatly acknowledged for all their contributions.

Finally, we want to thank those that have contributed to the proofreading of the final version of the manuscript: Daniel Eriksson, Erik Frisk, Erik Höckerdal (Scania), Mattias Krysander, Anders Larsson (Scania), Patrick Letenturier (Infineon), Oskar Leufvén, Tobias Lindell, Andreas Myklebust, Vaheed Nezhadali, Peter Nyberg, Andreas Thomasson, Frank Willems (TU/e and TNO Automotive), and Per Öberg.

Linköping, summer 2013
Lars Eriksson
Lars Nielsen

Series Preface

The heart of any automobile is the engine that converts stored energy into mechanical power. Taking that power and turning it into motion is the job of the driveline. The combination of the engine and the driveline are major defining elements of a vehicle. Almost certainly, when a consumer is planning on purchasing a high performance vehicle, engine and driveline specifications are the primary consideration. Historically, engine and driveline performance have significantly increased due primarily to technological innovations. Furthermore, the demands for higher performing vehicles that are fuel efficient and generate reduced amounts of emissions are being driven not only by the consumer market but by a wide spectrum of regulations worldwide. Thus, the need to fully understand the engine and driveline and their wide variety of configurations, such as spark ignition, diesel, electric hybrid and turbocharging, are critical for any professional in the automotive sector. This applies not only to automotive OEMs (Original Equipment Manufacturers) but also to the vast network of supplier companies that build and test every component that is integrated into the vehicle system.

Based on the rapid acceleration of engine and driveline technology, *Modeling and Control of Engines and Drivelines* presents a well-balanced discussion of the engine and powertrain including propulsion and engine fundamentals, modeling and control for both the engine and the driveline, and finally diagnostics and performance of propulsion systems. The text is designed as part of an advanced engineering course in engine and driveline systems and is part of the *Automotive Series* whose primary goal is to publish practical and topical books for researchers and practitioners in industry, and postgraduate/advanced undergraduates in automotive engineering. The series addresses new and emerging technologies in automotive engineering, supporting the development of more fuel efficient, safer, and more environmentally friendly vehicles. It covers a wide range of topics, including design, manufacture, and operation, and the intention is to provide a source of relevant information that will be of interest and benefit to people working in the field of automotive engineering.

Modeling and Control of Engines and Drivelines provides a thorough technical foundation for engine and driveline design, analysis, and control. It also incorporates a number of pragmatic concepts that are of significant use to the practicing engineer, resulting in a text that is an excellent blend of fundamental concepts and practical applications. The strength of this text is that it links a number of fundamental concepts to very pragmatic examples providing the reader with significant insights into engine and driveline design and operations. Not only do the authors provide both technical depth and breadth in this book, they also provide insight into some of the regulations that are driving the state-of-the-art in engine systems (e.g., emission standards), making the book a well-rounded reference for professionals in the field. It is a clear and concise book, written by recognized experts in a field that is critical to the automotive sector providing both fundamental and pragmatic information to the reader, and is a welcome addition to the Automotive series.

Thomas Kurfess
December 2013

Part One

Vehicle – Propulsion Fundamentals

1

Introduction

Customer needs and requirements from society have, together with a fierce competition among automotive manufacturers, had a tremendous effect on the development of our vehicles. They have evolved from being essentially mechanical systems in the early 1900s to the highly engineered and computerized machines that they are today. An important step has been the introduction of computer controlled systems that accelerate the development of clean, efficient, and reliable vehicles. Two trends are especially interesting for the scope of this book:

- Increased computational capabilities in vehicle control systems.
- New mechanical designs giving more flexible and controllable vehicle components.

These development trends are intertwined, as the development of new mechanical systems relies on the availability of more advanced controllers that can handle and optimally use these new systems. As a consequence, the design of vehicles is really evolving into co-design of mechanics and control. The tasks for such improved designs are numerous, but the main goals to strive for are:

- High efficiency, leading to lower fuel consumption.
- Low emissions, giving reduced environmental impact.
- Good driveability, providing predictable response to driver commands.
- Optimal dependability, giving predictability, reliability, and availability.

The goal of this book is to give insight into such new developments, and to do it in enough depth to show the interplay between the basic physics of the powertrain systems and the possibilities for control design. Having set the goals above, it is impossible to cover the field in breadth too. The text has to be a selection of important representatives. For example, two-stroke engines are not covered, since the usual four-stroke engine illustrates the general principles and by itself requires quite some pages to be described sufficiently.

Control systems have come to play an important role in the performance of modern vehicles in meeting goals on low emissions and low fuel consumption. To achieve these goals, modeling, simulation, and analysis have become standard tools for the development of control systems in the automotive industry. The aim is therefore to introduce engineers to the basics of internal combustion engines and drivelines in such a way that they will be able to understand today's control systems, and with the models and tools provided be able to contribute to the

Modeling and Control of Engines and Drivelines, First Edition. Lars Eriksson and Lars Nielsen.
© 2014 John Wiley & Sons, Ltd. Published 2014 by John Wiley & Sons, Ltd.
Companion Website: www.wiley.com/go/powertrain

development of future powertrain control systems. This book provides an introduction to the subject of modeling, analysis, and control of engines and drivelines. Another goal is to provide a set of standard models and thereby serve as a reference material for engineers in the field.

1.1 Trends

Modern society is to a large extent built on transportation of both people and goods and it is amazing how well the infrastructure functions. Large amounts of food and other goods are made available, waste is transported away, and masses of people commute to and from work both by private and public transportation. Transportation is thus fundamental to society as we know it, but there is increasing concern about its effects on resources and the environment. This is also stressed when considering the increasing demands in developing countries. To meet these demands there are many efforts toward making vehicles function as efficiently and cleanly as possible, and some of the major trends are

- downsizing
- hybridization
- driver support systems
- new infrastructure.

These will be briefly introduced below, after a section on the societal drive for care of our resources and environment.

1.1.1 Energy and Environment

Different standards and regulations have been the most concrete results that have come from concern for the environment. A perfect combustion of hydrocarbon fuels will result in CO_2 and water, whereas a non-perfect combustion results in additional unwanted pollutants. This means that the amount of CO_2 is a direct measure of the amount of fuel consumed, and a standard formulated in terms of CO_2 thus aims at restricting the use of fossil fuels. Worldwide standards are illustrated in Figure 1.1, illustrating that society is pushing the development of more fuel efficient vehicles. Standards and measures of control differ between regions, the USA, for example, uses a Corporate Average Fuel Consumption (CAFE) for manufacturers, while cars in Europe have a CO_2 declaration that is used for taxation of vehicles.

Another type of regulation is used to limit the emissions of important harmful pollutants. Examples are emissions of particulate matter (also called soot) and the gases carbon monoxide (CO), nitrogen oxides (NO and NO_2, collectively called NO_x), and hydrocarbons (HC). Legislators have made the levels that vehicles are allowed to emit increasingly stringent and Figure 1.2 shows the evolution for passenger cars in the USA.

Regulations like these in Figures 1.1 and 1.2 have been, and continue to be, drivers for better vehicles and have a decisive impact on technological development within the automotive area.

1.1.2 Downsizing

There are many ongoing developments to meet legislative requirements like those above, and one major trend in the search for solutions is downsizing. Downsizing has two meanings,

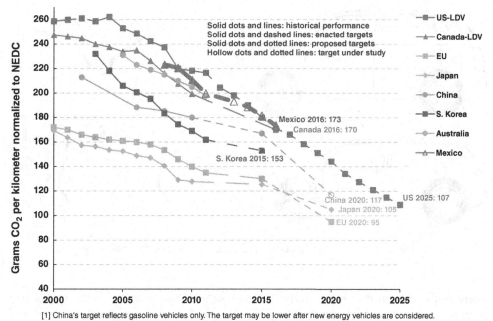

Figure 1.1 Global CO_2 emissions, historical data, and future standards. Reproduced with permission from The International Council On Clean Transportation

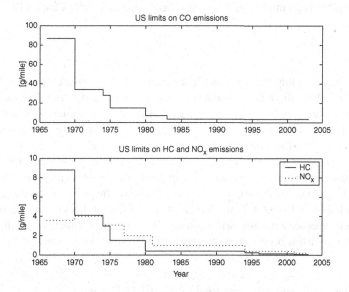

Figure 1.2 The evolution of federal emission regulations for carbon monoxide (CO), nitrogen oxides (NO_x), and hydrocarbons (HC) of passenger cars in the USA. At model year 2004 the Tier 2 standards started, specifying 10 bins where the manufacturers can place their vehicles, provided that they fulfill fleet average regulations. No data is plotted after 2004 due to the diversity of limits in the bins

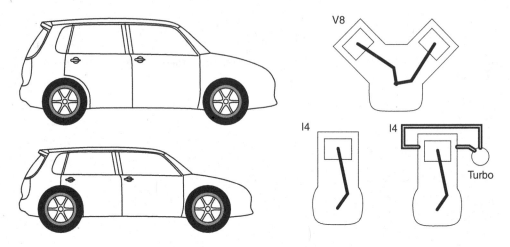

Figure 1.3   Downsizing of cars and engines to increase fuel efficiency

where one is that smaller and more lightweight cars need less fuel. The trends in this area cover new materials and new construction principles as well as customer acceptance of smaller cars. Another interpretation concerns the engine, where downsizing refers to having a smaller engine in the car that consumes less fuel. Downsizing is often used with turbocharging, where the smaller engine gets a boosted performance to come closer to that of a larger engine and improve customer acceptance. Both these ideas are depicted in Figure 1.3. The principle of downsizing engines is an important part of this book, see especially Chapter 8.

1.1.3 Hybridization

Downsizing is one path that leads to less fuel consumption and fewer emissions. Another path is hybridization, where there is an additional energy storage and retrieval in the car. Several ideas exist for storing and retrieving energy, and some candidates are to store energy as rotational energy in a fly-wheel, as pressure in an air tank, or as pressure in a hydraulic system. However, for now, electrification is the main line of development, where energy is stored electrically in a battery or in a super capacitor, and transformed to motion via electrical motors. Compared to traditional vehicles, hybrid vehicles are more complex since there are more components that should operate in harmony to achieve most of the promise of hybridization. This will be expanded on in Chapter 3, and a main theme is that, at the core of the solutions, the torques and velocities are the main variables to model and control; which means that the models and methodologies in this book can be directly applied to simulate and analyze hybrid systems.

1.1.4 Driver Support Systems and Optimal Driving

Fuel consumption and the amount of emissions are highly dependent on how a vehicle is driven. The fuel savings when driving optimally can be substantial compared to energy-unaware driving, so therefore there is a strong interest in systems that help the driver, or even replace the driver, when it comes to propulsion.

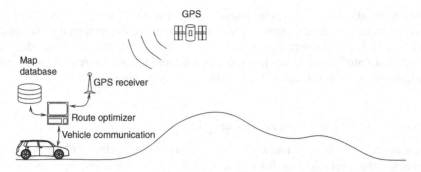

Figure 1.4   Depiction of a system for optimal driving regarding the upcoming topography of the road

A driver support system proposes speed and gear selections to the driver, and can also evaluate and educate a driver. There are also systems that can plan a fuel-optimal driving based on the topography of the road, that is using knowledge of the upcoming slopes of the road, as illustrated in Figure 1.4. The basis for such a system is positioning the vehicle using GPS, a map database used to read the upcoming road slopes, and on-board optimization algorithms that take control over propulsion. A number of names are given to these systems, such as Optimal driving, Look-ahead control, and Active prediction cruise control. The latter name reflects the fact that it is a natural extension of a conventional cruise control system.

New Infrastructure

Optimal driving as regards topography was made possible by the technological development of GPS and map databases. It would, of course, also be highly beneficial if driving could be optimal relative to all other circumstances like, for example, the traffic situation, other vehicles, and weather. To approach these potential benefits there is active development of vehicle-to-vehicle communication, road-side information systems, traffic systems, and on-line teleservices, such as weather and traffic reports. Such a situation is depicted in Figure 1.5. Some acronyms used are V2V for vehicle-to-vehicle, V2R for vehicle to road-side, and V2X as a generalization to any connection.

In addition to the infrastructure, the vehicle has its own sensors. These are both internal, regarding powertrain and vehicle motion dynamics, and external, like radars and cameras.

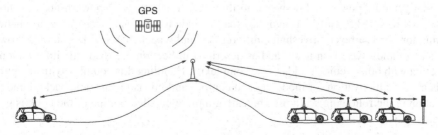

Figure 1.5   Illustration of a situation where each vehicle is provided with information from other vehicles, from road-side systems, and from teleservices such as GPS and weather information

In the future, the aim is to have superb situation awareness and planning potential within each individual vehicle, and the engineering task will be to utilize this potential in the best possible way. There is another benefit, with a system as sketched in Figure 1.5, besides making driving optimal. Information from other vehicles and from infrastructure providing road-side information, on-line weather, and traffic information, can also improve safety.

Integrated Propulsion and Powertrain Control

The situation in Figure 1.5 will make new functionality possible. One example, not far away, is platooning, where vehicles can drive close to each other to reduce the losses from air drag, see Section 2.2.3, and other more autonomous functionality will follow.

Eventually, all the aspects above will be part of truly integrated powertrain control based on the actual state within the powertrain, that contributes to a system that at every time instant can behave optimally.

1.1.5   Engineering Challenges

To sum up, transportation is crucial to society, but limited resources and environmental concerns have led to the need to find transportation solutions of the future. Luckily, new technological possibilities and developments have given many new possibilities, so there are now many trends constituting a vast plethora of challenging and interesting engineering tasks.

The full picture requires more than one book to cover, but one perspective is that all aspects come together in the question of optimal propulsion. The main scope of this book is to give the understanding and engineering tools for the powertrain that transforms energy to motion. The goal is to do this such that the systems of today are treated, but also so that a foundation is laid for approaching the engineering challenges of many years to come. To do this, a certain level of depth is needed, and our hope is that the reader will share a feeling of excitement about the challenges and the fun involved in exploring and developing future solutions.

1.2   Vehicle Propulsion

As seen in Section 1.1, there are many developments in transportation solutions for the future, and to be able to cope with these challenges the main focus in this book is the fundamental issue of efficiently transforming energy to motion without unwanted side effects such as pollution. This transformation is performed by the powertrain, which is the group of components that generate power and deliver it to the road. Illustrations of powertrains are shown in Figures 1.6, 3.1, 3.5, and 13.1, and they may include the engine, electrical motor, battery, transmission (or gearbox), driveshafts, differential, and wheels. As will be seen, the powertrain is a complex system in itself, and as described in Section 1.1 road-side information or interaction with other vehicles adds to the complexity. Handling this complexity in an optimal way is a strong motivation for modeling and control, and this is given some background and motivation in the following sections. Thereafter, a more detailed outline of the book is given in Section 1.3.

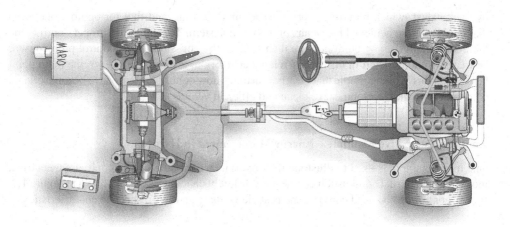

Figure 1.6 A sketch of a BMW 520D, touring, automatic, -08, that includes the driveline. This powertrain includes the engine, transmission (gearbox), propeller shaft, differential with final-drive, drive shafts, and wheels. Other components are: fuel tank, exhaust system, steering wheel, and suspension systems. Reprinted with permission from Mario Salutskij

1.2.1 Control Enabling Optimal Operation of Powertrains

The powertrain, with its components and with its external interactions, has to be coordinated into a single operational unit fulfilling a complex set of requirements. Hence, the need for control is natural, and potential and advantage is found in at least the following areas

- fulfilling legal requirements
- achieving performance
- handling complexity
- enabling new technology.

From the discussion above it should be clear that control is a strong enabler for the first three items. Regarding the fourth item, it is interesting to ask ourselves why so many advanced concepts, like supercharging, turbo, variable valve actuation, variable compression engines, and gasoline direct injection, are surfacing as commercial products. In fact, none of these concepts are new, even if they are sometimes presented so, but the novelty is instead that they can now with proper **control** achieve competitive functionality and performance. A well-known example is now used to illustrate this point.

An Illustrative Example – The Three-Way Catalyst

One important historic milestone was the introduction of the three-way catalyst that constituted a breakthrough in the reduction of emissions from a gasoline engine. The key step for successful application was the introduction and integration of a control system that continuously monitors the air–fuel mixture and modifies the fuel injection. This was necessitated by the catalyst, which requires a very precise mixture of air and fuel for optimal operation that could only be achieved by means of a control system. Together with proper controls the three-way catalyst now removes more than 98% of the emissions. This control problem will be treated in more detail in the engine-related chapters in the book. However, the main point here is that this is one example that clearly illustrates how control systems have become crucial components in the development of clean and efficient vehicles.

Another Illustrative Example – Energy Management in Hybrids

One more example is used to illustrate the importance of control. The torque of an electrical motor and an internal combustion engine have different characteristics, as shown in Figure 1.7. Proper control can be used combine the best elements from their respective characteristics.

High Ambitions Need Models

The ambitions for powertrain control are already high, and the demand for care in energy utilization and environment preservation will continue to develop toward optimal powertrain control. These societal drives are strong, and lead to striving to find really good designs from a performance perspective. To be able to handle these increasingly better and more complex systems, strong physical knowledge will be required, but it will also be necessary that this physical knowledge is provided in an efficient form for analysis and design. For this purpose, models are needed.

1.2.2 Importance of Powertrain Modeling and Models

This book covers modeling, control, and diagnosis of powertrains, with its main focus on models and model-based methods. In particular, much attention is given to modeling and models, and this choice has been made for two more reasons than its obvious use in model-based control.

Virtual Sensors

A first additional motive is seen by looking at the powertrain as the group of components that generate power and deliver it to the road, and the torque is thus fundamental to control.

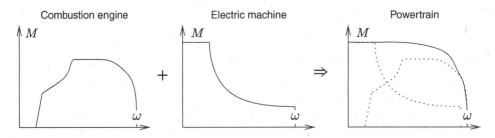

Figure 1.7 Illustration of control as an enabler for new functionality. Here, the example is about finding the best combination of an electrical motor and an internal combustion engine in a hybrid vehicle

One notes that the powertrain torque is not measured in current production systems, even though it is such a central variable. Thus, to be able to control this system, it is necessary for the system to have models that calculate (or estimate) the torque at various positions in the powertrain and especially the torque production from the engine. This generalizes to an important issue in mass produced vehicles: sensors cost money and cutting the cost of both the total system and of each component is of utmost importance. An additional sensor is not mounted unless it delivers a necessary input to the control system and, at the same time, is really worth its price. Models are therefore utilized to a high degree, instead of sensors, for determining interesting quantities in the system.

Systematic Build-Up of Knowledge

Secondly, models provide a foundation that can also be utilized in the development of future systems, one can say that they in a sense form a scientific basis for the control system design. Controllers and control architectures will change in the future, since these depend on the technical development of, for example, sensors and actuators. As an example, a particular control problem and its design to a large extent depend on what sensors and actuators are utilized, and if new better options become available and competitive the controller structure and control design can also be fundamentally changed. However, the physics of the energy conversion system does not change substantially, for example they follow Newton's and thermodynamic laws. Therefore, models that describe these system will also in the future provide a basis for analysis of system properties and future control designs.

1.2.3 Sustainability of Model Knowledge

Major constituents of modeling have developed since the introduction of the microprocessor in the 1970s and 1980s, but have developed with increased pace over the last 20 years. Many of the models presented in this book have received thorough experimental verification and have proved their usefulness in many existing designs. Therefore, it is our belief that these models, perhaps in new combinations but still comprising the same model components, will be the foundation for analysis and design for many years to come.

With these notes, about seeing modeling as the foundation for future development, it must also be mentioned that it is still important to analyze and understand current systems and controllers. This is because they give insight into current system designs and constitute design examples of how powertrain demands are formulated as control problems and how these are solved. Another aspect that this visualizes is the interesting interplay between thermodynamics, mechanics, and control that is seen in modern cars, and this is an interesting and dynamic area.

1.3 Organization of the Book

The core topics in this book are the modeling and control of powertrains, their components, and the interplay between these components. Models are provided for each system and for the integration between systems that are needed for successfully engineering a complete vehicle powertrain. In addition, it is also highlighted that systems should be designed such that they can be maintained and diagnosed over the vehicle lifetime, which is also an important engineering task in the development of control systems.

The text is organized into five parts: vehicles and powertrains, engine fundamentals, engine modeling and control, drivelines, and diagnosis. In the presentation of these subjects, measurements on real processes are used early in the treatment of different systems, and it is then shown how models are used as approximations of reality. For example: the process in the cylinder of a real gasoline engine (Otto engine) does not follow the ideal Otto cycle exactly, but the Otto cycle gives valuable insight into the engine's characteristics and properties. The main contents in each part will now be outlined in the following paragraphs.

Vehicle – Propulsion Fundamentals

The first part of the book gives an overview of vehicles and powertrains to set the framework for the rest of the chapters. The performance of a vehicle, regarding the motions coming from accelerating, braking, or ride, is mainly a response to the forces imposed on the vehicle from the tire–road contact. Chapter 2, Vehicle, gives sufficient background in these matters by providing models, so an engineer can study engines, motors, and drivelines in an complete vehicle setting. In Section 1.1 it was clear that there are many expectations of well-behaved vehicles, and in Chapter 2 this is further quantified by presenting legislative requirements and measures for consumer demands. Whereas Chapter 2 looks at the vehicle from outside, the following chapter, Chapter 3, Powertrain, continues the treatment by going inside the car to give a first overview of possible solutions. Already here there is a preliminary discussion on control structures for powertrain control.

Engine – Fundamentals

This second part summarizes important properties and basic operating principles of engines with respect to overall performance, limitations, and emissions. Chapter 4, Engine – Introduction, introduces basic engine geometries and quantities that are used to characterize the engine operating conditions and performance. Many of these appear as components or parameters in the models that are developed in later chapters.

Chapter 5, Thermodynamics and Working Cycles, covers the basics of the work production in a four-stroke engine operation and develops thermodynamic models for the process based on a thermodynamic foundation. The first sections are devoted to simplified thermodynamic processes, developing equations that both give insight into operating characteristics and can be used in models. Finally, Section 5.4 develops more detailed models that are often used for analyzing the effects of different design or control actions and optimizing set points for the controls.

Chapter 6, Combustion and Emissions, treats the combustion processes in spark ignited (gasoline) engines and compression ignited (diesel) engines as well as their characteristics. Further, the engine-out emissions and their treatment is summarized, giving a background for understanding the control goals for the engine with respect to emission.

Engines – Modeling and Control

Chapters 5 and 6 in the preceding part deal with work and emission production in the cylinder, and thus involve quantities that vary under one cycle, and the resolution of interest is in the region of one crank angle degree. The chapters in Part 3 on modeling and control treat the engine block, with the cylinders, as a system and develop component and system models that have longer time constants.

Chapter 7, Mean Value Engine Modeling, has as its theme *mean value engine modeling* and develops models for different components that are found in an engine. The timescales of these models are in the order of one to several engine cycles, and the variables that are considered are averaged over one or several cycles (i.e., the quantities are mean values over a cycle, giving the name mean value engine models). These models describe the processes and signals that have a direct influence on the control design. Another strong trend in engine development, namely downsizing and supercharging of engines, is treated in Chapter 8, Turbocharging Basics and Models, which gives a fundamental treatment of turbocharging and other variants of supercharging. The chapter leads to models for turbochargers and collects two complete turbocharged engine models, one gasoline and one diesel.

Generic components and tasks that are found in engine management systems are summarized in Chapter 9, Engine Management Systems – An Introduction. Control loops in spark ignited (SI) engines are treated in Chapter 10, Basic Control of SI Engines, covering both high level controllers, such as torque, air and fuel, and ignition control, and low level servo controllers such as throttle, waste gate, fuel injector, and so on.

Compression ignited (CI) engines are covered in Chapter 11, Basic Control of Diesel Engines, covering both high level controllers such as torque and gas flow control, and low level control, such as injection. Finally, Chapter 12, Engine – Some Advanced Concepts, describes some advanced engine concepts, such as variable valve actuation, variable compression ratio engines, and advanced feedback control. A theme of the topics in advanced concepts is that they rely on control systems in order to reach full utilization of their performance potential.

Driveline – Modeling and Control

From the prime movers (combustion engine or electrical motor) the driveline (clutch, transmission, shafts, and wheels) transmits the power for propulsion and is thus a fundamental part of a vehicle. Since the driveline parts are elastic, mechanical resonances may occur. The handling of such resonances is basic for functionality and driveability, but is also important for reducing mechanical stress and noise. Chapter 13, Driveline Introduction, introduces the nomenclature and defines the area of driveline control as a certain subarea of powertrain control. As a background to the coming chapters, it explains the physical background of unwanted vehicle behavior that results from inadequate driveline control. It clarifies the control tasks at hand, and gives a brief discussion on sensors and actuators. Chapter 14, Driveline Modeling, models the driveline and its components, providing descriptions of both how the engine is coupled to the wheels and how oscillations are caused by the elasticities found in, for example, the driveshafts. When describing the forces and torques on the wheels there is a connection back to Chapter 2, Vehicle, for descriptions of driving resistance. A systematic modeling methodology is used, and a set of driveline system models are developed with the purpose of giving a range of models that are suitable for analyzing different control problems.

Driveline control is treated in Chapter 15, Driveline Control, where, besides a general discussion on control formulations, the two main problem areas of speed control and torque control are given specific attention. Relating back to torque-based powertrain control in Chapter 3, Powertrain, both of these are examples where driveline control intervenes in the torque propagation structure with short-term demands. The two applications chosen to illustrate speed control and torque control respectively are *anti-surge control* and *driveline torque control for gear shifting*. The first application is important for handling wheel-speed oscillations, following from a change in accelerator pedal position or from impulses from towed trailers. The second application is used to implement automated gear shifting.

Diagnosis and Dependability

The availability of computing power in vehicles has also strongly influenced another field, namely diagnosis and dependability. Originally, the main driving force came from legislation requiring diagnostic supervision of any component or function that when malfunctioning would increase tail-pipe emissions by at least 50 %, the well-known On Board Diagnosis (OBD) requirements by the California Air Resource Board (CARB). Basically, there are observed variables or behaviors for which there is knowledge of what is expected or normal. The task of diagnosis is, from the observations and knowledge, to generate a fault decision, that is to decide whether there is a fault or not and also to identify the fault. Once a methodology to find faults or malfunctions has been developed then many new application areas open up. Chapter 16, Diagosis and Dependability, briefly introduces basic diagnostic techniques, and their wider use today is presented where the same techniques are used for safety, machine protection, availability, up-time, dependability, functional safety, health monitoring, and maintenance on demand. The consumer value is, for example, increased profit through dependability, or lower costs through maintenance on demand. Explicit examples of model based diagnosis are given where it is shown how the models that are developed in the book can also be used for diagnosis and dependability. These examples include important automotive examples. Finally an overview of OBDII is given.

2

Vehicle

The vehicle as a whole, and the situation it is used in, has to be considered when approaching vehicle propulsion. So, when aiming for insight and explicit tools for modeling and control of drivelines and engines, one has to take into account the external forces on the vehicle constituting the driving resistance, together with driver behavior and road characteristics. Further, many requirements are formulated for the complete vehicle, such as

- Fuel consumption, CO_2 and other emissions.
- Performance measures, for example acceleration.
- Driving feel.
- Diagnostics.

It is a complex set of requirements a car has to meet, and good design is very much about finding a good balance between them. One reason for the complexity is the origin of requirements, that may be

- Customer economy, including purchase, operation, and maintenance.
- Legal, such as for emissions.
- From society, for example the desire to reduce environmental impact.
- Good driveability, giving a predictable response to driver commands.

These requirements are typically formulated for the complete vehicle using external measurements on fuel consumption, tail-pipe emissions, acceleration, and so on.

The topic of this chapter is vehicle propulsion, different performance characteristics, and some details concerning their measurement. Sections 2.1 to 2.3 treat the basic equations for vehicle propulsion dynamics and lead to a set of useful models. Section 2.4 complements with driver and road models, so that a complete vehicle simulation is possible, as in Section 2.5. The rest of the chapter deals with characteristics and requirements. Section 2.6 treats performance measures, Section 2.7 fuel consumption, and Section 2.8 emissions. In the latter, the concept of a driving cycle is introduced.

2.1 Vehicle Propulsion Dynamics

Overall vehicle propulsion dynamics is the force balance, applying Newton's second law on a vehicle without looking inside it, that is without studying how the driving force is generated.

Modeling and Control of Engines and Drivelines, First Edition. Lars Eriksson and Lars Nielsen.
© 2014 John Wiley & Sons, Ltd. Published 2014 by John Wiley & Sons, Ltd.
Companion Website: www.wiley.com/go/powertrain

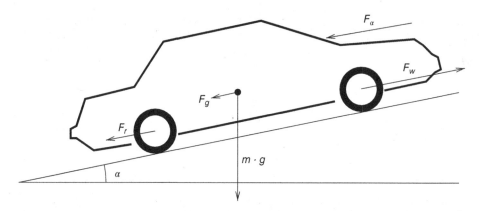

Figure 2.1 Illustration of the vehicle forces that act on the car body and that are relevant for longitudinal propulsion

The actual generation, and the characteristics of, this propulsion force, F_w, is the main topic in the rest of the book from the next chapter onward. Besides the propulsion force, the other two main forces on a vehicle are the driving resistance and the braking force. The driving resistance F_{DR} represents the sum of all external forces on the vehicle, and the braking force F_b represents all internal braking in the vehicle that neither stems from the engine nor the driveline, so F_b includes the usual brake system, but not forces due to negative torque from the engine while the engine is braking, nor losses in the powertrain. Given this definition, F_b will often be omitted when studying propulsion and powertrain behavior. Figure 2.1 shows the forces acting on the body of a vehicle with mass m and speed v. In the figure, the driving resistance F_{DR} is composed by the three most common components, air drag F_a, rolling resistance F_r, and gravitational force F_g.

Newton's second law in the longitudinal direction gives

$$m\,\dot{v} = F_w - F_{DR} - F_b \tag{2.1}$$

where the propulsion force F_w is acting between wheel and road. Decomposing the driving resistance F_{DR} into aerodynamic drag, rolling resistance, and gravitation, gives the typical equation for vehicle propulsion that is obtained from (2.1) by omitting the braking force and putting in the terms constituting driving resistance

$$m\,\dot{v} = F_w - F_a - F_r - F_g \tag{2.2}$$

2.2 Driving Resistance

The driving resistance, F_{DR}, includes many terms, the most important being aerodynamic drag, rolling resistance, and gravitation. The relative sizes of these main contributions depend on many factors, where, for example, the aerodynamic drag is strongly influenced by speed and gravitational force by vehicle weight. With this variability in mind, it is still worth looking at an example. Figure 2.2 illustrates where the energy produced by the engine goes for a 40-ton truck on a typical road. Such a figure would look different for a different vehicle with different driving, but the main conclusion would still typically be that losses due to aerodynamic drag, rolling resistance, and gravitation are all substantial, and larger than the total loss

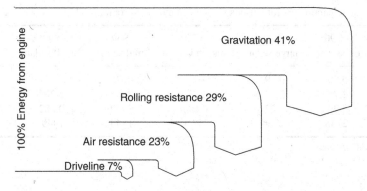

Figure 2.2 Distribution of the energy produced by the engine for a 40-ton truck on a typical road. Losses due to aerodynamic drag, rolling resistance, and gravitation are all substantial, and clearly larger than the total loss in the driveline

in the driveline. Further, it should be noted that the losses in potential energy due to gravitation are mainly due to braking going downhill. In an ideal situation where all potential energy could be recovered there would be no gravitational loss, and the ambition to recover at least some part of this potential energy is a driving force behind hybridization, as introduced in Section 1.1.3.

Aerodynamic drag, rolling resistance, and gravitation will be treated in the following sub-sections, and combined in models in Section 2.3. As a complement to the energy description in Figure 2.2, these models are compared as regards drag force in Figure 2.13.

2.2.1 Aerodynamic Drag

The aerodynamic force on a body is in physics approximated by the equation

$$F_a = \frac{1}{2} c_w A_a \rho_a (v - v_{\mathrm{amb}})^2 \tag{2.3}$$

where F_a is the air drag, c_w is the drag coefficient, A_a the maximum vehicle cross-section area, and ρ_a the air density. The term $v - v_{\mathrm{amb}}$ is the wind speed relative the vehicle, and it is the difference between the vehicle speed v relative to the ground and the ambient wind speed v_{amb} relative to the ground. To include the reverse, or when the wind is pushing the car, the sign has to be included in (2.3) by multiplying with $sign(v - v_{\mathrm{amb}})$. Modern midsize cars have $c_w \approx 0.3$ and frontal area $A_a \approx 2.2$ m^2, see Table 2.1 for examples of vehicle data.

A somewhat simpler expression is sometimes used by introducing C_w as the total effective drag, that is relating to (2.3) it is $C_w = \frac{1}{2} c_w A_a \rho_a$. If the wind is not blowing, that is $v_{\mathrm{amb}} = 0$, then the aerodynamic drag is

$$F_a = C_w v^2 \tag{2.4}$$

Drag Sources

In Figure 2.3, the main contributions to air drag are depicted together with very rough numbers for their relative contribution. They are: underbody 30%, wheel and wheel houses 25%, and vehicle shape 45%.

Table 2.1 Vehicle parameters for various vehicles in a model program

Model	XC90 crossover	S80 large sedan	V70 station wagon	S60 mid sedan	S40 small sedan	C30 small coupe
c_w [−]	0.36	0.29	0.31	0.28	0.31	0.28
A_a [m²]	2.75	2.34	2.23	2.27	2.20	2.18
$c_w A_a$ [m²]	0.99	0.679	0.691	0.636	0.682	0.610
Mass [kg]	2138	2186	1760	1544	1249	1347

Source: Volvo Cars web site.

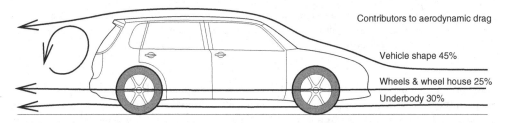

Contributors to aerodynamic drag

Vehicle shape 45%

Wheels & wheel house 25%

Underbody 30%

Figure 2.3 Main contributors to aerodynamic drag

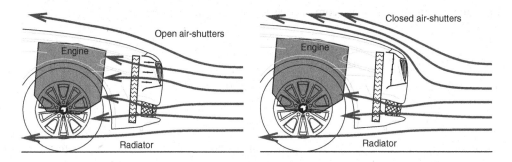

Figure 2.4 Active control of air-shutters

For a specific car, the values of the coefficients in (2.3) can be determined experimentally. However, note that there may be additional details contributing to aerodynamic drag, for example whether windows are open or closed.

2.2.2   Cooling Drag and Active Air-Shutters

An interesting aspect with consequences for aerodynamic drag is the fact that vehicles can now be equipped with active air-shutters so that cooling flow can be controlled. The purpose is to be able improve the thermal management of the engine, but the conclusion from the perspective of air drag is that it will vary depending on how open the air-shutters are. An example is seen in Figure 2.4.

The important modeling consequence is that the efficient aerodynamic drag will vary and (2.4) becomes

$$F_a = C_w(u_a)\, v^2 \tag{2.5}$$

where u_a is the control of the air-shutters. Numbers reported at the moment suggest that air drag can vary by 20%. There are also ideas concerning using valves to the wheel houses to be able to further control the cooling flow and thermal management. Such a system would also be covered by (2.5).

Connecting back to Section 1.1.4, optimal thermal management will be a part of optimal look ahead control, where the operation of the air-shutters will be controlled according to topography and situation.

2.2.3   Air Drag When Platooning

The above formulas for aerodynamic drag describe the situation when the vehicle is sufficiently far away from other vehicles. It is well known that air drag is reduced substantially if traveling behind another vehicle, and also that this effect is physically significant at the speeds of competing runners and bicycle competitors, that is already at speeds of 20–40 km/h. Since air drag grows as the square of speed the effect increases at typical vehicle cruising speeds. Even though this effect to some extent can be explained by thinking that the first vehicle puts the air in motion, that is gives value to v_{amb} in (2.3), it is better to look directly at some data as presented in Figure 2.5. The figure gives the reduction in effective air drag coefficient C_w for three vehicles in convoy as function of inter-vehicle distance. It is not surprising that the second and third vehicle in the convoy have strongly reduced air drag, but interestingly the first vehicle also gets a reduction when the distance moves below, in this case, 15 m. The physical explanation for this is that the drag caused by wake turbulence after the first vehicle is reduced thanks to the second vehicle. Note also that the sizes of the reductions imply that large fuel savings would be at hand if platooning could be utilized commonly.

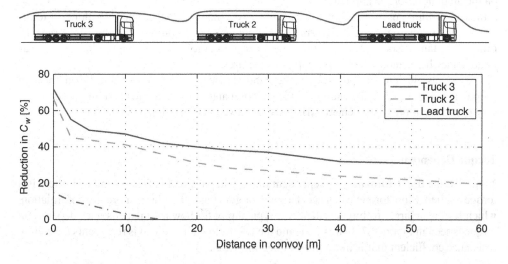

Figure 2.5   Reduction of effective air drag coefficient C_w for three vehicles in convoy

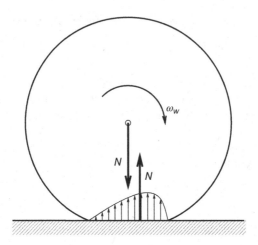

Figure 2.6   Rolling resistance comes from a force profile due to tire deformation

2.2.4   Rolling Resistance – Physical Background

Tire properties are another source of driving resistance. A pneumatic vehicle tire in contact with the ground and supporting the weight of a vehicle is neither round nor rolling. The real situation is depicted in Figure 2.6, where it can be seen that the tire is slightly deformed when it is in contact with the ground. The tire is somewhat compressed by the vehicle weight. Further, at the leading edge of a tire under the action of a driving torque, a front is built up and at the trailing edge there is an area of slipping friction. One consequence of the tire dynamics is that the contact patch between tire and ground is shifted in front of the wheel center, and as a consequence of that the resultant force acts in front of the wheel center, thus creating a resistance torque. Another consequence, mainly of the slipping, is that the rolling condition, $wr - v = 0$ is not fulfilled. Instead $wr - v > 0$, and how much bigger this is than zero depends on the amount of torque applied.

Instead of thinking in terms of forces forming driving resistance, one may think in terms of energy. The more the tire is deformed the more energy is required, and thus the larger the rolling resistance. Looking at the process as energy loss, it is natural that the lost energy has to go somewhere, and the result is heating of the tires.

A basic insight into the physical background sketched above is valuable when forming propulsion models, since depending on the situation and timescales involved it may be necessary to include more or less characteristics to get a valid description and analysis.

Torque Dependence

Naturally, the tire deformation is dependent on whether a wheel is free rolling or if there is a torque applied. More torque when accelerating, or going uphill, induces more tire deformation, which results in larger rolling resistance. A typical plot for how rolling resistance depends on torque is seen in Figure 2.7. In this plot, and the two following, the y-axis represents the rolling resistance coefficient that is introduced in Section 2.2.5.

Temperature, Pressure, and Velocity Dependence

Keeping the physical background of tire deformation in mind, it is clear that there are many relevant physical effects. The temperature of the tire will influence how flexible it is and will thus have a major influence on rolling resistance. The inflation pressure determines the initial deformation at rest, as in Figure 2.6, and thus the amount of deformation needed for each revolution. Examples of experimental data for temperature are shown in Figure 2.8a, where Figure 2.8b shows an example of evolution when driving. Whereas the dependence on torque, as depicted in Figure 2.7, can be handled as a look-up table capturing the plot, the dependence on temperature and pressure can instead be included in model parameters as described in Section 2.2.5, where pressure data is collected in the model (2.8) and Figure 2.10.

Regarding velocity dependence, one may think of a tire as consisting of many segments each acting as a spring and a damper. Since each element will be compressed more often with increasing speed, it is natural that the damping losses, that is the rolling resistance, should increase, and this is indeed the case. For normal tires the effect is small for usual cruising speeds, and the growth can be modeled as a linear function or a quadratic polynomial with low curvature. However, when approaching the rated tire speed the resistance grows very rapidly, and above rated speed the deformations are so large that the heat developing from them will damage the tire. The principal behavior is sketched in Figure 2.9.

2.2.5 Rolling Resistance–Modeling

As seen in the previous section, the forces and dynamics in the tire and in the contact between tire and ground have complex underlying physics, and we will now present some commonly used modeling equations. This section is about the rolling resistance and the next section, Section 2.2.6, treats formulas for tractive and braking force conditions. Keep in mind that these models are valid under certain conditions, where the timescale of interest is a main factor. For a study in vehicle dynamics, with a timescale of seconds, one set of models are sufficient, whereas for a driving mission, tire characteristics will change and, for example, long term thermal effects become important as will be treated in Section 2.2.7.

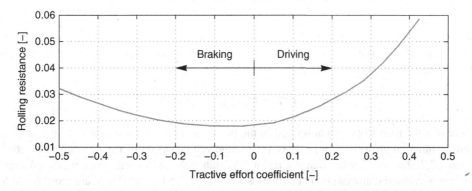

Figure 2.7 Rolling resistance coefficient as function of torque represented by the tractive force coefficient, F_w/N, where N is the normal force on the wheel

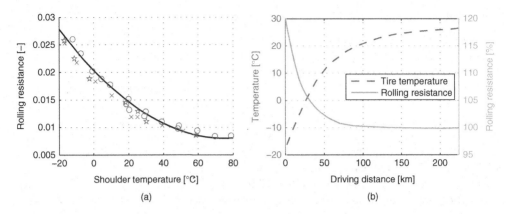

Figure 2.8   (a) Rolling resistance coefficient as function of tire temperature and (b) as function of driving distance

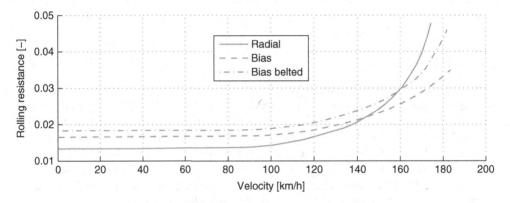

Figure 2.9   Rolling resistance coefficient as a function of speed. For low speeds there is a minor growth, but when approaching rated speed it increases rapidly and eventually damages the tire

The force, F_r, that is needed to overcome the rolling resistance is usually described with help of a *rolling resistance coefficient*, f_r, as follows

$$F_r = f_r(T, p, M, v, \ldots)\, m\, g\, \cos \alpha \tag{2.6}$$

where α is the road slope. As described in the previous section, f_r is function of tire temperature T, inflation pressure p, applied torque M, and vehicle speed v. The general trends from Figures 2.7, 2.8, and 2.9 are that the rolling resistance decreases with both increased pressure and increased temperature, but increases with torque and speed. There are more possible influences, and wet roads, for example, can increase f_r by 20%. However, this may already be captured by f_r in (2.6) since Gillespie (1992) contributes the fact that wet roads cool the tire.

Some approximate values of the rolling resistance coefficient are given in Dietsche (2011) and they are reproduced in Table 2.2.

Table 2.2 Approximate values for the rolling resistance coefficient f_r for different tires on different surfaces. Data adopted from Dietsche (2011)

Tire and road surface	Coefficient of rolling resistance
Pneumatic car tires on	
Large sett pavement	0.015
Small sett pavement	0.015
Concrete, asphalt	0.013
Rolled gravel	0.02
Tarmacadam	0.025
Unpaved road	0.05
Field	0.1–0.35
Pneumatic truck tires	
Concrete, asphalt	0.006–0.01
Strake wheels in field	0.14–0.24
Track-type tractor	
in field	0.07–0.12
Wheel on rail	0.001–0.002

Many empirical formulas that explain experimental data for the rolling resistance coefficient can be found in the literature, and to illustrate the situation some of the models will be summarized. These are typical models for short timescale behavior with application in vehicle dynamics or when thermal stationarity has been achieved. The model consists of a constant term and a speed dependent term. The following approximate relationships for the rolling resistance are given in Wong (2001), where v is given in km/h. The rolling resistance for a passenger car and truck with radial-ply and bias-ply tires are approximated by:

	Radial-ply	Bias-ply
Car	$f_r = 0.0136 + 0.40 \cdot 10^{-7} \, v^2$	$f_r = 0.0169 + 0.19 \cdot 10^{-6} \, v^2$
Truck	$f_r = 0.006 + 0.23 \cdot 10^{-6} \, v^2$	$f_r = 0.007 + 0.45 \cdot 10^{-6} \, v^2$

In Gillespie (1992) three models are summarized. The first is valid for lower speeds where the rolling resistance coefficient rises approximately linearly with speed

$$f_r = 0.01(1 + v/100) \tag{2.7}$$

where v is given in mph. The second comes from the Institute of Technology in Stuttgart and covers a larger speed interval and describes the rolling resistance coefficient as follows

$$f_r = f_o + 3.24 f_s \, (v/100)^{2.5} \tag{2.8}$$

where v is given in mph. The two coefficients depend on inflation pressure and are determined from a graph shown in Figure 2.10.

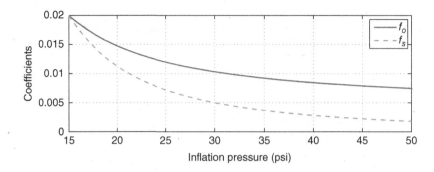

Figure 2.10   The pressure dependency in the rolling resistance coefficients used in (2.8)

Standard Model

As seen above, there is a spread in models for rolling resistance depending on the situation, where a model that aims at capturing the full behavior in Figure 2.9 needs a strong nonlinearity, whereas for a model for normal driving, that is the Figure 2.9b it is sufficient to use a linear model as in Gillespie (1992).

The following parameterization of the rolling resistance

$$F_r = m g (f_0 + f_s v) \tag{2.9}$$

will be used in this chapter and later in the chapters that cover modeling and control of powertrain and drivelines. A comment regarding (2.9) is that if there should be a larger growth than linear, then those effects can be captured by other terms when used in connection with the Standard Driving Resistance Model (2.28) as described in Section 2.3.2.

2.2.6   Wheel Slip (Skid)

Driven (or braked) wheels do not roll. Instead they rotate faster (or slower) than the corresponding longitudinal velocity. The difference is called *longitudinal slip*, and it is described by the *slip coefficient s*. Two different definitions of the slip coefficient are frequently used in the literature

$$s = \frac{r_w w - v}{r_w w} \qquad s' = \frac{r_w w - v}{v} \tag{2.10}$$

The first is most frequently used in textbooks on vehicle dynamics, while the second definition is used by the organization SAE. Vehicle *skid* usually refers to the conditions when the tractive effort is lost during braking.

To drive the vehicle forward and to brake the vehicle, a longitudinal force is needed in the contact patch between the vehicle and the road and this force depends on the slip. The tractive force is a function of slip and depends nonlinearly on the slip, its principle behavior is sketched in Figure 2.11 (where the first definition of slip is used).

The definition of the slip coefficient causes trouble for zero or low velocities. This can be seen in the definition of slip of (2.10) where s goes to infinity when w approaches zero (skidding with locked wheels) while s' goes to infinity when v approaches zero (wheel slip at standstill).

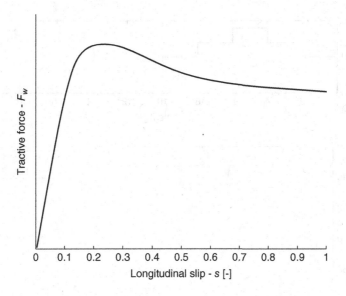

Figure 2.11   Tractive force, F_w, as function of slip s

When studying the conditions for driveline-control the normal situation is that the slip is small, that is for conditions close to the origin in Figure 2.11. As seen there, for small velocities and a small slip the tire–road contact force is well described by a linear dependence on the difference between the rotational and the longitudinal speed

$$F_w = m\,g\,k\,(r_w\,w - v) \tag{2.11}$$

In many cases, it is sufficient to connect the tire rotation to vehicle motion by using an assumption for *rolling condition*

$$r_w\,w = v \tag{2.12}$$

For the timescales relevant to vehicle handling, more models for the tire–road contact forces are given in standard books on vehicle and tire dynamics, see, for example, Gillespie (1992), Pacejka (2002), and Wong (2001).

2.2.7   *Rolling Resistance – Including Thermal Modeling*

A driving mission may start with tires at ambient temperature, say 10°C, and end up with tires at a significantly higher temperature, say for example 60°C. According to the general gas law, the pressure will increase. Relating to Figures 2.8 and 2.10, both temperature and pressure will cause the rolling resistance to vary during the mission, and end up in a significantly decreased value. The thermal behavior of a tire during a test cycle with different constant speeds can be seen in Figure 2.12. From the figure it is seen that the time constant, τ, is around half an hour. This is typical for a truck, and when experimentally determining stationary values one would drive for two hours before recording values. For a personal car the time scales involved are shorter and can be about 10 minutes.

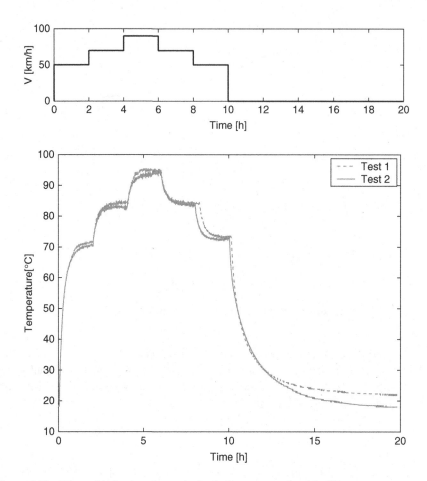

Figure 2.12   Thermal behavior of a truck tire during a test cycle with different constant speeds

It is also clear from Figure 2.12 that for stationary conditions, having kept a stationary speed, v_{sc}, long enough, the tire temperature converges to a corresponding stationary value, $T_{sc}(v_{sc})$.

Adding a Thermal Model

To capture the behavior in Figure 2.12 a thermal model is needed, and one approach is the following.

$$F_r = f_r(T, p, M, v, \ldots)\, mg \, \cos \alpha \qquad (2.13)$$

$$\frac{d}{dt}T = g(T, M, v, \theta_{amb}) \qquad (2.14)$$

$$\frac{p}{p_0} = \frac{T}{T_0} \qquad (2.15)$$

The basic idea is as follows. The coefficient, f_r, is left unchanged from (2.6). It is thus viewed as dependent on tire variables only. In addition to that there is a model for the tire temperature,

T, that in turn gives the tire pressure using the general gas law related to a reference state with subscript 0 that usually would be equal to ambient conditions. The parameter vector, θ_{amb}, is used to describe ambient conditions where parameters influencing tire temperature are air temperature, road temperature, sun radiation, and road conditions such as wet or not. The function g should capture all thermal effects on the tire including heat generated in the tire and heat losses to the environment. The heat generated in the tire can be modeled using the losses expressed by the difference between provided power, $M\omega$, and useful power, $F_w v$, but the heat dissipation from tires to environment is trickier to get right.

A detailed analysis of these phenomena is outside the scope of this book. Still, a fairly simple methodology can be found in Nielsen and Sandberg (2003). It is based on experimental data showing that a first-order linear system is an appropriate model. Then

$$\frac{d}{dt}T = \frac{1}{\tau}(T - T_{sc}) \tag{2.16}$$

With constant ambient conditions, the stationary temperature is a result of the heat produced by the losses in the tire, and it can be equivalently formulated in stationary velocity or stationary temperature, as described above in connection with Figure 2.12. The main steps are then to translate experimental data at stationarity to the equivalent stationary temperature T_{sc}, and to determine the time constant τ.

Parameterization of the Rolling Resistance Coefficient

The coefficient f_r in (2.13) can be obtained from Figures 2.8 and 2.10. One may fit algebraic expressions to these data, and just to illustrate the idea the following expression with separable variables is shown

$$f_r(T, p, M, v, \ldots) = C T^h p^d (mg)^\beta (a + bv + cv^2) \tag{2.17}$$

The exponents h, d need to be negative to capture the fact that the rolling resistance decreases with temperature and pressure. Another more direct approach is to make a look-up table from the data, as in Figures 2.8 and 2.10, and then to use interpolation.

2.2.8 Gravitation

The gravitational force, F_g, is straightforward. At a slope, it is the longitudinal component of the gravitational force, that is

$$F_g = mg \, \sin(\alpha(s)) \tag{2.18}$$

where α is the slope of the road. The slope of the road is dependent on where the vehicle is, so the position parameter is introduced, quite naturally, by the definition

$$\frac{ds}{dt} = v \tag{2.19}$$

The formula (2.18) is simple, but when used in a context of driving resistance, F_{DR}, it should be noted that in equations such as (2.1) and (2.2) the vehicle is modeled as a point mass. In reality, vehicles are extended objects, where a personal car is around 5 m and a truck can be

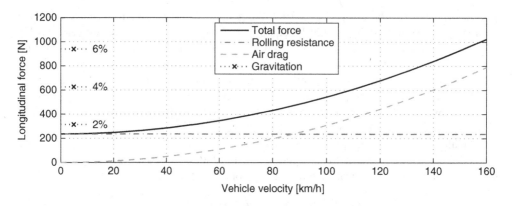

Figure 2.13   Driving resistance components as function of speed

around, say, 20 m. Then the question arises about which α to use, the momentary one for the center of gravity or an average slope for the extension of the vehicle. This will be further commented on in Section 2.4.2 on road modeling.

2.2.9   Relative Size of Components

We have now looked at the three main components of the driving resistance: F_{DR}, namely air drag F_a, rolling resistance F_r, and gravitational force F_g. Examples of numerical values were given in the previous sections, but it is illustrative to have a look at the relative size of the drag components. Keep in mind that speed is not the only underlying variable for driving resistance, but in many control situations or steady state evaluations it is sufficient to parameterize using speed, as Figure 2.13 illustrates.

2.3   Driving Resistance Models

Explicit models of the driving resistance, F_{DR}, can now be put together using the components from previous sections. The driving resistance F_{DR} represents the sum of all external forces on the vehicle, where the most important are aerodynamic drag, rolling resistance, and gravitation. Then

$$F_{DR} = F_a + F_r + F_g \tag{2.20}$$

The models are to be used in (2.1) which is

$$m\,\dot{v} = F_w - F_{DR} - F_b \tag{2.21}$$

or equivalently in (2.2), with omitted braking force, that is

$$m\,\dot{v} = F_w - F_a - F_r - F_g \tag{2.22}$$

As discussed before, the level of detail needed for F_{DR}, or its typical components F_a, F_r, and F_g, highly depend on the situation to be analyzed or simulated. Some common and useful examples of modeling, that will be used in the rest of the book, will be given now.

2.3.1   Models for Driveline Control

Many control problems, for example many driveline control problems, are on the time scale of seconds or less. Under these circumstances F_D is constant or slowly varying and then it is sufficient to model it as a constant load, commonly denoted by l. Then the model is as follows.

Model 2.1 Driving Resistance Model for Driveline Control

$$F_{DR} = l \qquad\qquad\qquad (2.23)$$

The driving resistance is modeled as a constant load.

Since the load is slowly varying compared variables of interest, in applications it will typically be estimated, for example by an extended Kalman filter. This will be exploited in Chapter 15.

Including Slope

A variation of the above model is obtained by including the slope α. One reason is that at low or relatively constant speed the other terms do not vary much on short timescales. On the other hand, the slope varies as the topology in combination with vehicle speed, and may also be known with some precision due to learned routes or using GPS, as in Section 1.1.4. Different parameterizations may be used but the following is used here.

Model 2.2 Driving Resistance Model for Driveline Control–2

$$F_{DR} = mg\left(f_0 + \sin(\alpha)\right) \qquad\qquad (2.24)$$

The driving resistance is modeled as a constant load and a varying slope.

The mass has been separated out from the constant part of the load. This is an arbitrary parameterization, but it can be motivated by the fact that the mass of the vehicle may be partly known or estimated by other means.

One application where the model is valuable is for heavy trucks operating in woods. Speeds are low but slopes vary. The ground may be slippery and torques are high going uphill. Then a missed gear shift may lead to a stop, and the truck may need to back down and try again. For torque control and gear shift control in such situations the above model is well suited.

2.3.2   Standard Driving Resistance Model

In many situations, the main purpose of a study or analysis is to gain insight into the main principles or to make comparative studies between concepts, and then absolute precision is not a main objective or one worth spending simulation time on. Here, it is typical to include all major components of F_{DR}, that is the components F_a, F_r, and F_g, but to restrict each to the basic model. The main example is as follows, where slip is neglected, and F_{DR} is described by the following quantities.

- F_a, the air drag, by

$$F_a = \frac{1}{2} c_w A_a \rho_a v^2 \tag{2.25}$$

where c_w is the drag coefficient, A_a the maximum vehicle cross-section area, and ρ_a the air density. However, effects from, for instance, open or closed windows will make the force difficult to model.
- F_r, the rolling resistance, by

$$F_r = mg\,(f_0 + f_S v) \tag{2.26}$$

where f_0 and f_S depend on, for instance, tires and tire pressure.
- F_g, the gravitational force, by

$$F_g = mg\,\sin(\alpha) \tag{2.27}$$

where α is the slope of the road.

Combining the components using (2.1) gives the following model.

Model 2.3 Standard Driving Resistance Model

$$F_{DR} = \frac{1}{2} c_w A_a \rho_a v^2 + mg\,(f_0 + f_S v) + mg\,\sin(\alpha) \tag{2.28}$$

The standard driving resistance contains the three major components and is used in many basic investigations.

Modeling and Parameter Estimation

Even though there is a choice of terms to include in the model, the Standard Driving Resistance Model includes the most relevant physical phenomena. From a modeling and estimation perspective, F_{DR} in this formulation includes terms that are constant, first and second power of velocity v, and dependent on slope α. Often, this model structure will be sufficient to get a good fit to experimental data, even though the interpretation of the coefficients would be slightly different. For example, if estimating the velocity term, this may of course also include viscous friction in the power train, but the estimated model will still be very useful for analysis and design. Still, one should be observant that no long-term dynamics such as thermal effects distort the model estimation.

2.3.3 Modeling for Mission Analysis

As stated above, the Standard Driving Resistance Model, can be tuned to experimental data so that it can be successfully used for system design or for relative studies between different concepts. However, sometimes it may desired to capture more precise values, for example to evaluate fuel consumption over a mission with precision in the order of percent. As seen in Figure 2.13, at normal cruising speed the air drag and the rolling resistance are of the same order, and from Figures 2.8 and 2.10 it is clear that this can influence the total driving resistance, and thus calculations of fuel consumption, by more than a percent. To get better precision the models can be combined in the following way.

Model 2.4 Mission Driving Resistance Model

$$F_{DR} = C_w(u_a)\,(v - v_{\text{amb}})^2 + f_r(T, p, M, v, \ldots)\,mg\,\cos\alpha + mg\,\sin(\alpha) \qquad (2.29)$$

$$\frac{d}{dt}T = g(T, M, v, \theta_{\text{amb}}) \qquad (2.30)$$

$$\frac{p}{p_0} = \frac{T}{T_0} \qquad (2.31)$$

The model is very similar to the standard driving resistance model, having the same three major components, but captures more effects, especially those on a longer timescale that vary over a mission. A feasible choice of the thermal model, g, is the linear first-order model in (2.16).

Driven and Non-Driven Wheels

For best precision, one has to treat the rolling resistance differently for driven or non-driven wheels, which means splitting the second term in (2.29) into two terms. To start with, there is a direct influence from the amount of torque utilized in relation to Figure 2.7. A short example illustrates that this is an effect of significance.

Example 2.1 Consider, somewhat unrealistically, two cars where one is two-wheel drive and one is four-wheel drive, but they are otherwise identical. The torque to overcome the driving resistance, F_{DR}, has for the 2WD to be obtained from the driven two wheels, whereas it for the 4WD it requires half the value for each of the driven wheels. Using Figure 2.7, consider a case which requires a tractive coefficient of 0.1 for the driven wheels in the 2WD, which will then require 0.05 for each wheel of the 4WD. The summed effective rolling resistance would then be $2 * 0.24 + 2 * 0.16 = 0.80$ for the 2WD, and $4 * 0.18 = 0.72$ for the 4WD. For a real 4WD, the gain is not that high since there are additional losses in the driveline, but the point is that it is a significant modeling issue if percent level absolute precision is needed.

Besides the direct torque dependent effect, there is also a clear difference in thermal behavior between driven and non-driven wheels, which is quite natural since the deformations due to torque are larger in the driven wheels, and the driven wheels will thus warm-up faster. At the beginning of the mission the decrease in rolling resistance due to this warm-up compensates for the increase in rolling resistance due to torque. To capture these aspects the thermal model has to be doubled, so that there is one for the driven wheels and one for the non-driven wheels.

2.4   Driver Behavior and Road Modeling

Now, let us return to overall vehicle propulsion dynamics, which means the basic (2.1), that we repeat for reference. Recall also Figure 2.1.

$$m\dot{v} = F_w - F_{DR} - F_b \tag{2.32}$$

In the previous two sections the driving resistance F_{DR} was treated, and it is now time to look at the other two forces, namely the propulsion force, F_w, and the braking force, F_b. Both these forces are the result of the driver acting on the accelerator pedal or the braking pedal. A remark relating to Section 1.1.4, is that support systems like cruise control will also influence the propulsion and braking forces, but since it is the driver that switches these support systems on and off, these can be included in driver behavior.

Aiming at simulation, a description of driver behavior is illustrated in Figure 2.14, see also Figure 2.16 for context. The actual situation in terms of driving mission and state variables are inputs to the driver model, and the outputs are the actuations on brake, gear, clutch, and gas pedal. These actuations are sometimes directly translated into a resulting force for the propulsion force, F_w, or the braking force, F_b. This view will be expanded on in coming chapters, where a force request on the vehicle will be propagated through the powertrain to result in, for example, a torque request from the engine – a structure that is called torque-based powertrain control.

2.4.1   Simple Driver Model

In the simplest driver model, the perspective is that the driver has a desired velocity, v_r, in mind, and adjusts the vehicle speed, v, by operating the gas pedal or brake pedal, which results in a force, F, that is actuated as a propulsion force, F_w, or a braking force, F_b, depending on sign. A standard way of achieving this is to use a PID controller, which means that the driver model is as follows.

Model 2.5 Simple Driver Model

$$F = PID(v_r - v) \tag{2.33}$$

Where F depending on the sign is the propulsion force, F_w, or the braking force, F_b.

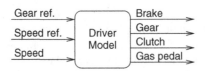

Figure 2.14   Driver model with its interfaces. Sometimes the outputs are translated into force requests

For many purposes in analysis or design, it is often sufficient to keep the vehicle in the operation range under study, or to make a transition from one state to another, and then it is usually appropriate to use the Simple Driver Model.

Driver behavior in its full extent is as complex as human behavior can be. For example, one may include human thresholds to different stimuli or changes in them, or include different driver modes by using state-machine models, for example. Sources for further reading are Bigler and Cole (2011) and Kinecke and Nielson (2005).

2.4.2 Road Modeling

As already discussed in Section 2.2.8, it is important to have the slope of the road, α, appropriately described. Available data is usually described in percent slope, where the relation to α is $\alpha = \mathrm{atan}(\%/100)$. The discretization varies but it could be a value for each meter or less if own data is used, or it could be 10 or 25 m if using a database service. Each value may have a resolution of 0.1%. A representation of such slope data leads to a piecewise constant function, as shown in Figure 2.15a. Using such road data in a simulation would typically lead to an overestimate of the fuel consumption. A better representation is shown in the same figure where the raw data has been filtered off-line with a zero phase filter with a filter constant that relates to vehicle length. The resulting road profile is shown in Figure 2.15b Not only analysis of fuel consumption can be influenced. Road slope information can potentially interfere with driveline control, since the step changes in the unfiltered version in Figure 2.15a, can potentially induce oscillations in simulations of driveline control (Myklebust and Eriksson 2012a).

Another aspect of a road model is its frequency content, where examples can be seen in Figure 2.15c. These are from three fairly long road segments (Koblenz–Trier, Södertälje–Norrköping, and Södertälje–Norrköping) that have been used in studies in optimal driving (Hellström et al. 2010). In the figure there are also the frequency plots of two shorter segments selected because they contain more variation. Quite naturally, they have more content at higher frequencies, but the main information is in low frequencies.

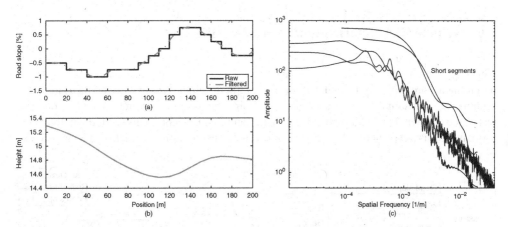

Figure 2.15 (a) road slope data, where the discretized slope data (0.25% at 10 m intervals) is presented as a piecewise constant function, and its filtered version (filter length corresponds to 25 m). The resulting road profile is seen in (b), (c) shows the frequency contents of three different roads, together with two selected shorter segments

Curved Roads

Road curvature is important for analysis of handling and many aspects like roll-over or risk of going off road. However, it also has some influence on longitudinal dynamics. When turning the steered wheels an angle, δ, the main effect is a lateral force, F_y, on the wheel. However due to the angle of the wheel there is a force component also in the longitudinal direction of the vehicle. This term is

$$2 F_y \sin(\delta) \tag{2.34}$$

where the factor 2 comes from the usual case of two steered wheels. For significantly curved roads, this term has to be added to the longitudinal dynamics of (2.32). An additional remark is that the rolling resistance is also influenced by handling, due to more complicated tire deformation, but the effect is small for regular maneuvers.

2.5   Mission Simulation

Now we have models for all the terms in (2.1), or equivalently in (2.32), and getting a complete simulation model is a matter of putting together these submodels as illustrated in Figure 2.16.

2.5.1   Methodology

The method illustrated in Figure 2.16, with an explicit driver model and perhaps road model, is called Forward simulation. There are alternatives, for example Inverse simulation, but before going into that it is advantageous to have introduced driving cycles, as will be done in Section 2.8, together with a discussion of tracking performance. It is also fruitful to have seen the torque based discussion in Chapter 3. We are therefore postponing the continued discussion on simulation methodology to Section 3.5.

2.6   Vehicle Characterization/Characteristics

As stated in the introduction to this chapter, a vehicle has to meet a complex set of requirements, and good design is very much about finding a good balance between them. These requirements are typically formulated for the complete vehicle, using external measurements such as fuel consumption, measurement of tail-pipe emissions, acceleration, and so on. The rest of the chapter deals with such characteristics and requirements. This section will treat performance measures, whereas Section 2.7 treats fuel consumption and Section 2.8 emissions.

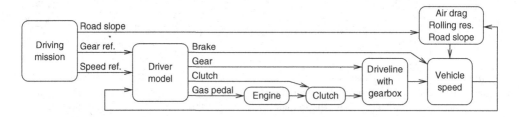

Figure 2.16   A complete mission simulation with explicit models for driver and road

2.6.1   Performance measures

Performance measures focus on the mechanical performance of a vehicle. To illustrate one basic performance measure, top speed is seen in Figure 2.17. This figure is a combination of Figure 4.4, with engine torque scaled for different gear ratios, using the driving resistance for a flat road from Figure 2.13.

There are many measures used to characterize vehicle performance, and a selection from different areas is listed below.

- Top speed. See Figure 2.17.
- Acceleration. Time for 0–100 km/h.
- Overtaking capability. Can be defined, for example as acceleration from 80 km/h to 120 km/h.
- Gradeability. Steepest slope a vehicle is capable of climbing at efficient speed.
- Responsiveness, where one example is the amount of turbo lag, see Chapter 8.
- Driveability, where one aspect is vehicle shuffle, or shunt and shuffle, see Chapter 13.

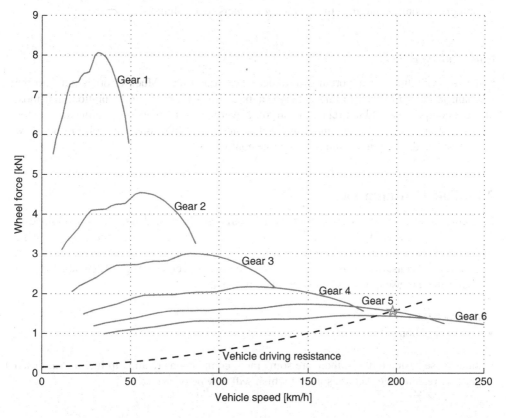

Figure 2.17   Torque diagram for a six-shift transmission with overdrive (the highest gear). Driving resistance, from Figure 2.13, as function of speed is also plotted. The top speed of the vehicle is marked, and note that it is not reached for the highest gear. The highest gear (overdrive) is instead used for economic driving

One may talk about hard or soft performance requirements. A hard requirement must be fulfilled and has a simple pass/fail condition. Take, for example, gradeability at a start from standstill. If starting at a certain gradient is not possible, then the vehicle cannot be used in that situation. Soft requirements are those that are important for a customer but for which there exists no hard fail limit. Rather, the value of the vehicle increases the better the requirement is met. For gradeability, this could mean maximum speed on steep gradients. The vehicle can be used even if the speed at a steep gradient is low, but a driver can rate the vehicle more highly if it can drive faster up a steep gradient. Such consumer values influence the price and the buy/no-buy decision of customers, and are thus important in vehicle design.

Driveability Measures

The whole concept of driveability, exemplified in the list above by turbo lag and vehicle shuffle, includes the basic performance characteristics in the previous section, but is also about the feel and smoothness of performance. This is an important topic in many places in this book, since it is an important part of control design. In Chapter 8 the principles of turbo engine design are treated, whereas Chapter 13 describes unwanted driveline behavior that may result from inappropriate control, and then the following chapters treat solutions.

Noise and Vibration

Noise and vibration are important for vehicle design and feel. Much effort is put into the mechanical design of cars to find the right compromises on noise and vibration, and most of these compromises do not relate to control. Nevertheless, there exist driveline and engine modes that are undesired but can be avoided or attenuated by proper control. Further, new regulations on drive-by noise influence engine control.

2.7   Fuel Consumption

Fuel consumption and CO_2 production in a car are closely related, as will be explained in Section 4.1.2, so recalling Figure 1.1 in Section 1.1 gives a motive for the focus on decreasing fuel consumption. Further, for the owner of a vehicle, the fuel consumption represents a cost. Therefore, fuel consumption is of primary concern in all aspects of design. There are different measures that are used, where the two most common are

- liters per 100 km
- miles per gallon.

Besides these basic numbers, there are some more comparisons relating to fuel that are used to get a more complete picture, some of which will now be presented.

2.7.1   Energy Density Weight

Why do we use diesel and gasoline as fuels? The main reasons are that they have a high energy to weight ratio and are readily available. As can be seen in Table 2.3, there is a vast difference

Table 2.3 Weight of one kilowatt hour for different energy sources

Storage medium	Weight/kWh
Lead acid	34 kg
Nickel cadmium	18 kg
Natrium sulfur	10 kg
Lithium ion	10 kg
Lithium composite	7 kg
Air zinc	4.5 kg
Gasoline	0.1 kg

in weight per kWh for different energy sources. The following example also illustrates how high the energy content in gasoline is.

Example 2.2 (Power fueling)   What is the energy flow through the hand when fueling a vehicle at a gas station with isooctane? The heating value for isooctane is $q_{LHV} = 44.3$ MJ/kg, and the density is $\rho = 0.69$ kg/dm^3. In a normal gas station it takes about 1 minute and 55 seconds to fill the tank with 55 dm^3 of fuel. The energy flow thus becomes

$$\dot{Q} = \frac{44.3 \cdot 0.69 \cdot 55}{115} = 14.6 \text{ MW}$$

This power is about twice that of the wind turbine "Enercon E-126," that has a rated power of 7.58 MW and was the largest in 2012. Another comparison is: to extract this power, 14.6 MW, using a 240 V electric system would require a current of 60 000 A!

These examples give an indication of how valuable hydrocarbon fuels are as energy carriers and also present us engineers with challenges when we have to find new solutions for the post-carbon era.

2.7.2 From Tank to Wheel – Sankey Diagram

Once the fuel is in the tank, the first question is: where does the energy go? An overall energy balance, based on the first law of thermodynamics, for a vehicle shows where the fuel's energy is consumed or disappears. The following list discusses where the energy contained in the fuel is consumed or dissipated during normal driving. The numbers given in the list below are only approximate, but they indicate the relative importance of each effect. An illustration is also given in the Sankey diagram in Figure 2.18.

Incomplete combustion: 1–5%. Energy is supplied to the engine by way of chemical energy in the fuel. Incomplete combustion, as mentioned earlier, comes from the fact that there is always some chemical energy which is not released during combustion and ends up as unburned HC, H$_2$, and CO in the exhaust. In a warmed-up SI engine 2–5% of the fuel's energy is not released.

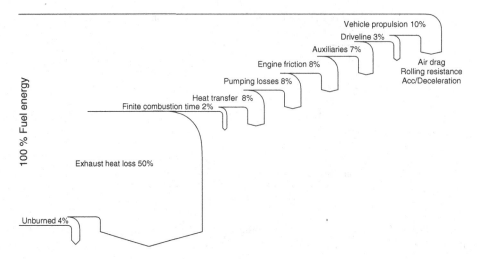

Figure 2.18 Sankey diagram illustrating a conventional car equipped with a gasoline engine. The numbers given in the diagram are only approximate, but it still illustrates where the fuel energy goes during normal driving of a conventional car

Ideal cycle: 50–55%. Figure 5.11 shows that an Otto cycle with $r_c = 10$ and a diesel cycle with $r_c \approx 20$ has an efficiency around 45–50%. The ideal thermodynamic cycle thus gives about 50–55% of the chemical energy directly to the exhaust.

Timing losses: 1–5%. The combustion takes a finite time, see Figure 5.15, which reduces the potential to produce work by a few percent.

Heat transfer: 5–15%. The heat transfer from the gas to the cylinder walls takes about 5–15% of the fuel's energy. With the losses described so far, the engine efficiency is about 30–35%.

Pumping losses: 4–10%. The engine's pumping is highly dependent on the engine's particular load point and can take around 4–10% of the fuel's energy. Therefore, the pumping wastes a third of the useful energy left. When studying the measured indicator diagram (the pressure-volume or p-V diagram) all losses discussed above are included, and the work produced is the indicated work. The real cycle can have an efficiency of around 20–27%.

Engine friction: 5–10%. The friction between the piston and walls, as well as that in the crank configuration. After the friction, the actual work that the engine produces can be measured, and at this point about 20% of the fuel's energy is available.

Auxiliaries: 5–10%. Servo pumps, electrical generators, air conditioners, and so on, are also connected to the engine and consume energy. After this equipment is connected, about 13% of the fuel's energy remains to drive the driveline.

Driveline: ~3%. The driveline, including the transmission, consumes a few percent of the fuel's energy. In the end, during normal driving conditions, about 10% of the fuel's energy is driving the vehicle in normal conditions.

2.7.3 Well-to-Wheel Comparisons

When comparing fuel consumption for vehicles of different types, say a conventional car and a hybrid car, it may be misleading to use only tank-to-wheel numbers. Instead, it is natural to

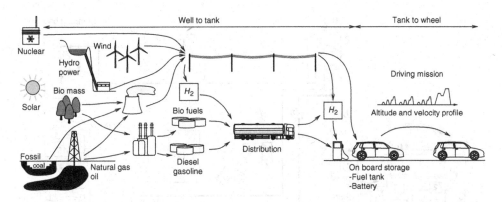

Figure 2.19   Depiction of well-to-wheel for a traditional car and for an electric or plug-in hybrid

compare well-to-wheel, as depicted in Figure 2.19. For example, for a hybrid car it matters what the original source of electricity is. If the electricity is generated from a mix of fossil energy, water power, or nuclear, then that is included in the well-to-wheel analysis.

2.8   Emission Regulations

In 1967, the Council of Europe provided the following definition of air pollution:

> *Air pollution occurs when the presence of a foreign substance or a large variation in the proportion of its components is liable to cause a harmful effect, according to the scientific knowledge of the time, or to create a discomfort.*

This definition captures an evolution of the notion of pollutant as knowledge expands. It also captures the fact that pollutants concern not only harmful emissions but also nuisances to humans.

In 1915 concerns were already being raised about the risk potential of automobile pollutants, and around 1945 vehicles were acknowledged as contributors to the "smog problem" in Los Angeles. The first regulations, enacted in California 1959, eliminated crankcase emissions (blow-by) and limited CO and HC. But it took until late the 1960s for the link between photochemical smog and engine emissions to be established. The first federal standards in the USA, under the Clean Air Act, applied to 1968 models and corresponded to the values set in California in 1960. Successive amendments converted the standard from pollutant concentration to pollutant mass per distance traveled, where the vehicle has to follow a certain test cycle.

Figure 1.2 and Table 2.4 show how the federal emission regulations have evolved and how stricter regulations have continuously been imposed. The table also shows some of the technical solutions that were developed to meet the amendments. In addition to exhaust emissions, evaporative emissions from the vehicle are also regulated.

2.8.1   US and EU Driving Cycles and Regulations

Emission regulations are formulated as maximum values on emissions from a complete vehicle driven in a vehicle dynamometer following a specified driving profile, called a test

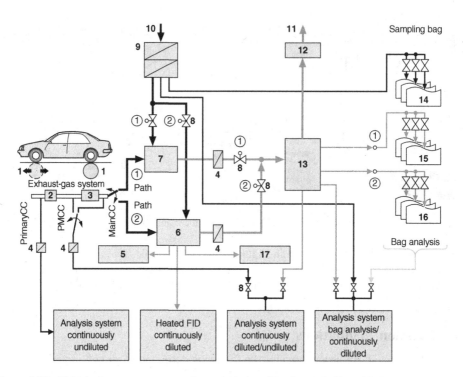

Figure 2.20   Vehicle dynamometer setup for emission certification. 1 Roller with dynamometer. 2 Primary catalytic converter. 3 Main catalytic converter. 4 Filter. 5 Particulate filter. 6 Dilution tunnel. 7 Mix-T. 8 Valve. 9 Dilution-air conditioner. 10 Dilution air. 11 Exhaust-gas/air mixture. 12 Blower. 13 CVS system (Constant Volume Sampling). 14 Dilution-air sample bag. 15 Exhaust-gas sample bag. 16 Exhaust-gas sample bag (tunnel). 17 Particulate counter. (1) Path for exhaust-gas measurement via Mix-T (without determination of particulate emission). (2) Path for exhaust-gas measurement via dilution tunnel (with determination of particulate emission). Source: Bosch Automotive Handbook 8e, Figure 1, page 510. Reproduced with permission from Robert Bosch GmbH

cycle. See Figure 2.20 for an example of a vehicle setup with emission sampling bags and analysis systems. A summary of the procedures and measurements can be found in, for example, Dietsche (2011). There are different test cycles in different countries. Together with the driving cycle, there are also specified limits on how good the tracking of the driving cycle must be during the experiment. The American FTP 75 cycle is frequently used and referred to and is outlined below. The FTP 75 test cycle and test layout are seen in Figure 2.21. The requirements on passenger cars relating to the FTP 75 cycle were shown in Table 2.4, for the years 1975 and later. It is also worth mentioning that the HC values for 1994, 1996, and 2001 cover only non-methane HC (NMHC). There are also more stringent requirements coming from California, as seen in Table 2.5.

In Europe, other cycles are used and the emissions are adapted to the size of the vehicles. One frequently used test cycle is the New European Driving Cycle (NEDC), which is shown in Figure 2.22. Emission regulations related to the NEDC are given in Table 2.6. Commercial vehicles, like trucks and so on, also have similar standards, but for heavy duty trucks and buses the standards are specified in terms of pollutants per produced work (i.e., g/kWh), instead of per distance traveled. An example of Euro I to Euro VI limits for a heavy duty diesel engine is shown in Figure 11.1.

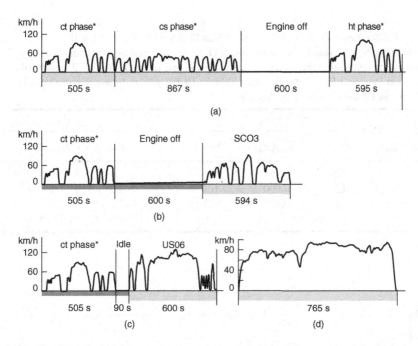

Figure 2.21   The US federal driving cycles for emission and diagnostic tests. Reproduced by permission of Robert Bosch Gmbh

Table 2.4   Evolution of federal emission regulation for passenger cars in the USA. The table also shows some of the technical solutions that have been applied to meet the amendments. Note that the emission levels are given in g/mile since these are US regulations. From model year 2004, Tier 2 standards started, specifying 10 bins where the manufacturers can place their vehicles, provided that they fulfill fleet average regulations. No data is given after 2004 due to the diversity of limits in the bins

year	CO g/mile	HC g/mile	NO_x g/mile	Methods
1966	87.0	8.800	3.60	Pre-control
1970	34.0	4.100	4.00	Retarded ignition, thermal reactors, and exhaust gas recirculation (EGR)
1974	28.0	3.000	3.10	Same as above
1975	15.0	1.500	3.10	Oxidizing catalysts
1977	15.0	1.500	2.00	Ox.cat. and improved EGR
1980	7.0	0.410	2.00	Improved ox.cat. and three-way catalysts
1981	7.0	0.410	1.00	Improved three-way catalyst and support material
1983	3.4	0.410	1.00	Continuous improvements
1994	3.4	0.250	0.40	Continuous improvements
1996	3.4	0.125	0.40	–
2001	3.4	0.075	0.20	–
2004	Tier 2 standards			–

Table 2.5 Californian emission limits for the "Clean Fuel Vehicle Fleet." LEV II is effective from model year 2004. NMOG stands for Non-methane Organic Gases (takes into account the ozone-generating potential of the various hydrocarbons in exhaust gas). TLEV – Transitional Low-Emission Vehicle; LEV – Low-Emission Vehicle; ULEV – Ultra-Low-Emission Vehicle; SULEV – Super Ultra-Low-Emission Vehicle

	LEV I			LEV II (2004–)		
Emission class	CO	NMOG	NO_x	CO	NMOG	NO_x
TLEV	4.2	0.156	0.6			
LEV	4.2	0.090	0.3	4.2	0.09	0.07
ULEV	2.1	0.055	0.3	2.1	0.055	0.07
SULEV				1.0	0.010	0.02

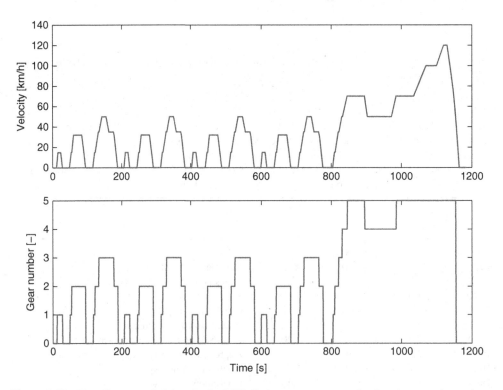

Figure 2.22 New European Driving Cycle (NEDC) for passenger cars that have a manual gear box with five gears. The cycle has total time 1 180 s, total distance 11 023 m, maximum speed 120 km/h, and average speed 33.6 km/h

Table 2.6 Example of EU emission standards for passenger cars, the values are given in g/km. (*) Only applicable for direct injected SI engines

Standard, year	Diesel				Gasoline					
	CO	NO_X	HC+ NO_X	PM	CO	THC	NMHC	NO_X	HC+ NO_X	PM(*)
Euro 1, 1992	2.72	–	0.97	0.14	2.72	–	–	–	0.97	–
Euro 2, 1996	1.0	–	0.7	0.08	2.2	–	–	–	0.5	–
Euro 3, 2000	0.64	0.50	0.56	0.05	2.3	0.20	–	0.15	–	–
Euro 4, 2005	0.50	0.25	0.30	0.025	1.0	0.10	–	0.08	–	–
Euro 5, 2009	0.50	0.180	0.230	0.005	1.0	0.10	0.068	0.060	–	0.005
Euro 6, 2014	0.50	0.080	0.170	0.005	1.0	0.10	0.068	0.060	–	0.005

Fulfilling these requirements is a strong driving force for development in engine management. It puts strong requirements on co-design of the engine and its control system. It is also stimulating the search for new sensors, for example to be able to directly measure the gas composition in the exhaust. This could lead to direct control of emission concentration.

Diagnostics

The emission requirements should not only be satisfied when the car is new. There are therefore also legal requirements on a diagnosis system that must detect and isolate faults that can increase emissions. More about this is found in Chapter 16.

3

Powertrain

The previous chapter, Chapter 2, looked at the vehicle from the outside and discussed its performance and requirements with a focus on the driving mission and longitudinal dynamics. This chapter takes up this thread and continues by going inside the vehicle to see how the longitudinal propulsion is generated. The system for doing this is the *powertrain*, which is defined as *the group of components that generate power and deliver it to the road*. An illustration of a powertrain is shown in Figure 1.6 and the powertrain seen there is typical in the sense that it includes engine, transmission (or gearbox), drive shafts, differential, and wheels. Relating to development goals, a powertrain with its control has as its main purpose to deliver the driver's desired power to the wheels, while at the same time fulfilling emission requirements, being efficient, and protecting the components from being harmed. This problem encapsulates numerous subsystems that have to be integrated into a single coordinated operational unit.

There are many possibilities and architectures for an automotive powertrain, and a number of possible solutions are given in Section 3.1. The main topic thereafter is the control issues of a powertrain, where Section 3.2 introduces the topic of vehicle propulsion control, and motivates the need for control structures in powertrain control. Then, a strategy named torque-based control is described in Sections 3.3 and 3.4. The chapter is concluded in Section 3.5 which includes simulation perspectives and in this way couples back to the models in Chapter 2. This lays a foundation for the thread of torque generation and control that continues throughout the book, and that will be followed up in the engine and driveline chapters.

3.1 Powertrain Architectures

There are many possible architectures regarding the powertrain. One classification is based on the topology regarding the choice of driven wheels, and the commonly used alternatives are

- Rear-wheel drive, see Figure 1.6 for an example.
- Front-wheel drive.
- Four-wheel drive, also called all-wheel drive.
- Rear-wheel drive with rear or mid-mounted engine.

The latter three are depicted in Figure 3.1.

Modeling and Control of Engines and Drivelines, First Edition. Lars Eriksson and Lars Nielsen.
© 2014 John Wiley & Sons, Ltd. Published 2014 by John Wiley & Sons, Ltd.
Companion Website: www.wiley.com/go/powertrain

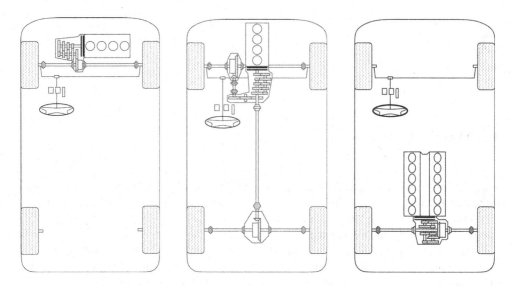

Figure 3.1 Three powertrain architecture topologies–from left to right–with front-wheel drive, with four-wheel drive, and with rear-mounted engine

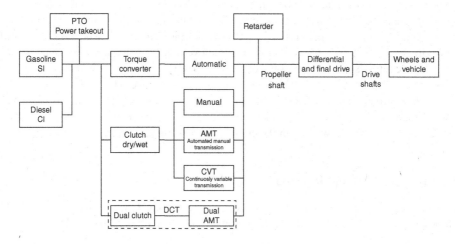

Figure 3.2 There are several options regarding components in a powertrain

Options for Components

Having decided on the topology with regard to what wheels to drive, there are still several choices for components. The engine may be gasoline or diesel, and there may be a torque converter or a clutch that can be either wet or dry. The transmission may be automatic, manual, automated manual (AMT), dual automated manual, or continuously variable (CVT). There may also be other components. Some of these choices are illustrated in Figure 3.2.

3.1.1 *Exhaust Gas Energy Recovery*

Another classification is based on whether or not there is a mechanism for recovering the energy leaving the engine cylinders as hot gases. As seen in Figure 2.18, this represents a substantial part of the energy. Some recovering alternatives (also illustrated in Figure 3.3) are

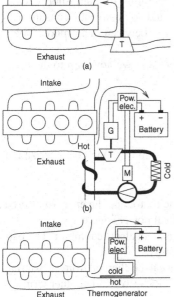

Figure 3.3 Recovery of exhaust energy by, from top: (a) turbocharging, (b) Rankine cycle, (c) thermoelectric effect. The components are: T–turbine, C–compressor, G–generator, M–motor

- Turbo, which will be treated in Chapter 8. A turbine is driven by the exhaust gases and a directly coupled compressor utilizes the energy in the engine intake.
- Turbo compound, which is similar to the above but the turbine has a direct mechanical coupling to the engine crankshaft.
- Rankine cycle-based recovery, where a heat engine (think in terms of a small steam engine) uses the exhaust energy to produce electricity. See, for example, Carlsson (2012), Ringler et al. (2009), Teng et al. (2007a,b).
- Thermoelectric generators based on the Peltier–Seebeck effect. Similar to the previous case, since the exhaust energy is used to produce electricity, but in this case directly. See, for example, Crane et al. (2001) and Stobart and Milner (2009).

Of these alternatives, the use of turbo aggregates is main stream, and will therefore be given a thorough treatment in Chapter 8. Turbo compound is commercially available on heavy trucks and will be discussed in the same chapter. The two latter are on an experimental level but have been evaluated to be promising. The future will show whether they are cost effective enough to become main stream. The drive in development of such powertrains is the natural wish to be as energy efficient as possible, but there are also driving forces from legislators that propose requirements on powertrain efficiency.

3.1.2 *Hybrid Powertrains*

A wide class of powertrains appeared on the market when it was opened up for alternative movers and energy storages. The current main trend is electrification, but a number of other alternatives are considered. A list of such other possibilities are

- Flywheel-based hybrid vehicles where energy is stored in a spinning inertia.
- Pneumatic hybrid vehicles where energy is stored as pressure in a gas tank.
- Hydraulic hybrid vehicles where a fluid, hydraulic oil, is used.

Many of these go under the notation of KERS–Kinetic energy recovery systems. These systems are currently mainly considered in vehicles with a substantial amount of stop-and-go action. One example would be a city bus, that when approaching a bus stop would use recuperation of braking energy to spin up a flywheel, or to compress gas in a gas tank. Then, when leaving the bus stop, the energy would be released and used for take-off-acceleration. Hydraulic hybrids have an additional area in which they are considered, and that is construction machines like wheel loaders or dumpers. Besides operating in many short cycles, these machines consume as much energy for loading, lifting, and so on, as they consume for motion. Therefore, hybridization that enables shifting of energy between these modes holds great promise.

Having seen some examples that are under current consideration, we now turn to the main trend, which is electrification.

3.1.3   Electrification

When we get to electrification the number of possibilities multiplies. One characterization depends on the way electric energy is accessed.

- Electric vehicle: has a powertrain that consists of an electric motor and the energy is stored in a battery. The battery is charged from an external electrical power source.
- Hybrid electric vehicle: having both a combustion engine and an electrical motor. The battery is charged by the engine or by recuperated braking energy.
- Plug-in hybrid electric vehicle: a hybrid electric vehicle, that can be charged from an external electrical power source. It is thus a combination of the previous two, and another term for this type of vehicle is range-extended electrical vehicle.
- Fuel-cell electric vehicle: yet another version of an electric vehicle where the electrical power is generated in a fuel cell.

Another classification for hybrid vehicles is the *degree of hybridization*, where the scale ranges from micro- and mild- to full-hybrid. The degree of hybridization is determined from the ratio between the maximum powers of the electric machine, P_{el}, and combustion engine, P_{ice}, and Figure 3.4 shows such a classification. The figure also places some commonly used

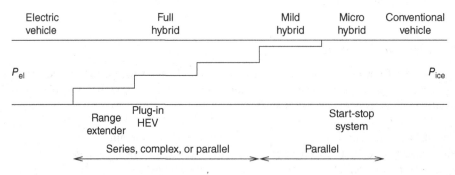

Figure 3.4   Illustration of the different types of hybrid vehicles and their relative use of power from electric machine, P_{el}, and internal combustion engine, P_{ice}

vehicle classifications on such a scale. Further, at the bottom of the figure, it is indicated that different architectures are used depending on the degree of hybridization, which naturally leads to a look at these architectures.

For the hybrid electric vehicle, having both combustion engine and electrical motor, the following main architectures for combining engine and electric motor exist:

- parallel hybrid
- series hybrid
- power split hybrid
- split axle hybrid.

Examples of architectures can be seen in Figure 3.5.

For all architecture alternatives, the combustion engine may be a traditional gasoline or diesel engine, or one may think of alternative engines as the combustion engine is operated differently to when it is the only mover. For all types, versions of supercharging may be considered, and finally, for all cases there are two alternatives of whether it is plug-in or not.

Sixteen Basic Types of Hybrid Electrical Vehicles

With the four main hybrid architectures in Figure 3.5, two main engine types (gasoline or diesel), and plug-in or not, there are sixteen basic hybrid electrical vehicles. One may add one more dimension regarding the electrical storage, that could be a battery or a super capacitor, or a combination of both, which would lead to another doubling of possible main architectures.

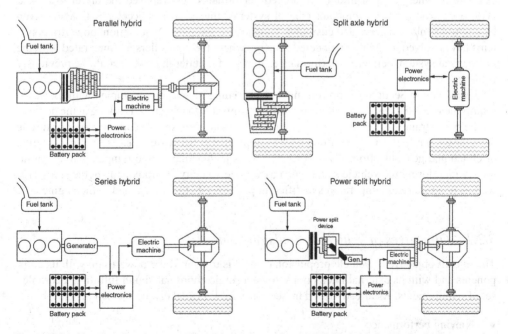

Figure 3.5   Sketches showing examples of the main architectures for hybrid electrical vehicles, namely: parallel hybrid, series hybrid, power split hybrid, and split axle hybrid

Design and Scope

There are of course many design issues when developing hybrid vehicles. The size and operation of the electrical storage differs, where a plug-in hybrid constructed to be used mainly in electrical mode would have a larger battery than a hybrid mainly supposed to recuperate braking energy. Also the type of electrical storage is chosen based on the fact that a battery can keep energy stored for long time, whereas super capacitors have fewer losses and are thus better for short-term energy storage. Further, a battery is operated at constant voltage, but for a super capacitor the voltage varies with the amount of stored energy. This leads to different requirements on the power electronics. Everything else in the powertrain also has to be dimensioned and matched. However, these aspects of physical design are not going to be the main topic here. Instead, we will present a control structure, termed torque-based control, that can handle all these cases.

3.2   Vehicle Propulsion Control

Historically, vehicle propulsion control was performed by the driver, who was in full control of the inputs to the mechanical systems. For example, the driver directly influenced the air and fuel delivery to the engine, which together with the clutch and gear selector were the main inputs to the powertrain. Originally, in the early 1900s for example on the T-Ford, the driver also directly controlled the spark advance. This was replaced by automatic spark advance using a mechanical system, where the distributor was mechanically adjusted by both the vacuum in the intake manifold and a centrifugal coupling to the crankshaft. This system mechanically controlled the ignition timing based on the operating condition, and it constitutes an early example of where control implied improved performance and relieved the driver of a task. With the introduction of computer control systems, this trend has accelerated. Also driven by environmental concerns and customer demands for low fuel consumption, powertrain systems have evolved to highly engineered systems where the controllers are integrated with the mechanical system, delivering performances that are far better than what could be previously achieved.

As was seen in Section 3.1, powertrains are becoming increasingly complex with increased requirements, and the goal of vehicle propulsion control is to meet this. Looking at the demand on the powertrain functionality, we start with the driver requesting a certain power or torque from the engine, which is done through, for example, gas pedal position, cruise controller input, or clutch and gear selection. The control system is then acting on these inputs while continuously monitoring the vehicle states, such as vehicle velocity, gearbox and engine speeds, as well as pressures and temperatures at different positions, both ambient and in the engine.

3.2.1   Objectives of Vehicle Propulsion Control

The overall objective of vehicle propulsion control is to control the powertrain, with its components and with its external interactions, to be a single coordinated unit fulfilling a complex set of requirements. As already listed in Section 1.2.1 this goal can be divided into

- achieving performance
- handling complexity
- enabling new technology.

These goals need to be mapped to a realization structure, where the main objectives are

- Control of all components in the powertrain.
- Coordination of all components in the powertrain.
- Keeping track of the state in the powertrain.
- Fulfillment of driver's request, while handling limitations.
- Distribution of resulting control actions to actuators.
- Integration of engine control functions in an implementation framework.

This list clearly illustrates how control systems have become crucial components in the development of clean and efficient vehicles. A comment when discussing powertrain control is that phenomena faster than the timescales relevant for control are not needed. For example, it is not necessary to go into detail on fast torque fluctuations due to individual ignitions, which means using the mean value concept, introduced in Chapter 1, that will be further expanded in Chapter 7.

3.2.2 Implementation Framework

The framework for implementing vehicle propulsion control is electronic control units (ECUs) with a real-time system as depicted in Figure 3.6a. A number of basic implementation functionalities need to be available. The different actuators need to be controllable from the ECUs. In the automotive area, the term drive-by-wire system is often used for this, and one example is electronic throttle control. Of course, several sensors for engine position tracking and so on are also needed. ECU hardware with hardware I/O is required, together with basic software and a real-time operating system. Today, it is quite common for function developers to use automatic code generation, often developed in symbiosis with their simulation environments, see Section 3.5.

Today there are usually several ECUs in a vehicle, and these units are connected together in a network. Figure 3.6b gives an example of such a system. Some nodes are related to vehicle propulsion control, but there are also several other functions.

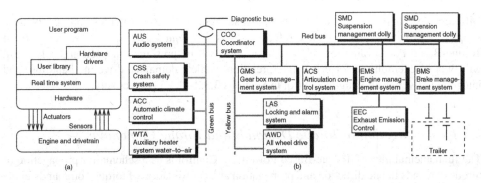

Figure 3.6 An implementation framework for vehicle propulsion control. (a) a schematic illustration of an electronic control unit (ECU). Several such ECUs are connected via one, or usually more than one, bus as depicted in (b)

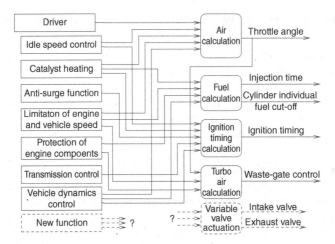

Figure 3.7 An example of a control system that has grown organically, and now has several functions influencing several controllers and actuators in complex ways–so complex that it may grow out of hand

3.2.3   Need for a Control Structure

When vehicle propulsion control systems started to evolve they grew organically, typically by adding new functions, where each function in isolation added value to the performance; but when combined with all other functions the complexity could grow out of hand. Reasons for this are, for example, that one function influences many actuator systems and that there are several cross-coupling effects. Coordination of limitations also became a major problem, with the effect that introduction and integration of new technologies could force major redesign of the whole system. Figure 3.7 gives a depiction of such a system. The conclusion drawn was that there was a need for a controller structure, and this now dominating structure is the topic of the next section.

3.3   Torque-Based Powertrain Control

Different design philosophies have been considered, and currently most producers have adopted a structure that is often referred to as a *torque-based structure*, and termed Torque-Based Powertrain Control. The basic idea will be presented in Section 3.3.1, and it will then be exemplified for driver interpretation, vehicle demands, driveline handling, and integrated driveline and engine control in Sections 3.3.2–3.3.5 respectively.

3.3.1   Propagation of Torque Demands and Torque Commands

The fundamental idea of Torque-Based Powertrain Control is as a scheme for propagation of torque demands in one direction and propagation and coordination of torque commands in the other direction. This basic idea is captured in Figure 3.8. Note that the figure has three parts. In the middle there is a depiction of a powertrain. At the top of the figure it is illustrated how torque and energy flow from the engine, via the powertrain to the wheels, to result in vehicle

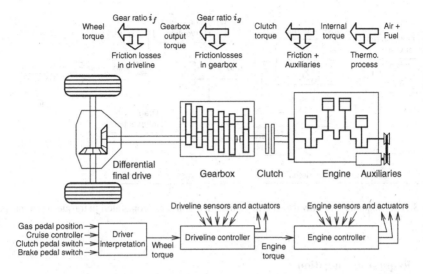

Figure 3.8 A vehicle powertrain showing the energy conversion path (top) and exemplifying a torque-based control structure (bottom). In the torque-based structure, a demanded wheel force or torque is propagated through the system to a demand on torque from the engine, and the engine controller then calculates the air and fuel delivery

propulsion. At the bottom of Figure 3.8, it is illustrated how the driver's demand is translated to a torque on the wheels, and this torque demand is then propagated in the control system to finally end up in a demand for torque from the engine.

Torque has Direct Physical Interpretation

One simple reason for choosing torque as the key variable is that it is directly related to the force at the wheels, which can be propagated through the powertrain components to the engine. Figure 3.8 shows how the torque-based calculations are propagated from the driver's input to the transmission and engine outputs. This is not the only reason, and the following is more important.

Handling of Complexity and Variants

Powertrains can differ significantly between vehicles in the same basic model series, which gives control engineers the delicate task of producing control systems for every configuration that gives the vehicle an integrated behavior. The purpose of the torque-based structure is to simplify the design process by giving well-defined interfaces between different subsystems. With the torque-based structure, the engine control design is decoupled from the driveline controllers and vehicle properties. The main points are thus

- Natural interfaces between components.
- Decoupling of the control tasks.
- Flexibility in change of components–engine, transmission, and so on.

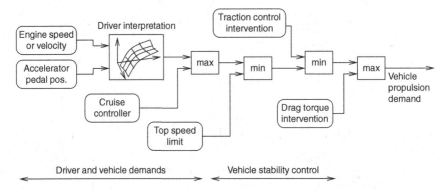

Figure 3.9 A schematic of how driver and vehicle demands are propagated in torque-based propulsion control

Torque Request Propagation

With the torque-based structure, the engine control design is decoupled from the driveline controllers and vehicle properties. There are limits in the driveline, like for example a maximum allowed torque that can be transmitted before it breaks. When the driver presses the accelerator and wants full acceleration, this request is, in the torque-based structure, transmitted to the driveline control that can limit the torque and pass the modified request to the engine. Then it is up to the engine to fulfill it and the engine controller does not need any knowledge about the driveline. If the design was reversed, so that the driver directly controlled the engine and it then controlled the driveline and so on, then the engine controller would need knowledge about the driveline and vehicle configuration to safely handle the driver request.

For each subsystem there are properties and controllers that deliver a torque request to the next link in the chain. The vehicle controller has functionality such as anti-spin control, cruise control, and so on. The driveline controller considers the torque request while handling internal torque limitations and driveline oscillations, and requests a torque from the engine. Finally, the engine uses the desired torque to determine the desired air and fuel flows to the engine while at the same time considering its limitations and emissions.

3.3.2   Torque-Based Propulsion Control – Driver Interpretation

As a first step going into a little more detail we will start in the lower left corner of Figure 3.8, namely with Driver interpretation. The function of that block is to translate all inputs, like gas pedal position, cruise control settings, gear selection, braking, and so on, into a torque demand on the wheels. In Figure 3.9 this is put into context, and the Driver interpretation is found to the left, while the arrow leaving it symbolizes the torque demand. The task of Driver interpretation is much more complex than many would guess at first, and it is typically a very large piece of code in production systems.

How this desired torque is propagated through the systems for driveline and engine, to finally be distributed at the engine, will now be exemplified in the following sections.

3.3.3   Torque-Based Propulsion Control – Vehicle Demands

The torque demand from the Driver interpretation has to be combined with other demands, limitations, and requests. For example, if it is larger than the request from the cruise controller then it should supersede it, otherwise not. There is a top speed limit that is to be respected. The way of handling these requests is to translate them to torque demands, and then use selectors as depicted to the left in Figure 3.9.

The right side of Figure 3.9 shows how vehicle demands enter. Traction control is a system that, when it detects that the driven wheels are spinning, reduces the torque on those wheels. Drag torque intervention restricts the maximum negative torque to prevent sliding. As seen in the figure, it is the same principle of using selectors in torque propagation when handling these vehicle demands.

3.3.4   Torque-Based Propulsion Control – Driveline management

Moving one step to the right in the bottom part of Figure 3.8 means that the torque request enters the module Driveline controller. The principle of torque-based control is again used. Figure 3.10 illustrates with the example of torque control for gear shifting. Actual methods for gear shifting intervention will be treated in Chapter 15. Regarding the last block to the right, Anti-surge, the background is that elasticities, found in the driveshafts of the driveline for example, can cause unwanted oscillations of the whole vehicle (called surge or vehicle shuffle). The prevention of this is called anti-surge and it will be explained and modeled in Chapters 13 and 14, while the control will be treated in detail in Chapter 15.

3.3.5   Torque-Based Propulsion Control – Driveline–Engine Integration

The propagation of the torque request from driveline management to the engine is similar to the examples above, and a schematic can be seen in Figure 3.11. The anti-surge function of driveline control is shown to the left of the dotted line that represents the interface to engine control. The reason is that the anti-surge function will be used to illustrate that the integration of driveline and engine can be based either on a feed forward filter or on feedback. In Figure 3.11 it shows up as a filter, but when describing feedback solutions in Chapter 15 additional signals are used as shown in Figure 15.2.

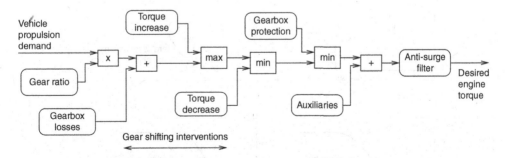

Figure 3.10   A schematic example of how driveline management for gear shifting is fitted into torque-based propulsion control

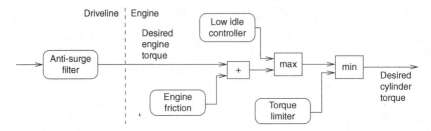

Figure 3.11 A schematic of engine management shown together with anti-surge, the final step of driveline management

Regarding the blocks of engine management, the losses due to friction are added, and there is an idle speed controller that ensures that the engine does not stall. There are also torque limitations, for example a smoke limiter in a diesel engine. The effect of torque limitations will be an interesting part of driveline–engine integration when looking at anti-surge in Chapter 15. The resulting torque is then the torque that is desired from the engine cylinders, and that the engine management system will generate using the available actuators. Models for friction and torque generation that are suitable for control implementation will be given in Chapter 7 and the basics of the torque control will be discussed in Chapters 10 and 11. Further, Chapter 9 goes into more detail, see for example Figure 9.3 and so on.

3.3.6 Handling of Torque Requests – Torque Reserve and Interventions

In addition to the structures in the previous section for propagating a torque request, it is important to have methodologies for handling limitations in response characteristics. In particular there are a number of circumstances that will modify a torque request. One obvious case is when a step in torque request cannot physically be obtained without some delay due to dynamics. Another example is intervention from other control units, for example the torque control during a gear shift. Four such situations are illustrated in Figure 3.12, where the signals are long-term torque demand, (immediately) available torque, and requested short-term torque. Each of these situations will now be discussed, and the discussion will also lead to the concepts of long-term torque demand (lead torque), short-term torque demand, and torque reserve. These concepts are introduced below, and are given a more precise treatment in Section 10.6.3.

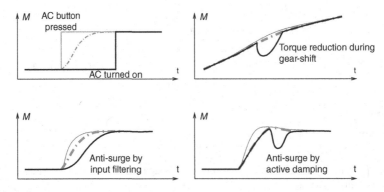

Figure 3.12 Four cases illustrating difference in time response between long-term torque demand (thin), available torque (dash-dotted), and requested short-term torque demand (thick)

Powertrain Preconditioning

Consider the upper left plot in Figure 3.12. It illustrates that the powertrain management waits some time after a torque request for the air conditioning equipment. In this way the available torque builds up during a period of time but is not used until a seamless start of the AC is possible.

Torque Reserve

The above example of powertrain preconditioning shows a build up of a difference between available torque and requested short-term torque. This difference is called torque reserve. The torque reserve is an important concept and needs to be managed, for example if there is a need for handling of fast requests or seamless mode transfers.

Long-Term Torque Demand (Lead Torque)

The example also shows that torque requests can be propagated in different ways in the torque-based structure. There are the usual demands called short-term torque demands, but there are also the torque demands, as in the AC example, that are called long-term torque demands. The situation is illustrated in Figure 3.13.

Long-term torque demand is also called lead torque, and is characterized by

- Usage: represents the future demand.
- Control of slower processes: air path, turbocharger.

Short-Term Torque Demand

Continuing with Figure 3.13, the short-term torque demand is the present torque request, that is used for

- Short interventions: anti-surge, gear-shift, and so on.
- Control of fast processes: ignition timing (SI), injection amount (DI).

For examples of these we now return to Figure 3.12.

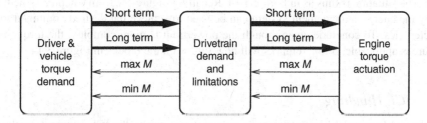

Figure 3.13   HOW torques are propagated for handling of response characteristics

Request Smoothing or Active Control

Consider the lower left plot and lower right plot Figure 3.12. The lower left plot illustrates the case where the anti-surge filter, see Figure 3.11, has smoothed the torque request in order not to excite oscillations. On the other hand, the lower right plot illustrates when active damping control is used instead. In that case the torque counteracts the oscillation based on feedback. It is also illustrated that the responsiveness is better in the latter case. See Chapter 15 for more information about this.

Torque Intervention

Finally, consider the upper right plot in Figure 3.12, that gives an illustration of the gear shifting intervention in Figure 3.10. As already mentioned, explicit examples of this will be given in Chapter 15.

Reporting Back the Resulting Torque

The fact that the resulting actually delivered torque is different from the demanded torque needs to be fed back in the control structure, and it has to be done for each step in reverse order so that each step in the structure knows the difference between its own demand and the result. This is also illustrated in Figure 3.13. The two main reasons are to

- Send feedback on achieved torque (or limits) to be used in the algorithms.
- Prevent integrator windup in controllers (for example the idle speed controller).

3.4   Hybrid Powertrains

An electric hybrid vehicle has an increased complexity in its powertrain as described in Section 3.1.3 and illustrated in Figure 3.5. Nevertheless, it is still a vehicle and when it comes to interpretation of, for example, driver intention, the same structure as in Figure 9.3 can be used, and the same can be said regarding the driveline. The major difference comes when leaving the driveline control where, for a traditional vehicle, this means a torque request to the engine as in Figure 3.11. For a hybrid vehicle, there is instead a choice of controlling how much of the requested torque should be provided by the combustion engine and how much should be provided by the electrical motor. The module responsible for this calculation is usually called the energy management system (even though torque division would also be a possible name). The situation is thus as in Figure 3.14. Recalling Figure 1.7 from Chapter 1, the torque split by the energy management system can be used to achieve the desired combined torque characteristics. To conclude, even though the powertrain is more complex, the torque-based structure is applicable. Some remarks will now follow on the blocks in Figure 3.14.

3.4.1   ICE Handling

The arrow in Figure 3.14 from the energy management unit to the ICE (internal combustion engine) is in principle no different from the corresponding arrow in Figure 3.11, so the same

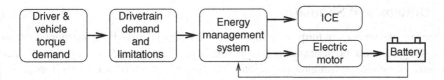

Figure 3.14 Torque-based structure for a hybrid vehicle. The torque request from driver interpretation and driveline management is split up by the energy management system (EMS), and propagated to the ICE and electrical motor respectively. The arrow from the battery to the EMS indicates the importance of battery management in powertrain control for hybrids

structure can be used. The parameters and strategies may be different, or very different, since the combustion engine may be operated differently when in a hybrid vehicle, but the control structure remains the same.

3.4.2 Motor Handling

The torque characteristics of an electrical motor were sketched in Figure 1.7, and the control of electrical drives is a well-established field both academically and industrially, see for example Fitzgerald et al. (2003) and Leonhard (1996). This book will not go into this field, but when deploying electrical motors in automotives, one has to consider some characteristics. There are very hard requirements on cost effectiveness and reliability, and some special functional requirements, for example to handle the fact that the battery voltage is more variable than the voltage from a power grid. Summing up, electrical machines and power electronics are established but have a special character in automotives.

3.4.3 Battery Management

The batteries in electrified vehicles are expensive and will for the foreseeable future continue to be an expensive component. The supervision and management of the battery system is thus a task of major importance. Existing battery packs can be composed of hundreds of battery cells connected together, and any individual cell can in the worst case scenario short the whole pack. Therefore a battery management system (BMS) needs to supervise the pack very carefully, and also manage the balancing of the cell voltages.

Preceding development of batteries in other areas, like mobile phones or computers, has considered environmental requirements that are simpler than the requirements in an automotive setting. It is also the case that requirements on energy density and power response drive the development in new battery chemistries, where questions about wear and aging are new but still need answers. This means that the development within battery management is intense. One major variable that is supervised is state-of-charge, SOC, and the arrow in Figure 3.14 going from the battery to the energy management system is used to represent the importance of battery management when deciding how to split the torque demand between the ICE and the electrical motor. This whole area of modulating the power outtake from the battery with respect to performance, thermal management, and aging, is an area of intense technological development.

3.5   Outlook and Simulation

Summing up, the strategy of torque-based control grew out of demands for a systematic method for handling complex powertrains, where it was desirable to be able to separate the design of vehicle, driveline, and engine controllers. The main characteristics of the Torque-Based Structure are

- Torque is a natural interface between driveline components.
- The Torque-Based Structure is physically organized.
- It improves transmission protection and compensates for auxiliaries.
- It has a modular structure with less interdependence between the functions.
- It is easy to realize additional torque interventions in a driveline.

More material for anyone interested in the background and historical development can be found in several sources, such as Dietsche (2011), Gerhardt et al. (1998), Heintz et al. (2001) and the book by Guzzella and Onder (2009).

3.5.1   Simulation Structures

The structure of torque-based control is a method for the systematic use of model-based control in vehicles. It turns out that a similar way of thinking is a sound basis for structuring modeling and simulation. It is natural to think in terms of physical entities, and a common way to formulate interfaces between model blocks is in terms of torques and velocities. These ideas will be exploited to a large extent in the modeling parts of this book, see for example the models in the engine and driveline chapters. The combined and complete models, in Figures 7.33 and 8.26, are combinations of component models with structured interfaces. Due to its functionality, this way of thinking about modeling has also resulted in the fact that tools for vehicle and powertrain simulation are structured and adapted to this way of thinking, and we will now continue with some more about simulation.

3.5.2   Drive/Driving Cycle

The concept of a drive cycle was introduced in Section 2.8 where it was used to specify the test cycle used in emission legislations. A drive cycle is a speed profile where speed is given as a function of time, and two examples were given in Figures 2.21 and 2.22. Often the term driving cycle is also used.

A drive cycle has several uses and the more important ones are

- as a test cycle in legislation
- as a representative usage of vehicles for vehicle design and testing
- specifically designed drive cycles to excite and test specific vehicle components or functions.

In the first and second case it is desired that the drive cycle represents realistic uses of a vehicle. With this goal in mind there are a number of ways to obtain a drive cycle. It can be

- a recording of an actual drive, as is the case for the FTP 75 cycle in Figure 2.21
- constructed, as is the case for the NEDC cycle in Figure 2.22
- obtained from a drive mission, road model, driver model, and vehicle model, as depicted in Figure 2.16.

When optimizing the behavior of a vehicle using only one specific drive cycle there is a risk that the calibration is overfitted to the specific cycle. The risk is that the vehicle behaves poorly in similar drive cycles, and a modern trend is therefore to seek methods that have a stochastic element but still generate drive cycles with the desired characteristics. One idea is to let the stochastic element be directed by statistics from large databases of driving data, so that the resulting drive cycle represents the type of driving selected in the database. One method, that is an evolution of the third item above using a Markov approach, can be found in Lee and Filipi (2011).

3.5.3   Forward Simulation

The direct method of forward simulation, using an explicit driver model, was described in Section 2.5 and illustrated in Figure 2.16. The corresponding schematic in Figure 3.15 shows the structure with a driving cycle as input to driver model that, for example, could be the Simple Driver Model described by (2.33).

3.5.4   Quasi-Static Inverse Simulation

Inverse simulation is an alternative approach using the driving cycle without an explicit driver model. To explain the basic idea, illustrated in Figure 3.16, we return to (2.32) which is the basic vehicle dynamics, and is recalled as

$$m\,\dot{v} = F_w - F_{DR} - F_b \qquad (3.1)$$

If the velocity, v, is given from a driving cycle then it can be differentiated. Then, with $m\,\dot{v}$ and F_{DR} known, it is easy to calculate the needed force $F_w - F_b$. This needed force can be recalculated to the torque at the wheels, which in turn can be propagated via the components of the powertrain to the torque from the engine. When the complete behavior has been computed, then, for example, the engine torque and speed can be used in an engine map to calculate fuel consumption. Note that once $\dot{v}$ is calculated, only static calculations remain, which motivates the name of this method, quasi-static inverse simulation.

3.5.5   Tracking

Tracking is a term used to describe how well a drive cycle is followed during a vehicle test or a simulation. In emission legislation there is an allowed tracking error resulting in a velocity

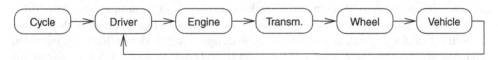

Figure 3.15 Schematic corresponding to Figure 2.16 with explicit use of the Simple Driver Model described by (2.33)

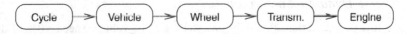

Figure 3.16 An illustration of quasi-static inverse simulation. There is no driver model since the drive cycle is used directly

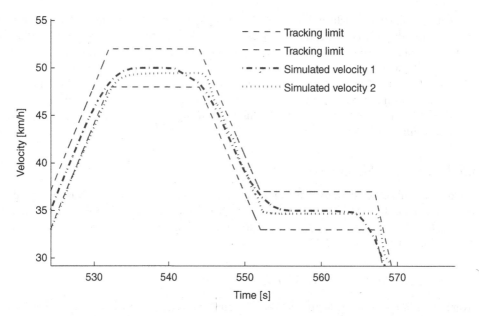

Figure 3.17 Two different simulations of NEDC. Both are within tracking limits but result in different fuel consumption

zone between a minimum and a maximum that needs to be kept. For a part of the NEDC cycle, in Figure 2.22, the allowed zone is illustrated in Figure 3.17. In the same figure, two simulations are presented. They are both within tracking limits, but are clearly different in behavior. They also result in different fuel consumption, even though they both fulfill the requirements of the test procedure.

Tracking within Limits and Tracking by Human Driver

The fact that drive cycle tracking is to be obtained within limits has a number of consequences. When using drive cycles for vehicle design, as indicated in Section 3.5.2, non-differentiable drive cycles and human tracking are of special interest and will be commented on now. Once again returning to the constructed NEDC cycle with its sharp corners in Figure 2.22, it is clear that a drive cycle cannot be expected to be continuously differentiable. This means that for models including additional dynamics, it is impossible to follow, or track, a drive cycle exactly, and hence the speed has to be controlled to track the drive cycle in the desired way. When drive cycles are used for emission legislation the cycle is tracked by a human driver within certain limits. It is therefore logical, in simulation too, to design driver models to track cycles within defined limits. As seen above, results in, for example, fuel consumption can differ between two different velocity trajectories that are both within prescribed limits, and it is thus important to study how a drive cycle is tracked and not just that the tracking is within limits.

3.5.6 Inverse Dynamic Simulation

Quasi-static inverse simulation uses only one state, that is, vehicle speed v(t), so that the force is determined by $F_w - F_b = G(v(t), \dot{v}(t))$. If more dynamics need to be added to the

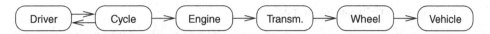

Figure 3.18 An illustration of inverse dynamic simulation. Compared to quasi-static inverse simulation, seen in Figure 3.16, an implicit driver model may be used, here seen to the left

vehicle model, then the inverse simulation strategy has to be extended. More states z have to be included, and they have to be obtainable from the velocity profile, which means that higher derivatives of the speed or the drive cycle may be needed. This can be formally written as $F_w - F_b = G(v(t), \dot{v}(t), z(t), \dot{z}(t)) = \tilde{G}(v(t), \dot{v}(t), \ddot{v}(t),...)$, where either additional states or higher derivatives may be used. Such simulation is called inverse dynamic simulation, and the idea is illustrated in Figure 3.18. The new block in the figure, Driver, is described below. To include the additional states, z, methodology is useful, and stable model inversion is a natural candidate, see Fröberg and Nielsen (2008). That paper gives examples where the vehicle dynamics is extended with engine dynamics, driveline dynamics, and gas flow dynamics for diesel engines. These examples are selected to represent important properties both regarding powertrain modeling but also mathematical properties, such as zero dynamics, resonances, and non-minimum-phase systems.

Driver Model

The driver model in inverse dynamic simulation, seen to the left in Figure 3.18, is based on the discussion in the Section 3.5.5. In practice, the driver model smooths the drive cycle like a human driver so that it is possible to track within limits. The smoothing of the drive cycle is in real life done by a test driver, but is done here by mathematical smoothing, which is interpreted as an implicit driver model. There are several possibilities for smoothing a drive cycle, and the shape of the smoothed speed profile is decided by the behavior of the driver model. One way is to filter the drive cycle with a standard linear low-pass filter with a relative degree of at least the same order as the system. This will ensure that the filtered drive cycle is sufficiently many times differentiable, and using a noncausal linear filter gives the driver look-ahead properties if that is desired.

Another way to smooth is to apply a convolution kernel on the drive cycle. Linear filtering of the drive cycle gives only asymptotically exact tracking, whereas a convolution kernel with compact support gives exact tracking on large parts of the drive cycle, as shown to the right in Figure 3.19. With a proper convolution kernel, the calculation of the derivatives does not require numerical differentiation, see Fröberg and Nielsen (2008). Instead, the derivatives are calculated by convolution using differentiation of the convolution kernel.

Thus, the smoothing of the drive cycle is interpreted as an implicit driver model, where the tracking behavior is specified in the velocity domain and is independent of the vehicle. In fact, this way of specifying tracking behavior can be argued to be closer to human behavior when performing drive cycle tests on a chassis dynamometer, since the tracking of a human driver is not specified by controller parameters but rather in terms of tracking smoothness. This is thus a general technique that also can be applied in forward simulation, Figure 3.15, which in that case means smoothing the drive cycle before feeding it as input to the explicit driver model.

Tracking Characteristics and Comparison

Drive cycle tracking depends on driver models for both forward and inverse dynamic simulation, and to compare tracking and behavior, different driver models are tested on the New European Drive Cycle, NEDC, see Figure 2.22. Figure 3.19 shows typical differences in tracking behavior between forward simulation and inverse simulation. For forward dynamic simulation, seen to the left, the driver model is explicit but the tracking is implicit since it depends on the driver model. For inverse dynamic simulation, seen to the right, the situation is that the driver model is implicit but tracking is explicit since it is specified directly.

The parameters of the driver model determine driver alertness, which, in this type of simulation, results in tightness of the cycle tracking. In Figure 3.19 the alertness parameters have been tuned to have comparable aggressiveness for both forward and inverse simulation.

3.5.7   Usage and Requirements

There are many important uses of drive cycle simulation and its applications. It is often used in concept studies, where examples are optimization over a design space of parameters, optimization of powertrain configuration, or design of powertrain control systems.

In these uses there are a number of requirements on the methodology. The main ones are

- set-up effort
- simulation time
- precision of solution
- tracking
- consistency in evaluation.

The first requirement is that it should not take too much time or other effort to set up a simulation. The requirement on simulation time is natural, and is especially important if many simulations are done, for example as part of an optimization loop. Precision and correctness are also natural and may be balanced against simulation time – if more time is spent a better solution may be obtained. As described above, the type of tracking influences the comparison between drive cycle evaluations, and this is also related to the subject of consistency. Consistency is a term to describe the idea that the comparison between two cases is fair, at least in relative order. For example, consider the case when a powertrain design is explored by

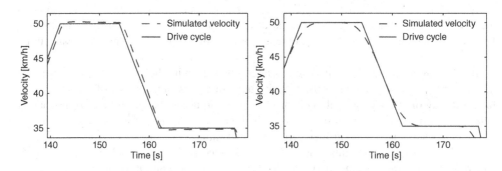

Figure 3.19   Tracking comparison for forward and inverse simulation

changing model parameters in the powertrain. Then the same driver model, that is the same controller, may give different tracking since the process has changed. When comparing a figure of merit, for example fuel consumption, the differences obtained in the exploration could then be due to the driver model resulting in better or worse tracking. The requirement on consistency is that even though there are imperfections and absolute errors, the order is consistent, that is the best also gives the best result and so on. To achieve consistency in an investigation it may be necessary to re-tune the driver model when the model is changing.

Some Tools

Because of the importance of drive cycle simulation, it is natural that there are several tools for the simulation of longitudinal vehicle models currently on the market. Advisor (Wipke et al. 1999) and Quasi-Static Simulation toolbox, QSS-TB, (Guzzella and Amstutz, 1999) are examples of quasi-static inverse tools, whereas PSAT (Rousseau et al. 2004), and V-Elph (Butler et al. 1999) are examples of forward dynamic simulation tools. All these tools use Matlab/Simulink.

3.5.8   Same Model Blocks Regardless of Method

It is clear that simulation is an important tool for propulsion analysis and design, and that there are different ways and methods of simulation. However, even though the methods vary, they are all built around model blocks with structured interfaces that describe different components of the driveline and engine, and the same model blocks can be used in the different methods. Existing or new powertrains can be modeled, analyzed, and simulated by combining the building blocks, and some examples can be seen in Figures 7.33 and 8.26. We conclude this chapter by recalling a point made already in Chapter 1. The models presented in this book have received thorough experimental verification and have proved their usefulness in many existing designs. Therefore, it is our belief that these models, perhaps in new combinations but still with the same model components, will be the foundation for analysis and design for many years to come. With that said, it is time to go into more model detail in the chapters to come.

Part Two

Engine – Fundamentals

4

Engine – Introduction

An internal combustion engine uses air and fuel, often based on hydrocarbons, to produce power and emissions, see Figure 4.1. To characterize and analyze the engine and its performance and energy conversion process, a set of relationships and parameters are defined. These parameters describe the air and fuel mixtures and also the engine geometry and performance measures.

Often, engines are described as gasoline or diesel engines, but there are many types of engines and also ways of classifying engines, see Heywood (1988) for a longer discussion. Here, they will be characterized based on the combustion process, and not fuel; into spark ignited (SI) engines and compression ignited (CI) engines. In SI engines the combustion is initiated with a spark plug and a flame propagates through the combustion chamber. Engines running on gasoline, natural gas, or bio-gas belong to this family. In CI engines fuel is injected into the cylinder and the mixture is ignited by the hot gases resulting from the compression. Engines running on diesel belong to this family. More details about SI and CI combustion and their characteristics are given in Chapter 6.

4.1 Air, Fuel, and Air/Fuel Ratio

A simplistic view of the engine operation is: the engine mixes air and fuel, the air and fuel mixture is combusted, and work and emissions are produced. The mixture of air and fuel is important for both work production and emissions. A characterization of the air and fuel mixture will follow here, while its influence on work and emissions is described in Chapters 5 and 6. First, let's look at the air.

4.1.1 Air

Air is a mixture of different gases and consists of oxygen, nitrogen, argon, carbon dioxide, water, and several other minor species. Their relative masses and concentrations are shown in Table 4.1, and some thermodynamic property data of air are collected in Appendix A. Oxygen is essential for an engine since it oxidizes the fuel and the oxidation process releases energy. The other gases are inert gases that have minor effects on the combustion but take up heat and space in the combustion chamber. A simple model for air is to consider it as being made up of

Modeling and Control of Engines and Drivelines, First Edition. Lars Eriksson and Lars Nielsen.
© 2014 John Wiley & Sons, Ltd. Published 2014 by John Wiley & Sons, Ltd.
Companion Website: www.wiley.com/go/powertrain

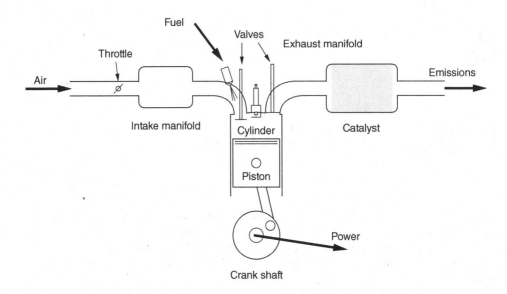

Figure 4.1   An engine takes air and fuel as input and generates power and emissions

Table 4.1   The table shows the constituents of dry air and their relative
concentrations. Water contents in air varies and, for example, a relative humidity
of 100% at 25°C has 2% water by mass

Constituent	Symbol	Molar mass	Volume [%]	Mass [%]
Oxygen	O_2	31.999	20.95	23.14
Nitrogen	N_2	28.013	78.09	75.53
Argon	Ar	39.948	0.93	1.28
Carbon dioxide	CO_2	44.010	0.03	0.05

only oxygen and nitrogen. Omitting the other molecules will introduce a small error, but this
error can be reduced if all other gases are lumped into nitrogen.

In 100 moles of air the total amount of nitrogen, argon, and carbon dioxide is 79.05 moles.
Considering all molecules except oxygen to be nitrogen will give that there are $\frac{79.05}{20.95} = 3.773$
nitrogen molecules for every oxygen molecule. This number of molecules is often referred to
as *atmospheric nitrogen* and it gives the following model for air with oxygen and the other
lumped molecules in the air.

Model 4.1 A Simple Model for Air

$$Air = O_2 + 3.773\,N_2$$

4.1.2   Fuels

The most important elements in fuels are hydrogen and carbon, and most fuels consist of
these, and in some cases there is also oxygen in the fuel (e.g., alcohols). Most liquid fuels

for vehicular engines are hydrocarbons, which exist in liquid phase at atmospheric conditions. There are sometimes also small portions of sulfur, water, and nitrogen in the fuels but these amounts can usually be neglected from a thermodynamic standpoint. However, they may have damaging effects on the emission treatment equipment. To keep the introductory discussion simple, we consider the fuel only to be an energy carrier.

Petroleum oils are complex mixtures of sometimes hundreds of different fuels, but the most important information for an engineer is the relative proportions of the elements, that is hydrogen, carbon, and so on. A frequently used reference fuel is isooctane C_8H_{18}, which is used when determining the octane number of a fuel, see, for example, Section 6.2.2.

The oxidizing reaction of a fuel releases heat, for example isooctane and oxygen gives

$$C_8H_{18} + 12.5\,O_2 \rightarrow 8\,CO_2 + 9\,H_2O + \text{Heat}$$

The energy that is released as heat from a fuel can be determined using, for example, a bomb calorimeter. Two fuel properties that are frequently used to quantify the amount of heat release are the higher heating value q_{HHV} and the lower heating value q_{LHV} of the fuel. The *higher heating value* is the amount of energy that the combustion of one unit of fuel can generate when the water among the combustion products is condensed to liquid phase. The *lower heating value* is the amount of energy that the combustion of one unit of fuel can release when the water in the products is in gaseous phase. Heating values for isooctane are $q_{HHV} = 47.8$ MJ/kg and $q_{LHV} = 44.3$ MJ/kg. The higher heating value is slightly greater since the phase transition for water from gas to liquid releases energy. In combustion engines, the exhaust gases are hot and thus the water is in the gaseous phase. Therefore, the lower heating value is used to describe the available energy in the fuel.

Another thing seen in the reaction is the relation between the fuel consumption (energy consumption) and CO_2, the more fuel we consume the more CO_2 is generated. For a given fuel, decreasing CO_2 means decreasing fuel consumption. Changing the fuel changes the CO_2 footprint. For example methane (essentially a natural gas) has a lower CO_2 footprint than isooctane (close to gasoline) while a renewable fuel such as ethanol produces CO_2 locally in the engine but does not give a net contribution to the atmosphere.

4.1.3 Stoichiometry and (A/F) Ratio

The *air/fuel ratio* (A/F) is defined as the ratio between mass of air m_a and mass of fuel m_f

$$(A/F) = \frac{m_a}{m_f}$$

A stoichiometric combustion reaction, between a general hydrocarbon fuel C_aH_b and air, produces only water and carbon dioxide and is written

$$C_aH_b + (a + \frac{b}{4})(O_2 + 3.773N_2) \rightarrow a\,CO_2 + \frac{b}{2}H_2O + 3.773\,(a + \frac{b}{4})N_2 \qquad (4.1)$$

This combustion reaction defines the stoichiometric proportions of air and fuel. By denoting the relative contents of hydrogen and carbon in the fuel $y = \frac{b}{a}$ and using the molecular weights for carbon, hydrogen, oxygen, and nitrogen we can derive an expression for the stoichiometric air/fuel ratio

$$(A/F)_s = \left(\frac{m_a}{m_f}\right)_s = \frac{(1 + y/4)(32 + 3.773 \cdot 28.16)}{12.011 + 1.008\,y} = \frac{34.56\,(4 + y)}{12.011 + 1.008\,y}$$

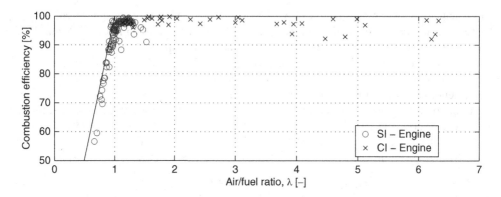

Figure 4.2   Combustion efficiency, η_c as a function of equivalence ratio ϕ. When the mixture becomes fuel rich the chemical energy in the fuel cannot be converted and the efficiency decreases. The line in the plot marks $\eta_c = \lambda$, which is $\min(1, \lambda)$ for the rich side. Data from Heywood (1988)

The stoichiometric air/fuel ratio varies between 13.27 and 17.23 for y varying between 1 (benzene) and 4 (methane). Usually the air/fuel ratio (A/F) is normalized with the stoichiometric mixture, and this is called the *air/fuel equivalence ratio* λ

$$\lambda = \frac{(A/F)}{(A/F)_s}$$

Another quantity is the (F/A) ratio, which is the inverse of the (A/F) ratio, and its normalized *fuel/air equivalence ratio* is denoted

$$\phi = \frac{(F/A)}{(F/A)_s} = \frac{1}{\lambda}$$

When there is excess air in the combustion ($\lambda > 1$) the mixture is referred to as *lean*. Then there will be excess air and some NO_x on the right side of Reaction (4.1). When there is excess fuel in the combustion ($\lambda < 1$) the mixture is referred to as *rich*. Under rich conditions the amount of air is insufficient for complete combustion of the fuel and the combustion efficiency decreases, see Figure 4.2.

A model for this behavior is to say that the fuel can not release its heat for rich conditions,

$$Q_{\text{heat}} = m_f \, q_{\text{LHV}} \, \eta_c(\lambda)$$

where the combustion efficiency η_c decreases when the mixture becomes rich. A simplified model for this behavior is $\eta_c(\lambda) \approx \min(1, \lambda)$. Furthermore, these rich mixtures produce unburned hydrocarbons HC and carbon monoxide CO that are a result of incomplete oxidation. This is treated more thoroughly in Section 6.4 that describes the mechanisms behind the emissions.

Model 4.2 A Simple Model for the Heat Release

$$Q_{\text{heat}} = m_f \, q_{\text{LHV}} \, \min(1, \lambda) \qquad\qquad (4.2)$$

4.2   Engine Geometry

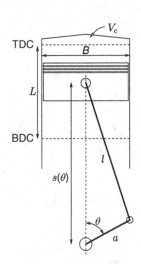

Figure 4.3 Definitions of the crankshaft, connecting rod, and cylinder geometries for an engine

The geometry of the piston, connecting the rod and crank is illustrated in Figure 4.3. The parameters shown in the figure and some important derived parameters are:

Cylinder bore	B	
Connecting rod length	l	
Crank radius	a	
Piston stroke	L	$= 2a$
Crank angle	θ	
Clearance (minimum) volume	V_c	
Displaced volume	V_d	$= \frac{\pi B^2 L}{4}$

The *displacement volume*, V_d, is the volume that the piston displaces in the cylinder and the definition here gives the volume for one cylinder. However V_d is often used to denote the total displacement of the engine, that is $V_D = n_{cyl}\frac{\pi B^2 L}{4}$ (where n_{cyl} is the number of cylinders in the engine). Here $V_D = V_d\, n_{cyl}$ will be used to denote the engine displacement.

The compression ratio, r_c, is an important parameter that influences the engine efficiency:

$$r_c = \frac{\text{maximum cylinder volume}}{\text{minimum cylinder volume}} = \frac{V_d + V_c}{V_c}$$

For spark ignited (SI) engines (also referred to as gasoline engines) the compression ratio usually lies in the range $r_c \in [8, 12]$. For diesel engines the compression ratio usually lies in the range $r_c \in [12, 24]$.

The instantaneous volume for the cylinder at crank position θ is given by

$$V(\theta) = V_c + \frac{\pi B^2}{4}(l + a - s(\theta))$$

where $s(\theta)$ is the distance between the crank axis and the piston pin

$$s(\theta) = a\,\cos\theta + \sqrt{l^2 - a^2\sin^2\theta}$$

In the expressions above the volume is specified in terms of the geometric parameters but the expressions are complex and hide some structure. It is possible to rewrite the expressions to

$$V(\theta) = V_d\left[\frac{1}{r_c - 1} + \frac{1}{2}\left(\frac{l}{a} + 1 - \cos\theta - \sqrt{\left(\frac{l}{a}\right)^2 - \sin^2\theta}\right)\right] \qquad (4.3)$$

and from this expression we see how the displacement volume V_d, the compression ratio r_c, and the $\frac{l}{a}$-ratio influence the cylinder volume.

4.3 Engine Performance

The focus here is on performance in terms of power, fuel consumption, and emissions but one needs to keep in mind that there are also other factors that are important for a driver and owner of a vehicle. For example: performance over the operating range, fuel consumption combined with fuel cost, noise and pollutants, as well as reliability, durability, and maintenance requirements. Now we turn to engine performance which is defined by:

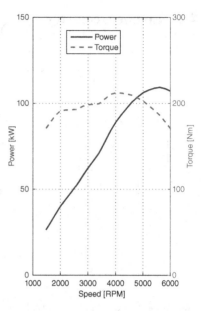

- The maximum power (or the maximum torque) available at each speed within the useful engine operating range.
- The range of speed and power over which engine operation is satisfactory.

Figure 4.4 shows the performance, maximum power, and torque, for a typical engine over the engine's operating range. Some other commonly used performance definitions are shown in the list below. These are also indicated in the figure.

Figure 4.4 Power and torque as a function of engine speed for a naturally aspirated Saab 2.3-liter SI engine

- Maximum rated power/torque: The highest power/torque an engine is allowed to develop for short periods of operation. (110 kW/212 Nm in Figure 4.4.)
- Normal rated power/torque: The highest power/torque an engine is allowed to develop in continuous operation. (Same as above for the engine in Figure 4.4.)
- Rated speed: The crankshaft rotational speed at which the rated power/torque is developed. (5500 resp 4000 RPM in Figure 4.4.)

4.3.1 Power, Torque, and Mean Effective Pressure

Engine torque is usually measured with a dynamometer, often referred to as an engine brake. The torque generated by the engine is denoted M_e (the symbol M is chosen for the torque to avoid confusion with the temperature T). The power P_b generated by the engine is

$$P_b = 2 \pi N M_e$$

where N is the engine speed measured in revolutions per second. The value of the power measured using a dynamometer is called *brake power* P_b. This power is the usable power delivered by the engine to the load, in this case a "brake." Note that torque is a measure of the engine's ability to do work, and power is the rate at which work is done.

Torque is a valuable measure of a particular engine's ability to do work but it depends on engine size. Another useful parameter is the mean effective pressure (MEP) which is defined as

$$\text{MEP} = \frac{\text{work produced per cycle}}{\text{volume displaced per cycle}} \tag{4.4}$$

The parameter obtained in this way is normalized with engine size and it has the unit (force per area), which is the same as pressure, thus giving its name. The mean effective pressure can be calculated based on the work that the engine produces in a dynamometer, then the parameter is called brake mean effective pressure: BMEP. BMEP is calculated by inserting the brake power or torque into the definition of MEP, giving

$$\text{BMEP} = \frac{P_b \, n_r}{V_d \, n_{cyl} \, N} = \frac{2 \, \pi \, M_e \, n_r}{V_d \, n_{cyl}} \tag{4.5}$$

here n_r is the number of crank revolutions in the complete power generation cycle ($n_r = 1$ for two-stroke engines and $n_r = 2$ for four-stroke engines).

The maximum brake mean effective pressure of good engine designs is well established and is essentially constant over a wide range of engine sizes. The actual BMEP that a particular engine develops can be compared using this norm, and the effectiveness with which the designer has used the engine's displacement volume can be assessed.

Typical values for BMEP are: for naturally aspirated spark ignition engines, maximum values are in the range 850 to 1050 kPa, at the engine speed where the maximum torque is obtained. Values at the maximum rated power are 10 to 15% lower. For turbocharged automotive spark ignition engines the maximum BMEP is in the range 1250 to 1700 kPa and depends on the pressure increase by the turbocharger. For diesel engines the BMEP lies in the ranges 700 to 900 kPa for naturally aspirated and 1000 to 1200 kPa for turbocharged. These values can be used to estimate either the power or torque from an engine of a given size or for design calculations to estimate the size of an engine where the power or torque requirement is known.

4.3.2 Efficiency and Specific Fuel Consumption

How efficiently the engine converts the fuel to useful work is another important aspect of the performance and two parameters are frequently used. The first is the fuel conversion efficiency which is the engine power divided by the fuel power $\dot{m}_f \, q_{LHV}$. This gives

$$\eta_f = \frac{\dot{W}}{\dot{m}_f \, q_{LHV}} = \frac{M \, 2 \, \pi \, N}{\dot{m}_f \, q_{LHV}} \tag{4.6}$$

the other frequently used parameter for efficiency comparisons is the specific fuel consumption (SFC), which is defined as the amount of fuel m_f per unit work output W, or equivalently fuel flow rate $\dot{m}_f$ per power $\dot{W}$,

$$\text{SFC} = \frac{m_f}{W} = \frac{\dot{m}_f}{\dot{W}} \tag{4.7}$$

The specific fuel consumption is often expressed in the units g/kWh, see Figure 4.5. From the definitions above one also sees that

$$\text{SFC} = \frac{1}{\eta_f \, q_{LHV}} \tag{4.8}$$

which shows that the SFC is inversely proportional to the efficiency. The specific fuel consumption is often related to brake power and denoted BSFC. In Figure 4.5 a measured BSFC *performance map* for a 2.3-liter passenger car engine is shown. A trend with an increasing BSFC for a decreasing load (output torque) is clearly seen, this is related to the part load efficiency.

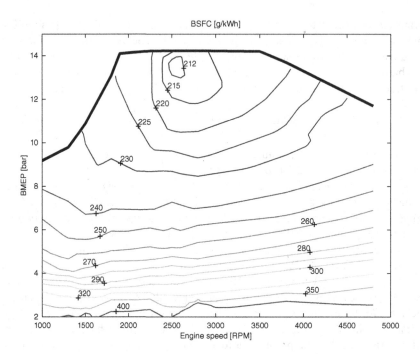

Figure 4.5 A performance map for a 2.3-liter SI engine equipped with a low boost turbocharging system. The map shows iso-lines for BSFC as a function of engine speed and load (BMEP), with the maximum load line marked. The lowest BSFC is attained near the maximum line. Note that the iso-BSFCs have different scales for the near optimum and low load regions

4.3.3 Volumetric Efficiency

The intake system consisting of air filter, carburetor and throttle plate (in SI engines), intake manifold, intake port, and inlet valve restricts the amount of air that the engine can induct. The measure used to measure the effectiveness of an engine's induction process is the *volumetric efficiency*, η_{vol}. The volumetric efficiency is defined as the volume flow rate of air into the intake system, $\dot{V}_a$, divided by the rate at which volume is displaced by the piston, $\dot{V}_d$,

$$\eta_{vol} = \frac{\dot{V}_a}{\dot{V}_d} = \frac{\dot{m}_a \, n_r}{\rho_{ai} \, V_d \, n_{cyl} \, N} \tag{4.9}$$

where ρ_{ai} is the air density in the inlet, $\dot{m}_a$ is the air mass flow, and $n_r = 1$ for a two-stroke engine and $n_r = 2$ for a four-stroke engine. An equivalent definition of the volumetric efficiency is to use the mass of air in one cylinder directly

$$\eta_{vol} = \frac{m_a}{\rho_{ai} \, V_d}$$

The definitions are equivalent but the mass of air in the cylinder is not easily determined.

The inlet density can either be taken as the intake system density (as made above) or as the atmospheric density. In the first case η_{vol} measures the efficiency of the pumping over the inlet valves and ports, while in the other it measures the efficiency for the whole intake. The volumetric efficiency for a naturally aspirated engine usually has a maximum of 0.8–0.9 but for an engine with well-tuned intake runners the volumetric efficiency can reach values above unity.

Important Notes of Caution Concerning Variables

Before proceeding it is necessary to give two *important* notes of caution:

1. The displaced volume V_d used in the equation above is that of one cylinder, but V_d is also frequently used for the total displaced volume of the engine which is $V_d\, n_{cyl} = V_D$ (as mentioned before).
2. The engine speed N, used in the equations above, is given in revolutions per second but expressions given in revolutions per minute (RPM) are also frequently used. Then there is a factor 60 included in the expressions.

It is usually clear from the context which volume or which engine speed is used, but be careful since their meaning is frequently interchanged in the literature.

4.4   Downsizing and Turbocharging

Engine downsizing, as mentioned in Section 1.1.2, is a concept used to achieve reduced loss through reduced engine size. The main virtue of having a small engine is that it has better efficiency at road loads than a larger one, which is due to the fact that a smaller engine operates closer to maximum load where the engine has its best efficiency, see Figure 4.5. The cause for this is that close to maximum load the relative effects of the losses, pumping and friction work, are smallest. In particular the friction work increases only slightly with load, while the pumping work decreases with load and this makes their relative importance decrease as the useful output is increased. More details, with models for torque generation and losses will be developed in Sections 5.2.3 and 7.9. The interested reader is recommended to study Soltic (2000) for more examples and analysis of different downsizing concepts for engines.

   To illustrate the behavior, let's consider a mid-size vehicle running at 90 km/h that can be fitted with two engines; one large 3.2-liter and one small 1.6-liter. With the gearbox the road load is translated to an engine speed of 2000 RPM and a torque of about 50 Nm. Figure 4.6 shows their efficiency and specific fuel consumption as function of load and has the point 90 km/h marked. At the cruising speed, the smaller engine operates at a point with better efficiency (nearer to maximum) and thus has a better fuel economy. As seen in the figure the small engine has a maximum torque of approximately 150 Nm while the bigger engine has a maximum torque of 300 Nm. Cruising at 90 km/h with the smaller engine requires more than a third of maximum torque while the bigger engine needs only one fifth. The torque reserve, from cruising to maximum, can be used to accelerate the vehicle, and it shows that the smaller engine naturally will give the car a lower acceleration performance. In summary, the upside of a small engine is lowered fuel consumption and the downside is lowered acceleration. One way of compromising between these is to use a supercharger or turbocharger and this will be introduced in the following section.

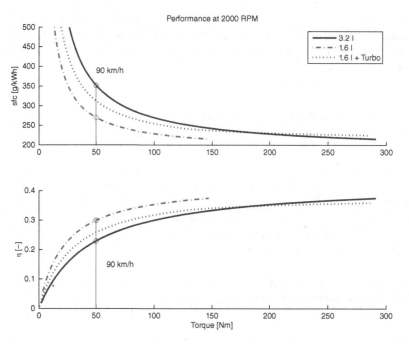

Figure 4.6 Comparison of the BSFC and efficiency at 200 RPM, for three engines: Solid – 6-cylinder 3.2-liter. Dash dotted – 4-cylinder 1.6-liter. Dotted – 4-cylinder 1.6-liter turbo charged. A mid-size vehicle cruising at 90 km/h would need about 50 Nm at 2000 RPM

4.4.1 Supercharging and Turbocharging

Supercharging is the collected name for several methods that increase the intake air density, that is methods that charge extra air to the cylinder, and one particular method is called turbocharging. The reasons why an engine is supercharged are coupled to engine power and torque. Engines are limited in power and torque by the amount of air that can be inducted into the cylinder, since the chemical energy in the fuel needs the oxygen in the air for combustion. (4.2) shows the total heat while the volumetric efficiency limits the amount of air in (4.9). Through an increased air density, the amount of air inducted into the cylinder can also be increased. Supercharging is often used on downsized engines to give fuel economy benefits.

In turbocharged engines, the energy from exhaust gases is used to compress the intake air and produce a higher intake manifold pressure. This is done by having a turbine in the exhaust system connected to a compressor in the intake system. The increased intake pressure results in an increased density and thus an increased amount of air into the combustion chamber, thereby increasing the power output that is available. A turbo also increases the intake air temperature, which can be cooled by an intercooler. The turbocharger speed and the resulting pressure after the compressor are controlled using either a wastegate or a variable geometry turbine (VGT).

Examples of Turbocharged Gasoline and Diesel Enignes

Figure 4.7 shows two turbocharged engines, one gasoline and one diesel. The main differences between them are the throttle on the gasoline engine and the exhaust gas recirculation (EGR)

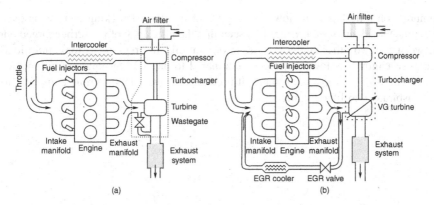

Figure 4.7 (a) A sketch showing the most frequently used configuration for a turbocharged SI engine. (b) The most common CI engine configuration

on the diesel engine. The gasoline engine has fuel injection in the port while the diesel has fuel injection directly into the cylinder. Furthermore, the turbine power is controlled using a wastegate on the gasoline engine while the diesel has a variable geometry turbine (VGT). The *wastegate* is a valve that bypasses the turbine, which results in less power to the turbine and as a result the turbocharger speed and intake pressure is decreased. VGT turbines change the flow characteristics of the turbine and can utilize the exhaust energy better since no mass flow is bypassed which gives better freedom in control and wider range for the engine. VGTs have moving parts that can have durability problems at high temperatures and therefore they are mainly used in diesel engines that have lower exhaust temperatures.

The compressor increases the pressure but also the temperature, which is undesirable since a higher temperature gives a lower charge density. Therefore an *intercooler* (sometimes also called *aftercooler*) is often utilized to decrease the temperature, and the charge density is increased which results in higher power output from the engine. An intercooler also gives better engine efficiency, since it can give the same mass flow through the engine for a lower intake pressure which in its turn requires less turbine pressure and thus reduces the pumping work of the engine.

In a gasoline engine an increased charge temperature can result in a higher knock tendency which can be detrimental (this will be discussed in Section 6.2.2). Different measures are taken to reduce the knock tendency, some are: reducing the compression ratio, cooling the intake air (intercooler), spark advance control, and wastegate control. One car can be equipped with two different 2.0-liter engines, one naturally aspirated and the other turbocharged with a low boost. The naturally aspirated engine has a compression ratio of $r_c = 10.1$ while the turbocharged engine has an intercooler and a compression ratio of $r_c = 9.0$. The power is increased by the turbo from 94 kW (128 hp) to 118 kW (160 hp).

Downsizing, Turbocharging, and Fuel Economy

We now return to the downsizing example in Figure 4.6. A turbocharger is fitted on the 1.6-liter engine, which increases the intake density and thereby the power and torque output. This design gives a compromise between the smaller engine's low fuel consumption and the bigger engine's power output. The drawback to raising the maximum intake pressure is that the compression ratio has to be lowered to guarantee against knocking conditions over the entire engine's load and speed range. It is seen that the efficiency of the turbocharged engine lies between the two other engines in the important cruising region. At maximum load the turbocharged engine has lower efficiency compared to the 3.2-liter engine (due to lowered compression ratio), but most importantly it produces a much higher torque compared to the naturally aspirated 1.6-liter engine.

5

Thermodynamics and Working Cycles

The operating principles of a four-stroke engine are studied in detail and used to illustrate the basic principles that govern the engine limitations and engine efficiency. The purpose is to give an in-depth understanding of the engine as a process that can be used as a foundation on which modeling, control, and diagnostics problems can be built. Furthermore it is also important to understand the compromises and limitations that are present. Following the course taken in Chapter 4, and focusing on the engine outputs, power generation, and emissions, this chapter describes and analyzes the operating principles and the mechanisms that govern power and efficiency, while the next chapter covers combustion and emissions.

5.1 The Four-Stroke Cycle

The spark ignited (SI) engine (sometimes also referred to as Otto, gasoline, or petrol engine) and the compression ignited (CI) engine (or diesel engine) are the most commonly used engines in automotive applications. Their working processes follow a four-stroke operating cycle that is illustrated in Figure 5.1. An important property which is fundamental for the engine operation and power generation is the in-cylinder pressure, p_{cyl}, because this pressure creates the force on the piston and thus the output torque of the engine. The in-cylinder pressure is displayed in Figure 5.2 together with the intake manifold pressure, p_{im}, and exhaust manifold pressure, p_{em}. The pressure trace plotted as a function of crank angle is also called the *indicator diagram*. The four strokes, that are related to the piston movement, are:

Intake (From top dead center (TDC) to bottom dead center (BDC).) The inlet valve is open and, while the piston moves downwards, the cylinder is filled with a fresh air/fuel charge from the intake manifold. Note that the cylinder pressure, p_{cyl}, during the intake stroke is close to the intake manifold pressure, p_{im}.

Compression (BDC-TDC) The air/fuel mixture is compressed to a higher temperature and pressure through the mechanical work produced by the piston. About 25° before TDC (BTDC) a spark ignites the mixture and initiates the combustion. The flame propagates through the combustion chamber and heat is added. The combustion continues into the expansion phase.

Modeling and Control of Engines and Drivelines, First Edition. Lars Eriksson and Lars Nielsen.
© 2014 John Wiley & Sons, Ltd. Published 2014 by John Wiley & Sons, Ltd.
Companion Website: www.wiley.com/go/powertrain

Expansion (TDC–BDC) The combustion finishes around 40° after TDC (ATDC). Work is produced by the fluid during the expansion stroke when the volume expands. Around 130° the exhaust port is opened and the blowdown process starts, where the cylinder pressure decreases as the fluid is blown out into the exhaust system by the higher pressure in the cylinder.

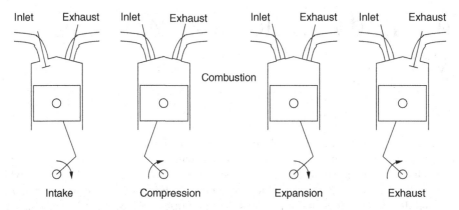

Figure 5.1 The four-stroke cycle in an internal combustion engine. The strokes are intake, compression, expansion, and exhaust of a four stroke

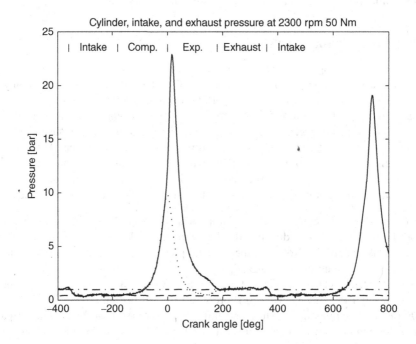

Figure 5.2 Cylinder pressure (solid), intake manifold pressure (dashed), and exhaust pressure (dash dotted), for a SAAB 9000 doing 80 km/h on slight uphill road in fourth gear. The four strokes – intake, compression, expansion, and exhaust – are displayed. The cylinder pressure for a motored cycle is also shown (dotted)

Exhaust (BDC–TDC) The fluid in the combustion chamber is pushed out into the exhaust
system. Note that the cylinder pressure, p_{cyl}, is close to the exhaust pressure, p_{em}. When the
piston reaches TDC the cycle is complete and the next cycle starts.

The data used to illustrate the process comes from an SI engine and the discussion thus refers
to the SI process. There are many similarities with the CI process and more details will be
covered in Sections 6.2 and 6.3.

The combustion is initiated by a spark discharge (also called ignition) at the spark plug. The
spark discharge usually comes between 0 and 40° before TDC. In Figures 5.2 and 5.3 the ignition event occurs around 25° BTDC. Around 5° BTDC the combustion pressure starts to rise
above the pressure for the motored cycle (which corresponds to the pressure from compression
only) and it reaches its maximum around 17° ATDC. When the pressure reaches its maximum
the pressure rise due to combustion is equal to the pressure reduction due to the expanding volume. The combustion still continues some 10–20°, burning the remainder of the mixture. The
cylinder pressure for the *motored cycle* is measured when the engine is pulled by an electric
motor without igniting the mixture, therefore the cylinder pressure follows only the compression and expansion and reaches its maximum at TDC (or ~1° BTDC due to heat transfer and
mass leakage).

Figures 5.2 and 5.3 also show another important principle in that the throttle affects the air
flow into the intake manifold and changes the intake manifold pressure and thereby also the
cylinder pressure. In Figure 5.2 the engine load is 50 Nm and the intake manifold pressure is

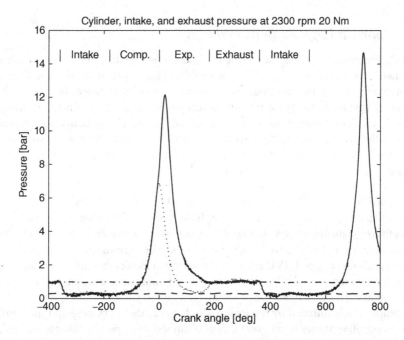

Figure 5.3 Cylinder pressure (solid), intake manifold pressure (dashed), and exhaust pressure (dash
dotted), for a SAAB 9000 doing 80 km/h on slight downhill road in fourth gear. The four strokes – intake,
compression, expansion, and exhaust – are displayed. The cylinder pressure for a motored cycle is also
shown (dotted)

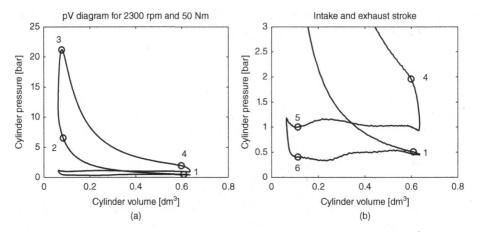

Figure 5.4 p-V diagram for the cycle shown previously in Figure 5.2. (a) shows the full cycle and (b) shows the intake and exhaust stroke enlarged. The marked point corresponds to: (1) Inlet valve closing (IVC). (2) Start of combustion. (3) Maximum pressure. (4) Exhaust valve opening. (5) Inlet valve opening. (6) Exhaust valve closing

p_{im} = 48 kPa, and in Figure 5.3 the engine load is 20 Nm with an intake manifold pressure of p_{im} = 30 kPa. This illustrates that the intake manifold pressure in an SI engine is related to the engine load, and can be used together with the engine speed to estimate the engine load.

5.1.1   Important Engine Events in the Cycle

Some of the important events in the engine cycle are shown in the p-V diagram (also referred to as the indicator diagram) in Figure 5.4, where the cylinder pressure shown in Figure 5.2 is plotted in relation to cylinder volume. The term indicator diagram, used for cylinder pressure traces plotted against crank angle revolution or cylinder volume, originates from the early mechanical cylinder measurement devices called *indicators*. The indicator was in some way connected to the cylinder so it could register the cylinder pressure, and it was also connected to either the crankshaft or the piston movement, so that pressure registration was made in relation to the crank angle rotation or the cylinder volume.

1. Inlet Valve Closing (IVC) usually occurs around 30–60° after BDC where the compression of the gas starts. The inlet valve closes after BDC in the compression stroke, this is in order to benefit from ram effects (the inertia of the gas moving in the intake manifold further adds some gas into the cylinder). Well-tuned intake runners can result in volumetric efficiencies greater than unity, η_{vol} > 1. IVC is one of the principal factors that influence the high-speed volumetric efficiency. It also effects the low-speed volumetric efficiency due to back flow into the intake (Heywood 1988).
2. The combustion is initiated by the energy discharge at the spark plug and the combustion starts propagating through the combustion chamber. The spark discharge occurs around 350° before TDC, and it has a direct influence on the engine efficiency and emissions. The spark timing will be dealt with more thoroughly in Section 10.6.
3. The pressure reaches its maximum around 17° after TDC and at this point the combustion is completed to 80–90%.

4. Exhaust Valve Opening (EVO) occurs around 50–60° before BDC and starts the blow-down process. When the exhaust valve opens, the hot gas expands out into the exhaust causing the cylinder pressure to decrease during the last part of the expansion stroke. The goal here is to reduce the cylinder pressure to the exhaust manifold pressure as quickly as possible.
5. Inlet Valve Opening (IVO) occurs around 10–25° before TDC. Engine performance is relatively insensitive to this point. It should occur sufficiently before TDC such that cylinder pressure does not dip early in the intake stroke.
6. Exhaust Valve Closing (EVC) occurs around 8–20° after TDC. At low loads with very low manifold pressures the timing determines the amount of residual gases that will be present in the combustion chamber. At high speeds and loads it influences the amount of burned gases that are exhausted.

More information about the valve events EVO, IVO, EVC, and IVC will be given in Section 12.1 on variable valve actuation. There are several parameters that can be controlled during this process that affect the performance of the engine. Some examples are:

- amount of air
- amount of injected fuel
- timing of combustion (ignition or injection)
- manifold pressure (using throttle and/or turbocharger)
- amount of burned gases in the cylinder (using exhaust gas recirculation (EGR))
- timing of the inlet valve opening and closing
- and timing of the exhaust valve opening and closing.

The first four are available on basically all production cars today, while the others have been used in advanced engines but are becoming more widely adopted.

5.2 Thermodynamic Cycle Analysis

Given a measured cylinder pressure that traces the work that has been produced during the cycle, W_i can easily be calculated by integrating the enclosed area in a p-V diagram

$$W_i = \oint p \, dV$$

To simplify the analysis of the cycle efficiency and the influence of different engine parameters, the pressure trace is plotted in a p-V diagram, Figure 5.4, and is modeled using cycles with similar properties and appearance. Figure 5.5 shows the p-V diagram of the cylinder pressure that was earlier displayed in Figure 5.4. Also shown in the figure are three ideal models of the engine cycle: Otto, Seiliger, and Diesel cycle. The Otto and Diesel cycles are just names of two specific model cycles and are not directly coupled to the operation of Otto or Diesel engines; at some operating conditions the Otto engine is best modeled as a Diesel cycle and sometimes the converse is better. However, it is the Seiliger cycle that best mimics the behavior of both engines.

An internal combustion engine is not a heat engine, if we strictly apply the thermodynamic definition, in the sense that it is not a closed system since it exchanges mass with the surroundings. A discussion of the differences between the ideal models and the measured pressure trace is performed later in connection with Figure 5.15.

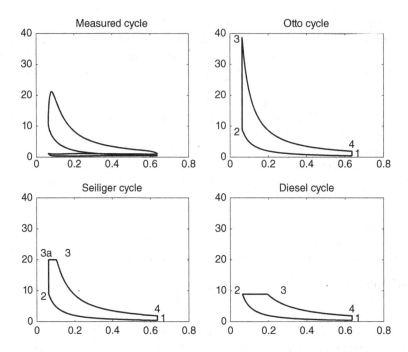

Figure 5.5   p-V diagrams for a measured cycle and for three other common cycles for the pressure

The work that is produced by the cycle is outlined by the events 1-2-3-4, in Figure 5.4. It is called *gross indicated work* $W_{i,g}$. When the engine passes through the events 4-5-6-1 it consumes work when it pumps out burned gases and inducts fresh mixture. The area in the p-V diagram that gives this work is called *indicated pumping work* $W_{i,p}$. The *net indicated work*, $W_{i,n}$, is the gross less the pumping work, that is $W_{i,n} = W_{i,g} - W_{i,p}$. The term indicated refers to the fact that the quantities are derived from the indicator diagrams, that is related to in-cylinder pressure. The first step in the analysis is to look at the gross indicated work, and the cycle efficiency for such cycles. Note that such a cycle is equal to a cycle where the engine operates at full load with wide open throttle. In such a case the inlet pressure p_{im} is at atmospheric pressure and equal to the exhaust pressure p_{em} which also is at atmospheric pressure, there fore the area enclosed by the points 4-5-6-1 is zero and can be left out. This is equivalent to disregarding the pumping losses, that is the work that the engine performs when it pumps out the burned gases and pumps in the fresh mixture.

5.2.1   Ideal Models of Engine Processes

For modeling and analysis, the engine operating cycle is divided into several separate processes that are modeled individually. The assumptions made in the modeling of the different processes are summarized in Table 5.1. In the ideal models the thermodynamic properties of the working fluid are considered to be those of an ideal gas with specific heats (c_p and c_v) constant through out the cycle. These models are usually referred to as *ideal gas standard cycles*. There is a change in c_p and c_v during the cycle and also before and after combustion. There is a small

Table 5.1 Table of thermodynamic properties for the different processes in the ideal gas standard cycle

Process	Model properties	Equations
Compression (1–2)	Isentropic (adiabatic and reversible)	$dq = ds = 0$
Combustion (2–3)	Adiabatic heat addition	
	a) Constant volume	$dq = c_v dT$
	b) Constant pressure	$dq = c_p dT$
	c) Partly constant volume and partly constant pressure	
Expansion (3–4)	Isentropic (adiabatic and reversible)	$dq = ds = 0$
Heat exchange (4–1)	Constant volume	$dV = 0$

change in the specific gas constant, R, due to the change in molecules and molecule weight before and after combustion. This small change is disregarded in the ideal gas standard cycles.

There are two parameters that are of great interest when comparing engine cycles. The first is the *thermal efficiency* or indicated fuel conversion efficiency,

$$\eta_{f,i} = \frac{W_i}{m_f \, q_{LHV}}$$

where m_f is the mass of fuel inducted into the engine during one cycle, q_{LHV} is the specific heating value of the fuel which represents the energy contents in one kilogram of fuel, and W_i is the indicated work. The indicated fuel conversion efficiency is a measure of how efficient the cycle is in converting the energy supplied by the fuel, $m_f \, q_{LHV}$, to mechanical work, W_i.

The second parameter is the *indicated mean effective pressure* (IMEP), which follows directly from the definition of MEP (4.4)

$$IMEP = \frac{W_i}{V_d}$$

The quantity IMEP is related to the indicated work per cycle, and is independent of engine size since it is normalized with the displaced volume V_d. The term indicated once again refers to the fact that the quantity is computed from the indicator diagram. There is a connection between IMEP and brake mean effective pressure (BMEP) which was the work that the engine actually produced (for a definition of BMEP see (4.5)). IMEP corresponds to the work produced inside the cylinder. Work is consumed by the friction between the piston and cylinder wall as well as friction in the connecting rod and crankshaft assembly. This friction work is calculated as the difference between the indicated work and the brake work

$$W_{fr} = W_i - W_b$$

The friction work can also be normalized with the engine size and gives an equivalent *friction mean effective pressure* (FMEP).

$$FMEP = IMEP - BMEP$$

A Closer Look at the Processes

The states and properties of the different processes that the engine cycle goes through are summarized here for convenience of referring to the equations later on. The derivation is made for a constant volume cycle but the modifications that are valid for the analysis of the other cycles are obvious and will be made in the section that derives the efficiencies. The final states of the processes in the cycle, 1-2-3-4-1, are defined in Figure 5.5.

Before we start analyzing the processes in the cycle we refresh two useful relationships from thermodynamics. First we have the difference between the specific heats c_p and c_v which for an ideal gas becomes

$$c_p - c_v = R$$

Then we also remember the definition of the ratio of specific heats

$$\gamma = \frac{c_p}{c_v}$$

Compression (1–2) The model is an isentropic (adiabatic and reversible) process. For adiabatic and reversible processes the first thermodynamic law states (see Table 5.1),

$$dq = du + dw = 0$$

The ideal gas law is stated as $pV = mRT$ and for unity mass the ideal gas law is expressed as $pv = RT$ where $v = \frac{V}{m}$ is the specific volume. Using the following relationships for reversible processes $du = c_v\, dT, dw = p\, dv$ we get

$$c_v\, dT + \frac{RT\, dv}{v} = 0$$

Dividing by T and integrating we get

$$c_v \ln\left(\frac{p\,v}{R}\right) + R \ln v = \text{const}$$

dividing by c_v, using $\frac{R}{c_v} = \frac{c_p - c_v}{c_v} = \gamma - 1$, and rewriting we get

$$\ln\left(\frac{pv}{R}\right) + \ln v^{\gamma-1} = \text{const}$$

Finally, since R is constant, we get

$$p\,v^\gamma = \text{const}$$

In summary, for a reversible adiabatic process for a perfect gas between states 1 and 2 we receive the following three relationships:

$$p_1 v_1^\gamma = p_2 v_2^\gamma \quad \text{or} \quad \frac{p_2}{p_1} = \left(\frac{v_1}{v_2}\right)^\gamma = r_c^\gamma \tag{5.1}$$

With the ideal gas law $p = \frac{RT}{v}$ we get

$$\frac{T_2}{T_1} = \left(\frac{v_1}{v_2}\right)^{\gamma-1} = r_c^{\gamma-1} \tag{5.2}$$

and

$$\frac{T_2}{T_1} = \left(\frac{p_2}{p_1}\right)^{1-1/\gamma} \tag{5.3}$$

Combustion (2–3) During the isochoric (constant volume) heating process, the fuel's chemical energy is converted to thermal energy

$$m_f\, q_{LHV} = m_t\, c_v\, (T_3 - T_2) \qquad (5.4)$$

where m_t is the total mass of the cylinder charge. The short-hand notation

$$q_{in} = \frac{m_f\, q_{LHV}}{m_t}$$

is used for convenience in the following text. With this notation, q_{in} is the specific energy contents of the cylinder charge, that is the energy contents of the charge per unit mass. The heat adding process can then be written

$$q_{in} = c_v\,(T_3 - T_2) \quad \text{or} \quad \frac{T_3}{T_2} = 1 + \frac{q_{in}}{c_v\, T_2} \qquad (5.5)$$

The isochoric process for an ideal gas, $\frac{p}{T} = \frac{R}{v} = \text{const}$, directly yields the following ratio for the pressure rise.

$$\frac{p_3}{p_2} = \frac{T_3}{T_2} \qquad (5.6)$$

Expansion (3–4) The expansion process is also an isentropic process as the compression and (5.1) to (5.3) are valid with notational changes.

Heat exchange (4–1) This is for cycles where the pumping during the gas exchange is not taken into account, that is Figure 5.5. In these cycles, heat is lost to the environment and the loss is $q_{loss} = c_v\,(T_4 - T_1)$.

5.2.2 Derivation of Cycle Efficiencies

Now we have gone through the processes in the ideal cycle, shown in Figure 5.5, and derived a set of equations for these, we can analyze the performance of the cycles. First we will look at the performance in terms of the efficiency, and then the work and indicated mean effective pressure.

Constant Volume Cycle – Otto Cycle

For the Otto cycle the heat is supplied from the fuel between points 2 and 3, $q_{in} = c_v\,(T_3 - T_2)$ and similarly the only heat loss is between 4 and 1 $q_{loss} = c_v\,(T_4 - T_1)$. Note that during the blow-down energy is lost, due to the hot gases that expand out into the exhaust manifold. The compression and expansion processes 1–2 and 3–4 are isentropic and therefore there is no heat flow during them. This yields the following expression for the thermal efficiency

$$\eta_{f,ig} = \frac{q_{in} - q_{loss}}{q_{in}} = \frac{c_v\,(T_3 - T_2) - c_v\,(T_4 - T_1)}{c_v\,(T_3 - T_2)} = 1 - \frac{(T_4 - T_1)}{(T_3 - T_2)}$$

Using the relationship for temperatures under isentropic compression and expansion, (5.2) we get the following $T_3 = T_4\, r_c^{\gamma-1}$ and $T_2 = T_1\, r_c^{\gamma-1}$. Inserting these expressions we finally get the Otto cycle efficiency

$$\eta_{f,ig} = 1 - \frac{1}{r_c^{\gamma-1}} \qquad (5.7)$$

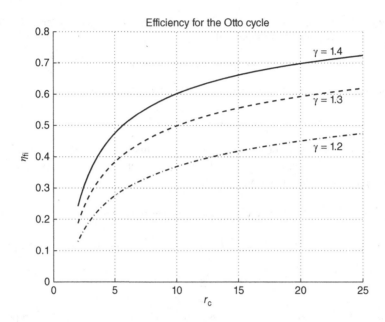

Figure 5.6   Efficiency of the Otto cycle as a function of compression ratio and ratio of specific heats γ

In Figure 5.6 the efficiency for the Otto cycle is plotted as a function of the *compression ratio* r_c for different values of γ. For an SI engine operating at stoichiometric mixture $\gamma = 1.3$ is a good approximation for the pressure data. In the figure it can be seen that the gain in efficiency decreases as the compression increases above 10.

Constant Pressure Cycle – Diesel Cycle

In the Diesel cycle, the heat is added in a reversible constant pressure process. The heat added during the process is $q_{in} = c_p (T_3 - T_2)$ and the heat loss is between 4 and 1 which is $q_{loss} = c_v (T_4 - T_1)$. The cycle efficiency is thus

$$\eta_{f,ig} = \frac{q_{in} - q_{loss}}{q_{in}} = \frac{c_p (T_3 - T_2) - c_v (T_4 - T_1)}{c_p (T_3 - T_2)} = 1 - \frac{(T_4 - T_1)}{\gamma (T_3 - T_2)}$$

For the isobaric process $dp = 0$, such as between 2 and 3, and for an ideal gas the following relationships hold:

$$\frac{v}{T} = \frac{R}{p} = \text{const}, \quad \frac{T_3}{T_2} = \frac{v_3}{v_2}$$

for the isentropic processes 1–2 and 3–4 we have $p_4 \, v_4^\gamma = p_3 \, v_3^\gamma$ and $p_1 \, v_1^\gamma = p_2 \, v_2^\gamma$. Dividing these with each other and noting that $v_4 = v_1$ and $p_2 = p_3$ yields $\frac{p_4}{p_1} = \left(\frac{v_3}{v_2} \right)^\gamma$. For the isochoric, $dv = 0$, process (4–1) we have

$$\frac{p}{T} = \frac{R}{v} = \text{const}, \quad \frac{T_4}{T_1} = \frac{p_4}{p_1}$$

The cycle efficiency can now be written

$$\eta_{f,ig} = 1 - \frac{1}{\gamma} \frac{T_1}{T_2} \frac{\frac{T_4}{T_1} - 1}{\frac{T_3}{T_2} - 1} = 1 - \frac{1}{r_c^{\gamma-1}} \frac{1}{\gamma} \frac{\left(\frac{v_3}{v_2}\right)^\gamma - 1}{\frac{v_3}{v_2} - 1}$$

Denoting $\frac{v_3}{v_2} = \beta$ finally gives

$$\eta_{f,ig} = 1 - \frac{1}{r_c^{\gamma-1}} \frac{\beta^\gamma - 1}{(\beta - 1)\gamma} \qquad (5.8)$$

Limited Pressure Cycle – Seiliger Cycle

By denoting $\frac{p_3}{p_2} = \alpha$ and $\frac{v_3}{v_2} = \beta$ we can derive the following efficiency for the Seiliger cycle

$$\eta_{f,ig} = 1 - \frac{1}{r_c^{\gamma-1}} \frac{\alpha\beta^\gamma - 1}{\alpha(\beta - 1)\gamma + \alpha - 1} \qquad (5.9)$$

The derivation of the Seiliger cycle efficiency is left as an exercise.

It is interesting to note that the Otto ($\beta = 1$) and Diesel ($\alpha = 1$) cycles are special cases of the Seiliger cycle. The parameter α introduced can be viewed as a parameter that gives the shape of the pressure trace and can be tuned to a real cycle.

Work and Mean Effective Pressure for Ideal Cycles

Above, we studied the efficiencies for the ideal cycles. These efficiencies can now be used directly to determine the gross indicated mean effective pressure IMEP_g and the gross indicated work $W_{i,g}$. The efficiency tells us how well the cycle utilizes the energy available in the fuel $m_f q_{LHV}$, and converts it to work

$$W_{i,g} = \eta_{f,ig} \, m_f \, q_{LHV}$$

Based on this expression, IMEP_g can directly be determined

$$\text{IMEP}_g = \frac{m_f \, q_{LHV} \, \eta_{f,ig}}{V_d}$$

5.2.3　Gas Exchange and Pumping Work

Up until now we have only studied the high pressure part of the cycles, these are also the work producing part of the cycles. To also include the pumping work (see Figure 5.7) the gas exchange process, with its state transitions, must be analyzed more carefully than above, where it was modeled only as a heat exchange state 4 to 1. The states and processes, 1-2-3-4-5-6-7-1, that are included in the thermodynamic cycle when accounting for pumping are shown in Figure 5.7.

Blow-down (4–5) The blow-down refers to when some of the hot gases in the cylinder are blown out into the exhaust and the cylinder pressure decreases down to $p_{cyl} = p_{em}$. The gas

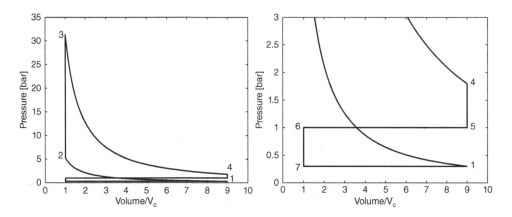

Figure 5.7 p-V diagram including intake and the exhaust stroke for part load (throttled) operation. The conditions were $p_{im} = 30\,kPa$ and $p_{em} = 100\,kPa$

that remains in the combustion chamber during the blow-down is assumed to undergo an isentropic expansion process. The connection between the pressure and temperature changes can be expressed as a state change of the fluid that can be derived by a similar derivation as (5.3). The result is

$$\frac{T_e}{T_4} = \left(\frac{p_{em}}{p_4}\right)^{1-1/\gamma} \tag{5.10}$$

Exhaust (5–6) The gases are pushed out into the exhaust system under constant pressure p_{em} and without heat transfer, that is $dT = 0$ which gives $T_6 = T_e$. The gas that remains in the combustion chamber at state 6 when the exhaust valve is closed is called the *residual gas*, and its mass is denoted m_r.

Inlet valve opening (6–7) When the inlet valve opens some of the gas expands into the intake manifold. During this phase it is assumed that the residual gas expands isentropically from p_{em} to p_{im} into the intake manifold which gives the following relationships for the residual gas temperature and residual gas volume

$$T_r = T_6 \left(\frac{p_{im}}{p_{em}}\right)^{1-1/\gamma} \qquad \text{and} \qquad V_r = V_c \left(\frac{p_{em}}{p_{im}}\right)^{1/\gamma} \tag{5.11}$$

Intake (7–1) The cylinder volume is filled with a fresh charge, m_{af}, at constant pressure, p_{im}, and the fresh charge mixes with the residual gas m_r in the combustion chamber. The mixing is modeled as an adiabatic mixing between the fresh charge and the residual gases, which is the same as $dU = 0$

$$U_1 = m_r\,c_v\,T_r + m_{af}\,c_v\,T_{im} \qquad \text{and} \qquad U_2 = (m_r\,c_v + m_{af}\,c_v)T_1$$

The energy balance together with the assumption that c_v is the same for burned and unburned gases gives an initial temperature T_1 that becomes the mass averaged temperature

$$T_1 = \frac{m_r\,T_r + m_{af}\,T_{im}}{m_r + m_{af}} \tag{5.12}$$

Pumping Work

During the exhaust and intake processes the piston produces work on the gases as burned mixture is pumped out, and the gases produce work on the piston as fresh mixture is pumped into the combustion chamber, respectively. To analyze the influence of the *pumping work* on the over all engine efficiency we also use the p-V diagram where the pumping phases of the combustion are considered. Figure 5.7 shows the p-V diagram for a cycle with intake and exhaust strokes. Assuming that the pressure equals the exhaust pressure, p_{em}, during the exhaust stroke and the intake manifold pressure, p_{im}, during the intake stroke, then the indicated pumping work can be derived as

$$W_{i,p} = p_{em}(V_5 - V_6) - p_{im}(V_5 - V_6) = (p_{em} - p_{im})(V_5 - V_6)$$

We can now determine the net indicated efficiency for the cycle as

$$\eta_{f,in} = \frac{W_{i,g} - W_{i,p}}{m_f\, q_{LHV}} = \eta_{f,ig}\left(1 - \frac{p_{em} - p_{im}}{IMEP_g}\right) \qquad (5.13)$$

where $IMEP_g$ is gross indicated mean effective pressure. The pumping work can also be expressed as a pumping mean effective pressure (PMEP), which becomes

$$PMEP = \frac{W_{i,p}}{V_d} = p_{em} - p_{im}$$

The net indicated mean effective pressure becomes

$$IMEP_n = IMEP_g - PMEP$$

With this notation for the pumping work, the indicated net efficiency can be expressed as

$$\eta_{f,in} = \eta_{f,ig}\left(1 - \frac{PMEP}{IMEP_g}\right)$$

The blow-down, exhaust, and intake processes and their thermodynamic properties will now be analyzed more carefully in the residual gas section below.

5.2.4   *Residual Gases and Volumetric Efficiency for Ideal Cycles*

During the gas exchange process, the combusted gas can not be exchanged completely, since the clearance volume is occupied by burned gases after the exhaust stroke and when the exhaust valve is closed. These residual gases are mixed with fresh mixture during the intake stroke. Additionally, back flow from the exhaust manifold can generate a higher amount of residual gases. Residual gases are sometimes also called internal EGR, since it is internal Exhaust Gas Recirculation (EGR).

Residual gases have two effects on the properties of the cycle. Firstly, there is the effect that it occupies a volume that cannot be filled with a fresh mixture of air and fuel, this reduces the volumetric efficiency. Secondly, the residual gases have a mass that influences the temperature rise from the combustion, see (5.4). There, the total mass in the combustion is the sum of air, fuel, and residual gas masses, that is

$$m_t = m_a + m_f + m_r$$

and when there is an increased residual gas mass the temperature after the combustion will decrease.

Volumetric Efficiency

Let's start by studying state 6 in the p-V diagram in Figure 5.7. In this state the pressure is p_{em} and the exhaust valve is closed. In this state the clearance volume V_c is completely filled with residual gases. When the intake valve is opened and the gases are expanded isentropically from p_{em} down to p_{im} along the path from state 6 to state 7 they occupy a volume

$$V_r = V_c \left(\frac{p_{em}}{p_{im}} \right)^{1/\gamma}$$

This means that some of the residual gases are expanded into the intake but they are then inducted into the cylinder during the intake stroke. The volume that the fresh air-and-fuel charge occupies at state 1 is now

$$V_{af} = V_1 - V_r = V_d + V_c - V_c \left(\frac{p_{em}}{p_{im}} \right)^{1/\gamma}$$

Now we can look at the volumetric efficiency, which is defined as the ratio between the volume of fresh air in the cylinder divided by the displacement volume, see the discussion before (4.9). The volumetric efficiency for the ideal cycle can thus be expressed as follows for

$$\eta_{vol} = \frac{V_a}{V_d} = \frac{V_a}{V_{af}} \frac{V_{af}}{V_d} = \frac{V_a}{V_{af}} \frac{V_d + V_c - V_c \left(\frac{p_{em}}{p_{im}} \right)^{1/\gamma}}{V_d} = \frac{V_a}{V_{af}} \frac{r_c - \left(\frac{p_{em}}{p_{im}} \right)^{1/\gamma}}{r_c - 1}$$

To get a more convenient expression for the ratio $\frac{V_a}{V_{af}}$ we have to look at the mixture of air and fuel and the molecule weights. The volumes that air and vaporized fuel occupy are

$$V_a = \frac{m_a R_a T}{p} = \frac{m_a \tilde{R} T}{M_a p}$$

and

$$V_f = \frac{m_f R_f T}{p} = \frac{m_f \tilde{R} T}{M_f p}$$

where M_a and M_f are the mole mass of air and fuel respectively. Some manipulations, where the definition of the air-to-fuel ratio (see Section 4.1.3) has been used, yield the final expression for the volumetric efficiency

$$\eta_{vol} = \underbrace{\frac{1}{1 + \frac{1}{\lambda (A/F)_s} \frac{M_a}{M_f}}}_{\approx 1} \frac{r_c - \left(\frac{p_{em}}{p_{im}} \right)^{1/\gamma}}{r_c - 1} \tag{5.14}$$

This expression shows how the volumetric efficiency depends mainly on the pressure ratio between the intake and exhaust. We also see that when the pressure ratio reaches the limit of

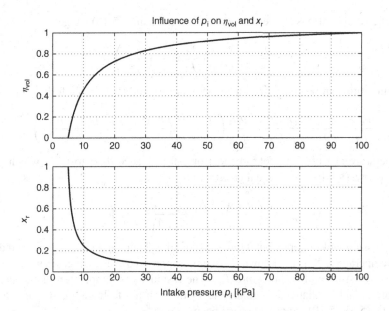

Figure 5.8 Residual gas and volumetric efficiency calculation for varying intake manifold pressures. In the calculations isooctane at stoichiometric conditions has been used and the conditions were: $p_{em} = 101.3$ (kPa), $T_{im} = 300$ (K), $q_{LHV} = 44$ (MJ/kg), (A/F) $= 14.6$, $\gamma = 1.3$, $R = 273$ (J/kg K), and $r_c = 10$. The residual gas fraction goes to unity and the volumetric efficiency goes to zero when the ratio p_{em}/p_{im} approaches r_c^γ, that is $p_{im} \rightarrow \frac{101.3}{10^{1.3}} \approx 5$ (kPa)

$\frac{p_{em}}{p_{im}} = r_c^\gamma$ the volumetric efficiency becomes zero and the engine cannot induct a new air and fuel mixture. In Figure 5.8 the volumetric efficiency is plotted as a function of intake manifold pressure, and the limit mentioned above is clearly seen.

Residual Gas Fraction

Given the residual gas mass, m_r, and the total mass of the gas in the cylinder, m_t, the residual gas fraction, x_r, is defined as

$$x_r = \frac{m_r}{m_t}$$

The residual gas fraction can be estimated using the ideal gas law and thermodynamic relationships for the engine cycle. In the following analysis only the Otto cycle is considered. The residual gas mass trapped when the exhaust valve closes is equal to the amount of gas at state 6, for which the mass can be calculated using the ideal gas law

$$m_r = \frac{p_{em} V_c}{R T_e}$$

The total mass of the charge m_t is determined by the ideal gas law in any of the states 1-2-3-4 and state 4 is chosen here since it simplifies the derivation. The total mass is

$$m_t = \frac{p_4 V}{R T_4}$$

Dividing the two equations above with each other and using (5.10) we get

$$x_r = \frac{m_r}{m_t} = \frac{V_c}{V} \frac{p_{em}}{p_4} \frac{T_4}{T_e} = \frac{1}{r_c} \left(\frac{p_{em}}{p_4} \right)^{1/\gamma}$$

The pressure ratio can be determined using (5.1), (5.6), and (5.5) in the following way

$$\frac{p_{em}}{p_4} = \frac{p_{em} \, p_{im} \, p_2 \, p_3}{p_{im} \, p_2 \, p_3 \, p_4} = \frac{p_{em}}{p_{im}} \left(\frac{1}{r_c} \right)^{\gamma} \frac{T_2}{T_3} \left(\frac{r_c}{1} \right)^{\gamma} = \frac{p_{em}}{p_{im}} \left(1 + \frac{q_{in}}{c_v \, T_2} \right)^{-1}$$ (5.15)

The temperature at state 2 is determined from state 1 according to (5.2) which gives the following expression for the residual gas fraction

$$x_r = \frac{1}{r_c} \left(\frac{p_{em}}{p_{im}} \right)^{1/\gamma} \left(1 + \frac{q_{in}}{c_v \, T_1 \, r_c^{\gamma - 1}} \right)^{-1/\gamma}$$ (5.16)

This expression looks simple, but it hides the fact that T_1 and q_{in} are functions of the residual gas. The temperature, T_1 is not known since it is composed by residual gases with temperature T_r and fresh mixture with temperature T_{im}. The residual gas temperature is the same as the temperature at state 7 that can be related to T_1 through the following chain of calculations, where (5.1), (5.5), (5.10), and (5.11) have been used

$$\frac{T_r}{T_1} = \frac{T_r \, T_5 \, T_4 \, T_3 \, T_2}{T_5 \, T_4 \, T_3 \, T_2 \, T_1} = \left(\frac{p_{im}}{p_{em}} \right)^{1 - 1/\gamma} \left(\frac{p_{em}}{p_4} \right)^{1 - 1/\gamma} \left(\frac{1}{r_c} \right)^{\gamma - 1} \left(1 + \frac{q_{in}}{c_v \, T_2} \right) \left(\frac{r_c}{1} \right)^{\gamma - 1}$$

Rewriting (5.15), inserting it into the equation above, and manipulating the expression, gives the following result

$$\frac{T_r}{T_1} = \left(1 + \frac{q_{in}}{c_v \, T_1 \, r_c^{\gamma - 1}} \right)^{1/\gamma}$$ (5.17)

Assuming that the mixing during the intake stroke is done under free expansion, $dq = 0$ and $dU = 0$, the temperature T_1 can be calculated from the relation

$$U_1 = m_r \, c_v \, T_r + m_{af} \, c_v \, T_{im} = U_2 = (m_r + m_{af}) \, c_v \, T_1$$

which gives

$$T_1 = x_r \, T_r + (1 - x_r) T_{im}$$ (5.18)

Finally we have to consider the input of heat to the system in which the influence of the residual gases is shown. The specific heat supplied to the system, q_{in}, from the fuel was defined earlier as

$$q_{in} = \frac{m_f \, q_{LHV}}{m_t} = \frac{m_f}{m_{af}} q_{LHV} \frac{m_{af}}{m_t}$$

The ratio $\frac{m_f}{m_{af}}$ is the ratio of fuel to fresh charge and is identified as $\frac{1}{1 + (A/F)} = \frac{1}{1 + \lambda (A/F)_s}$, and the ratio $\frac{m_{af}}{m_t}$ is recognized as $1 - x_r$. Inserting these we get the following expression for the specific energy contents in the combustion and temperature increase due to combustion

$$q_{in} = \frac{1 - x_r}{1 + \lambda (A/F)_s} q_{LHV} \quad \Rightarrow \quad \Delta T = T_3 - T_2 = \frac{q_{in}}{c_v} = \frac{1 - x_r}{1 + \lambda (A/F)_s} \frac{q_{LHV}}{c_v}$$ (5.19)

Given a certain fuel, thermodynamic tables can be used to get a value of q_{LHV}. Together with fuel mass this value gives the energy contents in the combustion chamber.

To determine the residual gas fraction the four equations (5.16) to (5.19) have to be solved to give the values for the four variables x_r, T_r, T_1, and q_{in}. Unfortunately, the system of equations cannot be solved analytically, but as they are phrased here they can be solved numerically by fixed point iteration techniques. The following example illustrates the iteration technique.

Example 5.1 (Calculation of residual gas fraction)   Find the residual gas fraction, for an engine running on isooctane C_8H_{18} and air with a stoichiometric mixture. During the engine operation the exhaust pressure is 101.3 kPa and the inlet air temperature is 20 °C and the intake manifold pressure is 30 kPa. (This is approximately the condition that the engine in Figure 5.3 was running under). The stoichiometric air/fuel ratio for the fuel is $(A/F) = 15.1$, the heating value for the fuel is $q_{LHV} = 44.6$ MJ/kg, the ratio of specific heats is $\gamma = 1.3$, and the specific heat is $c_v = 946$ J/kg·K. The engine has a compression ratio of 10.1.

To solve the problem we run the fixed point iteration scheme. First we have to assume an initial value for x_r so we get a starting point for the iteration, let's assume $x_r = 0$. Then we use (5.19) to determine a first value of $q_{in} = 2.77$ MJ/kg. With these initial conditions we can iteratively use (5.16), (5.19), (5.17), and (5.18) to determine the residual gas fraction. The result from the iterations is shown in the table below. The procedure has converged after six iterations and the residual gas fraction for this operating condition is 7.6%.

Iteration	(5.16) x_r [-]	(5.19) q_{in} [MJ/kg]	(5.17) T_r [K]	(5.18) T_1 [K]
1	0.0637	2.594	1114	345.3
2	0.0736	2.566	1177	358.1
3	0.0757	2.560	1192	361.1
4	0.0762	2.559	1196	361.8
5	0.0763	2.559	1197	362.0
6	0.0763	2.559	1197	362.0

The compression ratio r_c and the ratio between the exhaust pressure and intake pressure p_{em}/p_{im} both have a direct impact on the residual gas fraction. This is easily understood by considering the conditions that apply at states 6, 7, and 1 in the ideal cycle, Figure 5.7. State 6 gives the mass of the residual gas and state 7 gives how much the residual gas has expanded due to the lower pressure.

Figure 5.9 shows how the residual gas is influenced by r_c and p_{em}/p_{im}. The system of equations, (5.16) to (5.19), has been solved for different compression ratios and intake pressures, using isooctane fuel at stoichiometric conditions. It can be seen that an increased intake pressure decreases the residual gas fraction, and that an increased compression ratio also decreases the residual gas fraction.

It is also worth mentioning that the discussion above deals with theoretical cycles in which there is neither heat transfer nor any back-flow. However, it illustrates well the underlying principles that govern the engine operation.

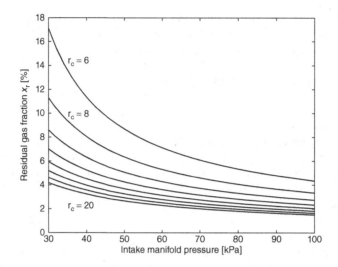

Figure 5.9 Residual gas calculation using isooctane at stoichiometric conditions. The conditions for the calculations are the same as in Figure 5.8 but here the compression ratio is also changed. The residual gas fraction decreases with increasing compression ratio and increasing intake pressure

5.3 Efficiency of Ideal Cycles

The most important results from the ideal gas cycles are summarized and described in this section. For proper cycle simulation and comparison some thermodynamic properties must be assumed, such as the energy contents of the mixture, but when comparing the efficiencies of the ideal cycles no assumptions (except for γ) have to be made. Changing the energy contents only changes the appearance of the cycle, that is the ratio p_3/p_2 (or v_3/v_2).

Ideal Gas Standard Cycles

In Figure 5.10, simulation results from the three standard cycles are shown. The figure illustrates how the three cycles deviate from each other when they all are supplied the same amount of chemical energy. For all three cycles the energy content of the fuel was assumed to be isooctane running at a stoichiometric mixture, which yields $q_{in} = 2.92 \cdot 10^6 (1 - x_r)$ and $\gamma = 1.3$. The cycle was generated with a residual gas fraction of $x_r = 7\%$, compression ratio of $r_c = 9$, and an intake manifold pressure of 100 kPa. Furthermore, for the limited pressure cycle the limit on the maximum pressure was set to $p_3/p_1 = 35$.

The Otto cycle reaches the highest pressure as expected since all combustion energy is used at TDC, the Seiliger pressure reaches the pressure limit of 3.5 MPa before it changes over to constant pressure, and the Diesel cycle has lowest maximum pressure. In the Otto cycle all chemical energy is consumed at TDC and used to raise the thermal energy of the gas in the cylinder. During the expansion its thermal energy is converted to mechanical work, so for the Otto cycle more thermal energy is converted to mechanical early in the expansion compared to a Seiliger cycle. Therefore, the pressure for the Otto cycle is lower than for the Seiliger cycle, late in the expansion. The same applies to a comparison between the Seiliger and Diesel cycles.

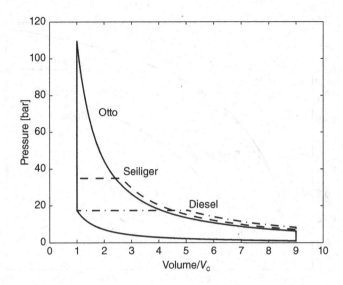

Figure 5.10 Simulation results for ideal gas standard cycles. Three p-V diagrams are plotted that show the behavior of the ideal cycles

Efficiency Comparison for Ideal Gas Standard Cycles

Efficiencies for the three ideal gas standard cycles are compared in Figure 5.11 for changed compression ratio r_c. The Otto cycle has highest efficiency, the Diesel cycle has lowest efficiency, and in between lies the Seiliger cycle. As upper limits for the limited pressure cycle, the pressure ratio p_3/p_1 has been chosen at 100, 70, and 40 bar. It follows naturally that the higher maximum pressure (and temperature) the engine tolerates the higher efficiency it achieves.

Studying Figure 5.5 it can be seen that the measured cycle is best depicted by the limited pressure cycle where part of the fuel burns at a constant volume and the rest burns at constant pressure. By placing realistic upper limits, ~ 10 MPa, on the maximum pressure for a cycle, the gain in efficiency is less when the compression ratio is increased over 10. The dotted line in Figure 5.11 represents a limited pressure cycle where the burning is constructed so that 70% of the energy is released at constant volume and 30% at constant pressure, this construction is maintained until the maximum pressure p_3 reaches 6.5 MPa. The choice 70/30% is made since it produces a cycle with a shape close to a measured cycle. The upper limit of 6.5 MPa is chosen as a realistic upper limit for the pressure so that the engine will not be destroyed by, for example, knock, see Section 6.2.2.

5.3.1 Load, Pumping Work, and Efficiency

The throttle angle and intake manifold pressure for SI engines have the function of controlling the engine load. As (5.13) shows, there is a loss in efficiency as the intake manifold pressure drops below the exhaust pressure. The engine load, or equivalently the indicated mean effective pressure, is used to illustrate the losses that come when a naturally aspirated operates at lower intake pressures. A set of cycle simulations has been performed in which the intake manifold pressure p_{im} was changed from $p_{im} = 1$ atm down until the indicated mean effective pressure

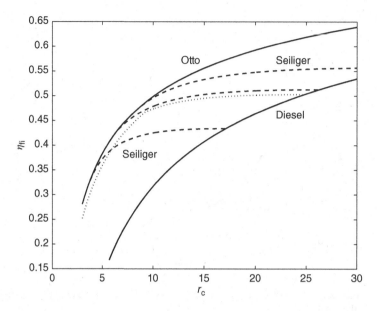

Figure 5.11 Efficiency comparisons for the three ideal gas standard cycles for $\gamma = 1.3$. The constant volume cycle is most efficient. For the limited pressure cycle, the pressure has been limited to $\frac{p_3}{p_1} = 100$, $\frac{p_3}{p_1} = 70$, and $\frac{p_3}{p_1} = 40$. A limited pressure cycle that has both a limit on the maximum pressure and a limit on the ratio, $\frac{p_3}{p_2}$, is also included. This cycle imitates the behavior of a real cycle, where the limit on the maximum pressure is imposed by the knocking and the ratio limit is imposed by the limited combustion rate

reached zero, which corresponds to an operating cycle that does not produce any work. During the simulation, efficiency and indicated mean effective pressure were recorded. To get a relative measurement of the load, IMEP is normalized by the maximum IMEP, denoted IMEP_{max}. The result is displayed in Figure 5.12 where the top plot shows that the efficiency decreases, when the load decreases as it has to overcome the pumping work.

In the bottom plot of Figure 5.12 the (indicated) specific fuel consumption (SFC) is shown in relation to the load. As the load decreases and the efficiency tends to zero the specific fuel consumption increases towards infinity, which follows from (4.8). Pumping work is one (friction is another) part of the explanation for the low efficiency at low loads of SI engines, compare the BSFC data along a single engine speed line in Figure 4.5. The trend is general with a higher SFC for lower loads and the best BSFC is attained close to the maximum load.

5.3.2 (A/F) Ratio and Efficiency

In the previous discussions on engine efficiency the influence of the air-to-fuel ratio has not been considered, this is because the influence of the mixture strength cannot be seen in the ideal gas models and therefore more complex models must be used. Changing the air/fuel ratio in the cylinder also changes the thermodynamic properties of the cylinder charge, for example γ. In the analysis of the impact of the equivalence ratio the standard cycles are completed with thermodynamic property data. The data is used to track the thermodynamic properties of the

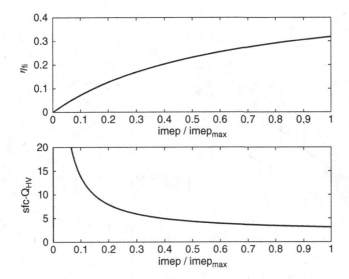

Figure 5.12 Efficiency for an Otto cycle with pumping losses as a function of engine load, with the load normalized by the maximum load. As load decreases from the maximum load the engine efficiency also decreases. Also shown is the (indicated) specific fuel consumption

mixture during the compression, combustion, and expansion more accurately than ideal gas standard cycles with their constant c_v and c_p. Changes in specific heats c_p and c_v are tracked for changes in mixture composition and temperature. Among others, the ratio of specific heats $\gamma = \frac{c_p}{c_v}$ depends nonlinearly on temperature. Such an analysis is called a *fuel–air cycle analysis* since it takes into account the properties of the fuel–air mixture. A fuel–air cycle analysis can be performed using, for example, the Chemical Equilibrium Program Package (CHEPP) (Eriksson 2005), but a full analysis is outside the scope of this text and the interested reader is referred to Heywood (1988, chapter 5) for details. However, the main results are important and are therefore summarized in Figure 5.13 and the following list.

- The compression ratio has the same effect on the efficiency as previously demonstrated in Figure 5.6. Changes in the equivalence ratio, λ, can be viewed as an appropriate change in γ, compare Figures 5.6 and 5.13.
- The efficiency increases as the mixture is becomes leaner, increasing λ. This occurs because the burned gas temperature decreases, which decreases the specific heats and thus increases the effective value of γ over the expansion stroke.
- A mixture with much fuel is called a rich mixture, the efficiency decreases because of lack of sufficient air for complete oxidation of the fuel. The lack of air more than offsets the effect of decreasing the burned gas temperature which decreases the mixture's specific heats.
- The increase in indicated mean effective pressure is proportional to the product $m_f \cdot \eta_{f,i}$. The fuel–air equivalence ratio ϕ is also proportional to m_f. The indicated mean effective pressure attains a maximum in the range $\phi \in [1.0, 1.1]$, that is slightly rich. This means that the maximum power output from an engine occurs for a slightly rich mixture.
- Variations in p_{im}, T_{im}, and x_r have only a modest effect on $\eta_{f,i}$. Their impact on IMEP is more substantial because IMEP depends directly on the initial charge density.

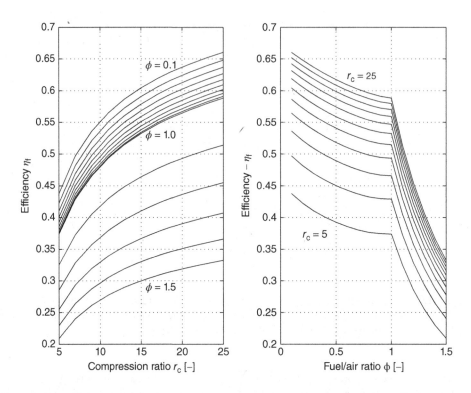

Figure 5.13 Efficiency of the constant volume cycle as a function of compression ratio and fuel–air equivalence ratio ϕ. The conditions were: isooctane fuel, $x_r = 0.05$, $p_{im} = 1$ bar, and $T_1 = 350\,K$

A p-V diagram from an engine operating close to optimum has an area of about 80% of the fuel–air cycle (Taylor 1985). This is often used for estimation purposes and in models for engine work and torque, see for example $\eta_{ig,ch}$ in the model (7.57).

SI Engines and Air/Fuel Ratio Sweeps

The fuel–air cycle analysis in the previous section indicates two things for the mixture of strength and efficiency. Firstly, when fuel enrichment is used ($\phi > 1$ or $\lambda < 1$) the efficiency drops, this is also seen in real engines. Secondly, efficiency increases monotonously with a leaner mixture, this is not seen in homogeneous SI engines. When the mixture is made leaner in an real SI engine, it reaches a limit where stable combustion can not be sustained and the torque and efficiency starts to decrease. Figure 5.14 shows two measured torque curves as a function of λ. In the data, marked by circles and a dashed line, λ was changed by varying the amount of fuel while maintaining the amount of air as constant. This experiment illustrates the fact that maximum torque and power are achieved when the mixture is slightly rich $\lambda \approx 0.9$ (i.e., $\phi \approx 1.1$). This effect can be utilized to increase the power and torque at full load; when the air flow has reached its limit the torque can be increased slightly by adding more fuel.

In the data in Figure 5.14, marked by crosses and a solid line, λ is changed by varying the amount of air while maintaining the amount of fuel as constant. With a constant amount of fuel the efficiency becomes proportional to the torque. As can be seen, the measured efficiency differs slightly in appearance compared to the results from the fuel–air cycle shown

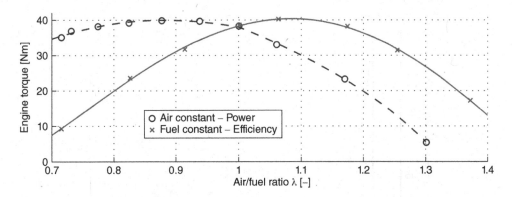

Figure 5.14 Torque as a function of λ measured at part load in an SI engine. The dashed line and circles represent a case when the amount of air is constant while the fuel is varied. This shows that the maximum torque and power of an engine can be attained for slightly rich mixtures. The solid line and crosses represent a case when the fuel was maintained as constant and the amount of air was varied. In this case the torque is proportional to the efficiency

in Figure 5.13. Maximum efficiency for an SI engine with a homogeneous mixture is attained around $\lambda = 1.1$.

There is a combustion stability limit for lean mixtures where the fuel no longer can be ignited and combusted, this is the reason for the reduction in torque when the mixture is made leaner in Figure 5.14. For very lean conditions the combustion reactions can no longer be sustained. This limit is called the *lean limit* and for engines running on homogeneous mixtures the lean limit lies around $\lambda \approx 1.5$ but is engine dependent. At the other end there is also a *rich limit* that occurs around $\lambda \approx 0.5$, where stable combustion cannot be sustained due to lack of oxygen.

5.3.3 Differences between Ideal and Real Cycles

Ideal cycles have been used to illustrate the operating principles and how different parameters effect output power and efficiency. Figure 5.15 shows one measured cycle and one Otto cycle. The discussion below summarizes the main differences between the ideal gas standard cycles and real.

1. The specific heats c_p and c_v for the fluid are not constant during the cycle. The ratio of specific heats for the unburned mixture is around $\gamma_u = 1.3$, and for the burned mixture $\gamma_b = 1.2$. During expansion, heat is transferred from the fluid to the combustion chamber walls which lowers the pressure during the expansion stroke. Measurements have shown that $\gamma_b = 1.3$ still gives a good fit for pressure and volume data during the expansion. Therefore, the choice of $\gamma = 1.3$ throughout the complete cycle is a good approximation. Additionally c_p and c_v also depend nonlinearly on the temperature, but these effects are often not considered under analysis and evaluation.
2. Compression and expansion are not isentropic processes, there are losses due to friction and heat transfer to the combustion chamber walls and to the coolants. The influence of heat transfer is not of importance in the compression stroke since the temperature difference between the walls and the fluid is not as great as in the expansion stroke when the mixture has burned and has a higher temperature. However, at the end of the compression stroke the

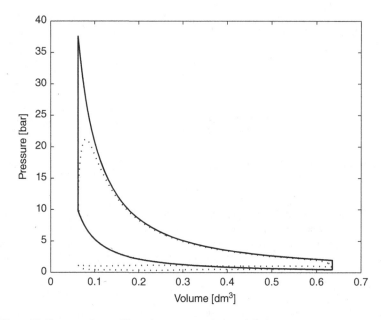

Figure 5.15 p-V diagrams for an SI engine compared to an ideal gas Otto cycle. The idealized processes, where one completes before the next starts, give the Otto cycle edges that are not found in a real cycle. Overall, and especially during compression and expansion, there are many similarities

measured pressure trace can lie slightly below the ideal gas cycle due to heat transfer, but this cannot be seen in the figure since it is too small. The slight increase in the measured pressure is explained under item 4. For the expansion stroke. the temperature of the gases is much higher and the heat transfer is more significant, but following the discussion in the previous item, the value of $\gamma = 1.3$ still gives a good fit to measured data during the expansion stroke.

3. The mixture composition changes during the combustion, which also changes the gas constant, R.

4. Time losses: the combustion is not instantaneous, it takes several crank angles to complete (~60°) and therefore there are losses during both compression and expansion. The major time losses occur early in the expansion since the combustion continues for several crank angle degrees after the piston has reached TDC. Late in the compression stroke, see Figure 5.15, the measured pressure trace starts to increase before the piston has reached TDC which is due to the fact that the mixture is ignited at approximately ~20° BTDC. Furthermore, it takes time for the combustion to complete and the measured cycle cannot reach the same high pressure as the Otto cycle.

5. The inlet and exhaust valve opening and closing are not instantaneous. Particularly, the exhaust valve opens 40–60° before BDC, which makes the pressure drop during the expansion stroke which reduces the expansion work.

6. Crevice effects. The crevices comprise 1–2% of the clearance volume. When the gas that enters the crevices assumes a temperature close to the temperature of the cylinder walls, which is significantly colder than the average gas, its density increases and the crevices can therefore include a very high percentage (up to 10%) of the cylinder charge (Gatowski et al. 1984).

7. Incomplete combustion. All the fuel present in the combustion chamber is not combusted. SI engines at warmed up conditions can have unburned hydrocarbons in the range of $1-3\%$ of the total fuel mass. Also, CO and H_2 contain $1-2\%$ of the energy from the fuel. The total combustion efficiency is therefore around 95% if there is enough air to complete the combustion reactions.

5.4   Models for In-Cylinder Processes

Cycles used as models in previous sections follow idealized processes, for example isentropic compression or instantaneous heat release at TDC, which allows us to develop analytical solutions and, most importantly, gain insight about the fundamental operating principles and limits. Due to the idealizations the p-V diagrams have, for example, sharp corners that are not seen in measurement data, compare Figure 5.5. Therefore, the next step is to develop more detailed models, capable of describing the measured pressure traces more accurately. These models incorporate effects such as finite burning time, as well as heat and mass transfers, giving models that cannot be solved analytically but instead can to be simulated using computers.

A cylinder pressure trace contains much information, it is, for example, influenced by the volume change, heat transfer, and combustion. Modeling these processes allows us to extract important information from the cylinder pressure, and one important element is when and how the combustion occurs in the cylinder. Using the pressure trace to extract the combustion profile in the cylinder is called *heat release analysis*. The analysis is performed within the framework of the first law of thermodynamics for the cylinder contents, and a commonly used model family is the single zone model.

5.4.1   Single-Zone Models

In a single-zone model, the cylinder contents are treated as a single gas with homogeneous pressure, temperature, and composition. Combustion is modeled as a heat adding process that increases the temperature and pressure of the gas. Mass and energy conservation laws are used to derive the basic governing equations for the models. The mass of the zone is denoted m and its differential is

$$dm = \sum_i dm_i \qquad (5.20)$$

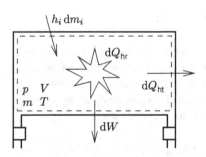

Figure 5.16 Control volume for a first law analysis of the gas in the cylinder. The boundary is dashed and the arrows indicate positive energy transfer directions

where dm_i is the mass transfer for flow i to the zone. A first law analysis of the processes in the cylinder, see Figure 5.16, is used to derive a differential equation for how the cylinder state evolves as a function of crank angle (or time). The result is the following expression

$$dU = dQ_{hr} - dW - dQ_{ht} + \sum_i dH_i$$

where extensive properties are used, that is, the mass is included. This prepares the equations so they will also be valid when considering mass flows to and from the cylinder. In the next step we insert the work, through volume expansion, done by the gas $dW = p \, dV$, and eliminate p with the ideal gas law $p \, V = m \, R \, T$. Differentiating the internal energy $U = m \, u$ and applying the chain rule gives

$$dU = m \, du + u \, dm = m \, c_v \, dT + u \, dm$$

which gives the following expression

$$m \, c_v \, dT + u \, dm = dQ_{hr} - \frac{m \, R \, T}{V} dV - dQ_{ht} + \sum_i h_i \, dm_i \tag{5.21}$$

Special care must be taken for the internal energy of the mass flows, $u \, dm$, these must have proper book-keeping for each flow $u_i^* \, dm_i$.

Solving (5.21) for the temperature differential gives

$$dT = -\frac{(\gamma - 1) \, T}{V} dV + \frac{1}{m \, c_v} \left\{ dQ_{hr} - dQ_{ht} + \sum_i (h_i - u_i^*) dm_i \right\} \tag{5.22}$$

which relates the temperature change to the volume change as well as mass and energy transfers. (5.20) and (5.22) constitute the basic equations for describing the contents and properties of a zone. These are the basis for many models that can be found in the literature, where the main differences lie in the submodels for heat release dQ_{hr}, heat transfer dQ_{ht}, and mass flow dm_i. In many models an equivalent formulation of (5.22) is used that is derived using the ideal gas law in standard form $p \, V = m \, R \, T$ and differentiated form $p \, dV + V \, dp = m \, R \, dT + R \, T \, dm$ to eliminate T and dT, giving an equation for the pressure differential instead of the temperature. One example of this is the much used model structure presented in Gatowski et al. (1984).

A Closed System Model

A first model for the closed system will now be derived. It is based on the assumption that the mass transfer is zero ($dm = 0$), that is considering the process during the time when the valves are closed and also assuming that there are no other mass transfers or leakages. This gives the following expression for the temperature differential

$$dT = -\frac{(\gamma - 1) \, T}{V} dV + \frac{1}{m \, c_v} (dQ_{hr} - dQ_{ht}) \tag{5.23}$$

which can be used to simulate the temperature during the closed part of the cycle when the heat release dQ_{hr} is given. The pressure can also be of interest and can be calculated from the ideal gas law $p = \frac{m \, R \, T}{V}$. As mentioned above, there is another frequently used formulation where (5.23) is rephrased into a differential equation for the pressure. This is done by using the ideal gas law in standard and differentiated form to eliminate dT and T, which gives

$$dp = -\frac{\gamma \, p}{V} dV + \frac{\gamma - 1}{V} (dQ_{hr} - dQ_{ht}) \tag{5.24}$$

If there is neither heat release, $dQ_{hr} = 0$, nor heat transfer $dQ_{ht} = 0$, then this equation describes the well-known pressure and volume connection of an isentropic process, see (5.1). When heat is released from combustion the pressure will rise above that of a pure compression and expansion process. How to determine and specify the heat release will be the next topic.

5.4.2   Heat Release and Mass Fraction Burned Analysis

Heat release analysis is a procedure that uses measured cylinder pressure traces to calculate the rate at which heat is released inside the combustion chamber. The basic idea is to study the pressure rise and deduct known effects such as volume change, heat transfer, and gas flows (such as flows to crevices and piston ring blow by) leaving the influence of combustion as the remaining effect that produces the pressure rise. The effects are deducted from the pressure rise based on a thermodynamic analysis of the process, and the most frequently used models are single-zone models.

Single-zone models are often used in two ways. One is the forward way, as in (5.24), where the heat release dQ_{hr} is provided as input which gives the pressure trace as output. Another way is to solve (5.24) for the heat release

$$dQ_{hr} = \frac{V}{\gamma - 1} dp + \frac{\gamma}{\gamma - 1} p \, dV + dQ_{ht} \tag{5.25}$$

and use a measured cylinder pressure trace p (and its derivative dp) as input to calculate the heat release rate dQ_{hr}. Cylinder pressure data is often sampled synchronously with the crank angle θ so that it is well synchronized with the volume, and it is then convenient to select θ as independent variable. The pressure trace $p_i = p(\theta_i)$ is available at the samples enumerated by i, and the pressure derivative needs to be calculated from it. It is calculated using

$$\frac{dp(\theta_i)}{d\theta} = \frac{p(\theta_{i+1}) - p(\theta_{i-1})}{\theta_{i+1} - \theta_{i-1}}$$

where it is important to use symmetric difference to avoid creating a phase lag between pressure, pressure derivative, volume, and volume derivative. To complete the analysis we also need the volume $V(\theta_i)$ and its derivative and these are determined directly from the volume function (4.3). To illustrate the procedure, two frequently used approaches will be used. The first is the *net heat release* method, that relies on a first law analysis, while the other is the classical Rassweiler–Withrow method.

Net Heat Release Method

In the net heat release method, developed in Krieger and Borman (1967), the heat transfer term dQ_{ht} in (5.25) is set to zero. The result from the heat release analysis of three cylinder pressure traces is shown in Figure 5.17. The Figure 5.17a shows the pressure traces $p(\theta)$, Figure 5.17b heat release rate traces

$$\frac{dQ_{hr}(\theta)}{d\theta} = \frac{V(\theta)}{\gamma - 1} \frac{dp(\theta)}{d\theta} + \frac{\gamma}{\gamma - 1} p(\theta) \frac{dV(\theta)}{d\theta}$$

and Figure 5.17c shows cumulative heat release traces. The heat release trace is calculated by integrating the heat release rate, starting from θ_{ivc} to the current angle

$$Q_{hr}(\theta) = \int_{\theta_{ivc}}^{\theta} \frac{dQ_{hr}(\alpha)}{d\alpha} d\alpha$$

It is not necessary to start the integration at IVC, an arbitrary point between IVC and the start of combustion can be used as starting point. The heat release profile is often specified as a function of crank angle with the help of a normalized curve called mass fraction burned,

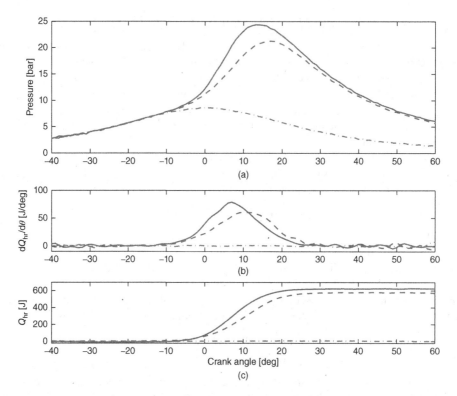

Figure 5.17 Heat release analysis performed for three cylinder pressure traces (solid – fast and more energy, dashed – less energy and slower, dash dotted – motored pressure). The tree plots are: (a) – cylinder pressure, (b) – heat release rate, (c) – cumulative heat release

$x_b(\theta)$, that starts at 0 before the combustion and ends at 1 when the combustion is finished, see Figure 5.18. Assuming that the mass fraction burned is directly proportional to the heat release gives a connection between the traces as

$$x_b(\theta) = \frac{Q_{hr}(\theta)}{\max Q_{hr}(\theta)} \qquad (5.26)$$

Rassweiler and Withrow's Classical Mass Fraction Burned Method

The classical method for determining the mass fraction burned x_b trace is the Rassweiler Withrow method (Rassweiler and Withrow 1938). The algorithm is based on the observation that when there is no combustion then the pressure and volume data can accurately be represented by the polytropic relation

$$pV^n = \text{constant}$$

where the exponent n for both compression and expansion gives a good fit to the data if it is in the range $n \in [1.25, 1.35]$. The pressure change, $\Delta p = p_{i+1} - p_i$, between two samples i and $i + 1$, is assumed to be made up of a pressure rise due to combustion, Δp_c, and a pressure rise due to volume change, Δp_v,

$$\Delta p = \Delta p_c + \Delta p_v \qquad (5.27)$$

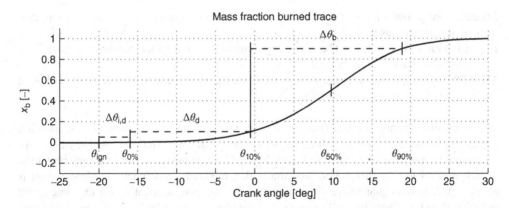

Figure 5.18 Mass fraction burned trace (solid curve) with combustion position and shape metrics: θ_{ign}–crank angle where ignition occurred, $\Delta\theta_{i,d} = \theta_{0\%} - \theta_{ign}$–ignition delay, $\Delta\theta_d = \theta_{10\%} - \theta_{0\%}$–flame development angle, $\Delta\theta_b = \theta_{90\%} - \theta_{10\%}$–rapid burning angle, and $\theta_{50\%}$–crank angle where 50% of the mass has burned

The pressures and volumes at the start and end of the interval, in the absence of combustion, are related by $p_i V_i^n = \hat{p}_{i+1} V_{i+1}^n$ which gives

$$\Delta p_v = \hat{p}_{i+1} - p_i = p_i \left(\left(\frac{V_i}{V_{i+1}} \right)^n - 1 \right) \tag{5.28}$$

The pressure rise due to combustion Δp_c can now be solved from (5.27). Assuming that the pressure rise due to combustion in the interval is proportional to the mass of mixture that burns, then the mass fraction burned at the end of the i'th interval can be summed up and thus becomes

$$x_b(i) = \frac{m_b(i)}{m_b(total)} = \frac{\sum_0^i \Delta p_c}{\sum_0^M \Delta p_c} \tag{5.29}$$

where M is the total number of crank angle intervals. Using (5.27) to (5.29) the mass fraction burned profile can be calculated.

5.4.3 Characterization of Mass Fraction Burned

When analyzing combustion, the heat release position and shape are often characterized using a set of metrics that are indicated in Figure 5.18 and defined in the list below. These metrics are used to characterize the rate of combustion and also the position for combustion.

Ignition timing θ_{ign} – The crank angle where the ignition occurs in SI engines. For CI engines this corresponds to the start of injection θ_{soi}.

Ignition delay $\Delta\theta_{i,d}$ – The crank angle interval from ignition or injection until the combustion starts or becomes visible in the x_b-trace. This is important in CI engines, but in SI engines the combustion flame is initiated by the ignition so there is no delay. However, in the early stages the flame front is so small that it does not become visible on the mass fraction burned trace, and this can be interpreted as a delay.

Flame development angle $\Delta\theta_d$ – The crank angle interval from start of combustion until 10% of the mass has burned.

Rapid burning angle $\Delta\theta_b$ – The crank angle interval during which combustion goes from 10% to 90% mass fraction burned.

MFB 50 $\theta_{50\%}$ – The crank angle where 50% of the mass has burned. This is often used as an indicator of the combustion position.

When modeling combustion, a functional description of the combustion process is often needed, and there are essentially three ways that have different variations. One way is to give a measured x'be-trace (e.g., obtained from heat release analysis) as input to the simulation model, another is to model the combustion with a parameterized function, while a third is to use a differential equation for the combustion propagation, accounting for the state in the cylinder. Here the discussion will be on the second alternative, and a frequently used equation is the Vibe (also called Wiebe or Wibe) model, described in Vibe (1970).

$$x_b(\theta) = \begin{cases} 0 & \text{if } \theta < \theta_0, \\ 1 - e^{-a\left(\frac{\theta-\theta_0}{\Delta\theta}\right)^{m+1}} & \text{if } \theta \geq \theta_0 \end{cases} \tag{5.30}$$

where the parameter θ_0 gives the start of combustion, $\Delta\theta$ and a are related to combustion duration, and m influences the shape of the burning profile. A small m positions the combustion early, see Figure 5.19 for an illustration. Measured mass fraction burned profiles have been fitted with $a = 5$ and $m = 2$ for SI engines in Heywood et al. (1979). One must note that the Vibe function is overparameterized, in particular a and $\Delta\theta$ cannot be uniquely determined (while the parameter m can).

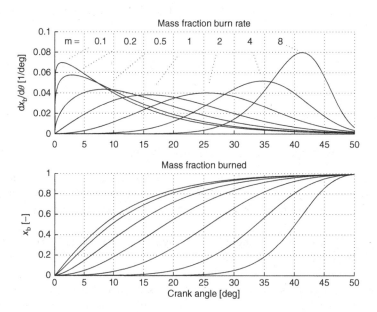

Figure 5.19 Shape of the Vibe function as a function of its parameter m. The other parameters are $\Delta\theta = 50°$ and $a = 5$

If the burn angles $\Delta\theta_d$ and $\Delta\theta_b$ are available, the Vibe parameters can be calculated from them. Due to the overparameterization, either $\Delta\theta$ or a must be specified for a unique solution, and by specifying $\Delta\theta$ beforehand (using for example $\Delta\theta = \Delta\theta_d + \Delta\theta_b$) the Vibe parameters can be calculated by

$$m = \frac{\ln\left(\frac{\ln(1-0.1)}{\ln(1-0.9)}\right)}{(\ln(\Delta\theta_d) - \ln(\Delta\theta_d + \Delta\theta_b))} - 1$$

$$a = -\ln(1 - 0.1)\left(\frac{\Delta\theta}{\Delta\theta_d}\right)^{m+1}$$

In CI engines with a single injection the combustion has two phases, a premixed burning and a main burning, and a frequently used approach for this is to use a sum of two Vibe functions (Merker et al. 2006). With multiple injections, more combustion phases are introduced and, depending on the number and their importance, these might also need to be modeled to adequately describe the combustion profile. A common approach is to model the mass fraction burned $x_b(\theta)$ with a sum of Vibe functions as

$$x_b(\theta) = \sum_i w_i\, x_{bi}(\theta), \quad \text{with } w_i \geq 0, \text{ and } \sum_i w_i = 1 \tag{5.31}$$

where each x_{bi}-function (5.30) has its own parameters a_i, m_i, $\Delta\theta_i$, and $\theta_{0,i}$.

5.4.4 More Single-Zone Model Components

Single zone models can give a good description of the cylinder pressure, and the quality depends on the submodels that are included. There are two components that are worth studying further, namely heat transfer and the description of the gas properties.

Gas to Cylinder Wall Heat Transfer

Heat losses from the combustion chamber to the walls are in the region of 20% of the supplied fuel energy, and need to be accounted for in many investigations. During the cycle, the majority comes from the expansion and exhaust stroke. In SI and CI engines, the major heat transfer path from the gas to the walls is through forced convection. Radiation from the gas to the walls is small in SI engines but in CI engines can contribute to 20–35% of the total heat transfer.

The rate of energy transfer by convection, which occurs in a direction perpendicular to the surface fluid interface, $\dot{Q}_{ht}$, is described using Newton's law of cooling

$$\dot{Q}_{ht} = h\, A\, \Delta T = h\, A\, (T - T_w) \tag{5.32}$$

where A is the surface area of the body which is in contact with the fluid, ΔT is the appropriate temperature difference between gas and surface, and h is the convection heat transfer coefficient. The most important task is to accurately predict the magnitude of the convection heat transfer coefficient. Since this quantity is a composite of both microscopic and macroscopic phenomena, many factors must be taken into consideration. For many flow geometries, h can be described by a Nusselt/Reynolds/Prandtl correlation

$$Nu = C\, (Re)^m\, (Pr)^n$$

see, for example, Appendix A.4 or Holman (2009) for heat transfer background.

Several expressions and correlations for determining h have been published for engines, see, for example Annand (1963), Eichelberg (1939), Zapf (1969). A much used method was proposed by Woschni in Woschni (1967), which essentially is a Nusselt–Reynolds number of the form $Nu = 0.035\,Re^m$, with $m = 0.8$ and several considerations for the terms in the Nusselt and Reynolds numbers. The final expression was set to

$$h = C_0\, B^{-0.2}\, p^{0.8}\, w^{0.8}\, T^{-0.53} \tag{5.33}$$

with $C_0 = 1.30 \cdot 10^{-2}$. The characteristic velocity w is modeled with two terms

$$w = C_1\, S_p + C_2\, \frac{V\, T_{ivc}}{V_{ivc}\, p_{ivc}}(p - p_m) \tag{5.34}$$

where the first describes the general gas movement and depends on the mean piston speed, S_p, while the second adds the effect of increased gas motion, due to combustion. The latter is modeled as the pressure rise over the motored pressure and is referred to a datum state at IVC (T_{ivc}, V_{ivc}, and p_{ivc}). The parameters, variables, and units are

B	cylinder bore [m]	S_p	mean piston speed [m/s]
p	cylinder pressure [Pa]	T	bulk temperature [K]
p_m	motored pressure [Pa]	V	volume [m^3]
h	heat transfer coeff [W/m^2 K]	w	characteristic velocity [m/s]

where parameters C_1 and C_2 depend on the phase in the engine cycle

	Gas exchange	Compression	Combustion and expansion m/s
C_1	6.18	2.28	2.28
C_2	0.0	0.0	0.00324

The Woschni heat transfer correlation was developed using a Diesel engine without swirl, and has become much used over the years for other configurations. It has also been modified to cover swirl (by adding a swirl term to S_p in (5.34) as well as new combustion concepts, see, for example, Chang et al. (2004), Hohenberg (1979), Sihling and Woschni (1979).

Newton's law of cooling (5.32) (with submodels (5.33)–(5.34)) gives the rate of the heat transfer $\dot{Q}_{ht}$ which is given per time unit. To simulate it in the crank angle domain, it is necessary to make a change of variable from time to crank angle as follows,

$$\frac{dQ_{ht}}{d\theta} = \frac{dQ_{ht}}{dt}\frac{dt}{d\theta} = \dot{Q}_{ht}\frac{1}{\omega_e}$$

where ω_e is the engine speed in rad/s.

Gas Properties

When analyzing the cylinder contents with the first law it is important to have a good description of the relation of work, temperature, and pressure to each other. This is by thermodynamic properties, c_p, c_v, that in the model equations are expressed using $\gamma = \frac{c_p}{c_v}$ and R. A sensitivity study of the impact of different model parameters on the cylinder pressure, in Klein (2007), pointed out γ as the most important variable. In an engine these depend on temperature and pressure, where γ decreases with temperature. A simple and effective model is to use a linear model in temperature (Gatowski et al. 1984)

$$\gamma(T) = \gamma_{300} - b\,(T - 300) \tag{5.35}$$

where $\gamma_{300} \approx 1.35$ is γ at $T = 300$ K, and $b \approx 7 \cdot 10^{-5}$ is the slope of the decrease with temperature. More advanced models for single zone models are investigated in Klein and Eriksson (2005).

5.4.5 A Single-zone Cylinder Pressure Model

All building blocks for a useful single-zone model are now available. So far the equations above have been described using differentials, which enables a modeler to select the independent variable when the implementation of the model is made. In many models, time is selected as independent variable but when the in cylinder processes are studied the crank angle θ is often selected as independent variable, as is done in the following model.

Model 5.1 Single-Zone Model with Combustion and Heat Transfer

---

An single-zone model that describes the cylinder contents from Inlet Valve Closing (IVC) to Exhaust Valve Opening (EVO) in crank angle domain is now developed. The starting point is IVC where the pressure, volume, and temperature are given, that is p_{ivc}, V_{ivc}, and T_{ivc}. The ideal gas law is used to determine the mass in the cylinder

$$m_t = \frac{p_{ivc}\, V_{ivc}}{R\, T_{ivc}} \tag{5.36}$$

which is constant during the interval. The functions for volume, volume change, gas temperature, specific heat ratio, and heat release rate are

$$V(\theta) = V_d \left[\frac{1}{r_c - 1} + \frac{1}{2}\left(\frac{l}{a} + 1 - \cos\theta - \sqrt{\left(\frac{l}{a}\right)^2 - \sin^2\theta} \right) \right]$$

$$\frac{dV(\theta)}{d\theta} = \frac{1}{2} V_d \sin\theta \left(1 + \frac{\cos\theta}{\sqrt{\left(\frac{l}{a}\right)^2 - \sin^2\theta}} \right)$$

$$T(\theta) = \frac{p(\theta)\, V(\theta)}{m_t\, R}$$

$$\gamma(T(\theta)) = \gamma_{300} - b\,(T(\theta) - 300)$$

$$\frac{dQ_{hr}(\theta)}{d\theta} = m_f\, q_{LHV}\, \eta_{co}\, \frac{dx_b}{d\theta}(\theta)$$

where $m_f\, q_{LHV}\, \eta_{co}$ is the amount of energy to be released and where $\frac{dx_b(\theta)}{d\theta}$ is the Vibe function described above. Using (5.24) with θ as an independent variable and the expressions above, gives the following ordinary differential equation for the pressure

$$\frac{dp(\theta)}{d\theta} = -\frac{\gamma(T(\theta))\, p(\theta)}{V(\theta)}\frac{dV(\theta)}{d\theta} + \frac{\gamma(T(\theta)) - 1}{V(\theta)}\left\{\frac{dQ_{hr}(\theta)}{d\theta} - \frac{dQ_{ht}(\theta)}{d\theta}\right\} \qquad (5.37)$$

that is solved numerically and gives the pressure as a function of crank angle, $p(\theta)$. In the expressions it is also necessary to have the motored pressure, which is equal to $p(\theta)$ before ignition and is then approximated with a polytropic process after ignition, giving the equation

$$p_m(\theta) = \begin{cases} p(\theta), & \theta \leq \theta_0 \\ p(\theta_0)\left(\frac{V(\theta)}{V(\theta_0)}\right)^n, & \theta > \theta_0 \end{cases}$$

where n is the polytropic exponent that approximates γ.

A comparison between this model and a measured cylinder pressure trace is shown in Figure 5.20. As can be seen in the figure, single-zone models can describe the in-cylinder pressure well and thus also the work production of the cylinder. This family of models are therefore well suited to analyzing the engine and combustion with respect to work production and engine efficiency.

Note that Model 5 has an equivalent heat release model that is obtained by solving (5.37) for $\frac{dQ_{hr}(\theta)}{d\theta}$ and providing the pressure traces $p(\theta)$ and p_m as input. This model has the same components, except for crevice flow, as the often used Gatowski model presented in Gatowski et al. (1984).

5.4.6   Multi-zone Models

When more detail about the cylinder contents or combustion is needed, for example to describe emissions or engine knock, other more detailed models are used. A common approach is to

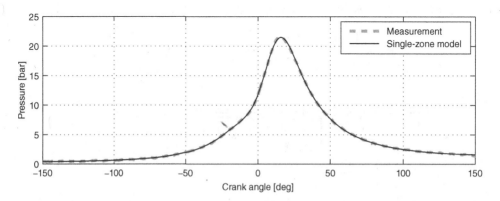

Figure 5.20 Comparison between the single-zone model (solid thin) and measured (shaded dashed) cylinder pressure trace from an SI engine, showing that the single-zone model can give a good description of the pressure development. Model tuning has been carried out using the method in Eriksson (1998)

divide the cylinder content into two zones, burned and unburned, and track the thermodynamic state and properties in each zone.

The fundamentals of the modeling are the same as for single zone, that is, energy and mass balances as well as the state equation (ideal gas law). There is a difference between the single- and multi-zone models in how the combustion is modeled. In single-zone models, combustion is modeled as a heat addition process, that is, dQ_{hr} is given as input, while in multi-zone models the combustion is modeled as a mass transfer between burned and unburned zones, where the mass element that burns is tracked with the absolute enthalpy of combustion reactants and products. The mathematical description given below for a general formulation for multi-zone models is based on the development discussed in Nilsson and Eriksson (2001).

In a multi-zone model, a system with N zones is considered, where the total system has a certain volume, pressure, and mass (V, p, and m). The pressure p is assumed to be homogeneous throughout the combustion chamber. Each zone has its own volume, temperature, mass (V_i, T_i, and m_i for zone i) and gas composition. The change in system volume and the mass transfer from zone i to zone j (dV and dm_{ij} respectively) are assumed to be known, and the mass of a zone can easily be determined by integrating the mass flows in and out of the zone. The changes of the remaining quantities (p, T_i and V_i, $i = 1, 2, \ldots, N$) are unknown and are to be determined.

The system consists of N zones and the sum of all V_i, $i = 1, 2, \ldots, N$, must thus be the same as the total volume, that is

$$\sum_i dV_i = dV \tag{5.38}$$

The energy balance for zone i is

$$m_i \, du_i + \sum_{j \neq i} u_i dm_{ij} = -dQ_i - dW_i + \sum_{j \neq i} dm_{ij} h_{ij} \tag{5.39}$$

In multi-zone models the gas composition can vary (foremost in burned zones) and the internal energy u_i can thus depend on both temperature and pressure

$$du_i(p, T_i) = \left(\frac{\partial u_i}{\partial p} \right)_{T_i} dp + \left(\frac{\partial u_i}{\partial T_i} \right)_p dT_i \tag{5.40}$$

The first term in (5.40) can be expanded, using the second law of thermodynamics, to

$$dU_i = T_i \, dS_i - p \, dV_i$$

and the Maxwell relations (see e.g., Finn (1998))

$$\left(\frac{\partial u_i}{\partial p} \right)_{T_i} = T_i \left(\frac{\partial s_i}{\partial p} \right)_{T_i} - p \left(\frac{\partial v_i}{\partial p} \right)_{T_i} =$$

$$= -T_i \left(\frac{\partial v_i}{\partial T_i} \right)_p - p \left(\frac{\partial v_i}{\partial p} \right)_{T_i} = -\frac{T_i^2}{p} \left(\frac{\partial R_i}{\partial T_i} \right)_p - T_i \left(\frac{\partial R_i}{\partial p} \right)_{T_i} \tag{5.41}$$

The second term in (5.40) is

$$\left(\frac{\partial u_i}{\partial T_i} \right)_p = \left(\frac{\partial h_i}{\partial T_i} \right)_p - R_i - T_i \left(\frac{\partial R_i}{\partial T_i} \right)_p \tag{5.42}$$

and according to the definition $\left(\frac{\partial h_i}{\partial T_i}\right)_p = c_{p,i}$. The energy balance (5.39) can with (5.40), (5.41), and (5.42) be written as

$$p\,dV_i + c_i\,dp + d_i dT_i = -dQ_i + \sum_{j\neq i}(h_{ij} - h_i + R_i T_i)dm_{ij} \tag{5.43}$$

where

$$c_i = -m_i T_i\left(\frac{T_i}{p}\left(\frac{\partial R_i}{\partial T_i}\right)_p + \left(\frac{\partial R_i}{\partial p}\right)_{T_i}\right)$$

$$d_i = m_i\left(c_p - R_i - T_i\left(\frac{\partial R_i}{\partial T_i}\right)_p\right)$$

The general state equation, based on the gas law, that allows a varying R, is $pV = m\,R(p, T)\,T$. In differentiated form it gives the last equation needed to get an unambiguous equation system for a multi-zone model

$$p\,dV_i + a_i\,dp + b_i\,dT_i = RT\sum_{j\neq i}dm_{ij} \tag{5.44}$$

where

$$a_i = V_i\left(1 - \frac{p}{R_i}\left(\frac{\partial R_i}{\partial p}\right)_{T_i}\right)$$

$$b_i = -m_i\left(R_i + T_i\left(\frac{\partial R_i}{\partial T_i}\right)_p\right)$$

(5.38), (5.43), and (5.44) make up a system of $2N + 1$ ordinary differential equations. In these equations, the right-hand side of the equality signs are known, while the differentials on the left-hand side are to be determined. The differentials on the left hand enter the equations linearly, and the system can thus be expressed

$$
\begin{bmatrix}
0 & 1 & 0 & \cdots & 1 & 0 \\
a_1 & p & b_1 & \cdots & 0 & 0 \\
c_1 & p & d_1 & \cdots & 0 & 0 \\
\vdots & \vdots & \vdots & \ddots & \vdots & \vdots \\
a_N & 0 & 0 & \cdots & p & b_N \\
c_N & 0 & 0 & \cdots & p & d_N
\end{bmatrix}
\begin{bmatrix}
dp \\
dV_1 \\
dT_1 \\
\cdots \\
dV_N \\
dT_N
\end{bmatrix}
=
\begin{bmatrix}
dV \\
R_1 T_1 \sum_{i\neq 1} dm_{1i} \\
-dQ_1 + \sum_{i\neq 1}(h_{1i} - h_1 + R_1 T_1)dm_{1i} \\
\vdots \\
R_N T_N \sum_{i\neq N} dm_{Ni} \\
-dQ_N + \sum_{i\neq N}(h_{Ni} - h_N + R_N T_N)dm_{Ni}
\end{bmatrix}
$$

which can be arranged as a matrix operation

$$\mathbf{A}(x)\,dx = \mathbf{B}(x, du) \tag{5.45}$$

where x is the state and du is submodels that drive the process, for example the volume function dV, heat transfer models dQ, and mass flow models dm that transfer mass between zones. The central component is dx, which is the vector with unknown differentials of the states

$$dx = [dp\ dV_1\ dT_1\ \cdots\ dV_N\ dT_N]^T$$

The differentials, or time/crank angle derivatives, of the state variables, dx, are determined by solving the system of linear equations (5.45). The result can expressed as

$$dx = \mathbf{A}(x)^{-1}\, \mathbf{B}(x, du)$$

where one needs to remember that a numerically better solution must be used in practice. When the derivatives dx have been determined, the states can thereafter be calculated by numerical integration. For more information on the numerics and applications of the multi-zone formulation see Brand (2005), Nilsson (2007), Nilsson and Eriksson (2001), and Oberg (2009).

5.4.7 Applications for Zero-dimensional Models

The models discussed in this chapter belong to a family of models referred to as zero-dimensional models, since they do not describe any spatial (dimension) dependence. These models have been used since the 1960s and have become an everyday tool for modeling and analyzing engine processes. Heat release analysis, described in the preceding section, is one example. They are used to analyze the effects of different control commands on engine performance and emissions.

In general terms, one can say that single-zone models are used when studying the cylinders as a component and it sees them as a unit that produces work or has other types of energy generation to and from them. Examples of applications are cylinder pressure and work production simulation, engine out temperature, and flows through the exhaust and inlet valves. Multi-zone models are used when there is a need to study things that occur in the cylinders and when there are effects that have a spatial dependence in the cylinders, for example separating the burned and unburned gases or even dividing them into more zones to describe their behavior. Examples of applications are knock modeling in SI engines, NO emission generation, and combustion in diesel engines.

6

Combustion and Emissions

The purpose of the engine is to combust a fuel, that is convert the chemical energy to thermal energy, and then transform it into mechanical work. The preceding chapter dealt with the work conversion while this chapter will look more into processes such as mixture preparation, combustion processes in SI and CI engines, and formation of pollutants or emissions that come as unwanted byproducts from the cylinder. The engine out emissions or pollutants can be reduced in exhaust gas treatment systems, and these are discussed in Section 6.5.

6.1 Mixture Preparation and Combustion

Spark ignited (SI) and compression ignited (CI) engines have different ways of mixing and combusting the fuel and they are often characterized as having homogeneous or stratified charge. With *homogeneous charge* it is meant that the gas in the cylinder constitutes a mixture that is homogeneous throughout the whole cylinder. In *stratified charge* the air–fuel mixture is not homogeneously mixed in the cylinder where, for example, 60% of the cylinder volume might be occupied by air and residual gases only, and the rest might be occupied by a air–fuel-residual mixture with varying λ. The homogeneity of the mixture depends on the mixture preparation, which is a collective name for mixture formation, mixture transport, and mixture distribution to the cylinders.

6.1.1 Fuel Injection

In SI engines, fuel metering was traditionally done using a carburetor, but it is now controlled using fuel injection systems. For fuel injection control and mixture preparation there are two major types of injection systems: sequential fuel injection and direct injection.

Sequential fuel injection systems have an injector for each cylinder and the injection is timed to the inlet valve opening events of the individual cylinders, that is they operate sequentially. The timing is often to terminate the injection just before the inlet valve is opened to avoid the emission of soot. The injections can be controlled individually and fuel supply can be individually cut off for each cylinder. A picture of a rail with fuel injectors in a sequential fuel injection system is shown in Figure 7.15. Historically, there has also been single point injection and multi point bank injection. In *single point injection* the fuel is injected onto the

Modeling and Control of Engines and Drivelines, First Edition. Lars Eriksson and Lars Nielsen.
© 2014 John Wiley & Sons, Ltd. Published 2014 by John Wiley & Sons, Ltd.
Companion Website: www.wiley.com/go/powertrain

throttle plate, while *multi point bank injection* systems have an injector for each cylinder but they inject simultaneously into the individual inlet pipes.

Direct injection systems inject the fuel directly into the cylinder. Historically, this technology was mainly used in CI engines, but is also becoming common in SI engines, and is often called gasoline direct injection (GDI). In CI engines the injection is timed near TDC to start the combustion, but for SI engines it occurs either during the induction stroke or during the compression to provide a more homogeneous mixture. One advantage with gasoline direct injection is that there is no fuel film buildup in the intake system, another is that the fuel economy can be improved by operating the engine extremely leanly using a stratified charge. When direct injection is used to create a stratified charge in SI engines, the throttle is opened fully which gives a high air flow, then a small amount of fuel is injected during the compression stroke. This results in a global λ much greater than 1, which would not be combustible if the mixture was homogeneous, but with the late injection the fuel has no time to be transported and mixed in the whole combustion chamber. Thus a stratified mixture is formed, with a local λ in the combustible region, $\lambda \in [0.8, 1.3]$. An improved fuel economy is achieved through reduced pumping losses, since the throttle is fully open.

6.1.2 Comparing the SI and CI Combustion Process

Figure 6.1a illustrates the combustion processes in an SI engine. The fuel is injected relatively early, either in the intake system in port injected engines or during intake or early in the compression stroke in direct injected engines. This gives the fuel time to mix with air and form a homogeneous charge before the spark ignites the mixture. Under normal conditions the combustion is ignited by a spark at the spark plug that gives a small flame kernel. The flame kernel increases in size and develops into a turbulent flame that propagates through the homogeneous air and fuel mixture in the combustion chamber. When the flame has propagated through the chamber it is extinguished (quenched) at the cylinder walls. The combustion process takes

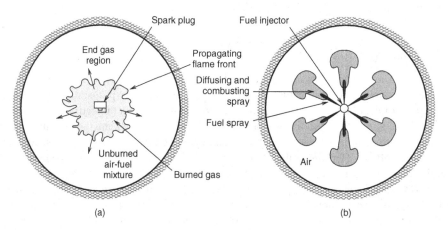

Figure 6.1   Top view of the cylinder, illustrating the differences between normal SI (a) and CI (b) combustion. In SI combustion, the charge is premixed and the spark starts a flame that propagates through the air–fuel mixture. In CI engines, the combustion starts after the fuel injection, as the fuel in the spray is heated and mixes with air

around 40–90° to complete, see the heat release trace in Figure 5.17, which corresponds to a flame propagation that is much slower than the speed of sound.

Figure 6.1b illustrates the combustion processes in an CI engine. The inducted air is compressed, and when it is time for combustion (some degrees before TDC) the fuel is injected directly into the combustion chamber. The fuel spray mixes with warm air and there is a delay from the start of injection to when the combustion reactions start. This occurs when the temperature is high enough and when there is oxygen available for the fuel to react with. Diesel engines thus operate with stratified charge since the fuel is injected directly into the cylinder and does not have time to mix before the combustion.

Engines that run with stratified charge, like diesels or gasoline direct injection (GDI) engines, can run with very high air-to-fuel ratios globally in the engine, but the combustion takes place in regions where λ is around 1. There are many variants of SI and CI engines, and the discussion here covers only the most common automotive configurations, the homogeneous SI engines and direct injected CI engines, for other types see, for example, Zhao (2010a,b), Heywood (1988) and Stone (1999).

6.2 SI Engine Combustion

Combustion rates in SI engines mainly depend on the laminar burning velocity and the turbulence. The *laminar burning velocity* depends on temperature, pressure, mixture strength (λ), and dilution (e.g., residual gases and EGR). For variations in mixture strength, the highest velocities are found for $\lambda \approx 0.9$ and it decreases with dilution (Heywood 1988). A higher temperature increases the velocity while a higher pressure decreases it. With little or no *turbulence* the flame propagates with a laminar flame front through the chamber and the laminar burning velocity sets the maximum. When turbulence increases, the flame front starts to wrinkle, which increases the effective area of the flame front and thereby the rate, at which the combustion consumes the air–fuel mixture is increased.

Engine operating conditions, like speed and load, influence these processes. Higher *engine loads* shorten the combustion duration which is due to: less residual gases, higher gas velocity (turbulence), and higher temperature. Higher loads also have higher pressures which decrease the velocity, but the other effects dominate, resulting in a faster combustion. Higher *engine speeds* increase the turbulence and thus the combustion rate when measuring it in seconds. Measuring the duration in crank angles shows a fairly flat trend in engine speed. In Eriksson (1999) the longest combustion duration (measured in crank angles) occurred for mid speeds around 2200 RPM, showing that turbulence for higher speeds increased the burn rate more than the speed shortened the time for combustion. Different models that describe trends in burn rates as a function of operating conditions have been presented, see Bayraktar and Durgun (2004), Blizard and Keck (1974), Csallner (1981) and Lindström et al. (2005) for some examples.

6.2.1 SI Engine Cycle-to-Cycle Variations

Cycle-to-cycle variations are always present in SI engines. Figure 6.2 shows ten consecutive cylinder pressures that clearly show the variations. The engine was operated at steady state conditions with all controllable parameters held constant, that is engine speed, throttle angle, fuel injection, and all temperatures and pressures were held constant. Thus, the variations in

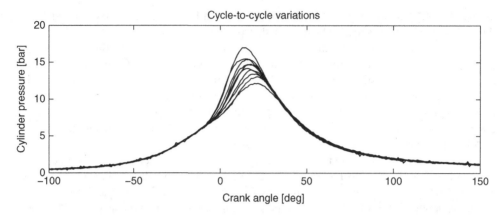

Figure 6.2 The variations in the combustion process and cylinder pressure are always present in SI engines. Cylinder pressure measurements from ten consecutive cycles have been measured where the engine is operating under steady state conditions

the figure are cycle-to-cycle variations and not environmental or controller variations. Three causes for the variations in the rate of how fast the mixture burns have been found:

- Variations in the gas motion in the cylinder. Changes in gas motion influence the combustion propagation.
- Variations in the amount of fuel, air, and recycled gases cause the amount of energy in the cylinder to vary from one cycle to another.
- Spatial variations in the concentration of air, fuel, and recycled gases. There is one spatial distribution within the cylinder for one cycle and another spatial distribution the next. The distribution close to the spark plug is particularly important for the early flame development and propagation.

The cycle-to-cycle variations in cylinder pressure impose limits on the engine operation. The engine controllers are tuned so that they have good performance in general, but the extreme cases, that can occur in the cycle-to-cycle variations can impose restrictions on the engine operation. For example, at high load conditions the fastest burning cycle is most probable to cause the engine to knock (see Section 6.2.2), due to their higher temperatures. Thus, in many cases the extreme cycles are the restricting ones and set the limit for performance.

6.2.2   Knock and Autoignition

Knock is a fundamental problem in spark ignited engines that can easily destroy an engine if it is allowed to continue.*Knock* is associated with the noise transmitted through the engine structure when there is a spontaneous ignition of a portion of the end-gas. The part of the air−fuel mixture that is in front of the flame front and has not yet burned is called the *end-gas*. When a spontaneous combustion process occurs, there is an extremely rapid release of much of the chemical energy in the end-gas, see Figure 6.3. The duration of this release is in the

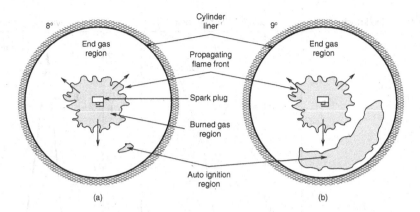

Figure 6.3 Cylinder seen from above at 8° and 9° ATDC, illustrating a combustion with autoignition. A spark plug has initiated the combustion and produced a flame that propagates through the chamber. At 8° (a) an autoignition occurs and it releases the heat much more rapidly than normal combustion, at 9° (b) it has consumed a large portion of the end-gas

order of 1°. Such a rapid process causes very high local pressures which in turn causes a propagation of pressure waves across the combustion chamber. This is illustrated in Figure 6.3 where a large portion of the end-gas is consumed rapidly, while not much happens in the flame front between 8° and 9° ATDC.

Knocking is seen in the cylinder pressure as oscillations that have an amplitude that decays with time and that often occur near the maximum pressure. Figure 6.4 shows three cylinder pressure traces: one with normal combustion, one with slight knock, and one with intense knock. The rapid and local pressure rise causes a pressure wave to propagate through the combustion chamber and excites the resonance modes of the cylinder. It is the resonance in the combustion chamber that produces the oscillatory signature in the cylinder pressure. The frequencies of the oscillations, associated with knocking condition, depend on the resonance modes of the combustion chamber and the first fundamental usually lies in the range 5 to 10 kHz.

The main theory for explaining the origin of knock is the *autoignition* theory. It states that when the fuel–air mixture in the end-gas region is compressed to sufficiently high pressures and temperatures, the fuel starts to oxidize spontaneously in parts or all of the end-gas region, see Figure 6.3. When the autoignition occurs this further heats the end-gas, leading to a combustion process that escalates and gives a very rapid energy release. An indication of this is seen in Figure 6.4, where the cylinder pressure increases steadily up until 8° ATDC and then there is a rapid acceleration of the burning.

Severe knock, if it is allowed to continue, can significantly damage the engine, and even if it does not damage it or is not severe it is a disturbing source of noise. The mechanisms that inflict damage on the engine through knock are thought to depend on heat transfer where the oscillations influence the thermal boundary layer at the cylinder wall. Severe knock can lead to engine failure in minutes Heywood (1988). Engine knock tendency can be controlled using the ignition timing, where a later ignition gives lower end-gas temperatures and thus lower knock tendency. This will be discussed more in Section 10.6.1.

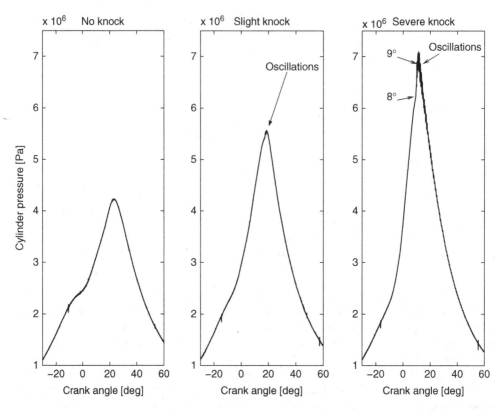

Figure 6.4 Three cylinder pressure traces from cycles where: no knocking, slight knock, and severe knock has occurred. In the rightmost plot a rapid pressure increase is visible, between 8° and 9° ATDC, coming from the heat release associated with the knock

6.2.3 Autoignition and Octane Number

Knocking is also related to fuels and their ability to start their oxidation process. The octane number, in particular, is a measure of how resistant the fuel is against knock. The octane number scale is based on two hydrocarbons: by definition *normal heptane* (n-C_7H_{16}) has an octane number of zero and *isooctane* (C_8H_{18} 2,2,4-trimethylpentane) has an octane number of 100. Figure 6.5 shows the chemical composition of these two hydrocarbons. Generally it can be said for paraffins that the longer the chain the higher the knock tendency, which is also true

Figure 6.5 The composition of the hydrocarbons n-heptane and isooctane, only the positions of the carbon molecules are shown since they define the structure

Table 6.1 Engine operating conditions for octane number determination

	Research	Motor
Engine speed	600 rpm	900 rpm
Ignition timing	13° BTDC	19–26° BTDC
	fixed	varies with r_c
Inlet temperature	52°C (125°F)	149°C (300°F)
Inlet pressure	1 atm	
Humidity	0.0036–0.0072 kg/kg dry air	
Coolant temperature	100°C	
Air to fuel ratio	Adjusted for maximum knock	

for these two molecules. A short or compact molecule is more knock resistive than a long molecule.

A special single-cylinder engine is used to determine the octane number of a fuel. The engine has two carburetors that it is possible to switch between and it is possible to change the compression of the engine while it is running by raising or lowering the cylinder. One carburetor contains the fuel that is studied and the other a reference fuel mixed from n-heptane and isooctane. The engine is run at a specified state, where intake pressure, intake temperature, humidity, coolant temperature, engine speed, ignition timing, and air–fuel ratio are specified. See Table 6.1 for the conditions for *Research Octane Number* (RON) and *Motor Octane Number* (MON) tests. When the engine is running at the reference conditions, the compression ratio is increased until the engine runs under slight knocking conditions with the fuel under test. Then the reference fuel is blended from isooctane and n-heptane until the reference mixture has the same knock intensity as the fuel under test. If the reference fuel has a mixture of 5% n-heptane and 95% isooctane the octane number for the tested fuel is 95.

This method is performed on a single-cylinder engine under specified and constant conditions, and does not always predict how a fuel will behave in an automotive engine under a variety of speed loads and weather conditions. MON is often lower than RON for normal fuels since it uses more severe operating conditions. Therefore, there exist several measures for further quantifying fuel properties and one is *fuel sensitivity*

$$\text{fuel sensitivity} = \text{RON} - \text{MON}$$

which describes how the knocking tendency of a fuel depends on the operating conditions. Another is the *anti-knock index* (AKI),

$$\text{AKI} = \frac{1}{2}(\text{RON} + \text{MON})$$

sometimes also called *pump octane number* (PON), which is the result of attempts to find a *road octane number* (RdON=a RON + b MON + c) that would have conditions similar to those of a vehicle on the road. RdON was found to be between the RON and MON with parameters $a \approx b \approx 0.5$ and $c \approx 0$ which is close to the AKI. Around the world AKI is used, for example in the USA, Canada, and Mexico, while RON is used in Europe, Asia, and Australia.

A vehicle–engine combination has an *octane number requirement* (OR) which is defined as the minimum fuel octane number that will resist engine knock over the complete load and speed range of the engine. With the knock control (see Section 10.6.1) available in modern engines, the recommended OR can, for example, be 98 RON, but the engine can tolerate 95 RON with a power and torque loss at high loads, since the engine control system can adapt the ignition to avoid knocking.

Fuel Additives

All gasoline fuels contain additives that are intended to improve fuel quality in several ways. The octane number can be increased by adding anti-knock agents, which increase the octane number to a lower cost than changing the fuel's hydrocarbon composition by refinery processing. The most effective anti-knock element known is lead, and it is stable and soluble in fuels in the form of lead alkyls. The lead alkyl, tetraethyl lead $(C_2H_5)_4Pb$, was first introduced in 1923. Since the 1970s, when air pollution and exhaust gas after-treatment through catalysts became an issue, effort has been put into developing lead free gasolines, using other chemical additives that are more environmental friendly after combustion. This is because the lead has a toxicological effect in urban areas and because the lead deposits on the catalyst and decreases its efficiency.

Compression Ratio and Knock

Knock occurs when the end-gas temperature is high enough to promote autoignition. The compression ratio is directly related to the knocking tendency through its influence on the temperature of the unburned gases, see also the discussion above concerning octane number determination. Engine knock and the octane number of the fuel give an upper limit on the compression ratio for an SI engine, which in most cases lies slightly above $r_c = 10$. Currently, naturally aspirated automotive SI engines have compression ratios around $10-12$ while turbocharged SI engines have slightly lower compression ratios around $8-10$.

6.3   CI Engine Combustion

Compression ignition (CI) combustion is initiated by the *start of injection* (SOI) and is characterized by three phases; ignition delay, premixed combustion, and mixing controlled combustion. These are shown in Figure 6.6. When the diesel fuel is injected it must be transformed from a cold liquid to a vapor and be heated so that it can autoignite. This time from the start of injection to start of combustion is called the *ignition delay* $\Delta\theta_{i,d}$ (see also Section 5.4.2). When this vapor–fuel mixture is at or above the autoignition temperature it will ignite and combust and this phase is called the *premixed* combustion phase. Then the fuel in the main body of the spray will mix with the air and burn during the *mixing controlled* combustion phase. In the CI engine, the fuel follows a path from a fuel rich spray with $\lambda \ll 1$ to a global lean mixture where there are sufficient amounts of air to oxidize the fuel. More about the diesel spray, its oxidation and emissions, will be given in Section 6.4.

6.3.1   Autoignition and Cetane Number

The ignition delay in a diesel engine influences how much of the fuel burns in the premixed phase, a longer ignition delay gives more premixed fuel that results in a more pronounced

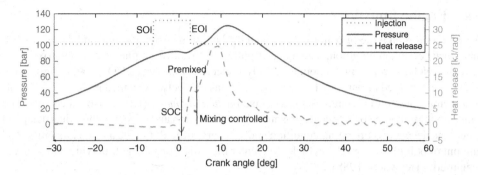

Figure 6.6 Injection timing with start of injection (SOI) and end of injection (EOI), cylinder pressure, and heat release in a heavy duty diesel engine. The signals show the three phases: (i) Ignition delay, which is from SOI to SOC. (ii) Premixed combustion, from SOC over the first peak in the heat release. (iii) Mixing controlled combustion from the end of the first peak to the end of combustion

Table 6.2 Engine operating conditions for cetane number determination

Engine speed	900 rpm
Injection timing	13° BTDC
Inlet temperature	65.6°C (150°F)
Coolant temperature	100°C
Injection pressure	10.3 MPa

premixed burning. Under extreme conditions this can emit a knocking sound, referred to as *diesel knock*. The ignition delay depends on the operating conditions and on the fuel quality, that for a diesel fuel is measured with the *Cetane Number* (CN), which is defined in a similar way as the octane number. A special pre-chamber diesel engine with adjustable compression ratio is run at the conditions specified in Table 6.2. The compression ratio is adjusted so that the ignition delay is 13° with the fuel to be examined. Then the fuel is changed to a blend of the two reference fuels n-hexadecane ($C_{16}H_{34}$ also called cetane), which is defined to have CN = 100, and heptametylnonane (HMN), which is defined to have CN = 15. Initially a less stable compound (α-methylnaphtalene $C_{11}H_{10}$) was used to define CN = 0 but this fuel was substituted for HMN and the scale was adjusted so that the blend determines the cetane number in the following way

$$CN = \%\text{n-hexadecane} + 0.15\ \%\text{HMN}$$

Octane and cetane numbers can be seen as their opposites. A high ON implies high resistance to autoignition, while a high CN fuel autoignites easily. This is also illustrated in the approximate relation between the cetane and octane numbers

$$ON = 120 - 2\ CN \qquad \Longleftrightarrow \qquad CN = 60 - 0.5\ ON$$

where the minus sign shows the opposite trends of the octane and cetane numbers.

6.4 Engine Emissions

Air pollution is an issue of major interest, and the emissions from gasoline and diesel engines have received much attention. The exhaust pollutants from SI engines are mainly oxides of nitrogen (NO and very small amounts of NO_2, collectively known as NO_x), carbon monoxide (CO), and unburned or partially burned hydrocarbons (collectively named HC). The main polluting emissions from CI engines are NO_x and *particulate matter* (PM), also called *soot*. The exact amounts that are produced during and after combustion and appear in the exhaust depend on engine design and operating condition and it is very difficult to give an exact quantity. The amount of pollutants from the cylinder that enter the exhaust have the following approximate magnitudes Heywood (1988)

	SI engines		CI engines	
NO_x	500–1000 ppm	$\sim$ 20 g/kg fuel	$\sim$ as SI	$\sim$ 20 g/kg fuel
HC	3000 ppm	$\sim$ 25 g/kg fuel	600 ppm	$\sim$ 5 g/kg fuel
CO	1–2 %	$\sim$ 200 g/kg fuel	very small	
Soot	very small		$\sim$ 2–5 g/kg fuel	

where HC is normalized to C_1. It must be noted that the data are engine-out emissions from engines of the 1970s and 1980s. Modern diesel injection and combustion systems have reduced these emissions significantly. Direct injected gasoline engines have also appeared, and with stratified charge have been shown to emit some soot.

There are also emissions from other sources than from the engine exhaust, for example unburned hydrocarbons come from piston blow-by gases and fuel evaporation through vents in the tank. Therefore, measures are taken that effectively minimize these emissions. Examples of measures are ventilating the crank case and returning the gases to the cylinder and also venting the fuel tank through a carbon canister which is purged by air during the engine operation, see Section 10.5.3.

6.4.1 General Trends for Emission Formation

Engine-out emissions are the result of complex processes and there is not a complete understanding of when, where, and how all chemical species are formed in the engine. This is an active research area combining engine experiments and modeling with thermodynamic and reaction kinetic models. The most detailed models involve calculation of all the chemi-kinetic reaction paths that occur when a fuel is oxidized, which is a complex task even for simple fuels. For example the model in Hunter et al. (1994) that describes combustion of methane CH_4 in air includes 40 species and 207 reactions, while a model for propane oxidation (Hoffman et al. 1991) includes 493 reactions. Considering that normal fuels are blends of different, more complex, hydrocarbons, highlights the complexity of such modeling tasks.

By studying the species concentration at chemical equilibrium one can gain some insight into when and why certain species are formed in an engine without the need to simulate the complex reaction paths. For example, the emissions depend strongly on the (A/F) ratio, see Figure 6.10 for emissions from an SI engine, and equilibrium compositions are useful for understanding the basic trends with (A/F) ratio, temperature, and pressure.

Chemical Equilibrium Compositions

Chemical equilibrium is achieved when the rate of change of concentrations goes to zero for all species. For a constant temperature and pressure system, chemical equilibrium can be determined using Gibb's free energy $g = h - sT$. In order to find chemical equilibrium for the following global reaction,

$$\frac{\phi}{8+18/4} C_8H_{18} + O_2 + 3.773N_2 \rightarrow$$

$$x_1CO_2 + x_2H_2O + x_3N_2 + x_4NO + x_5CO + x_6O_2 + x_7CH_4 + x_8OH + x_9O +$$

$$x_{10}C_2H_2 + x_{11}C_2H_4 + x_{12}HCN + x_{13}N + x_{14}H_2 + x_{15}NO_2 + x_{16}H$$

then Gibb's free energy is stated based on the components $g = \sum x_i \tilde{g}_i$ and the value for x_i that minimizes g is searched for by an optimization procedure. For more information about chemical equilibrium see for example, the textbooks Borman and Ragland (1998) or Turns (2000) or the CHEPP package (Eriksson 2005).

The equilibrium composition for varying (A/F) ratios $\lambda = \frac{1}{\phi}$ is shown in Figure 6.7. There are three regions that exhibit different characteristics. For *lean* mixtures, $\lambda > 1$, the excess air produces high concentrations of molecules containing oxygen, O_2, NO, OH, and O. It is lean mixtures that have the highest NO emissions. For *rich* mixtures, $0.32 < \lambda < 1$, there is not enough air to fully oxidize the fuel and higher amounts of CO, H_2, and H are generated. For *excessively rich* mixtures with carbon/oxygen ratios C/O > 1 (which corresponds to $\lambda < 0.32$ for isooctane) there is an increase in hydrocarbon compounds. At these rich conditions, with too little oxygen, soot is formed. In real engines, the critical C/O ratios for soot formation are from about 0.5 to 0.8 Heywood (1988).

Temperature has a strong influence on the composition and this is shown in Figure 6.8. For the lean mixture at low temperatures the major species are, N_2, CO_2, H_2O, and O_2. When the temperature is increased, *dissociation* of these species occurs, resulting in increasing numbers of NO, OH, CO, H_2, O, and H. The pressure has also an effect on the concentration, but this is mainly at higher temperatures where a higher pressure reduces dissociation.

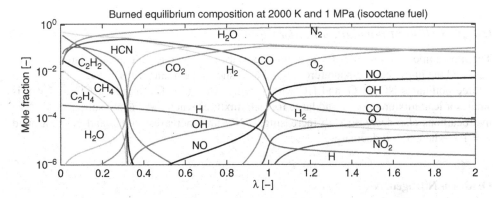

Figure 6.7 Equilibrium composition as function of air–fuel ratio (λ) at 2000 K and 1 MPa after combustion of isooctane in air, generated with CHEPP (Eriksson 2005). Note the logarithmic scale on the y-axis

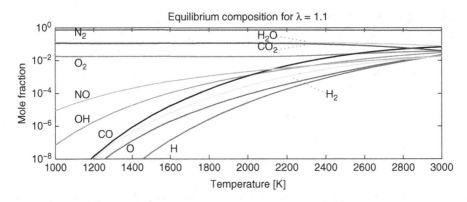

Figure 6.8 Species concentrations for varying temperatures with $\lambda = 1.1$. Note that N_2, CO_2, and H_2O concentrations decrease with increasing temperature, which is due to the dissociation that produces other species at the expense of these

Adiabatic Flame Temperature and Mixture Strength

Figures 6.7 and 6.8 showed that the composition depends on temperature and (A/F) ratio in a complex way. With the aid of chemical equilibrium and flame temperatures, species concentrations can be calculated and evaluated for simple processes. The following example is illustrative: a mixture between isooctane C_8H_{18} and air is combusted at constant pressure and without heat loss (i.e., isobaric and adiabatic combustion). For such a combustion, the enthalpy for the reactants $h_{re}(T_{re})$ is equal to the enthalpy for the products $h_p(T_p)$. With the initial temperature and pressure known, the temperature for the products, called adiabatic flame temperature, can be found using the equation $h_{re}(T_{re}) = h_p(T_p)$. Figure 6.9 shows the adiabatic flame temperature T_p and equilibrium composition when isooctane is adiabatically combusted at isobaric conditions. Things to note here are: firstly, the highest flame temperature occur at slightly rich mixtures, $\lambda = 0.95$. Secondly, NO does not peak at the highest temperature nor at the leanest side but at slightly lean mixtures, $\lambda = 1.14$, which is close to what can be measured experimentally, see, for example, Figure 6.10.

6.4.2 Pollutant Formation in SI Engines

The most important variables for determining spark ignition engine emissions are the (A/F) ratio and the in-cylinder temperature. Figure 6.10 shows the main trends of the

three pollutants NO_x, CO, and HC with varying A/F ratio. The trend for CO is that it is small for lean mixtures $\lambda > 1$ and large for rich mixtures when there is not sufficient air. The hydrocarbons also decrease with increasing λ. NO first increases with λ and peaks around $\lambda = 1.1$, whereafter it decreases.

Oxides of Nitrogen, NO_x

NO_x is the collective name for nitrogen monoxide NO and nitrogen dioxide NO_2. The nitrogen dioxide, NO_2, formed in an SI engine is small compared to NO, with maximum NO_2 concentrations reaching about 2% of the NO concentration. However, later oxidation of NO in the

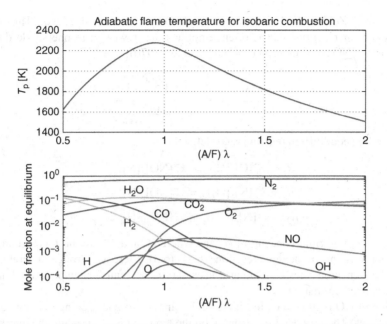

Figure 6.9   Concentration for adiabatic and isobaric combustion of isooctane (C_8H_{18}). Initial conditions for the combustion: $T_{re} = 293$ K and $p = 1$ atm

atmosphere gives NO_2 which, in the presence of ultra violet light, reacts with hydrocarbons to form photochemical smog.

Three mechanisms for forming nitrogen monoxide have been identified, and they are called prompt, nitrogen oxide, and thermal. The prompt mechanism

$$CH + N_2 \rightarrow HCN + N$$

occurs in the flame when the fuel is combusted, and can be important when there is fuel-bound nitrogen. The nitrogen oxide mechanism is

$$N_2 + O + M \rightarrow N_2O + M$$

that later decomposes to NO. This can dominate at low temperatures or very lean mixtures. Normally in SI engines the most dominant mechanism is the thermal formation, which is frequently modeled using the extended Zeldovich mechanism

$$O + N_2 \rightleftharpoons NO + N \qquad (6.1)$$

$$N + O_2 \rightleftharpoons NO + O \qquad (6.2)$$

$$N + OH \rightleftharpoons NO + H \qquad (6.3)$$

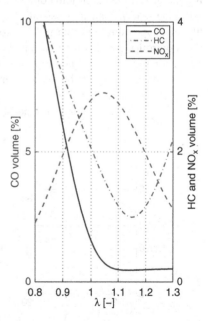

Figure 6.10   Engine-out emissions for different (A/F) ratios, measured on a 5-cylinder 2.4-liter SI engine at 2500 RPM and 65 Nm

originating from Zeldovich (1946) with (6.3) added by Lavoie et al. (1970). This is a much studied system of reactions that is often simplified and rewritten to one single differential equation for the NO concentration (see Lavoie et al. (1970))

$$\frac{d[NO]}{dt} = 2\,R_1\,\frac{1 - \left(\frac{[NO]}{[NO]_e}\right)^2}{1 + \frac{[NO]}{[NO]_e}\frac{R_1}{R_2+R_3}} \tag{6.4}$$

where R_i are the equilibrium reaction rates defined by

$$R_1 = k_1^+[O]_e[N_2]_e = k_1^-[NO]_e[N]_e \tag{6.5}$$

$$R_2 = k_2^+[N]_e[O_2]_e = k_2^-[NO]_e[O]_e \tag{6.6}$$

$$R_3 = k_2^+[N]_e[OH]_e = k_3^-[NO]_e[H]_e \tag{6.7}$$

where $[\cdot]_e$ are equilibrium concentrations and where k_i^+ and k_i^- are the forward and reverse reaction rate constants of the three reactions (6.1)–(6.3). There is a strong temperature dependence in the formation rates k_i, making it necessary to account for temperature dynamics in the analysis of the formation.

Formation of NO is illustrated in Figure 6.11 using a two-zone setting, having burned and unburned zones. The rate depends strongly on the temperature, and while the temperature is high NO is formed and recombined quickly. During the expansion stroke the reaction rate goes down with the temperature and the reactions freeze, giving NO concentrations significantly above equilibrium concentrations (a factor 20 higher at 130°). As a simple rule of thumb, NO is generated if the maximum combustion temperatures go above 1800 K.

Engine load has an indirect influence on the NO formation, because a higher load has a higher temperature and more NO is formed. Beside the load dependence, NO formation is also controlled by the (A/F) ratio and attains its maximum at slightly lean conditions, as can be seen in Figure 6.10. Spark timing also influences NO and a later spark timing gives less NO since it reduces the peak burned gas temperatures.

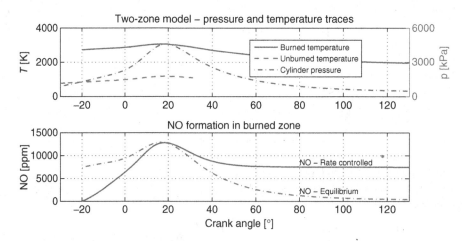

Figure 6.11 Two-zone calculations, for a high load condition, with NO production in the burned zone where rate controlled NO is compared to equilibrium NO. The reaction rates k_i^+ are from Heywood (1988) and the engine operates at full load, $\lambda = 1$, and $\theta_{ign} = -20°$

NO emissions can be reduced with exhaust gas recirculation (EGR), where a portion of the exhaust gas is cooled and transferred back to the intake. The presence of burned gas in the combustion chamber reduces the amount of oxygen and the burned gas temperatures, which thus reduces the NO formation. Adding recycled gases to the SI engine has the drawback that it decreases combustion stability by increasing cycle-to-cycle variations. SI engines can tolerate about 20–30% dilution by EGR before the misfire limit is reached, and with EGR the combustion temperature is lowered, which helps to reduce the emissions of nitrogen oxides NO_x.

Carbon Monoxide, CO

The carbon monoxide generation is mainly controlled by the air/fuel ratio. This is indicated by the combustion reaction: if there is excess fuel in the combustion there is not enough oxygen to fully oxidize the fuel so CO forms instead CO_2. Figure 6.10 shows how the engine-out CO is affected by the equivalence ratio λ. As the mixture becomes rich $\lambda < 1$ the CO increases steadily, and as the mixture becomes lean the level of CO remains fairly constant around 10^{-3} mole fraction. Under rich conditions, the amounts of CO in the exhaust is close to what chemical equilibrium calculations show. However, the mechanisms that generate CO under lean conditions are more complex than those that can be evaluated by studying equilibrium, and chemi-kinetic mechanisms (similar to those for thermal NO) must be utilized.

Hydrocarbons, HC

Figure 6.10 shows the influence of (A/F) on the hydrocarbon HC emissions. The trend is a steady decrease with increasing λ until the lean limit is approached and a stable combustion can no longer be maintained. For those conditions, the flame is quenched before the air–fuel mixture is fully consumed. There are several mechanisms that contribute to the HC emissions and they are all related to the fact that fuel, in one way or an other, evades combustion. The four main mechanisms are described below.

Crevice effects: During compression and the earlier stages of flame propagation the increasing cylinder pressure forces some of the air–fuel mixture into the crevices, that is narrow volumes connected to the combustion chamber, such as between the piston, rings, and cylinder wall. Most of the gases that enter the crevices are unburned gases and they also remain uncombusted since the crevices are too narrow for the flame, and thus combustion reactions, to enter. During expansion and blow-down, the cylinder pressure decreases and the unburned hydrocarbons, entrained in the crevices, expand back into cylinder chamber and mix with the bulk gas, and then exit the cylinder with the other gases during the exhaust stroke.

Flame quenching: At the combustion chamber walls there is a small quench layer (< 0.1 mm) that is cooled by the wall and the flame is quenche before it reaches the wall. When the walls are clean it has been shown that the unburned mixture in these layers is burned up quickly, but it has also been shown that porous deposits increase the hydrocarbon emissions.

Oil layer absorption: The oil layers present in the combustion chamber absorb hydrocarbons during compression that are later released during expansion and exhaust. In this way, some of the hydrocarbons escape combustion and exit the combustion chamber unburned.

Poor combustion quality: A final source of unburned hydrocarbons is incomplete combustion where the flame is quenched by local conditions in the bulk gas. Such events can occur under very lean or very diluted mixture conditions. This can be seen in Figure 6.10 where the HC emissions increase when the mixture becomes very lean.

6.4.3   Pollutant Formation in CI Engines

The pollutants from CI (diesel) engines are mainly nitrogen oxides NO_x and soot or particulate matter (PM). Pollutant formation depends to a large extent on local properties, such as mixture strength and temperature, in the spray controlled mixing and combustion. A conceptual model for diesel engine combustion with soot and nitrogen monoxide NO emissions has been presented in Dec (1997) and Dec and Canaan (1998) and the main results are summarized here. Cylinder pressure, injector needle lift, and heat release traces for the experiment are given in Figure 6.12. An engine with optical access to the combustion chamber was used, so the events in the combustion chamber could be recorded and analyzed. Results are summarized as a sequence of sketches, Figure 6.13, showing the temporal evolution of a reacting diesel fuel jet from the start of injection up through the first part of the mixing–controlled burn. The sketches show that the liquid-phase fuel spray reaches its maximum penetration length just a few degrees after the start of injection and before the premixed ignition occurs. Fuel vapor is formed in front of the spray tip and along the sides of the spray. The first indication of chemical activity is a weak blue light of chemiluminescence that is found along the sides and later also around the entire downstream region of the vaporized spray.

Soot Formation

The premixed combustion starts with fuel breakdown in the leading solid portion of the vaporized jet, which takes place at λ values of 0.25–0.5. Small sized soot particles are formed downstream of the rich burning premixed flame. In addition, the particle size increases along the line of flow, soot is thus formed in the inner parts of the spray plume as a result of rich premixed burning. The precursor for soot formation, Poly aromatic Hydrocarbons (PAH), was detected in the same region where the soot was formed.

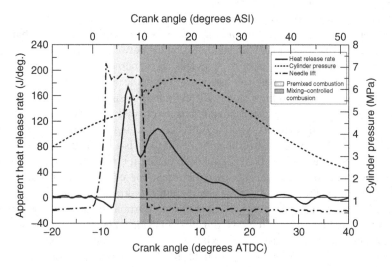

Figure 6.12   Cylinder pressure and heat release traces for the engine in Dec (1997). The heat release drops initially, when the fuel evaporation takes energy from the combustion chamber, and then rises when the combustion starts to occur. (*Source*: Dec, J.E. 1997. Reproduced with permission of SAE)

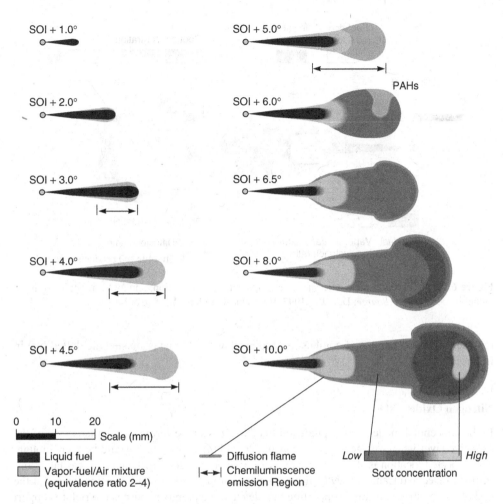

Figure 6.13 Conceptual model for spray development, soot formation, and NO formation from the model in Dec (1997). (*Source*: Dec, J.E. 1997. Reproduced with permission of SAE)

Figure 6.14 shows a developed reacting fuel jet during the mixing–controlled burning. It explains the sequence of events a fuel package follows as it moves from the injector downstream through the mixing, combustion, and emissions-formation processes. The rich premixed burning and soot forming inner parts were covered with a thin sheath of diffusive combustion. In this layer, the products from the premixed combustion were further oxidized. The diffusive combustion was detected by measuring the OH radical that is mainly formed close to stoichiometric conditions.

In summary, the most important characteristic of the model is that the combustion takes place in two steps. Firstly, a premixed and fuel rich flame where soot and other partly oxidized products start to form. Initially the formed soot particles are very small, but increase in size as they go downstream. Oxidation of soot and other products from the rich premixed combustion takes place on the rich side of the thin, close to stoichiometric, diffusive layer. This layer

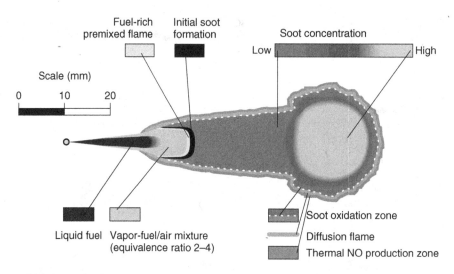

Figure 6.14 Conceptual model for spray development, soot formation, and NO formation from the model in Dec (1997). (*Source*: Dec, J.E. 1997. Reproduced with permission of SAE)

represents the second step of combustion and is shown as the layer outside the dotted line in Figure 6.14.

Nitrogen Oxides, NO_x

In the conceptual model, it was predicted that NO_x should be formed on the lean side of the diffusive layer where oxygen would be available and the temperature would be high enough. The expected location is shown as the outermost layer in Figure 6.14, and this was later confirmed in Dec and Canaan (1998). It was shown that the start of NO formation occurs at the end of the premixed burning, supporting the idea that the premixed burn is too rich to support NO formation.

Carbon Monoxide, CO

Carbon monoxide is usually small in CI engines, since the engine is operating leanly with enough oxygen to fully oxidize the fuel to CO_2. When the engine is operated at lean conditions and late injection timings CO can increase.

Hydrocarbons, HC

Hydrocarbons are only produced in low quantities in CI engines. The mechanisms through which the HC evades combustion in SI engines are not present in CI engines. There is no premixed fuel that can enter crevices nor is the oil layer adsorption present. One important source for HC is the small holes in the fuel injector, called sac volume, which contain unburned fuel that later enters the chamber during the expansion stroke. Some HC emissions can also be produced either in cold engines or for excessively rich mixtures.

6.5   Exhaust Gas Treatment

So far the discussion has been concentrated on emissions directly from the engine. These levels of pollutants are too high compared to the limits imposed by the legislators, see Section 2.8, and thus have to be reduced by treating the gases in the engine exhaust system. This is achieved by using one or a mixture of catalysts, filters, and traps that store emissions. After this there is a short discussion on catalyst material and general properties.

Catalyst design with materials and their blend selection, for achieving an efficient and long-term stable catalyst, is a complex area. The active surfaces of the catalysts are usually made of combinations of three precious metals: platinum (Pt), palladium (Pd), and rhodium (Rh). Each of the precious metals is used for a specific purpose, but are best served if kept separate from the others (Kummer 1986). These precious metals are scarce, with catalyst production being a major consumer. Chemical formulation depends both on the desired performance and current precious metal prices. Metal loading is typically of the order 0.2% in weight of Pt or Pd and 0.04% in weight of Rh and the relative ratio between the species differs significantly, for example the Pt/Pd ratio varies from 15/1 to 1/3, see Degobert (1995) for more information.

6.5.1   Catalyst Efficiency, Temperature, and Light-Off

The *conversion efficiency* is determined in terms of the mass flow rate $\dot{m}$ of the constituents, that is CO, NO_x, and HC. With HC as example, the definition is

$$\eta_{cat,HC} = \frac{\dot{m}_{HC,in} - \dot{m}_{HC,out}}{\dot{m}_{HC,in}} = 1 - \frac{\dot{m}_{HC,out}}{\dot{m}_{HC,in}}$$

Catalyst efficiencies are determined from conditions before and after the catalyst, such as those shown in Figure 5.13.

The efficiency of a typical catalytic converter depends directly on the temperature of the converter. The *light-off temperature* is used to describe the temperature where the catalyst reaches 50% efficiency, which for oxidizing catalysts lies around 150–30 0°C. Figure 6.15 shows the efficiency and light-off temperature for three different catalysts, in the promoted Pt catalyst Rh has been used as promoter. The Pt/Rh combination lights off at lower temperatures than the Pt/Pd, Pd/Rh, and Pt/Pd/Rh systems. The light-off temperature increases with aging.

The cold start emissions are a major source of pollutants since during the engine warm-up period, before the catalyst has reached its light-of temperature, the catalyst is very inefficient and all emissions just pass through directly to the atmosphere. Figure 6.16 shows a cumulative plot of the emissions under a drive cycle, and it is clear that the bulk of the vehicle emissions are generated under the first 120 seconds, before the catalyst has reached the light-off temperature.

Too high temperatures can also inflict permanent damage that reduces its conversion efficiency. Pt is especially sensitive. The temperature of the exhaust gases decrease as distance to the engine is increased. Therefore, there is an obvious design conflict in where the converter should be positioned: if it is placed near the engine it will reach the light-off temperature sooner, reducing the cold-start emissions, but then it might be damaged by the high exhaust temperatures at full load. And the converse is true for a catalyst placed far from the engine.

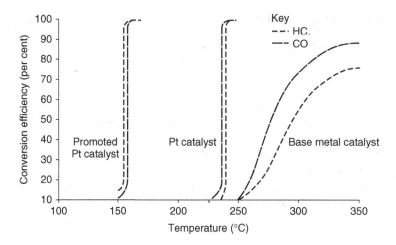

Figure 6.15 Light-off temperature for different catalysts. The promoted Pt catalyst has Rh as promoter. (*Source*: Stone, R. 1999. Reproduced with permission of Johnson Matthey)

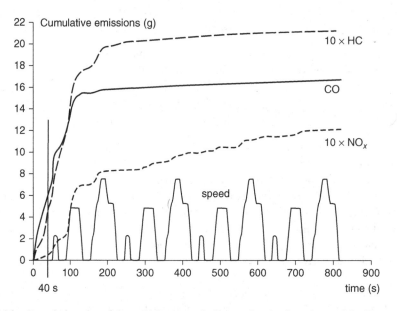

Figure 6.16 Cumulative plot of the vehicle-out emissions under the first phase of the European drive cycle. The majority of the emissions are generated under the first 120 seconds of the cycle, before the catalyst has reached the light-off temperature. (*Source*: Degobert, P. 1995. Copyright Editions Technip)

Some catalysts change characteristics with temperature and, for example, promote one reaction in one temperature region and another in a higher region. So monitoring and managing the temperature becomes an important task for the control system.

For example, one measure taken to avoid catalyst overheating in SI engines, is to enrich the air–fuel mixture at high loads, which results in a colder exhaust temperature that spares the catalyst, but on the other hand it reduces the efficiency of both the engine and catalyst.

6.5.2   SI Engine Aftertreatment, TWC

The three-way catalyst (TWC) is the most common catalyst on gasoline engines today, it received its name because it removes all three pollutants simultaneously. If the engine is operated at stoichiometric conditions there will be a sufficient amount of reduction gases to remove NO and enough O_2 to oxidize CO and HC.

Figure 6.17 shows the efficiency of the three-way catalyst in relation to the A/F ratio and also the emissions before and after the catalyst are shown. In a very narrow band around $\lambda = 1$ the catalyst has its best conversion efficiency. The width of the narrow band lies around 0.7% and is too narrow to be achieved by carburetors, but with fuel injection based control systems these requirements can be achieved. The introduction of the TWC was a key point for the introduction of feedback control in SI engines, as this is necessary for achieving the narrow tolerances on λ.

The catalyst performance shown in Figure 6.17 is for steady flow operation, but there are also important storage dynamics occurring. Three-way catalysts consist of Cerium oxide and precious metals, the precious metals make the reactions happen and Cerium oxide stores oxygen. Cerium (Ce), first used in the mid-1980s, now typically forms a large proportion of the washcoat, around 30% in weight. Cerium oxide $Ce\,O_2$ (ceria), and to some extent the precious metals, provide the TWC with an oxygen storage capability. Ceria stores and releases oxygen through $Ce\,O_2 \leftrightarrow Ce_2\,O_3$, where under lean conditions, ceria stores excess oxygen from the exhaust gas and oxygen catalyzed by the precious metals. If there is room to store oxygen, NO_x is also reduced to nitrogen. During rich cycles, cerium oxide releases stored oxygen to oxidize CO and HC to CO_2 and water.

Due to the oxygen storage capability in the catalyst, λ may go outside of this narrow band for short periods as long as the mean value is close to 1. Experimental data shows that there is a widening of the band if there is a cyclic variation in the fuel flow, however the maximum efficiency is slightly lowered compared to when there are no fluctuations. The influence of the

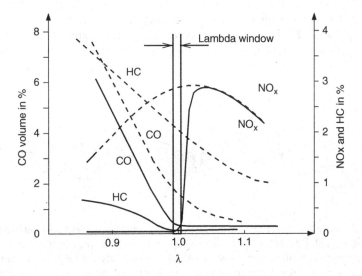

Figure 6.17   Emissions before (dashed) and after (solid) the catalyst. The narrow band around $\lambda = 1$, in which there is good reduction of emissions, is called the lambda window

fluctuation frequency is that frequencies around $1-2$ Hz are most effective and can increase the width of the band to about 5 %. Accurate control of the catalyst includes monitoring and control of the oxygen stored in the catalyst.

Particulate Matter from GDI Engines

In direct injected SI engines, soot or particulates can be produced, foremost during cold starts and transients, and therefore it has been proposed that future legislation for SI engines should also have limitations on particulate emissions. As a result, there is a need for particulate filters in SI engines, called gasoline particle filters GPF, and there are ongoing efforts to study and optimize the particulate filter and the TWC see, for example, Richter et al. (2012) and Zhan et al. (2010) for some investigations.

6.5.3 CI Engine Exhaust Gas Treatment

A wide range of exhaust gas treatment systems are used in the diesel engine, where different catalysts are combined with particulate traps or filters, see Figure 6.18 for an example. Beside filters and catalysts, the CI engine is often equipped with exhaust gas recirculation (EGR) systems for reducing the engine-out NO_x.

Diesel Oxidation Catalyst – DOC

Oxidizing catalysts require sufficient amounts of oxygen to oxidize CO and HC, which can be supplied either by running the engine leaner than stoichiometric or pumping additional air into the exhaust. Saturated carbons are hardest to oxidize, while increasing molecule weight makes the oxidizing easier. A diesel oxidation catalyst (DOC) uses the elements palladium and/or platinum as catalysts to convert hydrocarbons and carbon monoxide into carbon dioxide and water. Most DOCs operate with around 90% efficiency.

In the DOC, CO and HC react with the oxygen in the lean mixture instead of the oxygen carried in NO_x. Therefore, a DOC does not reduce the NO_x emissions significantly, but it oxidizes NO to NO_2 which has the positive benefits of increasing the NO_x conversion efficiency

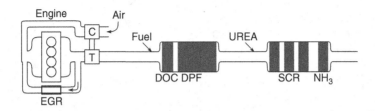

Figure 6.18 Examples of components in a diesel engine gas treatment in a heavy duty engine, see, for example, Charlton et al. (2010) for details. The components are: Compressor (C), Variable Geometry Turbine (T), Exhaust gas recirculation (EGR), Diesel oxidation catalyst (DOC), Diesel particulate filter (DPF), three blocks with Selective catalytic reduction (SCR) catalysts, and last a fourth block with an Ammonia slip catalyst (NH_3). There is control of fuel injection for the DOC and Urea injection for the SCR

of an SCR and improving low temperature regeneration of a DPF. Some DOCs can also store NO_x—Glover et al. (2011) and Millo and Vezza (2012)—at low temperatures when the SCR's efficiency is low. For diesel engines, it is therefore used as a component in the total system, see Figure 6.18 with DPF and an LNT or SCR to improve the reduction of particulate matter and NO.

Another function of the DOC is that it can act as a catalytic burner to raise the temperature, for example to enable a particulate filter regeneration. Fuel is injected either directly into the exhaust, as in Figure 6.18, or very late during the expansion or exhaust stroke, to provide the air mixture with hydrocarbons that will burn (oxidize) in the catalyst and heat the gas.

Lean NO_x Catalyst, LNC

There is not yet any sufficiently efficient catalyst available for automotive use that efficiently promotes the reduction reaction $2\,NO \rightarrow N_2 + O_2$. Therefore, the most efficient de-NO_x catalysts use reducing agents. In the lean NO_x catalyst (LNC) NO reduction catalysts use the CO or H_2 present in the exhaust. Under rich conditions, NO can be reduced, since there is an excess of the reducing species over the oxidizing species. Reduction of NO by CO or H_2 can, for example, be accomplished by base metal catalysts (CuO, NiO) in the temperature range 350 to 600°C.

Lean NO_x trap, LNT

A lean NO_x trap (LNT), also called a NO_x storage catalyst (NSC), stores NO_2 in metal oxides deposited in the converter, where the storage is in the form of metal nitrates $NO_2 + MeO \rightarrow MeNO_3$. This thus requires that NO is converted into NO_2, which is achieved by having an DOC (or oxidizing catalyst material) directly in the LNT. After some time, the metal oxide sites are filled and conversion efficiency drops, then the trap needs to be regenerated. Regeneration is performed by feeding a rich mixture to the catalyst, which has the effect that the metal nitrates become unstable and decompose through $2\,MeNO_3 \rightarrow 2\,MeO + 2\,NO + O_2$. Finally, CO and HC in the rich mixture acts as reduction agents, converting NO into N_2 and CO_2 and H_2O. The storage phase takes about 30–300 seconds while the regeneration takes 2–10 seconds, depending on the operating condition (Dietsche 2011).

Selective Catalytic Reduction, SCR

In selective catalytic reduction (SCR), a reducing agent is added to the flow to the catalyst. This agent selectively prefers to oxidize with NO_x instead of O_2, which is present in the lean gases. For this purpose, ammonia NH_3 has proven to be highly selective and efficient, but since ammonia is toxic a non-toxic carrier called urea $(NH_2)_2CO$ is used. It is distributed as a urea/water solution, either called Diesel Exhaust Fluid (DEF) or by the brand name AdBlue. Solution concentrations by mass are 32.5% urea and 67.5% deionized water, which is selected since it has lowest mixture freezing point (-11°C). Further, DEF expands 7% and the tanks need to be designed to accommodate this expansion.

The urea solution is injected upstream of the SCR catalyst, see Figure 6.18, and is decomposed to ammonia, which reduces NO_x in the catalyst. In this there is a balance, on one side if too little urea is injected the NO_x will not be reduced, on the other side if too much urea is injected NH_3 will slip past the SCR. To remedy this, the last segment in the exhaust treatment system in Figure 6.18 is inserted to reduce the *ammonia slip* from the SCR. Catalyst temperature has an effect on the NH_3 storage, where more ammonia can be stored when the catalyst is cold. Consequently, a transient that heats the catalyst will cause a release of NH_3 that must be counteracted with less urea injection to avoid ammonia slip, the converse happens in a cooling transient. All in all, this illustrates the requirement of careful control, especially during engine transients.

Diesel Particulate Filter, DPF

A diesel particulate filter (DPF) is used to efficiently eliminate particulate matter (PM) (soot) and has an efficiency of about 80–100%. Passive and active regeneration is used to keep the filters clean, since a clogged DPF reduces conversion efficiency and increases the fuel consumption through an increased back pressure.

Sensors in the exhaust system combined with soot generation models are used to determine when the filter has accumulated a given amount of PM, which triggers an active regeneration mode. Soot is the result of partially combusted fuel, and therefore it is natural to burn it out of the filter, for which a temperature of about 600°C is needed. During active regeneration, fuel is introduced into the exhaust stream, either by injecting fuel into the combustion chamber during the exhaust stroke or by a dedicated fuel injector in the exhaust, and burned to increase the heat of the DPF. Usage of active regeneration is a trade-off, too frequently impairs the fuel economy while too seldom clogs the DPF and impairs the driveability and fuel economy.

Passive regeneration can also be achieved with NO_2, that has a soot oxidizing effect in the temperature region 300–450°C and can therefore be used to continuously regenerate DPF. The DOC is located upstream of the DPF to generate NO_2, and it also needs to be warm enough to generate NO_2. Finally, the DPF can also self-regenerate under high engine load conditions if sufficiently high temperatures are reached.

6.5.4 Emission Reduction and Controls

As is evident from the presentation above, the reduction of polluting components, from engine and exhaust gas relies on the existence of and functions in the control system. For example, the three-way catalyst in the SI engine, with its stringent demands on λ for optimum efficiency, requires a closed loop control system. The performance of the EGR system, with nonlinear effects, needs a well-designed control system. The SCR system and urea dosing are also precarious control problems. Finally, the mode switching and regeneration of LNT and DPF systems also relies on a control that uses sensors and models to determine the regeneration. These topics will be discussed more in Chapters 10 and 11 on SI and CI engine control respectively.

Part Three

Engine – Modeling and Control

7

Mean Value Engine Modeling

The previous engine chapters covered combustion fundamentals and in-cylinder processes and developed models that are used for analysis and understanding of the complex phenomena that influence engine performance, efficiency, and emissions. These in-cylinder models describe parameters that vary under one cycle, and the resolution of interest is in the order of one crank angle degree, that is frequencies up to about 5–15 kHz. The models described in this chapter are called *mean value engine models* (MVEM) and have applications in analysis and design of control and diagnosis systems.

Definition 7.1 Mean value engine models *are models where the signals, parameters, and variables that are considered are averaged over one or several cycles.*

Mean value models, therefore, describe variations slower than an engine cycle and have a frequency range of about 0.1–50 Hz. Since the models have control and diagnosis applications, they are also often referred to as *control oriented models*. Mean value models have been used since the 1970s, but the term MVEM is attributed to Elbert Hendricks and appeared as mean value model (MVM) in Hendricks (1986). There is a rich literature on MVEM models that has followed the intense development of engine control systems; similar model components with the same basic equations for components, but that differ in parameters or submodels for efficiencies and so on. The model family also goes under the name of *filling and emptying models* (this will be clear from Sections 7.5 and 7.6) and also belongs to the family of zero-dimensional models, since no spatial dimensions are included in the models.

   Modeling for control and diagnostics leads naturally to a sensor and actuator perspective, since the sensors and actuators are interfaces between the engine hardware and engine management system. This chapter therefore starts with an overview of sensors and actuators in Section 7.1. Further simulation of engines and their subsystems is an important engineering tool, used in the development process of engine control system development, and Subsection 7.1.2 discusses how component modeling can be used to compile system models. Gas flow is an important part of the models and Sections 7.2 to 7.6 go through the gas path components, while Section 7.7 covers fuel injection and its interaction with the gas path. Torque generation is the topic in Sections 7.8 and 7.9, with in-cycle torque variations in the former and MVEM torque in the latter. Thermal effects with engine-out and exhaust temperatures as well as intercoolers are modeled in Sections 7.10, 7.11, and 7.12. Finally, the mechanics and modeling of the electronic throttle body is given in Section 7.13.

Modeling and Control of Engines and Drivelines, First Edition. Lars Eriksson and Lars Nielsen.
© 2014 John Wiley & Sons, Ltd. Published 2014 by John Wiley & Sons, Ltd.
Companion Website: www.wiley.com/go/powertrain

Turbocharging is important, and Chapter 8 is devoted turbocharger modeling. That chapter ends by compiling two complete engine models as examples, this is achieved by collecting component models from Chapters 7 and 8.

7.1 Engine Sensors and Actuators

A sketch of an SI engine with examples of sensors and actuators that are used for control and diagnosis is shown in Figure 7.1. The sensors and actuators are also described in Table 7.1.

There are three control inputs shown: ignition angle θ_{ign} – controlling the efficiency. Fuel injection – controlling the fuel amount and thereby the (A/F) ratio and load. Throttle torque M_{th} – controlling the throttle angle α which in turn controls the amount of air and thereby load. It also influences the (A/F) ratio. In many models the throttle angle is considered to be directly controlled without taking into account the servo, since its dynamics can be faster than that of the other components. For example, in the air flow models that will be developed, the throttle angle α is seen as the input.

Modeling of CI engines follows the same principles as SI engines, with the main difference being that the load is controlled with fuel injection (in the cylinder) instead of a throttle. Essentially, all automotive CI engines have turbochargers and a first thorough example of a CI engine model will therefore come in Chapter 8, when turbocharger models have been treated. The interested reader can for example look at Figure 4.7 to see the similarities between the gas flow components in SI and CI engines.

7.1.1 Sensor, System, and Actuator Responses

The goal with control oriented models is to develop models that can describe the system response from actuator to sensor, as well as relations between sensors. Before going into

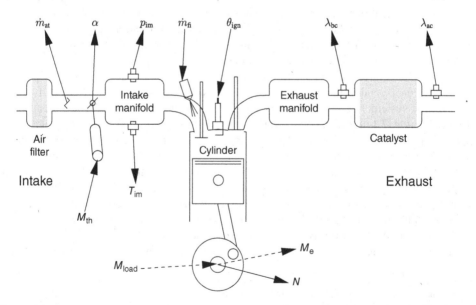

Figure 7.1   An overview of some of the sensors and actuators that are used for controlling and diagnosing SI engines

Table 7.1   Table with notation and variables used in Figure 7.1. Engine output torque and load are not measured on production engines, but they are important variables in the modeling and are therefore included in the list

Sensors		Actuators	
Air mass flow past throttle	$\dot{m}_{at}$	Throttle servo torque	M_{th}
Throttle angle	α	Fuel mass injected	m_{fi}
Intake manifold pressure	p_{im}	Ignition angle	θ_{ign}
Intake air temperature	T_{im}		
λ before catalyst	λ_{bc}		
λ after catalyst	λ_{ac}		
Engine speed	N		
Engine output torque	M_e	Engine load	M_{load}

equations, we start by looking at some measured signals to see what kind of system responses and timescales one sees in engines.

Figures 7.2 and 7.3 illustrate the air and fuel dynamics of the fuel preparation. Figure 7.2 shows the manifold filling dynamics, which are also called air intake dynamics. A step input in the throttle angle, α, produces a change in the air flow past the throttle, $\dot{m}_{at}$ and also a change in the intake manifold pressure, p_{im}, and this also increases the engine torque. Figure 7.2 displays an experiment where a sequence of step changes in the throttle angle are performed, the first one occurs at $t = 10$ s. When the throttle is opened air rushes past the throttle to fill the intake manifold and increase the pressure. The air flow past the throttle peaks just after $t = 10$ s and then decreases as the manifold pressure increases. A few seconds after the throttle step, the

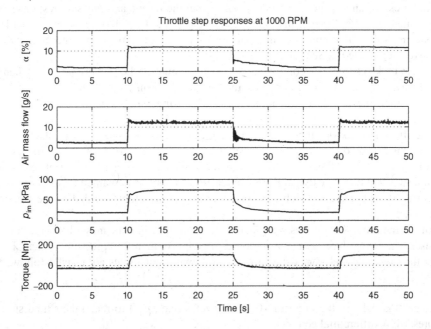

Figure 7.2   A step input on the throttle angle results in an increased flow of air through the throttle, which builds up the intake manifold pressure, p_{im}. The flow peaks slightly after the throttle steps, which indicates an air rush past the throttle. The oscillations in flow after 25 s are a result of compressor surge

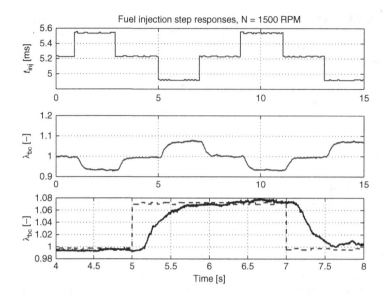

Figure 7.3 Step responses for the amount of fuel injected and the response in lambda. When fuel injection increases, λ decreases. In the bottom plot the signals have been enlarged and the fuel injection (dashed) has been scaled and flipped, showing more clearly that the time delay and dynamics are present in the system from t_{inj} to λ_{bc}

system has reached steady state conditions again. In the figure, one can see that the response from throttle angle to mass flow is fast, and that there is a dynamic from throttle to intake manifold pressure and torque. One can also note that the torque and pressure responses have similar time constants. Finally, just after the closing step, at $t = 10$ s, there is an oscillation in the air mass flow, that comes from a phenomenon called compressor surge. Surge will be discussed and modeled in Section 8.6.5.

Let's also look at some data concerning the fuel dynamics of the fuel preparation. Changing the fuel injection time without changing the throttle position changes the (A/F) ratio. The amount of fuel injected into the engine is metered using a pulse width modulated signal, the longer the fuel injection valve is open, the more fuel is injected. Figure 7.3 shows step responses from injector opening time t_{inj} to the measured A/F ratio before the catalyst.

The response in λ_{bc} due to the changed fuel injection time is not instantaneous. First there is a delay of about 0.2 s, and then it takes about another 0.4 s for it to reach steady state conditions. This indicates that there are dynamic phenomena that cannot be explained by just considering the amount of air inducted (which was constant in the experiment) divided by the amount of fuel injected.

From the measurements it was seen that some quantities, like air mass-flow $\dot{m}_{at}$, have fast responses while other have dynamic responses, like intake manifold pressure p_{im} and λ. Coupled to the timescales relevant for control, the following guidelines are usually followed for MVEMs.

- Changes that take in the order of 10–1000 cycles (e.g., p_{im}) to reach their final state are expressed by differential equations.
- Changes that are faster (e.g., $\dot{m}_{at}$) are expressed by static relations.

- Changes that are slower (e.g., coolant and ambient temperatures) are expressed using constants.

7.1.2   Engine Component Modeling

The following sections will collect the models for some important subsystems that are present in most MVEMs. The subsystems that are modeled are engine processes as well as sensors and actuators. In conjunction with the model descriptions, some properties and problems related to engine control problems are also pointed out. In the model descriptions, the time dependence, variable t, is omitted when there is no chance of misinterpretation. The majority of all components are applicable to both CI and SI engines and no difference will be made between the model components.

Engines are thermodynamic machines and therefore the thermodynamic laws are to a large extent used in the modeling process. When we study the flow system in engines, the transport and conservation of both mass and energy are the essential physical processes and these appear as a theme in the models. A component based approach will be followed in the text where there is a subdivision of the models into *volume* components, that have storage of mass and energy, and *flow* components, that transport mass and energy between the storing components, see Figure 7.4 for an illustration. Each component needs information from the surrounding components to determine its state and output, this is illustrated by the arrows going both up and down. Examples of flow components in MVEMs are: air filter, throttle, cylinders, catalyst, and mufflers, as well as the compressor and turbine in a supercharged or turbocharged engine. Examples of volume systems are the intake manifold, exhaust manifold, and pipes that have finite volume. Using a strategy of modeling the system with flow components in series with volume components, as in Figure 7.4, gives a structured way of modeling which enables reuse of components and easy extension to build models for complex engine systems, see for example, Eriksson et al. (2002b) for a discussion on this and the extension to turbocharged engines. The next step is now to dive into models for flow components and continue with volume models.

7.2   Flow Restriction Models

An engine consists of several components that affect the gas flow, and we first turn to components like the air filter, intercooler, exhaust system, and throttle that are collectively called

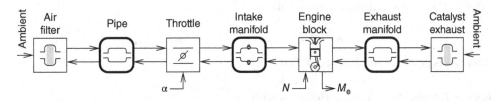

Figure 7.4   Illustration of component-based modeling for the air path in an engine, compare with Figure 7.1. Examples of *flow* components in MVEMs are: air filter, throttle, cylinders, and catalysts (also compressor and turbine in a turbocharged engine). Examples of *volume* components are intake manifold, exhaust manifold, and pipes between components that have finite volume

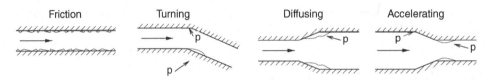

Figure 7.5 Example showing different occasions where pressure losses occur in a flow system. Adverse pressure gradients occur at several places, some of them are marked with a "p" in the figure, these gradients give rise to the pressure losses

flow restrictions. When a gas flows through these there are pressure drops over them, and some examples of occasions where pressure losses occur are shown in Figure 7.5. More examples and the theory behind the losses are given in text books that cover fluid mechanics, see, for example, Miller (1990) or Massey (1998).

Flow restriction models are divided into different categories that represent the physical process and the system behind the equations. One classification of the flow in restriction is whether the fluid can be modeled as incompressible or compressible. Models for incompressible flow capture flows with velocities up to Mach numbers 0.2–0.3, which for air at ambient conditions corresponds to velocities up to 70–100 m/s. For higher velocities, the fluid often needs to be modeled as being compressible. Another subdivision of the incompressible fluid is into the two subcategories laminar and turbulent flow, where the Reynolds number, *Re*, (see Appendix A.3 for a definition) is used to determine if it is turbulent or laminar. Example A.1 in the Appendix indicates that most flows in engines are turbulent and incompressible. However, in some places, like in the throttle and in other control valves, the flow needs to be treated as a compressible fluid.

Mass Transport in Flow Restriction Models
Flow restrictions are, in MVEMs, used to describe the mass flow through a component, and the pressures in the surrounding pipes or components are considered given for the flow restriction. The structure of the equations is that the mass flow is formulated as a function of the upstream and downstream pressures and the upstream density, that is

$$\dot{m} = f(p_{us}, p_{ds}, \rho_{us}, \ldots)$$

There is only one mass flow, since a flow restriction model is modeled as having no mass storage and the mass flows before and after the restriction are thus the same. Another result of this assumption is the following equation, that relates the mass flow $\dot{m}$ to the mean flow velocity U for up- and downstream conditions

$$\rho_{us} A_{us} U_{us} = \dot{m}_{us} = \dot{m}_{ds} = \rho_{ds} A_{ds} U_{ds} \tag{7.1}$$

where A_x is cross-section area and ρ_x is density.

Energy Transport in Flow Restriction Models
Energy is transported through the restriction and the temperature change is related to the energy change. For many restriction models, neither work nor heat is added or removed and therefore the energy is conserved. This follows since the first law of thermodynamics states that the energy is conserved, therefore the energy at the upstream conditions becomes the same as for

the downstream conditions, which is expressed in the following equation

$$H_{\text{us}} + KE_{\text{us}} + PE_{\text{us}} = H_{\text{ds}} + KE_{\text{ds}} + PE_{\text{ds}}$$

Here H is the enthalpy, KE kinetic energy, and PE is the potential energy. In engines, the changes in kinetic and potential energy can be neglected, and therefore the enthalpy is constant. In these models we consider the fluid to be an ideal gas, and for ideal gases the internal energy (and enthalpy) is a function of temperature only. Therefore the temperatures upstream and downstream are the same and the equation for the temperature of the flow is simply

$$T_{\text{flow}} = T_{\text{us}} \tag{7.2}$$

which gives the enthalpy flow

$$\dot{H}_{\text{ds}} = \dot{H}_{\text{us}} = \dot{m}_{\text{us}} \, c_{\text{p}} \, T_{\text{us}} \tag{7.3}$$

These relations apply to the systems where there is no heat or work transport to or from the restriction, which is a good assumption for many systems. However, there are systems, like the intercooler (heat exchanger) in a turbocharged engine, where there is significant heat transfer and then the temperature and enthalpy are not described by (7.2) and (7.3) and need to be modeled in more detail.

7.2.1   Incompressible Flow

For incompressible flows, there are essentially two different models for the flow (or pressure loss). The first model equation is for *laminar flow*. Experimental investigations have shown that the pressure loss $\Delta p = p_{\text{us}} - p_{\text{ds}}$ for flows through porous media and laminar flow in pipes are proportional to the flow velocity, $\Delta p \propto U$. This proportionality can, with (7.1) be expressed as a function of mass flow

$$\Delta p = K_{\text{la}} \frac{R \, T_{\text{us}}}{p_{\text{us}}} \dot{m}$$

where geometric and fluid properties are lumped into K_{la} that is seen as a tuning constant and $\frac{R \, T_{\text{us}}}{p_{\text{us}}}$ accounts for the inlet density. This is the basis for the laminar flow model and an equivalent formulation that expresses the flow as function of pressure ratio is used in component-based mean value models. Additionally, the gas temperature of the flow is also needed, which follows from (7.2). This results in the following model.

Model 7.1 Incompressible Laminar Restriction

$$\dot{m}(p_{\text{us}}, T_{\text{us}}, p_{\text{ds}}) = C_{\text{la}} \frac{p_{\text{us}}}{R \, T_{\text{us}}} (p_{\text{us}} - p_{\text{ds}}), \qquad T_{\text{flow}} = T_{\text{us}}$$

where $C_{\text{la}} = 1/K_{\text{la}}$ is a tuning constant.

The second, and more important, model is that of fully developed *turbulent flow* that can be used to describe several components such as: area changes, pipe bends, viscous flow, and so on.

The pressure loss is proportional to the density and the square of the fluid velocity through the restriction. This gives the following pressure drop

$$\Delta p = K_{tu,1}\, \rho_{us}\, U^2 = [\text{using } (7.1)] = K_{tu,2}\frac{\dot{m}^2}{\rho_{us}} \tag{7.4}$$

where $K_{tu,1}$ and $K_{tu,2}$ are constants that depend on geometric and fluid properties and the density can be expanded as $\rho_{us} = \frac{p_{us}}{R\, T_{us}}$ for gases. It is also important to consider the calculation causality, where we in engine systems use the pressure drop over the component as the driver for the flow through it. Therefore, the pressure is viewed as the input and the flow as output, which gives the following model for the mass flow.

Model 7.2 Incompressible Turbulent Restriction

$$\dot{m}(p_{us}, T_{us}, p_{ds}) = C_{tu}\sqrt{\frac{p_{us}}{R\, T_{us}}}\sqrt{\Delta p}, \qquad T_{flow} = T_{us} \tag{7.5}$$

Here C_{tu} is a tuning constant that depends on geometry and fluid properties.

An aspect that is of practical importance, when using this model for simulation, is the fact that the derivative of (7.5) approaches infinity when Δp approaches zero. The function thus does not satisfy the Lipschitz condition, which is important because it is connected to the existence and uniqueness of a solution to the ordinary differential equation. Therefore, certain precautions have to be taken when using the model in simulation. A simple and often used solution is to modify the function in a region close to $\Delta p = p_{us} - p_{ds} = 0$ so that it receives a finite derivative. Often a linear portion is used, resulting in the following model.

Model 7.3 Incompressible Turbulent Restriction with Linear Region

---

$$\dot{m}(p_{us}, T_{us}, p_{ds}) = \begin{cases} C_{tu}\sqrt{\frac{p_{us}}{R\, T_{us}}}\sqrt{p_{us} - p_{ds}}, & \text{if } p_{us} - p_{ds} \geq \Delta p_{lin} \\[2mm] C_{tu}\sqrt{\frac{p_{us}}{R\, T_{us}}}\frac{p_{us} - p_{ds}}{\sqrt{\Delta p_{lin}}}, & \text{otherwise} \end{cases} \tag{7.6}$$

$$T_{flow} = T_{us} \tag{7.7}$$

where the linear region is in $p_{us} - p_{ds} \in [0, \Delta p_{lin}]$, and C_{tu} is a tuning constant that depends geometry and fluid properties.

This extension is also motivated by the physics, since the flow is laminar for low flow velocities. When the flow velocity increases it becomes turbulent, and this gives a transition; see Ellman and Piché (1999) for an example of a smooth transition.

Examples of Incompressible Turbulent Restrictions

Most gas flows encountered in the pipes in engines are incompressible and turbulent, and therefore model (7.5) (or (7.6)) is frequently used to describe components in models of engine systems. This model is easy to tune and validations show that it also gives an accurate description of the flow. This is illustrated in the example below.

Example 7.1 (Mass flow models for air filter, intercooler, exhaust system) Consider the following three components in an engine: air filter, intercooler, and exhaust system (including the catalyst and muffler). How well does the model for an incompressible flow restriction (7.4) describe the pressure losses over (or flows through) the components?

Figure 7.6 shows the result when the models are adjusted to experimental data from a turbocharged engine. As can be seen, the simple model with one adjustable parameter gives an accurate description of the pressure loss (see Eriksson et al. (2002b) for a discussion).

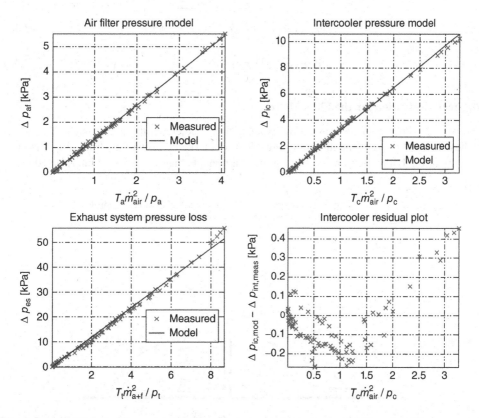

Figure 7.6 Validation of flow models for three components in a turbocharged SI-engine. The agreement between model and measurement is very good

7.2.2 Compressible Flow

Now we turn to models for components like the throttle and other control valves where the gas velocity is high and compressible fluid models are needed. Flows passing components where the cross-section area is small and where the pressure difference over the component is large are well described by isentropic *compressible flow*. The mass flow, $\dot{m}$, depends on the opening area, the density before the throttle, as well as the pressure ratio $p_r = \frac{p_{ds}}{p_{us}}$. We directly give the equation for the mass flow and give a motivation of its properties. The interested reader can turn to, for example, Heywood (1988, Appendix C) for a derivation of the equation. The equation for mass flow through a venturi is

$$\dot{m} = \frac{p_{us}}{\sqrt{RT_{us}}} \cdot A \cdot C_D \cdot \Psi(p_r) \tag{7.8}$$

where A is the area, C_D is a discharge coefficient that depends on the shape of the flow area, and $\Psi(p_r)$ is

$$\Pi(p_r) = \max(p_r, \left(\frac{2}{\gamma + 1}\right)^{\frac{\gamma}{\gamma-1}}) \tag{7.9a}$$

$$\Psi_0(\Pi) = \sqrt{\frac{2\gamma}{\gamma - 1} \left(\Pi^{\frac{2}{\gamma}} - \Pi^{\frac{\gamma+1}{\gamma}}\right)} \tag{7.9b}$$

$$\Psi(p_r) = \Psi_0(\Pi(p_r)) \tag{7.9c}$$

Physical Motivation for the Flow Model

The pressure drop over the restriction expands and accelerates the fluid, and it reaches its highest velocity close to the throat. Figure 7.7 shows a sketch of the flow lines and illustrates the flow through a restriction, explaining the components that are included in (7.8) and (7.9).

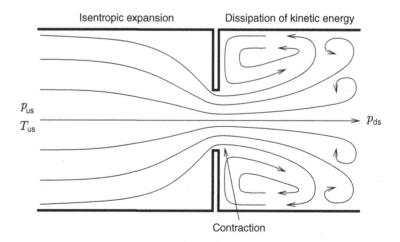

Figure 7.7   A restriction where the pressure drop over the restriction accelerates the flow. At the throat, the flow lines are contracted and the effective flow area A_{eff} becomes smaller than the real area $A_{eff} = C_D A$. After the throat, the kinetic energy in the flow is dissipated to turbulence

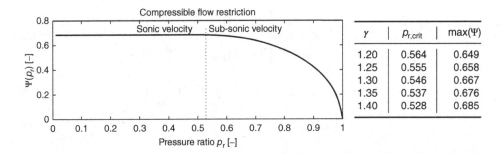

Figure 7.8 Shape of the $\Psi(p_r)$ function and the influence of γ, dotted line shows the critical pressure ratio $p_{r,\text{crit}}$. The table next to the plot shows how the critical pressure ratio and the maximum value of $\Psi(p_r)$ changes with γ. It can be seen that γ only has a minor influence on $\Psi(p_r)$

In particular $\Psi(p_r)$ in (7.8) determines the fluid velocity, which is limited by the sonic velocity. The velocity at the throat is determined by the pressure ratio, and the value where the flow reaches sonic velocity is called the critical pressure ratio

$$p_{r,\text{crit}} = \left(\frac{2}{\gamma+1}\right)^{\frac{\gamma}{\gamma-1}}$$

where γ is the ratio of specific heats. When the pressure ratio is below $p_{r,\text{crit}}$ the sonic velocity is reached. Figure 7.8 shows the function $\Psi(p_r)$ for $\gamma = 1.4$. The ratio of specific heats influences the shape of the function but, as can be seen in the table next to the plot, it has only a minor influence. In the model, Ψ gives information about the flow velocity, and the terms $\frac{p_{\text{us}}}{\sqrt{RT_{\text{us}}}} \Psi(p_r)$ in (7.8) describe the density and velocity in terms of inlet conditions. At the throat the flow lines contract slightly more than the geometric area A, and this contraction is described by the discharge coefficient C_D. A and C_D jointly determine the effective flow area in the restriction. After the throat, the kinetic energy is dissipated to turbulence.

This model, (7.8) and (7.9), gives a good description of the mass flow through several components such as the throttle, wastegate, and EGR valve. When using this model, the same precautions have to be taken as for the incompressible and turbulent model, since it does not fulfill the Lipschitz condition for $p_r = 1$ either (note that $p_r = 1 \iff \Delta p = 0$). One way to overcome this problem is to use a linear function around $p_r = 1$, this is seen in Figure 7.8 where the linear region goes from 0.99 to 1.00.

The isentropic flow also preserves the energy, and the temperature of the fluid that leaves the restriction will thus be the same as that of the upstream. The isentropic flow model now can be formulated in the following way.

Model 7.4 Compressible Turbulent Restriction

$$\Pi\left(\frac{p_{ds}}{p_{us}}\right) = \max\left(\frac{p_{ds}}{p_{us}}, \left(\frac{2}{\gamma+1}\right)^{\frac{\gamma}{\gamma-1}}\right) \tag{7.10a}$$

$$\Psi_0(\Pi) = \sqrt{\frac{2\gamma}{\gamma-1}\left(\Pi^{\frac{2}{\gamma}} - \Pi^{\frac{\gamma+1}{\gamma}}\right)} \tag{7.10b}$$

$$\Psi_{li}(\Pi) = \begin{cases} \Psi_0(\Pi) & \text{if } \Pi \leq \Pi_{li} \\ \Psi_0(\Pi_{li})\dfrac{1-\Pi}{1-\Pi_{li}} & \text{otherwise} \end{cases} \tag{7.10c}$$

$$\dot{m}(p_{us}, T_{us}, p_{ds}, A) = A\, C_D\, \frac{p_{us}}{\sqrt{RT_{us}}}\, \Psi_{li}(\Pi(\frac{p_{ds}}{p_{us}})) \tag{7.10d}$$

$$T_{\text{flow}} = T_{us} \tag{7.10e}$$

where the linear region is defined by $\frac{p_{ds}}{p_{us}} \in [\Pi_{li}, 1]$.

A Note on Linear Regions and Simulation

In the models (7.6) and (7.10) above, it has been commented that a linear region around $\Delta p = 0$ should be added so that these functions fulfill the Lipschitz condition. The size of this linear region is not easy to determine beforehand (one way to determine it was given in Ellman and Piché (1999)). An indication is: if oscillations are present in the simulated mass flow of an MVEM component that operates at steady state and thus should be smooth, then a linear region is needed or it should be made larger.

7.3   Throttle Flow Modeling

There are two different aspects that can be modeled for a throttle, the first is the throttle motion and the second is the air mass flow through the throttle. The latter is important for the engine air flow and intake pressure build-up. The most frequently used throttles are of butterfly type, see Figure 7.9 for an example, where the angle of the throttle plate controls the flow area. Throttles are either directly controlled by the driver through, for example, a mechanical wire coupling or controlled by the EMS with an electric servo motor. The latter is the standard in automotive engines today, since it provides high flexibility in determining the behavior of the vehicle by designing different profiles for the throttle behavior.

In the flow models developed below the throttle angle α is seen as the input. The mechanical system, describing the throttle motion and the resulting α, will be described and modeled later in Section 7.13.

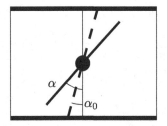 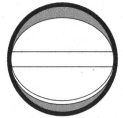

Figure 7.9   (a) Throttle body with definitions of the angle where the throttle is closed α_0 and the actual throttle angle α in relation to the throttle housing. (b) The two moon crescent shaped areas show the flow areas

7.3.1 Throttle Area and Discharge Coefficient

Figure 7.9 shows a throttle body and defines some of the parameters that control the throttle opening area, and thus the air mass flow. The throttle area is usually expressed in terms of the throttle angle α in relation to the throttle body, and the angle where the throttle is closed α_0. For a throttle, both the opening area and the shape of the flow area changes with throttle angle, which means that the area and the discharge coefficient are functions of α, that is $A_{th}(\alpha)$ and $C_{D,th}(\alpha)$.

For most throttles there is a small air flow through the throttle $\dot{m}_{at,0}$, when $\alpha = \alpha_0$, that comes from areas caused by imperfections in the throttle. The flow area can thus be expressed as

$$A_{th}(\alpha) = A(\alpha) + A_L$$

where A_L represents a leakage flow area which is present at $\alpha = \alpha_0$, when the throttle is closed. It is suitable to model the flow through the throttle as a compressible fluid ((7.8) and (7.9)) since the flow velocity becomes very high in the narrow regions. This results in the following model for the air mass flow through a throttle.

Model 7.5 Throttle Mass Flow

$$\dot{m}_{at}(\alpha, p_{us}, p_{ds}, T_{us}) = \frac{p_{us}}{\sqrt{RT_{us}}} A_{th}(\alpha) C_{D,th}(\alpha) \Psi_{li}(\Pi) \tag{7.11a}$$

$$\Psi_{li}(\Pi(\frac{p_{ds}}{p_{us}})) = [(7.10a) \ to \ (7.10c)] \tag{7.11b}$$

$$T_{flow} = T_{us} \tag{7.11c}$$

For simple geometries, the area $A(\alpha)$ can be determined by geometric considerations of the throttle housing, throttle plate, and throttle plate shaft, however the geometries around the plate can be arbitrarily complex. A simple area function is attained if we make the following assumptions: (i) disregard the throttle plate shaft, (ii) the plate is infinitely thin, (iii) the pipe is circular with diameter D, (iv) the throttle is closed at α_0. Then the plate is elliptic with a major axis of length $D/\cos(\alpha_0)$. For this simplification the area becomes

$$A_{th}(\alpha) = \pi D^2 \left(1 - \frac{\cos(\alpha)}{\cos(\alpha_0)}\right)$$

which is simple but has errors at large throttle openings when the shaft is important. Including the shaft gives a more elaborate expression for the area, see for example (Heywood 1988, Eq. (7.18)). Nowadays the throttle housings are designed and manufactured to shapes that no longer are straight pipes, which makes it even more difficult to derive the throttle area from the geometry. Simple parameterizations that approximate the area are to use, for example, a trigonometric representation or a polynomial

$$A_{th}(\alpha) = A_0 + A_1 \cos(\alpha) + A_2 \cos^2(\alpha) + \cdots \tag{7.12}$$

$$A_{th}(\alpha) = A_0 + A_1 \alpha + A_2 \alpha^2 + \cdots$$

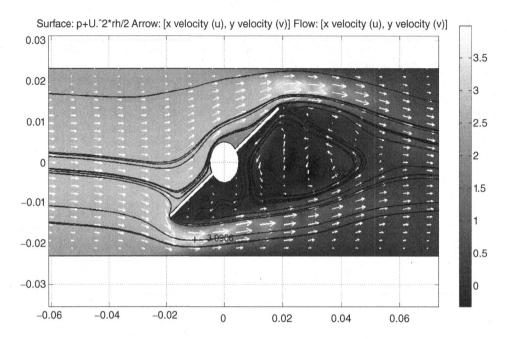

Figure 7.10 Two-dimensional flow over a throttle plate, with an angle of $\alpha = 50°$. Static pressure is shown in gray scale and arrows show the velocity vectors. A significant contraction of the flow lines is shown after the plate. (Picture from Krysander (2000).)

If three terms are not sufficient then it is easy to add new terms to the expression (sin terms can also be included). When the throttle is closed, at $\alpha = \alpha_0$, the idealized throttle area is zero, but due to production tolerances and material roughness there is a small leakage area which either should be included in the model or added separately, $A(\alpha) = A_L + A_{th}(\alpha)$. After the geometric area is determined, tests in either a flow bench or an engine have to be made to determine the leakage area A_L and the discharge coefficient.

It is difficult to model the discharge coefficient $C_{D,th}$ from the geometry, which is problematic since it has a significant impact on the effective area, see Figure 7.10 that clearly shows the contraction. The general difficulty is clearly illustrated in (Blair 1999, Ch. 3) where a whole chapter is devoted to the problem of determining the discharge coefficient. $C_{D,th}$ depends on the angle, pressure ratio, as well as the throttle plate Reynolds number, $C_{D,th}(\alpha, p_r, Re)$. $C_{D,th}$ is therefore often determined using measurements. Extensions of the model have also been proposed that include the pressure ratio (or pressure difference) in the discharge coefficient. Another extension, presented in (Hendricks et al., 1996), proposes two parallel flows and changes to the $\Psi(p_r)$ function.

Since models for both area and discharge coefficient are difficult to obtain, it is often beneficial to lump them together and determine the product $A_{eff}(\alpha) = A_{th}(\alpha) C_{D,th}(\alpha)$ experimentally and then use a parameterized model, such as (7.12).

7.4   Mass Flow Into the Cylinders

The air mass flow passing the inlet valves, $\dot{m}_{ac}$ (the subscript ac refers to air and cylinder), out of the intake manifold into the cylinders depends on many parameters, but the engine speed, N, intake manifold pressure, p_{im}, and temperature, T_{im}, are most important. The parameter volumetric efficiency, η_{vol}, is used to describe the effectiveness of the engine's ability to induct new air into the cylinders, see Section 4.3.3. The volumetric efficiency depends on several engine parameters, like engine speed, intake manifold pressure, fuel evaporation, coolant temperature, and so on, that is $\eta_{vol}(N, p_{im}, \ldots)$. At steady state the air mass entering the intake manifold is equal to the mass of air inducted into the cylinder, and the volumetric efficiency can be determined in tests using measurements with the air mass flow meter over the operating range. The submodel for the volumetric efficiency is then determined from the engine tests, and examples of such models will be described in Section 7.4.1.

With the volumetric efficiency model given, the mass flow into the cylinders can be determined by using the definition for volumetric efficiency (4.9), and solving for the $\dot{m}_{ac}$. The air density, ρ_{ai}, is given by the ideal gas law for intake conditions, that is $\rho_{ai} = \frac{p_{im}}{R\,T_{im}}$, which gives the following mass flow model.

Model 7.6   Engine Mass Flow

$$\dot{m}_{ac}(N, p_{im}, T_{im}) = \eta_{vol}(N, p_{im}, \ldots)\frac{V_D\,N\,p_{im}}{n_r\,R\,T_{im}} \qquad (7.13)$$

which uses a submodel for the volumetric efficiency $\eta_{vol}(N, p_{im}, \ldots)$.

Note that this model gives the flow for the whole engine, and individual cylinders are not considered, so in this case V_D is the displaced volume for the whole engine. The air mass flow is thus parameterized by the volumetric efficiency, which can be determined using the air mass flow sensor when running the engine at steady state operation. This parameterization, using the volumetric efficiency, handles variations in intake system density. The following sections will look into parameterizations for η_{vol}, the first are some simple black box approachesm while those that follows incorporate some of the physics behind the air induction process. Another possibility is to represent the mass flow using a lookup table (also called a map), this will be described in Section 9.3.1.

7.4.1   Models for Volumetric Efficiency

Several models for volumetric efficiency have been proposed in the research literature on mean value engine modeling, and range from black box models to physically based models. We start with some examples of black box parameterizations of $\eta_{vol}(N, p_{im})$, where the most simple is a single constant while others that are frequently used are second-order polynomials in pressure and engine speed. Even higher-order polynomials are often used. Some examples are:

$$\eta_{vol} = c \qquad\qquad\qquad\qquad\qquad\qquad\qquad\qquad\text{"The constant"}$$

$$\eta_{vol} = c_0 + c_1 \sqrt{p_{im}} + c_2 \sqrt{N} \qquad\qquad\qquad\qquad\qquad\text{A simple relation}$$

$$\eta_{vol} = \eta_{v,0} + \eta_{v,N1} N + \eta_{v,N2} N^2 + \eta_{v,p1}p_{im} \qquad\quad\text{Hendricks and Sorenson (1990)}$$

$$\eta_{vol} = s(N) + \frac{y(N)}{p_{im}} \qquad\qquad\qquad\qquad\qquad\qquad\text{Jensen et al. (1997)}$$

Where $s(N)$ and $y(N)$ are weak functions of N in the last expression.

Pressure, Fuel, and Heat Transfer, Dependency

Now we turn to the physical processes behind the air induction. As a first step, we disregard the influence of the engine speed and concentrate on effects such as the displacement of fresh gases by residual gas, heating of the gas through the walls and ports as well as cooling by the evaporating fuel. Influences of the residual gases in the cylinder were discussed and illustrated in Chapter 5 where the following expression was derived

$$\eta_{vol}(p_{im}, p_{em}, \lambda) = \underbrace{\frac{r_c - \left(\frac{p_{em}}{p_{im}}\right)^{1/\gamma}}{r_c - 1}}_{\text{residual gas}} \cdot \underbrace{\frac{1}{1 + \frac{1}{\lambda\,(A/F)_s}\frac{M_a}{M_f}}}_{\text{molecule displ.}\approx 1} \qquad (7.14)$$

This covers the effect of the expansion of the residual gases from the exhaust conditions to intake conditions. Molecule displacement has only a minor effect on the volumetric efficiency, while heat transfer has a more significant effect. To get a good agreement with the experimental data, a speed dependent factor can simply be added to the model

$$\eta_{vol}(p_{im}, p_{em}) = C_{\eta_{vol}}(N)\frac{r_c - \left(\frac{p_{em}}{p_{im}}\right)^{1/\gamma}}{r_c - 1}$$

Other effects, like charge cooling from fuel enrichment, can also be handled, see (Heywood 1988, Ch. 6) for a derivation of a general model. An SI engine normally operates at $\lambda = 1$, but on some occasions fuel enrichment is used to protect the catalyst. In Andersson and Eriksson (2004) a model is proposed that describes the variation as λ deviates from $\lambda = 1$

$$\eta_{vol}(p_{im}, p_{em}, \lambda, T_{im}) = C_{\eta_{vol}} \frac{r_c - \left(\frac{p_{em}}{p_{im}}\right)^{\frac{1}{\gamma}}}{r_c - 1} \cdot \underbrace{\frac{T_{im}}{T_{im} - C_1(1/\lambda - 1)}}_{\text{Charge cooling}} \qquad (7.15)$$

where only one extra constant C_1 is added. The constant C_1 depends on fuel properties, like the enthalpy of evaporation, see Andersson and Eriksson (2004) for details. This augmentation of the model improved the model agreement; without charge cooling effect the relative error was 10% and with the charge cooling effect it was 3%.

Engine Speed Influence

The engine speed has an influence but this depends on the engine design, and for each case it must be explored whether or not it should be included. Using a structure of the form shown below gives good capabilities for also capturing the speed variations

$$\eta_{vol}(N, p_{im}, p_{em}, \lambda, T_{im}) = C_{\eta_{vol}, N1}(N) \cdot \eta_{vol}(p_{im}, p_{em}, \lambda, T_{im}) + C_{\eta_{vol}, N2}(N)$$

where $C_{\eta_{vol}, N1}(N)$ and $C_{\eta_{vol}, N2}(N)$ are weak functions of N, and $\eta_{vol}(p_{im}, p_{em}, \lambda, T_{im})$ come from the models above, for example (7.15). When considering the engine speed dependence, it is difficult to draw conclusions of a general nature; an engine with tuned intake runners can require a fourth (or even sixth) order polynomial to describe the resonating effects in the intake runners, while a turbocharged engine that has other design considerations might only have minor variations in η_{vol} with engine speed. Including variable valve actuation gives a more complicated situation, see Section 12.1 for a discussion.

Including EGR with Oxygen and Burned Gas Fractions

When exhaust gas recirculation (EGR) is used, burned gases are fed into the intake system and displace air. Therefore, when EGR is present, the gas flow to the cylinders is not just air. A common approach to handle this is to use the volumetric efficiency to determine the total mass flow to the cylinders $\dot{m}_{cyl}$, and complement this with a model that includes the air and recirculated masses in the intake system. The EGR fraction is defined as

$$x_{egr} = \frac{m_{egr}}{m_{tot}} = \frac{m_{egr}}{m_{air} + m_{egr}}$$

where the masses are those in the intake system (but can in the general case be those in any system). This can now be used to determine the mass flows of air and EGR into the cylinders

$$\begin{pmatrix} \dot{m}_{ac} \\ \dot{m}_{egr} \end{pmatrix} = \begin{pmatrix} x_{air} \\ x_{egr} \end{pmatrix} \dot{m}_{cyl} = \begin{pmatrix} 1 - x_{egr} \\ x_{egr} \end{pmatrix} \dot{m}_{cyl}$$

The same procedure can also be applied to more detailed gas composition models. To refine this further, one can note that if an engine runs leanly, as a diesel engine does, then the EGR also carries oxygen. The opposite is less common, but in case of rich mixtures there is CO and HC in the EGR gas, which can be seen as a deficit of oxygen. There are two approaches to track the oxygen. One is to track the *oxygen fraction*, which is defined as the

$$X_O = \frac{m_O}{m_{tot}} \tag{7.16}$$

where m_O is the mass of oxygen in the volume. When there are several flows $\dot{m}_i$, with different oxygen fractions $X_{O,i}$, into a volume, the oxygen fraction in the volume will change. To describe the change a differential equation can be derived by differentiating (7.16) and using the ideal gas law. With $m_{tot} = \frac{pV}{RT}$ the expression becomes

$$\frac{dX_O}{dt} = \frac{RT}{pV} \sum_i (X_{O,i} - X_O)\dot{m}_i \tag{7.17}$$

The other approach is to track the *burned gas fraction* X_B, which separates out the burned gases from the unburned air in the EGR. The definition and differential equation becomes

$$X_B = \frac{m_B}{m_{tot}} \quad \Rightarrow \quad \frac{dX_B}{dt} = \frac{RT}{pV} \sum_i (X_{B,i} - X_B)\dot{m}_i \tag{7.18}$$

To give some examples, air has $X_{Bair} = 0$ and $X_{Oair} = \frac{32}{32+3.773 \cdot 28.16}$ (recall Model 4.1), while a gas combusted at $\lambda = 1$ has $X_B = 1$ and $X_O = 0$.

7.5 Volumes

Up until now, the MVEM components have been static and expressed as static relations. Now we turn to some of the models that describe the dynamic behavior of the system, and these are described with ordinary differential equations.

The intake and exhaust manifolds, as well as many other volumes in the engine system, can be viewed as a thermodynamic control volume that stores mass and energy, see Figure 7.11. The thermodynamic control volume has a fixed volume V and its storage has filling and emptying dynamics. The increase or decrease of the mass is determined by the inlet and outlet air mass flow, $\dot{m}_{in}$ and $\dot{m}_{out}$. Mass and energy conservation are fundamental and will be used to derive the differential equations for the dynamics in the volume.

The difference between the inflow and outflow of mass ($\dot{m}_{in} - \dot{m}_{out}$) directly gives the rate of change in mass in the system

$$\frac{dm}{dt} = \dot{m}_{in} - \dot{m}_{out} \tag{7.19}$$

Energy is also conserved and stored in the system. Considering the energy, there might be heat transfer $\dot{Q}$ but there is no mechanical work transfer to the control volume, see Figure 7.11. Energy is also transferred to and from the system through the in- and outflows. For the open system that we are considering, the first law (energy conservation) gives the following rate of change of the internal energy

$$\frac{dU}{dt} = \dot{H}_{in} - \dot{H}_{out} - \dot{Q} \tag{7.20}$$

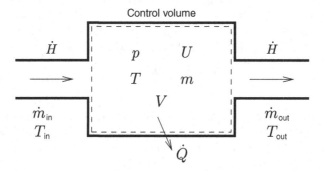

Figure 7.11 The control volume is marked with dashed lines and the pressure, temperature, and mass states are shown in it. There are two flows across the boundaries. There is no volume change (V is a constant), except those from the flow, so there is no mechanical work exchange

This kind of volume is called a well-stirred reactor since the incoming enthalpy is instantaneously spread out over the whole volume. It is also called a lumped model since it lumps all spatial information into just one point.

Mass and energy are difficult to measure, so these equations are usually manipulated to express the system states in measurable quantities like temperature and pressure, see the isothermal and adiabatic models below. Before we go into this, we make the following assumptions:

- the gas is ideal (i.e., $pV = mRT$)
- c_p and c_v are constant

With these assumptions, the temperature can be determined directly from the internal energy and the mass through

$$U = m\,u(T) = [c_v\text{-constant}] = m\,c_v\,T \tag{7.21}$$

and the pressure can be determined from the ideal gas law

$$pV = mRT \tag{7.22}$$

Furthermore, the enthalpy flows are given by

$$\dot{H}_{in} = \dot{m}_{in}\,c_p\,T_{in} \quad \text{and} \quad \dot{H}_{out} = \dot{m}_{out}\,c_p\,T_{out} \tag{7.23}$$

where the temperature of the outflowing gas is the same as that in the volume, that is $T_{out} = T$. (7.19) to (7.23) give a system of equations that can be solved to give all information about the system, for example the temperature and pressure. Note that if the flow is reversed, the sign of the mass flow also changes, and in addition the flow temperature must also be changed accordingly. Any number of in- and outflows to the volume can be added by adding their enthalpy flows.

Two formulations of the volume model occur frequently in the literature, one is called the *isothermal model* and the second is called the *adiabatic model*. In these the differential equations are formulated so that the pressure and temperature are states in the models, and they will be discussed below. A comparison of the difference between the models is shown in Section 7.6 in Figure 7.14.

Isothermal Model

The first and most frequently used model is the *isothermal* model, which is named after the assumption that there is no temperature change in the system $T = T_{in} = T_{out}$. In this way the temperature is determined. To derive the differential equation for the pressure, the ideal gas law $pV = mRT$ is differentiated, which gives

$$\frac{dp}{dt} = \frac{RT}{V}\frac{dm}{dt}$$

Inserting the mass conservation (7.19) yields the following model for the pressure dynamics in a volume.

Model 7.7 Isothermal Volume

The dynamic equation for the pressure development is

$$\frac{\mathrm{d}p}{\mathrm{d}t} = \frac{R\,T}{V}(\dot{m}_{\mathrm{in}} - \dot{m}_{\mathrm{out}}) \tag{7.24}$$

and the temperature is constant $T = T_{\mathrm{in}} = T_{\mathrm{out}}$.

This model has good capabilities of describing the pressure dynamics in the intake and exhaust systems. This model is, as all models, a simplification and the energy conservation has been neglected through the assumption that the temperature is constant. The assumption that the intake manifold temperature is constant is not always valid during fast transients. For example, the temperature can rise for a short time when the gas in the intake system is compressed after a fast opening of the throttle. A more detailed model is therefore described below.

Adiabatic Model

The next volume model is called the *adiabatic* model, where the name adiabatic was adopted because the heat transfer in the energy equation was often set to zero. This model is derived by rewriting the state equations for mass (7.19) and internal energy (7.23). As a first step, the internal energy is rewritten so that the temperature is selected as a state. To get the temperature differential, the internal energy (7.21) is differentiated, which gives the relationship $\frac{\mathrm{d}U}{\mathrm{d}t} = \frac{\mathrm{d}(m\,u)}{\mathrm{d}t} = u(T)\frac{\mathrm{d}m}{\mathrm{d}t} + m\frac{\mathrm{d}u}{\mathrm{d}t}$ and this is inserted into (7.20) which gives

$$m\,c_{\mathrm{v}}\frac{\mathrm{d}T}{\mathrm{d}t} = \dot{m}_{\mathrm{in}}(h(T_{\mathrm{in}}) - u(T)) - \dot{m}_{\mathrm{out}}(h(T_{\mathrm{out}}) - u(T)) - \dot{Q} \tag{7.25}$$

where $T_{\mathrm{out}} = T$. As the next step, the definition of enthalpy and the assumption of ideal gases are used which gives $h(T) = u(T) + RT$. Furthermore, applying the assumption that c_{p} and c_{v} are constant and rearranging the terms in the equation above gives the temperature derivative

$$\frac{\mathrm{d}T}{\mathrm{d}t} = \frac{1}{m\,c_{\mathrm{v}}}[\dot{m}_{\mathrm{in}}c_{\mathrm{v}}(T_{\mathrm{in}} - T) + R(T_{\mathrm{in}}\dot{m}_{\mathrm{in}} - T\,\dot{m}_{\mathrm{out}}) - \dot{Q}]$$

This equation and the mass balance give a complete model for the state of the volume. From the mass and temperature states, the pressure in the volume can be calculated with the ideal gas law. This gives one formulation of a model for the volume.

Model 7.8 Adiabatic Volume with Temperature and Mass States

The differential equations that describe the thermodynamic state (m and T) are

$$\begin{cases} \frac{\mathrm{d}m}{\mathrm{d}t} = \dot{m}_{\mathrm{in}} - \dot{m}_{\mathrm{out}} \\[2mm] \frac{\mathrm{d}T}{\mathrm{d}t} = \frac{1}{m\,c_{\mathrm{v}}}\left[\dot{m}_{\mathrm{in}}c_{\mathrm{v}}(T_{\mathrm{in}} - T) + R\,(T_{\mathrm{in}}\dot{m}_{\mathrm{in}} - T\dot{m}_{\mathrm{out}}) - \dot{Q}\right] \end{cases} \tag{7.26a}$$

and the outputs from the volume are the states and pressure

$$\begin{cases} m = m \\ T = T \\ p = mRT/V \end{cases}$$ (7.26b)

Using mass and temperature as states gives a compact expression as seen above, but it is necessary to add an extra output equation to get information about the pressure. An equivalent formulation can be derived where the only difference is that the equations are formulated so that pressure and temperature become the state variables instead of mass and energy. The aim is to formulate the model in terms of differential equations for pressure and temperature. When the mass, pressure, and temperature are allowed to change the differentiated ideal gas law becomes

$$V\frac{dp}{dt} = RT\frac{dm}{dt} + mR\frac{dT}{dt}$$

Eliminating the mass with the ideal gas law, inserting (7.19), and rewriting the equation, we get the following differential equation for the pressure

$$\frac{dp}{dt} = \frac{RT}{V}(\dot{m}_{in} - \dot{m}_{out}) + \frac{p}{T}\frac{dT}{dt}$$

that in its turn depends on the pressure differential equation $\frac{dT}{dt}$. In summary, the differential equations for the adiabatic model of a volume become as below.

Model 7.9 Adiabatic Volume with Pressure and Temperature States

---

The differential equations for the states (T and p) are

$$\begin{cases} \frac{dT}{dt} = \frac{RT}{pVc_v}\left[\dot{m}_{in}c_v(T_{in} - T) + R(T_{in}\dot{m}_{in} - T\dot{m}_{out}) - \dot{Q}\right] \\ \frac{dp}{dt} = \frac{RT}{V}(\dot{m}_{in} - \dot{m}_{out}) + \frac{p}{T}\frac{dT}{dt} \end{cases}$$ (7.27)

where the result of the first equation is plugged into the second. The outputs are the states (T and p). If the mass is also required then it can be determined with the ideal gas law $m = \frac{pV}{RT}$.

Setting $\dot{Q} = 0$ gives an adiabatic model, which is the source for the name in the literature, however the heat transfer can be included to give a more general model. Note that Models 7.9 and 7.8 are equivalent formulations of the mass and energy conservation equations, and have only a difference in selection of thermodynamic state.

Of the two volume models, the isothermal model is frequently encountered in literature because it is simple to implement and is able to capture the pressure dynamics. The adiabatic model has also been used by several authors, but the name adiabatic was made popular in Chevalier et al. (2000) and a descriptive overview of the adiabatic model is given in Hendricks (2001). A comparison of the isothermal and adiabatic models will be made at the end of Section 7.6 that covers the intake manifold model.

7.6   Example – Intake Manifold

Now we have reached the point in modeling where we have enough components to model the intake system with throttle, intake manifold, and engine air mass flow, see Figure 7.12. The intake system is important because it determines the transient dynamics of naturally aspirated engines, and here we will now look at the intake manifold pressure p_{im} for a transient in throttle angle α.

Example 7.2 (Intake Manifold Dynamics)   The task is to model the intake manifold system of a naturally aspirated SI engine, shown in Figure 7.12. The system consists of a butterfly throttle with a servo, intake manifold, and air mass flow to the engine cylinders. Validate the model against measured data for a step response in throttle angle.

We start with the throttle, which is controlled by a servo. A reference signal α_{ref} is fed to the servo which controls the throttle plate angle, and the resulting angle is measured as the output voltage α from a potentiometer on the throttle plate. The throttle servo is modeled using a first-order system with time constant τ_α, that is $\alpha = \frac{1}{1+s\,\tau_\alpha} \alpha_{ref}$. Throttle air mass flow is modeled using (7.11), where we assume that pressure and temperature before the throttle are equal to the ambient conditions $p_{us} = p_{amb}$ and $T_{us} = T_{amb}$. Downstream we have intake manifold pressure p_{im}. As effective area A_{eff} a polynomial in $\cos(\alpha)$ is used.

Next, we model the air flow to the cylinders, using (7.13), where a first-order polynomial in p_{im} and N is used for volumetric efficiency.

The intake manifold dynamics is modeled using an isothermal model (7.24), with intake temperature equal to ambient, $T_{im} = T_{amb}$. To summarize the model, we have the following equations

$$\alpha = \frac{1}{1+s\,\tau_\alpha} \alpha_{ref}$$

$$A_{eff}(\alpha) = A_{th}(\alpha)C_{D,th}(\alpha) = A_0 + A_1\,\cos(\alpha) + A_2\cos^2(\alpha)$$

$$\dot{m}_{at}(p_{amb},p_{im},T_{amb},\alpha) = \frac{p_a}{\sqrt{RT_a}} A_{eff}(\alpha)\Psi_{li}(\Pi(\frac{p_{im}}{p_a}))$$

$$\Psi_{li}(\Pi(\frac{p_{im}}{p_a})) = [(7.10a)\ to\ (7.10c)]$$

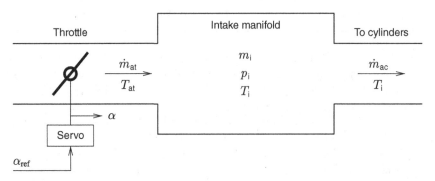

Figure 7.12   A schematic drawing of the intake manifold as a volume that is filled by the throttle flow and emptied by the air flow to the cylinder

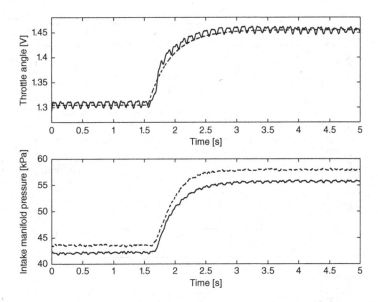

Figure 7.13 Evaluation of the isothermal model, (7.24), of the intake manifold dynamics. Solid line – measurements. Dashed line – model output

$$\dot{m}_{ac}(N, p_{im}, T_{amb}) = \eta_{vol}(N, p_{im}, \ldots) \frac{V_D N p_{im}}{n_r R T_{im}}$$

$$\eta_{vol}(N, p_{im}) = \eta_0 + \eta_1 p_{im} + \eta_2 N + \eta_3 N p_{im}$$

$$\frac{dp_{im}}{dt} = \frac{R T_i}{V_i}(\dot{m}_{at}(\alpha, p_{amb}, p_{im}, T_{amb}) - \dot{m}_{ac}(N, p_{im}, T_{im}))$$

Finally, we compare how well this model can describe the throttle motion and the measured pressure trace. In Figure 7.13, a model for a throttle servo (top figure) with the corresponding throttle movement is validated, furthermore the manifold pressure model is also validated. The validation verifies that the model with its assigned parameters captures the pressure build-up and has the correct response time and appearance to a change in throttle angle.

This example shows that the isothermal model with one state is sufficiently good for capturing the behavior of the pressure dynamics.

Comparison of Isothermal Adiabatic Control Volumes

In Figure 7.14, the isothermal and adiabatic control volume models are compared using the example of an intake system. The comparison is made for the intake system, shown in the example above, where all parameters are the same and the only difference is the volume model assumption. Two step changes are made in the throttle angle, one closing and one opening, and the pressure and temperature in the control volume are compared for the two models.

The comparison shows that the pressure dynamics is slightly faster for the adiabatic model, but there is not a big difference. The isothermal model has no temperature dynamics while

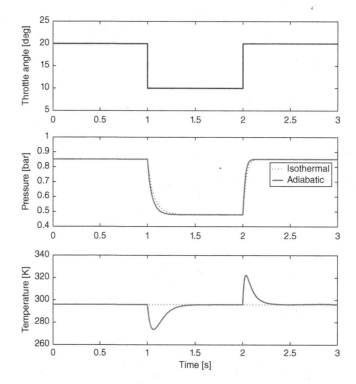

Figure 7.14 Comparison between the isothermal and adiabatic model. Dotted line – isothermal model. Solid line – adiabatic model. The pressure dynamics is slightly faster in the adiabatic model, but there is no big difference. No temperature dynamic is visible in the isothermal model

the adiabatic model shows the temperature excursions as the gases in the control volume are expanded and compressed with the closing and opening steps. At steady state there is no difference between the models.

7.7 Fuel Path and (A/F) Ratio

The fuel path is here defined as the parts: fuel tank, fuel pump(s), fuel rail, injectors, and mixture preparation. During the mixture preparation, it enters the gas path, consisting of air and burned gases with residual or recirculated exhaust gases, and is no longer an independent process. In fact, fueling has a strong influence on the gas path in EGR and turbocharged engines, and to be able to describe the engine system dynamics they need to be treated together. In the sections to come the fuel path will be discussed and then the joint gas path, the main focus is on models for SI engines.

7.7.1 Fuel Pumps, Fuel Rail, Injector Feed

In SI engines, there are two options for the fuel pump and delivery: a speed pump with a pressure regulator and return line, or a more advanced controllable pump with pressure sensor and controller, and no return line. In CI engines the high fuel pressures are generated by two pumps, a low pressure and a high pressure pump, that produce the pressure levels needed. The output of the fueling model is to describe the fuel pressure over the injector nozzle. In port fuel injected SI engines, the combination of fuel pump and pressure regulator, connected to the rail and intake system, produces a stable pressure Δp_{reg} and ensures that the injections become repetitive and accurate. The model is simply

$$p_{rail} = p_{im} + \Delta p_{reg}$$

In direct injected engines where the pressure is significantly higher, it becomes necessary to account for the compressibility of the fuel. Pressure fluctuations and wave propagation, that follow after injector openings, often occur in the high pressure rails. These influence the injector characteristics (see Figure 7.17) but are so fast that they are outside the range of mean value engine models. Examples of fuel rail and injector modeling are found in Chiavola and Giulianelli (2001), Hu et al. (2001), Woermann et al. (1999).

7.7.2   Fuel Injector

The fuel flow into the cylinder is measured and controlled through an electrically controlled valve. A fuel injector (see Figure 7.15) consists of a solenoid, valve, and spring, where a current in the solenoid opens the injector and the spring closes when the current is shut off. The fuel injector is opened once each cycle, and the closing time for the injector (end of injection) is usually timed to occur just before or after the inlet valve opening.

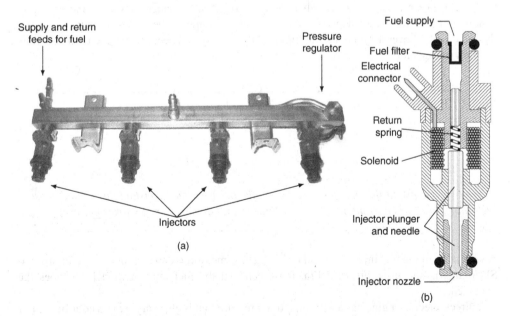

Figure 7.15   (a) A picture of a fuel rail with four injectors and a pressure regulator, (b) A cutout of a fuel injector showing the key components, solenoid, valve needle, and spring

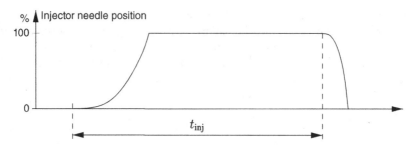

Figure 7.16   A pulse with duration, t_{inj}, is used to control fuel injector opening using the needle lift. As illustrated in the sketch, the needle movement is not instantaneous. When the pulse is applied the coil is magnetized and the injector starts to move, and after some time the injector becomes fully open. When the pulse is lowered, then the magnetization is removed and the spring pushes the needle to the closed position

The amount of fuel metered by the injection valve is proportional to the square root of the pressure over the valve (see (7.4)) and the opening time, t_{inj}, see Figure 7.16, less a lumped opening and closing time.

$$m_{fi} = C_0 \sqrt{\rho_{fuel} \, \Delta p_{inj}} \; (t_{inj} - t_0(U_{batt})) \tag{7.28}$$

The opening and closing times of the valve, t_0, depend on the battery voltage U_{batt}, since the opening of the valve depends on the current through the solenoid and thus the voltage, while the closing depends on force from the spring. When the injector model is applied to an engine, the pressure dependence is usually omitted since the pressure regulator, see Figure 7.15, maintains a constant pressure over the valve ($\Delta p_{inj} = \Delta p_{reg}$), independent of the engine operating condition. Furthermore, the density of the fuel is also assumed to be fairly constant. Averaging the injected fuel over a cycle, and multiplying by the engine speed, yields the total fuel flow through the injector

---

Model 7.10   Simple Fuel Injector Mass Flow Model

$$\dot{m}_{fi} = \frac{N \, n_{cyl}}{n_r} \, m_{fi} = N \, C \, (t_{inj} - t_0(U_{batt})) \tag{7.29}$$

where C is a constant that includes pressure influence (assumed constant due to the pressure regulator), injector constant, number of cylinders, and the number of strokes per cycle.

A validation of the fuel injector model (7.29) against measurement data for a port fuel injected SI engine, is shown in Figure 7.17a. It is seen that the fuel injector model describes the data well.

A direct injected engine has a high injection pressure that is also varying and controlled. For these it is necessary to have a more detailed model for the injector and the injected fuel mass. An example of such data is shown in Figure 7.17b, and it seen that a simple linear model is

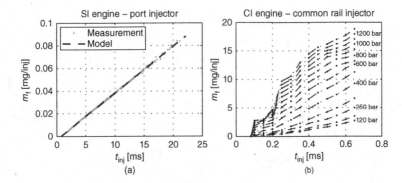

Figure 7.17 The pulse width, t_{inj}, to the fuel injector determines the time that the fuel injector is open and thereby the amount of fuel that is injected. (a) Port fuel injector for an SI engine, where the injector pressure is controlled by the pressure regulator, (b) Common rail injector data for a CI engine (from Woermann et al. (1999)), where the EMS controls the pressure depending on the operating point. Note that the data is monotonous in injector opening time and rail pressure, but varies considerably

not sufficient. Instead, the data presented in the figure can be included in a look-up table and used as a model.

$$m_f = f(t_{inj}, p_{rail}, p_{cyl}, \ldots) \tag{7.30}$$

In diesel engines, that can have multiple injections during a cycle, there can also be inter-pulse dependencies, due to the fact that one injection reduces the pressure locally in the fuel rail which in turn can influence the next injection.

7.7.3   Fuel Preparation Dynamics

Fuel delivered from the injector must be transported into the cylinder and go through a phase change to the gaseous phase so that it can be combusted. Evaporation cools the air flow which has two side effects, it increases the volumetric efficiency, see Section 7.4.1, and also reduces the knock tendency, see Section 6.2.2. Here we will focus on the delivery of fuel mass to the cylinders.

Direct Injected Engines

In direct injected engines, the fuel is delivered inside the cylinder and it is immediate in the sense of MVEM. For a direct injected engine, the mass of fuel delivered to the cylinder is identical to the injected.

Model 7.11   Fuel Mass in the Cylinder for Direct Injected Engine

---

$$m_{tc} = m_{fi} \qquad and \qquad \dot{m}_{tc} = \dot{m}_{fi}$$

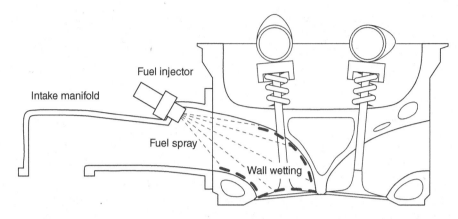

Figure 7.18 Illustration of the wall wetting effect. A fraction X of the injected fuel hits the walls and valve and builds up a fuel film, while the rest $(1 - X)$ enters the cylinder. The fuel on the walls evaporates, with a time constant of τ_{fp}, and enters the cylinder

Port Fuel Injected Engines

In a port fuel injected engine, only a fraction of the injected fuel is inducted into the cylinder directly. Some fuel is deposited on the intake walls either as a film or as puddles. Figure 7.18 shows a schematic picture of the wall wetting phenomenon. Wall wetting has been proposed as an explanation for the dynamic response from a change in the fuel injection time, t_{inj} to the measured λ. The other part of the explanation is the dynamics in the exhaust gas mixing and sensor dynamics, which is discussed below.

The fuel-film dynamics has been extensively studied in the literature, see, for example, Aquino (1981), Curtis et al. (1996), Hendricks et al. (1992), and Hires and Overington (1981), and one popular formulation of the model is the one in Aquino (1981). A fraction X of the injected fuel, $\dot{m}_{fi}$, is deposited on the wall and forms a *fuel film* (sometimes also called a fuel puddle), while the rest, $(1 - X)$, mixes with the air. The fuel in the film evaporates with a time constant τ_{fp} and mixes with the air. Evaporation is assumed to be proportional to the area of the film, which in turn is assumed to be proportional to the mass in the film, m_{fp}, this is an assumption of a thin film where the fuel mass is spread out.

It is illustrated in Figure 7.18 and described as follows:

- The rate of change in fuel film mass, $\frac{dm_{fp}}{dt}$, is the deposited mass, $X\dot{m}_{fi}$ minus the mass evaporated from the film, $\frac{1}{\tau_{fp}}m_{fp}$.
- The fuel flow into the cylinder, $\dot{m}_{fc}$, is the sum of the part that goes directly from injection, $(1 - X)\dot{m}_{fi}$, and the part that is evaporated, $\frac{1}{\tau_{fp}}m_{fp}$.

These two effects give the model for the fuel flow to the cylinders

---

Model 7.12 Wall Wetting Model (Aquino)

The dynamics for the wall wetting and fuel transport system is

$$\frac{dm_{fp}}{dt} = X\dot{m}_{fi} - \frac{1}{\tau_{fp}}m_{fp} \tag{7.31}$$

and the output equation for the fuel flow to the cylinders is

$$\dot{m}_{fc} = (1 - X)\,\dot{m}_{fi} + \frac{1}{\tau_{fp}} m_{fp} \tag{7.32}$$

where X and τ are model parameters.

The parameters X and τ depend on the engine state as well as the fuel properties. Even if everything around the engine is kept constant, the parameters still change with the operating condition. An explanation is that the deposition and evaporation depend on the air flow passing the fuel deposits. A simple model for this is to let the parameters be static functions of operating condition,

$$X = X(N, p_{im}) \tag{7.33}$$

$$\tau_{fp} = \tau_{fp}(N, p_{im}) \tag{7.34}$$

The fuel film dynamics is most pronounced during cold starts and becomes less pronounced at warmed-up conditions and is also of less importance at high engine speeds. Examples of more detailed physical models for the wall wetting and evaporation processes can be found in, for example, Curtis et al. (1996), and Locatelli et al. (2004).

In Cylinder (A/F) Ratio

In the preceding sections the air mass flow and fuel mass flow to the cylinders were modeled, and we now have all the components for modeling the (A/F) ratio and lambda for the engine.

Model 7.13 Cylinder Lambda

$$\lambda_c = \frac{\dot{m}_{ac}}{\dot{m}_{fc}} \frac{1}{(A/F)_s} \tag{7.35}$$

From the components in the model and the equations in the previous section, some of the challenges related to (A/F) control in SI engine can be explained. In order for the engine controller to know how much fuel to inject, $\dot{m}_{fi}$, it must know the air flow into the cylinder, $\dot{m}_{ac}$, as well as the contribution to and from the fuel film. This will be discussed in Section 10.4.

Exhaust Gas Recirculation (EGR) and λ_O

To track λ when there is exhaust gas recirculation (EGR), the cylinder flow $\dot{m}_{cyl}$ should be separated into air and burned gases. For example, if an engine runs leanly, as a diesel engine, then the EGR also carries oxygen. With the total cylinder flow $\dot{m}_{cyl}$ determined from the volumetric efficiency, one can define a normalized oxygen/fuel ratio,

$$\lambda_O = \frac{m_O}{m_f} \frac{1}{(O/F)_s} = \frac{\dot{m}_{cyl}\,X_{Oim}}{\dot{m}_{fc}\,(O/F)_s} = \frac{\dot{m}_{cyl}\,(1 - X_{Bim})}{\dot{m}_{fc}\,(A/F)_s} \tag{7.36}$$

where the oxygen fraction (7.16), or the equivalent burned gas fraction, has been used. In the expression $(O/F)_s$ is the stoichiometric oxygen/fuel ratio, defined as $(O/F)_s = (A/F)_s X_{Oair} = (A/F)_s \frac{32}{32+3.773\cdot28.16}$. Note that in stationary conditions we have $\lambda_O = \lambda$. An application of λ_O is to track the limits on λ for smoke generation in diesel engines during transients.

7.7.4   Gas Transport and Mixing

The distance from the injection point to the sensor position and the reciprocation of the engine causes a variable transport delay, τ_d, that limits the performance of closed loop controllers. By looking at the transport delay from the fuel injection to the measured λ, shown in Figure 7.3, it can be split into two different parts: the first part comes from the transport through the engine cycles and depends on the engine speed. This is because the air-fuel mixture is drawn into the engine, compressed, and expanded, which results in a time delay of $\sim \frac{180+180+180}{360\,N}$ s. The second part comes from the exhaust processes with the blow-down and the exhaust stroke and finally transport in the exhaust system. This component depends on both the engine speed and the engine load.

These different delay contributions are aggregated into a single time delay, $\tau_d(N, p_{im})$. Measurements of the variable transport delay on an engine have shown that it is often sufficiently accurate, for simulation purposes, to model the transport delay with a function of just the engine speed (Bergman, 1997), that is $\tau_d(N)$. The time delay, τ_d, (and also dynamics) from injection to measurement position poses another challenge for closed loop fuel-injection control since it influences the phase margin and thus limits the feedback gain.

Another effect that occurs while the gas travels from the cylinder to the λ-sensor is that the gases mix along the pipe. This gives rise to a dynamic effect that smooths out the variations in the exhaust gases. This can be modeled as a low pass filter with a mixing time constant τ_{mix}, which together with the time delay gives the following model for the exhaust gas, for example, at the lambda sensor.

Model 7.14   Exhaust Gas Lambda

$$\frac{d}{dt}\lambda_{eg}(t) = \frac{1}{\tau_{mix}}(\lambda_c(t - \tau_d(N)) - \lambda_{eg}(t)) \tag{7.37}$$

When looking downstream on the exhaust side of the engine, the mixing time constant and time delay increase with the distance from the cylinders. More about the gas mixing dynamics can be found in, for example, Locatelli et al. (2003).

7.7.5   A/F Sensors

One or more λ-sensors, see Figure 7.1, are used to determine the air/fuel ratio in the exhaust of an SI engine. Figure 7.19 shows two sensor examples: the finger sensor and the planar sensor. The finger sensor belongs to a class called EGO-sensors, which is an acronym for Exhaust Gas Oxygen-sensor. The planar sensor is heated and belongs to the family called Heated EGO

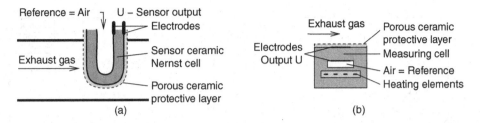

Figure 7.19 Cross-section drawings of two lambda sensors. (a) A finger type sensor, mounted in the exhaust. (b) A planar type sensor with heating. Note that air is used as the reference gas for both sensors

(HEGO) sensors. Heating enables the sensor to both start earlier and have temperature control. These two sensors are basically switch-type λ sensors, giving information about only fuel rich or fuel lean mixtures, by sensing the presence of oxygen in the exhaust. There are also sensors that give a continuous output, and they are referred to as wide range λ sensors or Universal EGO (UEGO) sensors.

The λ-sensor response is not instantaneous. It has a response time coupled to the oxygen diffusion in the sensor. A simple and often used model that captures this effect, is a first-order system with time constant τ_λ (Chang et al., 1995; Onder et al., 1997).

Model 7.15 Linear λ-Sensor Dynamics

$$\frac{d}{dt}\lambda_s(t) = \frac{1}{\tau_\lambda}(\lambda_{eg}(t) - \lambda_s(t)) \tag{7.38}$$

where $\lambda_{eg}(t)$ is gas mixture strength at the sensor.

The model describes the λ-value in the sensor element, and a modern sensor has a time constant of about 10–20 ms.

(A/F) Ratio and Sensor Voltages

Air-to-fuel ratio sensors measure the oxygen contents in the exhaust gas, as it depends strongly on λ, and they use a Nernst cell as sensing element. The principle is that at the surface of the sensor element there is a change of oxygen atoms to oxygen ions

$$O_2 + 4\,e^- \leftrightarrow 2\,O^{2-}$$

and the oxygen ions are carried through the sensor thimble which gives rise to a potential that follows the Nernst equation

$$U = \frac{k\,T}{4\,e} \ln \frac{p(O_{2,a})}{p(O_{2,e})} \tag{7.39}$$

where k is the Boltzmann constant, T temperature, e elementary charge, and the partial pressure of oxygen in the atmosphere and exhaust are $p(O_{2,a})$ and $p(O_{2,e})$ respectively. Note that

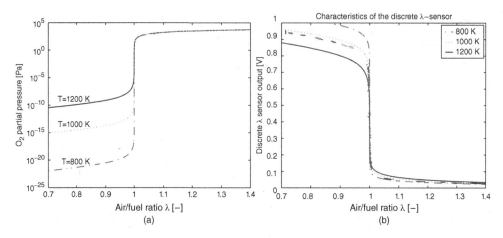

Figure 7.20   (a) Variations in the oxygen partial pressure at equilibrium as a function of variations in λ and temperature (calculated with CHEPP: Eriksson (2005)). (b) The lines represent the voltage, according to the Nernst equation, for the variations in λ and temperature. Dots are engine measurements using a HEGO sensor

air is used as the reference gas (see Figure 7.19) and thus provides the reference partial pressure of oxygen $p(O_{2,a})$. Figure 7.20 illustrates the EGO and HEGO sensor components, where Figure 7.20a shows the oxygen partial pressure at equilibrium in exhaust gases, as a function of temperature and λ. Figure 7.20b shows the resulting voltage from the Nernst equation (7.39) which is compared to a set of measurements from a λ sensor on an engine. The EGO and HEGO sensors, that are used in many production cars, only give information about whether the output is greater or smaller than $\lambda = 1$, see Figure 7.20, and this can be approximated by the following relay output

$$\lambda_{ego} = \begin{cases} 1 & \text{if} \quad \lambda_s < 1 \\ 0.5 & \text{if} \quad \lambda_s = 1 \\ 0 & \text{if} \quad \lambda_s > 1 \end{cases}$$

UEGO Sensor Characteristics
The wide range λ-sensor, UEGO, has a characteristic where the voltage depends nonlinearly on λ, and example characteristics are shown in Figure 7.21. Modern wide range λ sensors, built on planar technology, are based on two Nernst cells, where one is used as an oxygen pump and the other is used as a sensing element (Dietsche, 2011). The design is such that the pumping current depends on λ, and the direction of the current switches at $\lambda = 0$. The λ-sensor is sensitive to the absolute pressure, which changes the pump current proportionally to the pressure. As a result there is no change at $\lambda = 1$ (where the current is 0) but an increased sensitivity for richer, and foremost for leaner, mixtures. This becomes important if the λ sensor is used for control of lean air/fuel ratios in CI engines or as an oxygen/EGR sensor in the intake system.

Furthermore, planar sensors (both EGO and UEGO) have time constants τ_λ that are short enough for individual pulses from the cylinders to be detected. A spread in λ between cylinders,

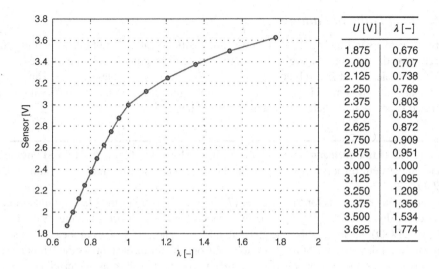

U [V]	λ [−]
1.875	0.676
2.000	0.707
2.125	0.738
2.250	0.769
2.375	0.803
2.500	0.834
2.625	0.872
2.750	0.909
2.875	0.951
3.000	1.000
3.125	1.095
3.250	1.208
3.375	1.356
3.500	1.534
3.625	1.774

Figure 7.21   Sensor characteristics for a UEGO λ-sensor. Calibration points are marked with circles, and the function $f(x)$ is obtained through linear interpolation between the points.

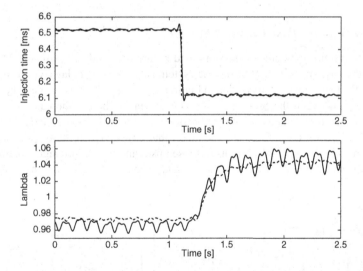

Figure 7.22   Validation of the wall wetting dynamics, λ is measured by a continuous sensor *UEGO*. Solid lines are the measured inputs and outputs and dashed lines are model inputs and outputs. The variation superimposed on the λ-signal comes from one cylinder that runs leaner than the other three

caused by uneven air-charge distribution or individual variations in injectors, for example, can therefore be detected. Cylinder individual pulses are seen as a ripple on the measured λ-signal (solid) in Figure 7.22, and examples of detection and control are presented in, for example, Cavina et al. (2010), and Grizzle (1991).

7.7.6   Fuel Path Validation

The two λ-sensors in Figure 7.1, λ_{bc} and λ_{ac}, are positioned at different places and thus have different time delays from the fuel injection to the sensor response: these are denoted τ_{bc} and τ_{ac}. All components that are included in the fuel path have now been described and an example that validates this system and model can now be developed. Here, the fuel path validation is performed from the fuel injector to a sensor before the catalyst.

Example 7.3 (Fuel path model and validation)   Model the fuel path of an SI engine and validate the system response from a step in the metered fuel to the measured lambda using a UEGO.

Combining the equations for fuel injector (7.29), fuel film (7.31) and (7.32), transport delay, and λ-sensor dynamics (7.38) yields the dynamic system that we are studying. A continuous λ-sensor (UEGO) is used (in this case the switch above is not used). This model now describes the dynamics of the system from injector to λ-sensor. The time constant for the lambda sensor is taken from the supplier's data sheet, parameters for the fuel film are determined from the step response, and the time delay is also taken from the step response. Figure 7.22 shows a validation of how well the model approximates the fuel dynamics from t_{inj} to the λ-sensor.

7.7.7   Catalyst and Post-Catalyst Sensor

In the exhaust system there are two sensors, and before the catalyst has reached its light-off temperature the gases pass untreated and the system acts as a time delay (with some gas mixing) between the two sensors. This is seen in Figure 7.23a where there is a time delay (and some small dynamics) between the pre- and post-catalyst sensors before the catalyst has reached operational temperature. When the catalyst becomes operational, it stores and releases oxygen as described in Section 6.5.1. This has the effect that oxygen concentrations downstream of the catalyst are not only delayed versions of those upstream, but also change dynamically. See Figure 7.23b where the catalyst acts as a buffer and attenuates the oscillations in λ.

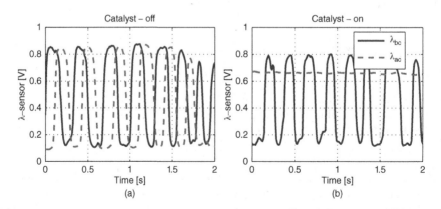

Figure 7.23   Responses in the pre-catalyst and post-catalyst sensors (switch type) for non-working catalyst (a) and a working catalyst (b). When the catalyst is working (b) the oscillations in the incoming gas are attenuated due to the oxygen storage capability in the catalyst

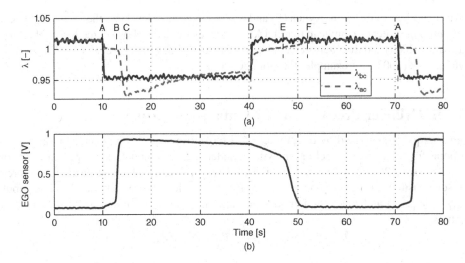

Figure 7.24   (a) Responses in two λ-sensors, one before the catalyst λ_{bc} and one after λ_{ac}. (b) The corresponding trace for the switch type (EGO) sensor after the catalyst

The responses are nonlinear in nature due to saturation of oxygen storage and the fact that the catalyst can become deactivated when there is not enough oxygen available. In addition, the sensor is not only sensitive to oxygen and pressure but also to, for example, hydrogen. Figure 7.24 illustrates the situation using a sequence of steps around $\lambda = 1$. Before the instant where the wide-range λ-sensor before the converter, λ_{bc}, switches from lean to rich (A), the engine has been running leanly and the amount of stored oxygen in the catalytic converter is thus high and the degree of catalyst deactivation is low. Directly after λ_{bc} switches from lean to rich (between A and B), λ_{ac} stays close to stoichiometric even though λ_{bc} becomes rich, since the large amount of oxygen in the catalyst compensates for the lack of of oxygen in the incoming gas. λ_{ac} does not start to fall until the converter is out of excess oxygen (B). Then λ_{ac} drops to its richest value (C).

During the interval between (B) and (C), the oxygen storage level decreases even more to compensate for the lack of oxygen in the incoming gas. At the same time, the degree of catalyst deactivation is increased slightly due to the incoming rich gas that cannot be oxidized due to the lack of excess oxygen. Hence, the engine is running richly and there is a high number of vacant sites on the converter's surface, which promotes hydrogen generating reactions. The increasing amount of hydrogen makes the λ_{ac} show a richer value than the true one. This explains the big difference between λ_{ac} and λ_{bc} at (C).

In the interval between (C) and (D), where both λ_{bc} and λ_{ac} show rich values, the degree of catalyst deactivation is continuously increased since there are more species that need to be oxidized than available oxygen. Hence, the number of vacant sites is decreased, and the hydrogen generation is inhibited. This affects λ_{ac}, which shows an increasingly leaner value.

When λ_{bc} switches back to lean (D), λ_{ac} stays close to the stoichiometric level in the same way as when λ_{bc} switched to rich. At first, the excess oxygen in the incoming gas oxidizes the species that have been gathered in the catalytic converter during the rich period and caused the catalyst deactivation. Then, the excess oxygen in the incoming gas is adsorbed in the converter and increases the level of stored oxygen again. λ_{ac} starts to increase when the catalytic converter is close to full with oxygen (E) and reaches its leanest level (F), then stays at the lean level until λ_{bc} switches from lean to rich (A). The measurements above come from Johansson and Waller (2005) and the interested reader is referred to Auckenthaler et al. (2002, 2004) and Peyton Jones (2003), for example, for more details.

7.8   In-Cylinder Pressure and Instantaneous Torque

Engine torque is generated by the pressure and piston force in the cylinders, and as a starting point for the torque model an analytic cylinder pressure model, developed in Eriksson and Andersson (2003), is described. It is based on the ideal cycles that describe compression, combustion, and expansion processes for one cycle. The model uses MVEM inputs $(p_{im}, p_{em}, T_{im}, \ldots)$ and gives the cylinder pressure at an arbitrary crank angle θ as a static function

$$p_{cyl}(\theta) = f(\theta; p_{im}, p_{em}, T_{im}, \theta_{ign}, \Delta\theta_d, \Delta\theta_b, x_r, T_r \ldots)$$

where the crank angle denotes the crank dependency while the rest of the inputs are parameters to one cycle. The model is thus expressed as a function of crank angle without the need to solve a differential equation for the cylinder pressure curve, which is the main difference from the single- and multi-zone cylinder models, described in Section 5.4.

Figure 7.25 shows the idea behind the model and outlines the calculation steps and structure of the cylinder pressure model. It is divided into parts that follow the strokes: exhaust, intake, compression and expansion, and an interpolation that gives smooth transitions between them. The following list highlights the important model components:

- The compression process is modeled as a polytropic process, with exponent k_c. With a properly selected polytropic exponent, the compression with its heat transfer can be well approximated. Heat transfer is thus implicitly included in the model. The compression stroke is determined from the intake system conditions.
- Similarly, the expansion asymptote is also well described by a polytropic process with exponent k_e. The expansion stroke is determined from the compression stroke and combustion conditions, providing a reference point for expansion temperature and pressure that is calculated using a constant-volume combustion process.
- The concept of pressure ratio management of Matekunas (1983, 1984, 1986) provides a way to interpolate smoothly from compression to expansion. The pressure ratio is similar to the mass fraction burned profile and the Vibe function (5.30), for example, can be used in the interpolation.
- The gas exchange is treated as follows. During the intake stroke EVC–IVC, the pressure is approximated by the intake manifold pressure. During the exhaust stroke EVO–IVO the pressure is approximated by the exhaust manifold pressure. Between the phases, during the valve overlap, the pressure can be determined through an interpolation using, for example, a sine function.

7.8.1   Compression Asymptote

The compression process can be modeled with good accuracy by a polytropic process. Such a process is described by a polytropic exponent k_c and a reference value at one reference

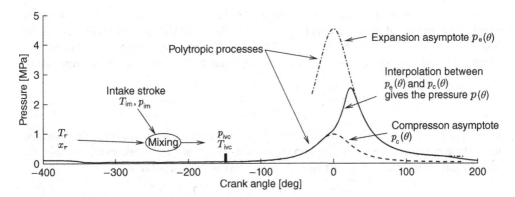

Figure 7.25 The model is based on the compression pressure, the expansion asymptote, and an interpolation between these. Initial conditions are determined from the intake conditions and the residual gases

point. One point that can be used as reference is the inlet valve closing (IVC), which gives the following expressions for the compression pressure and temperature

$$p_c(\theta) = p_{ivc}\left(\frac{V_{ivc}}{V(\theta)}\right)^{k_c} \qquad T_c(\theta) = T_{ivc}\left(\frac{V_{ivc}}{V(\theta)}\right)^{k_c-1} \qquad (7.40)$$

These traces describe the cylinder pressure and temperature up to the point of combustion. The temperature model is also necessary since it has a direct impact on the expansion pressure.

Determination of Initial Pressure

The manifold pressure gives a good indication of the initial pressure for the compression stroke. However, pressure drops over valves, and tuning effects in the intake runners also have an influence on the pressure. As a first approximation a reference condition just before IVC is used to determine the initial pressure

$$p_{ivc} = p_{im}(\theta_{ivc}) \qquad (7.41)$$

The crank angle for inlet valve closing θ_{ivc} is not exactly known due to production tolerances, it can also be used as a tuning parameter to compensate for pressure drops over the valves and so on. This can be extended with more detailed submodels, for example an affine correction in engine speed was added in Eriksson and Andersson (2003),

$$p_{ivc} = p_{im}(\theta_{ivc}) + c_1 + c_2\,N$$

which improved the accuracy of the compression pressure. Here, c_1 and c_2 are parameters that have to be determined, which increase the model flexibility but also its complexity.

Determination of Initial Temperature

It is more difficult to determine the fluid temperature at inlet valve closing compared to the pressure, since it is influenced by heat transfer and residual gases that are difficult to measure and determine. The air in the intake manifold is heated from T_{im} to T_a by the hot valves and the locally high heat transfer coefficients in the cylinder. Fuel is added and undergoes an evaporation which also influences the temperature. By considering the energy equation with a lumped process for heating, evaporation, and mixing, the initial air/fuel mixture temperature can be stated as

$$T_{af} = \frac{m_a c_{pa} T_a + m_f c_{pf} T_f - m_f h_{v,f} + Q}{m_a c_{pa} + m_f c_{pf}}$$

where $h_{v,f}$ is the vaporization enthalpy for the fuel and Q is the heat added to the fresh mixture. These are both difficult to determine without detailed measurements. In the cylinder, the residual gases and fresh charge are mixed and the mixture temperature is

$$T_{ivc} = \frac{m_{af} c_{paf} T_{af} - m_r c_{pr} T_r}{m_{af} c_{paf} - m_r c_{pr}}$$

During the gas exchange, prior to mixing, the residual gases have also been cooled down by heat transfer to the walls.

Simplifying the Temperature Model

The outlined models for the heating, evaporation, and mixing processes are complex and contain several variables that have to be determined, and therefore some simplifications are made. Neglecting heat transfer gives $T_{af} = T_{im}$, and assuming that c_p are the same for residuals and fresh air and fuel mixtures, yields

$$T_{ivc} = T_{af} (1 - x_r) + x_r T_r \tag{7.42}$$

where the residual gas fraction is defined as

$$x_r = \frac{m_r}{m_a + m_f + m_r}$$

The residual gas fraction x_r and temperature T_r can be seen as tuning parameters, but submodels can also be used, for example, either based on an ideal Otto cycle or using one of the procedures outlined in Heywood (1988) or Mladek and Onder (2000). Residual gases are important for the model since they directly influence the initial temperature T_{ivc} (7.42) and will also be shown to influence the expansion pressure (see (7.40) and (7.45)). Their importance is illustrated by a numerical example: $x_r = 0.07$, $T_{im} = 298$ [K], $T_r = 1000$ [K] yields an initial temperature of $T_{ivc} = 347$ [K], which is a 16% increase from T_{im}.

7.8.2   Expansion Asymptote

The asymptotic expansion process is also modeled as polytropic, with exponent k_e.

$$p_e(\theta) = p_3 \left(\frac{V_3}{V(\theta)} \right)^{k_e} \qquad T_e(\theta) = T_3 \left(\frac{V_3}{V(\theta)} \right)^{k_e - 1} \tag{7.43}$$

Here we need to determine V_3, p_3, and T_3, that refer to state 3 in the ideal Otto cycle. This is done by going from state 2 to state 3 in the pV diagram, top right plot in Figure 5.5. The temperature increase is

$$\Delta T_{\text{comb}} = \frac{m_{\text{f}} \, q_{\text{LHV}} \, \eta_{\text{c}}(\lambda)}{c_{\text{v}} \, m_{\text{tot}}} = \frac{(1 - x_{\text{r}}) \, q_{\text{LHV}} \, \eta_{\text{c}}(\lambda)}{(\lambda(A/F)_{\text{s}} + 1) \, c_{\text{v}}} \tag{7.44}$$

where the fuel conversion efficiency $\eta_{\text{c}}(\lambda)$ comes from Figure 4.2, which takes the effect of varying air-to-fuel ratios by considering the effect that the fuel mass has on the temperature increase. Exhaust gas recirculation, EGR, can also be included, this enters the equations in the same way as the residual gas and influences both the initial temperature T_{ivc} and the dilution x_{r}. The thermodynamic properties of the fluids (i.e., c_{v}, k_{c}, and k_{e} of burned and unburned gases) depend on λ, but this effect is often neglected in simpler models. All in all this gives the following expressions for the temperature and pressure after the combustion

$$T_3 = T_2 + \Delta T_{\text{comb}} \qquad p_3 = p_2 \frac{T_3}{T_2} \tag{7.45}$$

where p_2 and T_2 are determined from (7.40).

7.8.3   Combustion

The combustion part is produced by interpolating between the two asymptotic pressure traces, $p_{\text{c}}(\theta)$ and $p_{\text{e}}(\theta)$, using the mass fraction burned $x_{\text{b}}(\theta)$. As an interpolation function one can use the well-known Vibe function of Vibe (1970), see (5.30), which gives the following expression for the pressure

$$p_{\text{cyl}}(\theta) = (1 - x_{\text{b}}(\theta)) \cdot p_{\text{c}}(\theta) + x_{\text{b}}(\theta) \cdot p_{\text{e}}(\theta)$$

From the end of combustion to EVO the pressure follows the expansion asymptote, (7.43), and burn duration can be approximated by

$$\Delta\theta \approx 2\theta_{\text{d}} + \theta_{\text{b}} \tag{7.46}$$

using flame development angle, θ_{d}, and fast burn angle, θ_{b}. It is worth pointing out that this approach can be seen as the inverse of the pressure ratio management that has been investigated in great detail in Matekunas (1983, 1986), and Sellnau and Matekunas (2000).

Combustion Phasing

The ignition timing and combustion phasing influence the final pressure, p_3, for example a cycle with late combustion has higher pressure than the optimal (see, e.g., Figure 7.28). This effect can be accounted for by phasing the ideal Otto combustion using the burn angles. The position for the combustion θ_{c} is chosen to be at TDC if the calculated position for 50% mass fraction burned, $\theta_{50\%}$, is at its optimal value, $\theta_{50\%,\text{opt}}$. If the $\theta_{50\%}$ position deviates from its optimum, the angle θ_{c} is set to that deviation in CAD. This is expressed as

$$\theta_{\text{c}} = \theta_{50\%} - \theta_{50\%,\text{opt}} \tag{7.47}$$

$$\theta_{50\%} = \Delta\theta_{\text{d}} + \frac{1}{2}\Delta\theta_{\text{b}} \tag{7.48}$$

where $\theta_{50\%,\text{opt}}$ is a model parameter. In the original work (Eriksson and Andersson, 2003) the optimal value for the 50% mass fraction burned was set to 8°, but it is a tuning parameter and can be set closer to TDC. The phasing couples the mass fraction burned trace to the single θ_c that phases the combustion of the ideal Otto cycle and thus defines state 2, and volume for state 3 to

$$V_2 = V_3 = V(\theta_c), \quad p_2 = p_c(\theta_c), \text{ and } T_2 = T_c(\theta_c) \tag{7.49}$$

These enter (7.43) and (7.45) that define the expansion asymptotes.

7.8.4   Gas Exhange and Model Compilation

After the exhaust valve has opened, the blow-down phase begins and the pressure approaches the pressure in the exhaust system. For this phase, an interpolation scheme can be used, as was mentioned above. The same applies for the valve overlap as well as the inlet valve closing. Interpolation between phases can, for example, be performed using the cosine function as follows

$$x_i(\theta, \theta_0, \theta_1) = 0.5 \left(1 - \cos\left(\pi \frac{\theta - \theta_0}{\theta_1 - \theta_0} \right) \right), \quad \theta \in [\theta_0, \theta_1] \tag{7.50}$$

A complete cycle can now be modeled using the following expressions.

Model 7.16 Analytical In-cylinder Pressure Model

The complete model is a compilation of (7.40) to (7.49) with the interpolation described by the Vibe equation (5.30) and interpolation function (7.50). One cycle is modeled by

$$p_{\text{cyl}}(\theta) = \begin{cases} p_{\text{im}} & , \theta_{\text{evc}} \leq \theta < \theta_{\text{int}} \\ p_{\text{im}} \left(1 - x_i(\theta, \theta_{\text{int}}, \theta_{\text{ivc}}) \right) + p_c(\theta) \, x_i(\theta, \theta_{\text{int}}, \theta_{\text{ivc}}) & , \theta_{\text{int}} \leq \theta < \theta_{\text{ivc}} \\ p_c(\theta) & , \theta_{\text{ivc}} \leq \theta < \theta_0 \\ p_c(\theta) \left(1 - x_b(\theta) \right) + p_e(\theta) \, x_b(\theta) & , \theta_0 \leq \theta < \theta_{\text{evo}} \\ p_e(\theta) \left(1 - x_i(\theta, \theta_{\text{evo}}, \theta_{\text{exh}}) \right) + p_{\text{em}} \, x_i(\theta, \theta_{\text{evo}}, \theta_{\text{exh}}) & , \theta_{\text{evo}} \leq \theta < \theta_{\text{exh}} \\ p_{\text{em}} & , \theta_{\text{exh}} \leq \theta < \theta_{\text{ivo}} \\ p_{\text{em}} \left(1 - x_i(\theta, \theta_{\text{ivo}}, \theta_{\text{evc}}) \right) + p_{\text{im}} \, x_i(\theta, \theta_{\text{ivo}}, \theta_{\text{evc}}) & , \theta_{\text{ivo}} \leq \theta < \theta_{\text{evc}} \end{cases} \tag{7.51}$$

It has a set of tuning parameters and inputs that have physical interpretations, and they are given in Table 7.2. See also Figure 7.25 where a graphical illustration of the model was given.

7.8.5   Engine Torque Generation

Engine torque is the result of the in-cylinder pressure, that produces a force on the piston that in turn is translated to a torque from the connecting rod and crank mechanism. The pressure in the crank casing is close to atmospheric, and opposes the in-cylinder pressure. In an engine with several cylinders, the torque contribution from each cylinder j is added to the crankshaft, in (7.51) the pressure model is related to the TDC of the individual cylinders, therefore the

Table 7.2 Tuning parameters and inputs for the analytic cylinder pressure model

	Tuning parameters		Inputs
c_p	specific heat at combustion	p_{im}	intake manifold pressure
k_c	polytropic coefficient for compression	T_{im}	intake manifold temperature
k_e	polytropic coefficient for expansion	$\Delta\theta_d$	flame development angle
q_{LHV}	fuel heating value	$\Delta\theta_b$	fast burn angle
T_r	residual gas temperature	θ_{SOC}	start of combustion angle
x_r	residual gas fraction	λ	normalized air to fuel ratio
θ_{ivc}	inlet valve closing angle		

pressure and crank lever need to be phased with cylinder individual offsets θ_j^0. The sum is the instantaneous indicated torque

$$M_{e,i}(\theta) = \sum_{j=1}^{n_{cyl}} (p_{cyl,j}(\theta - \theta_j^0) - p_{amb}) \, A \, L(\theta - \theta_j^0) \tag{7.52}$$

Note that the product of the area and crank lever is equivalent to the volume derivative with respect to the crank angle, that is $A \, L(\theta) = \frac{dV(\theta)}{d\theta}$ with $V(\theta)$ from (4.3). It is important to note that the model describes the indicated torque, and friction is not yet present in the model. Friction will be added to the engine torque model in the coming section and discussed in more detail in Section 7.9.3.

The result from adding the cylinder pressures and calculating the instantaneous indicated torque is shown in Figure 7.26, where the cylinder pressure for a four-cylinder engine is shown

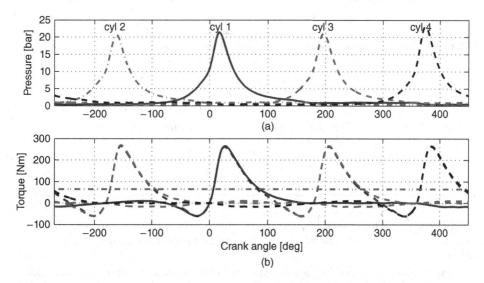

Figure 7.26 (a) Cylinder individual pressures in a four-cylinder engine. The firing order among the cylinder is also seen, being 1-3-4-2. (b) (thick dashed line) – indicated torque, (thin solid and dashed) – torque contributions from the individual cylinders, (dash-dotted) – cycle averaged indicated torque, which is the indicated torque in a mean value engine model

together with the cylinder individual instantaneous torques as well as the total torque. The combustion in the cylinders in a multi-cylinder engine follows a certain order, which is controlled by the crankshaft and valve openings. This is referred to as the firing order. In the engine shown here the firing order is 1-3-4-2, which gives the following angles $[0, 3\,\pi, \pi, 2\,\pi]$ for the crank offset θ_j^0 in (7.52).

This model takes MVEM inputs and produces instantaneous pressures and torque without the need to solve the differential equation as is done in the zonal models in Section 5.4. This model is thus well suited to applications where the driving torque and its pulsations are of interest while the details of the in-cylinder processes are of less importance. See Andersson and Eriksson (2009), Eriksson and Nielsen (2003), Johansson (2012), Larsson and Schagerberg (2004), and Scarpati et al. (2007) for some example applications that use the model.

7.9   Mean Value Model for Engine Torque

Engine torque is an important output that governs the engine and vehicle performance and this section develops one simple and one more advanced model for it. In this section, the goal is a mean value model, meaning that the instantaneous engine torque, $M_{e,i}(\theta)$, that fluctuates with the different cylinders, pumping and work generation are averaged out. Friction also needs to be subtracted from the torque that is produced in the cylinders, that is it is the following torque we are interested in

$$M_e = \frac{1}{4\pi} \int_0^{4\pi} M_{e,i}(\theta)d\theta - M_f$$

A simple model is obtained by studying experimental data, for example data from an SI engine is shown in Figure 7.27. In this case, the engine operates with nominal conditions and it is seen that BMEP depends almost linearly on the intake manifold pressure. Therefore, the following simple affine model (linear with offset) gives a good approximation of the BMEP.

Model 7.17   Affine Torque Model from Intake Manifold Pressure

$$BMEP(p_{im}) = -C_{p1} + C_{p2}\, p_{im} \qquad (7.53)$$

where C_{p1} and C_{p2} can be determined from experimental data. For the data in Figure 7.27 $C_{p1} = 3.5\,10^5$ and $C_{p2} = 12.5$ give a good fit. The engine torque then becomes

$$M_e(p_{im}) = \frac{BMEP(p_{im})\, V_D}{n_r\, 2\,\pi} \qquad (7.54)$$

For CI engines a similar affine relation can be obtained between mass of fuel injected, m_{fi}, and BMEP (or torque). A frequently encountered name for these affine models is Willans line models.

The model above is a good first approximation, but there are many other factors beside p_{im} that influence the torque, for example λ and ignition angle θ_{ign}. To capture such effects a more refined model is needed. The torque produced by the engine, M_e, is therefore modeled using three components for the production and consumption of work: gross indicated work per cycle

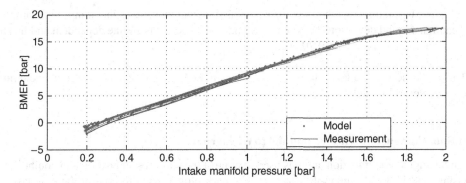

Figure 7.27 BMEP plotted as a function of intake manifold pressure. The engine has been operating at its nominal operating conditions with respect to λ and θ_{ign}. In the plot there is also validation data for a torque model, based upon (7.55)

$W_{i,g}$, pumping work $W_{i,p}$, and friction work consumed by the engine components as well as some auxiliary devices W_{fr}. The engine torque model is expressed using the three different models for work production and consumption in the following way.

$$M_e = \frac{W_e}{n_r \, 2 \, \pi} = \frac{W_{i,g} - W_{i,p} - W_{fr}}{n_r \, 2 \, \pi} \tag{7.55}$$

The three submodels will be discussed and described below.

7.9.1 Gross Indicated Work

The indicated gross work is coupled to the energy that comes from the delivered fuel and can be modeled as a function of fuel mass, fuel heating value, operating conditions, and

$$W_{i,g} = m_f \, q_{LHV} \, \tilde{\eta}_{ig}(\lambda_c, \theta_{ign}, r_c, \omega_e, V_d) \tag{7.56}$$

To specialize this model, we can make the simplifying assumption that the gross indicated efficiency can be written as a product of efficiencies that account for the different effects that lead to losses. The following equation illustrates the effects

$$\tilde{\eta}_{ig}(\lambda_c, \theta_{ign}, r_c, \omega_e, V_d) = (1 - \frac{1}{r_c^{\gamma-1}}) \cdot \min(1, \lambda_c) \cdot \eta_{ign}(\theta_{ign}) \cdot \eta_{ig,ch}(\omega_e, V_d) \tag{7.57}$$

and gives a good approximation while including the following losses:

- $(1 - \frac{1}{r_c^{\gamma-1}})$, represents the efficiency for an ideal Otto cycle – even the most efficient cycle has losses.
- $\min(1, \lambda_c)$, captures the fact that the energy in the fuel mass cannot be fully utilized when there is a rich mixture, see Figure 5.13 where the efficiency drops for $\phi = \frac{1}{\lambda} > 1$.
- η_{ign}, describes the combustion phasing losses, that is when the SI ignition timing or CI injection timing is not optimal, this will be discussed below.

- $\eta_{ig,ch}(\omega_e, V_d)$, lumps together the other combustion chamber losses including, for example, heat transfer and deviations between ideal and real cycles. It is engine dependent and lies in the region around $\eta_{ig,ch} \in [0.70, 0.85]$.

This model serves as a good starting point for the modeling and it can be refined by including more detailed submodels.

Ignition, Combustion Timing, and the p-V Diagram

In an CI engine, the injection angle is used to control the start of combustion and combustion phasing, and in an SI engine the spark at the spark plug is used to control the combustion phasing. Both angles can be denoted θ_{ign} (but in the CI case the angle is often denoted α), and ignition data for SI engines will be used to illustrate the principles.

The ignition positions the combustion in relation to the crankshaft rotation, or equivalently the volume change. Figure 7.28 shows pressure traces and mass fraction burned curves for six different ignition timings together with their p-V diagrams. In the figure, the positioning of the combustion is clearly shown, the left-most mass fraction burned trace had its ignition at $32°$BTDC and the right-most trace at $7°$BTDC. The spacing between the traces is $5°$. From the p-V diagram it is seen that the combustion position influences the indicated work and thus the engine efficiency. An early combustion results in an early pressure build-up and lower pressure during the later part of the expansion stroke. A too early pressure build-up, that occurs before TDC, counteracts the piston motion and lowers the output power. Additionally, it has a lower pressure during the expansion phase, which results in a lower output power. A later combustion results in a later pressure build-up but higher pressure during the expansion stroke. With a too late ignition, the gains of higher pressure during the later part of the expansion stroke cannot compensate for the losses early in the expansion stroke, that is the loss area in the upper left corner of the p-V diagram becomes larger than the area gained in the expansion. The low expansion stroke pressures for cycles with early combustion have two sources from the

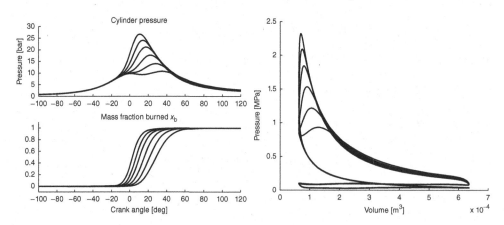

Figure 7.28 Left: Cylinder pressure and mass fraction burned for six different ignitions at 2300 RPM, and 50 Nm, showing that the ignition timing controls the pressure development in the combustion chamber. Right: The corresponding p-V diagram showing how the ignition influences the enclosed area in the p-V diagram.

physics: firstly, the internal energy that comes from the combustion has been converted to work, compare the p-V diagram in Figure 7.28 with those for the ideal gas cycles in Figure 5.10. If combustion occurs late, the thermal energy has not been converted to work, which gives higher temperatures and pressures during the expansion stroke. Secondly, an earlier combustion gives more heat transfer to the combustion chamber walls, due to both higher temperatures and longer exposure time. This contributes to the lower temperatures and pressures during the expansion for earlier ignitions.

Ignition Timing and Efficiency

For each operating point there is an optimal position for the ignition timing, $\theta_{ign,opt}$, and it depends on many operating parameters like engine speed, fuel, mixture strength, dilution and other parameters $\theta_{ign,opt}(\omega_e, m_f, \lambda, x_r)$. When studying the ignition efficiency η_{ign} it is seen that the efficiency has a strong dependence on the deviation from the optimal ignition timing. Experimental data on η_{ign} for an SI engine is shown in Figure 7.29. The maximum point is related to the maximum brake torque (MBT) and the optimum ignition angle is called *MBT timing*. The interesting region in the figure is for optimal and delayed ignitions (the maximum and the region to its right), since earlier ignitions generate higher pressures, temperatures, and more emissions that can be detrimental for the engine. Each dot is one ignition timing for one operating point and the data is the collection of ignition sweeps for several operating points. Figure 7.29 shows that the ignition efficiency is independent of the operating point in the interesting region. Therefore, a model for the ignition efficiency can conveniently be expressed in terms of the deviation from the optimum.

$$\eta_{ign} = \eta_{ign}(\theta_{ign} - \theta_{ign,opt}(\omega_e, m_f, \lambda, x_r)) \tag{7.58}$$

This model says that the ignition efficiency follows the deviation from the optimal ignition timing and is independent of the operating condition. When θ_{ign} deviates from the optimal value, the efficiency decreases and a quadratic function can be used as a first approximation

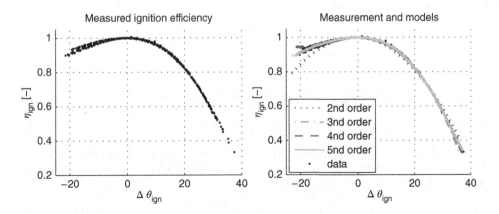

Figure 7.29 (a) Ignition fish hooks for an SI engine showing, the ignition efficiency $\eta_{ign} =$ IMEP$_g$/ max(IMEP$_g$) as function of $\theta_{ign} - \theta_{ign,opt}(\omega_e, m_f, \lambda, x_r)$. (b) Second- to fifth-order polynomials fitted and compared to measured efficiency

to describe the effect.

$$\eta_{ign} = 1 - C_{ig,2} \cdot (\theta_{ign} - \theta_{ign,opt}(\omega_e, m_f, \lambda, \ldots))^2$$

This model also says that the decrease is symmetric, which is a simplification, a better approximation is attained if a third (or higher) order polynomial is used. To the right, in Figure 7.29, a set of polynomials of order $i \in [2, 3, 4, 5]$ have been fitted to the ignition sweep data.

$$\eta_{ign}(\omega_e, m_f, \lambda, \ldots) = 1 - \sum_{i=2}^{N} C_i \cdot \left(\frac{\theta_{ign} - \theta_{ign,opt}(\omega_e, m_f, \lambda, \ldots)}{100} \right)^i \qquad (7.59)$$

Table 7.3 Ignition efficiency polynomial coefficients for (7.59)

N	C_i			
2	4.316			
3	3.217	4.722		
4	3.439	8.157	−10.47	
5	3.012	7.063	7.672	−36.95

Polynomial coefficients C_i are identified for the data, and they are given in Table 7.3. Naturally, the fifth-order polynomial gives best agreement over the whole interval of data, but already a third order gives a fairly good agreement, see Figure 7.29.

7.9.2 Pumping Work

The pumping work accounts for the work consumed in the breathing process. It is described by the pumping mean effective pressure, see Section 5.2.3.

$$W_{i,p} = V_D \, \text{PMEP}(p_{em}, p_{im}, N, \ldots) = V_D \, (p_{em} - p_{im})$$

In the second step above, a simplification is made that PMEP can be described by the intake and exhaust manifold pressures. This doesn't take the pressure losses over the valves nor the blow-down and valve overlap processes into account, many of these are most pronounced at high speeds and mass flows, see Nilsson et al. (2008). This simple model often serves well as a first approximation, but more effects related to the valve events might be needed at higher speeds or in engines with variable valve actuation.

7.9.3 Engine Friction

The friction is modeled using the friction mean effective pressure

$$W_{fr} = 2 \pi n_r M_{fr} = V_D \, \text{FMEP}(N, T_e, p_{im}, \ldots)$$

One example of a model for the friction mean effective pressure is the ETH model, see Inhelder (1996), and Stöckli (1989). The model is derived from data published in papers about engine friction, and the expression is as follows

$$\text{FMEP} = \xi_{aux} \cdot [(0.464 + 0.0072 \, S_p^{1.8}) \cdot \Pi_{bl} \cdot 10^5 + 0.0215 \cdot \text{BMEP}] \cdot \left(\frac{0.075}{B} \right)^{0.5}$$

where ξ_{aux} is the load from the auxiliary devices which is around $1.3-1.4$. Π_{bl} is the boost layout that comes from the effect that supercharging has on the dimensions of the bearings, a naturally aspirated engine (that isn't boosted) has $\Pi_{\text{bl}} = 1$. S_{p} is the mean piston speed and it is included to describe the speed dependence of the friction. B is the bore. The included term BMEP captures the effect that load has on the rubbing friction. One minor disadvantage of this model is that it requires the BMEP to be known before the FMEP can be calculated. However, there are two solutions: one is to phrase the equations so that BMEP can be solved from the model, the other is to approximate BMEP using a model. One way to do this is to use the intake manifold pressure (as was done in (7.53)) and another is to use the air-mass captured in the cylinder.

There are also other friction models, for example a simple model is to use only a polynomial in engine speed

$$\text{FMEP} = C_{\text{fr},0} + C_{\text{fr},1} \frac{60\,N}{1000} + C_{\text{fr},2} \left(\frac{60\,N}{1000} \right)^2$$

where $C_{\text{fr},0} = 0.97 \cdot 10^5$, $C_{\text{fr},1} = 0.15 \cdot 10^5$, and $C_{\text{fr},2} = 0.05 \cdot 10^5$ (Heywood, 1988). Here FMEP is given in Pa. A comprehensive model that accounts for engine bearing dimensions is developed in Patton et al. (1989). This model was later extended in Heywood (2003), with models for the effect that the temperature has on the friction. An example of a friction map, showing the speed and temperature dependence for a 14.4-liter diesel truck engine, is given in Figure 7.30.

Relation Between Pumping and Friction Work

Figure 7.31 shows the useful work and losses for an SI engine. The relative importance of pumping and friction losses are largest when the load is low. Note: at zero load the pumping and friction are of approximately equal size. A diesel engine has to overcome the friction which also results in low part load efficiency, but it does not have a throttle that restricts the air and

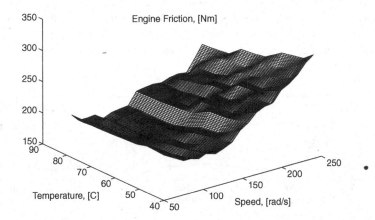

Figure 7.30 Engine friction for the 14.4-liter truck as a function of engine speed and engine temperature. Data from Pettersson and Nielsen (2000)

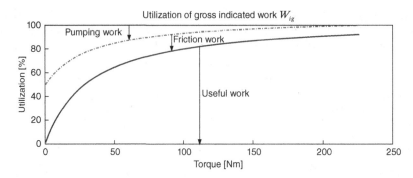

Figure 7.31 Illustration of how friction, pumping, and useful work utilize the net indicated work as a function of engine load

gives rise to the pumping work. Consequently, the diesel engine is more efficient at part load when compared to an SI engine.

7.9.4 Time Delays in Torque Production

In the engine and torque production there are small but inevitable delays between the events that control the combustion and the torque that is produced. Figure 7.32 illustrates the three instants where (i) the fuel is injected, (ii) air is inducted, and (iii) ignition occurs. The torque production is also shown and there are delays until the torque is produced. The torque production occurs during the expansion, where the torque maximum occurs at 30° ATDC and the center of gravity of the torque production is at 55° ATDC. Engine speed N has a direct influence on the delay and the figure gives the following approximate time delays from the events: fuel injection $\tau_{M,f} = \frac{500}{360\,N}$ s, air induction $\tau_{M,a} = \frac{330}{360\,N}$ s, and ignition $\tau_{M,\mathrm{ign}} = \frac{80}{360\,N}$ s, where engine speed is given in revolutions per second.

It is important to note that here there is only the delay from the event to the torque production. In an engine control system, one also needs to add a calculation and synchronization time to the time delay from when command is being calculated to when it occurs on the engine actuator.

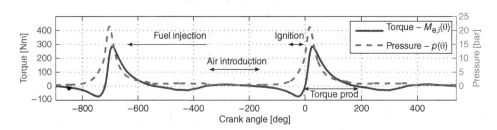

Figure 7.32 Cylinder pressure and instantaneous torque for two cycles. The events that control the fuel, air, and ignition in a port fuel injected SI engine are marked. Fuel injection is marked with one directional arrow, illustrating the direction where the value can be open. Injection is often timed to end at IVO, but the length of the injection in a crank angle depends on engine speed and fuel mass t_{inj}, at full load and maximum engine speed the injector can be open almost 720°

7.9.5   Crankshaft Dynamics

At this point, a complete torque model is available and this will interact with the rest of the powertrain to give the engine speed. Basically Newton's second law gives the engine speed, which is the following equation for the rotational dynamics of the engine crankshaft with inertia J_e,

$$J_e \frac{d\omega_e}{dt} = M_e(p_{im}, N, \lambda, \theta_{ign}, T_m) - M_{load} \qquad (7.60)$$

when the load torque M_{load} (from either the engine dynamometer or the driveline) is given. It should be noted that the engine interacts with the driveline and a complete driveline model might be necessary for simulating the engine speed correctly in a driving scenario. In many cases the engine model delivers the driving torque to the rotational parts of the powertrain, and the powertrain model then delivers the rotational speed to the engine, see for example the driveline models in Chapter 14 that include the engine inertia. For the MVEM, the engine speed becomes an exogenous input to the engine model and its subsystems. Engine models that have engine speed as an input are frequently encountered and are used in, for example, Wahlström and Eriksson (2011a).

Model 7.18 A Naturally Aspirated SI Engine

A complete engine model can be derived by compiling the equations in the preceding sections. A simplification of the mean value engine model is shown as a block diagram in Figure 7.33. The block diagram shows the essential flow paths of mass and, to some extent, energy, but does not show all dependencies; for example the engine speed also influences some components. To get the full model the following components and equations are used

- *Throttle air mass flow is discussed in Section 7.6.*
- *Intake manifold filling and emptying dynamics with air flows and pressure build up are discussed in Section 7.6.*
- *Fuel dynamics, including the direct term and the fuel film build up, is described in Section 7.7.3.*
- *Torque generation and crankshaft dynamics are discussed in Section 7.9.*
- *Finally, the lambda sensor dynamics is described in Section 7.7.5.*

This model can be utilized in many ways when studying naturally aspirated engines, it can, for example, be used as a simulation model, a model for analysis of the dynamics, or as the design or validation model for different controllers. One controller in particular is the idle speed controller that can be analyzed and studied simulated with the system above.

7.10   Engine-Out Temperature

When modeling the turbine or catalyst input temperatures, it is also necessary to have a model for the engine output temperature, T_{eo}. Several operating parameters affect this temperature, like ignition timing, λ, speed, and load. This is a system that is complex and difficult to model since it depends on many parameters and the measurement situation is also difficult due to

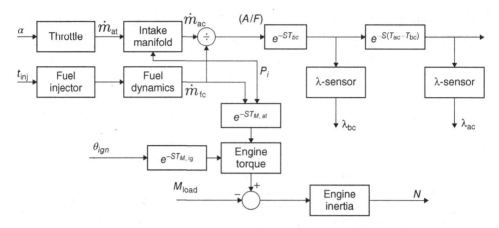

Figure 7.33 Schematic picture of the subsystems that are included in the MVEM for a naturally aspirated engine. Note that the figure is a simplification for the illustration of the essential connections and therefore not all signals and dependencies are shown

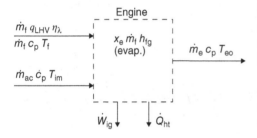

Figure 7.34 First law analysis of the engine focusing on the exhaust temperature. Air and fuel are delivered to the engine. The energy is split to: evaporate fuel, produce work on the piston $\dot{W}$, heat transfer to coolant $\dot{Q}_{ht}$, and exhaust energy

heat transfer and highly varying flow conditions. A first law analysis of the engine as an open system, see Figure 7.34, gives insight into what affects the temperature.

Figure 7.34 shows that air and fuel deliver energy to the system, where the fuel carries energy both as thermal energy, $\dot{m}_f c_p T_f$, and chemical energy $\dot{m}_f q_{LHV} \eta_\lambda$. Here η_λ describes that it is not possible to release all the fuel's chemical energy at rich conditions. The delivered energy is spent on: evaporation of a fraction of the fuel $x_e \dot{m}_f h_{fg}$, work production on the piston $\dot{W}_{ig}$, heat transfer to the coolant system $\dot{Q}_{ht}$, and thermal energy that passes through the exhaust valves $\dot{m}_e c_p T_{eo}$.

With these components, the exhaust enthalpy and temperature is described by the following equation

$$(\dot{m}_{ac} + \dot{m}_f) c_p T_{eo} = \dot{m}_{ac} c_p T_{im} + \dot{m}_f c_p T_f + \dot{m}_f q_{LHV} \eta_\lambda$$

$$- \underbrace{x_e \dot{m}_f h_{fg} - \dot{m}_f q_{LHV} \tilde{\eta}_{ig}(\lambda_c, \theta_{ign}, r_c, \omega_e, V_d, n_{cyl})}_{\dot{W}_{ig}} - \dot{Q}_{ht} \qquad (7.61)$$

where $\dot{m}_e = \dot{m}_{ac} + \dot{m}_f$ and (7.56) for $\dot{W}_{ig}$ have been used. The three most important effects are: (i) Fuel enrichment with $\lambda < 1$ in SI engines reduces the temperature through the evaporation energy and increased mass. (ii) Going to very lean or very EGR diluted mixtures (as can be the case in CI engines) reduces the temperature, since there is a smaller fraction of fuel mass that heats the air and/or EGR mass. (iii) A reduced engine efficiency $\tilde{\eta}_{ig}(\dots)$ increases the exhaust temperature, and one way to achieve this is to move the combustion from the optimum position. The effect is illustrated in Figure 7.35, where the cylinder temperature is shown for

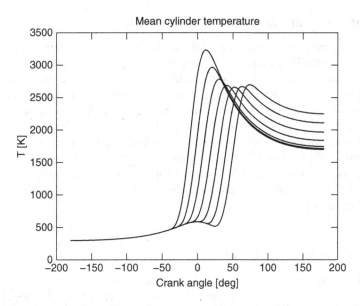

Figure 7.35 Simulation of the cylinder mean gas temperature for different spark advances, using Model 5.1. A later ignition timing gives higher temperature at EVO, which gives higher exhaust temperatures

different ignition timings. A later ignition timing (or fuel injection timing in CI engines) gives a higher temperature at the end of the expansion which gives a higher exhaust temperature. With variable valve timing engines, even more ways of influencing the work production and exhaust temperature are possible. The discussion above gives some information about the effects that play important roles in the determination of the exhaust temperature. See Ainouz and Vedholm (2009), Keynejad and Manzie (2011), Roth and Guzzella (2010) for some examples of exhaust temperature models. With this introduction and outlook, we now turn to a special case.

A Simple Model for a Special Case
If one only considers standard SI engines that run at MBT timing with $\lambda = 1$, then the exhaust temperature can be modeled using an affine function in mass flow $\dot{m}$. The model is:

Model 7.19 Simple Engine-Out Temperature Model

$$T_{eo}(\dot{m}) = T_{cyl,0} + \dot{m} \cdot K_t \tag{7.62}$$

where T_{eo} is the temperature of the fluid delivered by the cylinder to the exhaust.

In Eriksson (2003), a two-zone cylinder pressure simulation is used to study the exhaust temperature and it supports this choice of simple model structure. Experimental data for engine exhaust temperatures, at the turbine inlet, show that this simple model gives satisfactorily good agreement.

7.11   Heat Transfer and Exhaust Temperatures

Heat transfer is an inherent process in engines where gases that flow through pipes and other components are cooled or heated. In many engine components, the heat transfer is of minor importance and can be neglected in the modeling. However, there are occasions when these effects dominate and have to be accounted for, like for example in heat exchangers and exhaust manifold. This section summarizes some of the heat transfer phenomena that influence the gases flowing in engines, and describes models for the temperature change of a gas in a pipe. A general equation for the temperature change in a pipe is first developed in Section 7.11.1, then the attention is turned to the exhaust system where the heat transfer is very high. The different modes of heat transfer and their relative importance in the exhaust system are studied in Section 7.11.2 and a set of temperature models, that are suitable to use together in MVEM, are compiled in Subsection 7.11.3.

7.11.1   Temperature Change in a Pipe

A pipe with the geometry of length L and diameter d is studied, in which there is heat transfer from the fluid to a surrounding of constant temperature T_s, see Figure 7.36 for a sketch of the situation. The temperature of the fluid that enters the pipe is T_i, and the temperature of the fluid leaving the pipe is T_o. The heat flux (heat flow per unit area) from the fluid at position x to the surroundings, with temperature T_s, follows Newton's law of cooling $\dot{q} = h\,(T(x) - T_s)$ with heat transfer coefficient h. Under these conditions it is possible to derive the following differential equation for the temperature along the pipe, see Eriksson (2003).

$$-\frac{dT}{dx} = \frac{h(x)\,\pi\,d_i}{\dot{m}\,c_p(T)}(T - T_s)$$

This differential equation can be solved for the gas temperature at position x. With the boundary conditions from the inlet the temperature solution becomes

$$T(x) = T_s + (T_i - T_s)\,e^{-\frac{h\,\pi\,d_i}{\dot{m}\,c_p}x} \tag{7.63}$$

At the outlet, $x = L$, the temperature is

$$T_o = T_s + (T_i - T_s)\,e^{-\frac{h\,\pi\,d_i\,L}{\dot{m}\,c_p}} = T_s + (T_i - T_s)\,e^{-\frac{h\,A_i}{\dot{m}\,c_p}} \tag{7.64}$$

where $\pi\,d_i\,L = A_i$ is the inner wall area of the pipe.

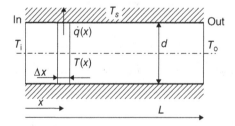

Figure 7.36   Sketch of the pipe that shows the heat transfer situation used for the derivation of the differential equation

The total heat transfer from the gas in the pipe to the surroundings can now be determined by integrating the local heat transfer $\dot{Q}_i(x) = \pi\,d_i\,\Delta x\,h_{cv,i}\,(T(x) - T_s)$ over the pipe length. Inserting (7.63) and integrating yields

$$\dot{Q}_i = \pi\,d_i\,h_{cv,i} \int_0^L (T_i - T_s)\,e^{-\frac{h_{cv,i}\,\pi\,d_i}{\dot{m}\,c_p}x}\,dx$$

$$= \dot{m}\,c_p\,(T_i - T_s)\,(1 - e^{-\frac{h_{cv,i}\,\pi\,d_i\,L}{\dot{m}\,c_p}})$$

This expression can be rewritten to an expression similar to Newton's law of cooling

$$\dot{Q}_i = \underbrace{\frac{1 - e^{-\frac{h_{cv,i}\,A}{\dot{m}\,c_p}}}{\frac{h_{cv,i}\,A}{\dot{m}\,c_p}}\,h_{cv,i}\,A}_{h_{g,i}}\,(T_i - T_s) \tag{7.65}$$

where a generalized internal heat transfer coefficient, $h_{g,i}$, is defined. (7.63) and (7.64) describe the temperature change for both heating and cooling conditions in a pipe, and together with (7.65) they are central equations in the temperature models that will follow.

7.11.2 Heat Transfer Modes in Exhaust Systems

This section describes and compares the different heat transfer modes in the exhaust system with the aid of a resistor analogy. The resistor analogy is a simplification, but provides a convenient basis for performing a comparison. A sketch of the different heat transfer modes occurring in an exhaust system is shown in Figure 7.37. Table 7.4 summarizes the relative importance of the different heat transfer modes in and around an exhaust pipe. For radiation, the values given are for the equivalent heat transfer coefficient h_{rad}, defined in Eq. (A.4). From the table, it is clear that conduction through the pipe wall is very high and acts as a short circuit which has motivated the development of dual wall exhaust manifolds. Conduction to the engine block depends highly on the length of the exhaust pipe but it has, none the less, a high influence on the total amount of heat transferred. Finally, the radiation is of the same size as the external heat transfer coefficient, which shows that it is necessary to take the radiative heat transfer into account, see Eriksson (2003) and Wendland (1993) for more discussions on radiation.

Table 7.4 Comparison of the different heat transfer modes in the exhaust, from Eriksson (2003)

Heat transfer mode	$[\text{W/K m}^2]$
Internal convection	70–150
Wall conduction	10^4
Conduction to engine	20–30
Ext. forced convection	20–35
Ext. nat. convection	10
Ext. radiation	10–35

7.11.3 Exhaust System Temperature Models

Based on (7.64), three different models for the temperature drop in a pipe will be described. These models are lumped parameter models with an input–output relation for the temperature

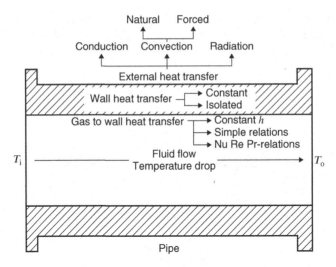

Figure 7.37 Heat transfer and temperature model components in the exhaust pipe temperature are: engine-out temperature, temperature drop, heat transfer from the gas, wall conduction, and heat transfer from the wall to the surroundings. The arrows represent different ways of modeling the phenomena

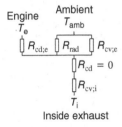

Figure 7.38 Resistor analogy of the heat transfer paths in the exhaust

change in a pipe section. All three models are based on the same components, see Figure 7.37, that is input temperature, temperature drop of the fluid, and different models for the heat transfer modes. The different possibilities for modeling each part are indicated by the arrows in Figure 7.37. The models were developed and investigated in Eriksson (2003) and more details and validations can be found therein.

The different heat transfer modes were discussed and compared in Section 7.11 and they are included in the model using the resistor analogy, see Figure 7.38. The resistances are the reciprocals of the heat transfer coefficients $R = \frac{1}{h}$. Compared to the exhaust gas and wall temperatures, the engine temperature T_e is close to ambient T_{amb}. The engine temperature is thus approximated to be equal to the ambient temperature $T_{amb} = T_e$, giving one single temperature for the sinks and a total heat transfer coefficient between T_i and T_{amb}

$$\frac{1}{h_{tot}} = \frac{A_i}{A_o}\frac{1}{h_{cv,i}} + \frac{1}{h_{cd}} + \frac{1}{h_{cv,e} + h_{cd,e} + h_{rad}} \quad (7.66)$$

Normally the conduction h_{cd} is so large, see Table 7.4, that it effectively acts as a short circuit and is neglected. Furthermore, the ratio between the inner and outer pipe area can be approximated by $\frac{A_i}{A_o} \approx 1$, the contribution from these terms are smaller than the general uncertainty in the heat transfer coefficients. An analysis of measurements shows that it is necessary to include the conductive heat transfer into the engine block, $h_{cd,e}$, this is also reported in Shayler et al. (1997). In the expression above, radiation is included by an approximation but it can also be included explicitly, see Eriksson (2003).

Static Temperature Model

A static temperature model can now be compiled from the results above. The heat transfer modes are included in the total heat transfer coefficient h_{tot}, see (7.66), and it is assumed that it is independent of T and x. Furthermore it is assumed that there is no conduction in the wall along the flow direction, so that the heat transfer is from the gas to constant ambient conditions with a constant heat transfer coefficient. The heat flux is then $q(x) = h_{tot} (T(x) - T_{amb})$, and the assumptions leading to (7.64) now hold with $T_s = T_{amb}$. The model becomes.

Model 7.20 Temperature Drop in an Exhaust Pipe

Model from the input temperature T_i to the output temperature.

$$\frac{1}{h_{tot}} = \frac{1}{h_{cv,i}} + \frac{1}{h_{cv,e} + h_{cd,e} + h_{rad}} \tag{7.67}$$

$$T_o = T_{amb} + (T_i - T_{amb}) e^{-\frac{h_{tot} A}{\dot{m} c_p}} \tag{7.68}$$

where the heat transfer coefficients in (7.67) are modeled with submodels from Appendix A.4.

The two assumptions stated above are analyzed in Eriksson (2003), showing that the model well approximates the temperatures in the exhaust system. Furthermore, a comparison and validation of the models are shown in Figure 7.39.

Model Validation

Temperature measurements on and in the exhaust system from three different engines have been used to build and validate the models. The engines used for modeling and validation and the data available from them are shown in Table 7.5. The following model validations are made

- Temperature model from engine to a measurement sensor. The model uses the linear model (7.62) for the engine output temperature.
- Temperature model for the temperature drop between two sensors in the exhaust system. The measured temperature at the first sensor is used as input to the model that predicts the next sensor output.

Evaluation 1: Models for engine-out and temperature drop, are shown in Figure 7.39a, c, d. The data shows that a linear model for the engine-out temperature T_i-gives a good fit to measured data, when the engine operates close to nominal conditions ($\lambda = 1$ and optimal ignition timing).

Evaluation 2: Temperature drop between two sensors are shown in Figure 7.39b and d. The agreement between modeled and measured temperature drops between sensor 1 and 2 is excellent.

Engine C is equipped with a dual wall exhaust manifold. A dual wall is not explicitly included in the model, but the model still gives good agreement with the measured data. The engine has one gas temperature sensor close to the exhaust port and one just before the precatalyst. There

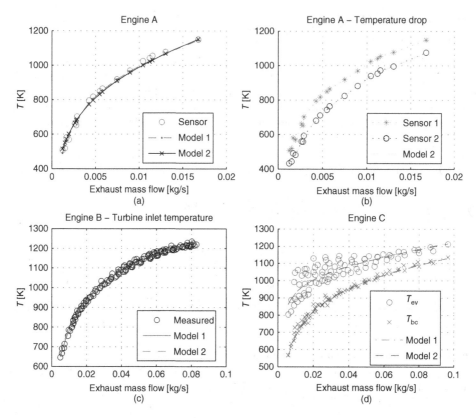

Figure 7.39 Evaluation of the stationary exhaust heat transfer models for Engine A. (a) – model for the first temperature sensor, (b) – model for the temperature drop between two sensors. Evaluation of the stationary exhaust heat transfer model for Engine B, (c) and C (d) Engine B has one sensor just before in front of the turbine inlet and engine C has two sensors, one inserted close to the exhaust valve and one before the precatalyst

Table 7.5 The engines used for modeling and validation and the data that is available from them. Stat–stationary data. Dyn–dynamic data. Surf–Number of surface temperature sensors. Gas–Number of gas temperature sensors

Engine	Short description	Stat	Dyn	Surf	Gas
A	0.36 l pressure wave supercharged engine	yes	yes	4	2
B	2.3 l turbocharged engine with standard exhaust manifold	yes	no	0	1
C	3.2 l naturally aspirated engine with dual wall exhaust manifold	yes	no	0	2

are high variations in the measured temperatures close to the port which are due to a speed dependence and the unsteady flow from the port. When measuring farther from engine the influence from engine speed is no longer seen and the mass flow dominates.

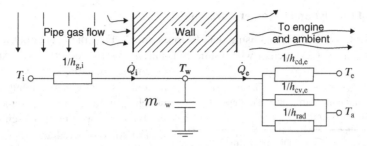

Figure 7.40 Electrical analogy for the dynamic wall temperature, where the heat capacity of the wall stores the heat

Dynamic Exhaust Temperature Model

The model above covers stationary conditions, but the derivation can easily be extended to include the wall temperature dynamics and the associated gas heat transfer. It is assumed that the wall temperature varies with time but is the same along the pipe, and the situation is illustrated by the electrical analogy in Figure 7.40. The temperature development is described using the energy balance, where the difference between the internal and external heat transfer rates $\dot{Q}_i - \dot{Q}_e$ drives the temperature change of the wall that has mass m and specific heat capacity c_w.

Model 7.21 Dynamic Temperature Model

The dynamic temperature model is expressed as

$$\dot{Q}_e = A[h_{cv,e}\,(T_w - T_{amb}) + F_v\,\epsilon\,\sigma\,(T_w^4 - T_{amb}^4) + h_{cd,e}\,(T_w - T_e)$$

$$\dot{Q}_i = h_{g,i}\,A\,(T_i - T_w)$$

$$\frac{dT_w}{dt}\,m_w\,c_w = \dot{Q}_i(T_w, T_i) - \dot{Q}_e(T_w, T_{amb}, T_e)$$

$$T_o = T_w + (T_i - T_w)\,e^{-\frac{h_{cv,i}\,A}{\dot{m}\,c_p}}$$

where $h_{g,i}$ is determined from (7.65), c_w is the specific heat capacity of the wall material, and m_w is the wall mass. For a circular pipe the wall mass is given by $m_w = \rho_w \frac{\pi\,(d_o^2 - d_i^2)}{4}\,L$. The ordinary differential equation above includes the heat transfer and wall temperature under dynamic conditions.

For all engines, the following values gave good fits to the measurement data. External heat transfer coefficient lie in the range of $h_e = h_{cd,e} + h_{rad} + h_{cv,e} \approx 100$ (W/m² K). Internal heat transfer coefficients $h_{cv,i}(\dot{m})$ were calculated using experimental Nu-Re-Pr relations, for Engine A and B the correlation from Eriksson (2003) was used, and for Engine C Meisner and Sorenson (1986) was used.

Validation of the Dynamic Temperature Model

The dynamic model is validated using data from engine A. First the engine-out temperature is modeled using (7.62) together with a time constant of 15 s, that describe, for example, the engine temperature dynamics. This temperature is then given as input to Model 3 which is used for the gas temperature in the exhaust pipe. Finally the sensor dynamics is also considered and Table 7.6 shows the sensor dynamics for different sensor diameters and flow conditions. The sensor used had a diameter of 1 mm and therefore a time constant of 0.6 s is used for the sensor dynamics. Figure 7.41 shows the results from the dynamic tests on Engine A.

Table 7.6 Time constants for the Thermocoax sensors, used in the measurements. The sensors had diameters of 0.5, 1.0, and 1.5 mm

Thermocoax	Diameter [mm]				
Time constant for	0.25	0.5	1	2	3
Flowing air [s]	0.1	0.2	0.6	2	3
Non flowing air [s]	0.4	1.2	3	8	12

The measured and modeled gas temperatures are shown in the top of the figure, while the four wall temperature sensors and the modeled wall temperature are seen in the bottom of the plot. Both gas temperature and wall temperatures are well described by Model 3. It is also seen that the three wall temperatures differ from each other and that it is coldest closest to the engine while one of the middle sensors is the warmest.

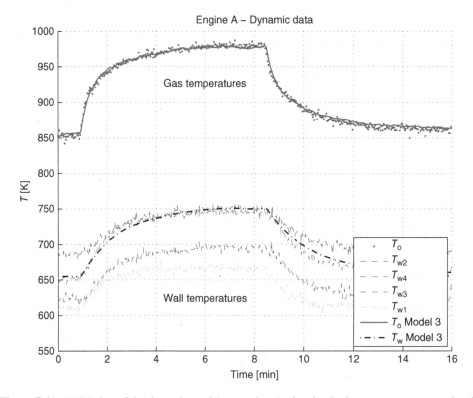

Figure 7.41 Validation of the dynamic model on engine A, showing both gas temperatures and wall temperatures. Note that there is only one lumped wall temperature in the model

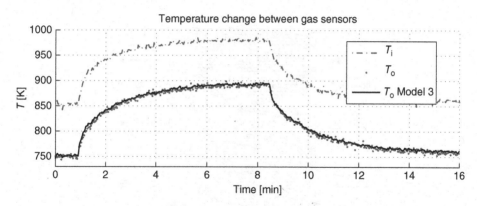

Figure 7.42   Validation of the dynamic model on engine A for response between gas temperature sensors 1 and 2

In the second test, sensor 1 is used as input to the model that is used to describe the temperature at sensor 2. The data is shown in Figure 7.42 which shows that the model describes the data accurately.

A Note on Pipe Wall to Engine Conduction

For all engines, included in the model validations above, the total external heat transfer coefficient $h_e = h_{cv,e} + h_{cd} + h_{rad}$ had to be set to values that are higher than that postulated by radiation and external convection only. This extra amount of heat transfer seen in the data is a strong indication that there is significant conduction from the wall into the engine block. The validations performed above show that the model can incorporate and capture such conditions too.

7.12   Heat Exchangers and Intercoolers

In turbocharged engines a heat exchanger, called an intercooler, is used after the compressor to cool the compressed air back to near ambient temperature. It increases air density, which increases both the maximum power output and engine efficiency, for a given maximum intake pressure in both SI and CI engines. In SI engines it is mainly used for knock reduction, which leads to the fact that a higher engine compression ratio can be tolerated. In engines the most common heat exchanger for charge cooling is the *cross flow intercooler* with both fluids unmixed, which is shown in Figure 7.43. There are also gas to water intercoolers, especially for EGR (exhaust gas recirculation) systems.

To achieve efficient cooling of the charge air, the tubes in the intercooler need to be rather thin so that the air is exposed as much as possible to the cooling medium. The result is that some of the gain in intake air density is lost. Therefore, beside the obvious need to model the outlet air temperature, a model for the pressure head loss is also needed and such a model was described and validated in Example 7.1.

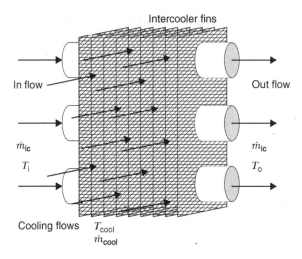

Figure 7.43 Heat exchanger of the type: cross flow, both fluids unmixed. The total flow $\dot{m}_{\mathrm{ic}}$ is split up between the different tubes in the heat exchanger and the fins make the transport of heat between the fluid more efficient. Cross flow means that the inner flow $\dot{m}_{\mathrm{ic}}$ and outer flows $\dot{m}_{\mathrm{cool}}$ are perpendicular to each outer

7.12.1 Heat Exchanger Modeling

For all practical purposes of heat exchangers (intercoolers) in combustion engines the flow rate of the cooling fluid $\dot{m}_{\mathrm{cool}}$ is greater than the mass flow, $\dot{m}_{\mathrm{ic}}$, through it. This suggests, Holman (2009), that the following equation can be used to measure the effectiveness

$$\varepsilon = \frac{T_{\mathrm{i}} - T_{\mathrm{o}}}{T_{\mathrm{i}} - T_{\mathrm{cool}}}$$

The intercooler effectiveness depends on the operating conditions, like temperatures and flow conditions, that is in MVEM sense it is treated similarly to the volumetric effectiveness. Solving this equation for T_{ic} yields the desired expression for the intercooler outlet temperature, expressed in terms of the temperatures and the intercooler effectiveness

$$T_{\mathrm{o}} = T_{\mathrm{i}} - \varepsilon(\dot{m}_{\mathrm{cool}}, \dot{m}_{\mathrm{ic}}, \ldots)\,(T_{\mathrm{i}} - T_{\mathrm{cool}}) \tag{7.69}$$

T_{cool} is the temperature of the cooling medium, in this case ambient air, $T_{\mathrm{cool}} = T_{\mathrm{a}}$. In order to determine the intercooler outlet temperature, a model for the intercooler effectiveness, $\varepsilon(\dot{m}_{\mathrm{cool}}, \dot{m}_{\mathrm{ic}}, \ldots)$, is needed. Below, two models of different complexity are presented; a regression model and the standard NTU-model (see e.g., Holman (2009) for a derivation).

Regression Effectiveness Model

The physics of heat transfer is built into the NTU-model but its structure is complex, which motivates the search for other models. Another approach to model ε is to use the heat transfer considerations made when developing the NTU-model as an indicator of what parameters effect the efficiency and use them as regressors in a linear regression model. Choosing the mean tube temperature and air mass flow together with the ratio of the air mass flow and the cooling air mass flow, the following model for ε may be stated

$$\varepsilon = a_0 + a_1 \left(\frac{T_c + T_{cool}}{2} \right) + a_2 \dot{m}_{air} + a_3 \frac{\dot{m}_{air}}{\dot{m}_{cool}} \tag{7.70}$$

This model based on linear regressors will be called the REG-model.

NTU Model

A derivation and discussion on how to apply the NTU-model to an automotive intercooler is given in Bergström and Brugård (1999). The model equations are arranged as follows

$$\varepsilon = 1 - e^{\frac{e^{-C N^{0.78}} - 1}{C N^{-0.22}}}$$

$$N = \frac{UA}{c_{p,air}\, \dot{m}_{air}} = \frac{K}{c_{p,air}} \dot{m}_{air}^{-0.2} \mu_i^{-0.5} \tag{7.71}$$

$$\mu_i = 2.3937 \cdot 10^{-7} \left(\frac{T_c + T_{cool}}{2} \right)^{0.7617}$$

$$C = \frac{\dot{m}_{air}}{\dot{m}_{cool}}$$

The unknown constant K is determined from a least squares fit to measured data, and as can be seen, the model is expressed in variables that are easily determined. The grouping $\frac{UA}{c_{p,air}}$ is called the *number of heat transfer units* (NTU), which gives the name NTU-model.

Validation of Temperature Models

The REG-model and NTU-model are validated in Figure 7.44 with respect to how they describe efficiency. Further validations for the temperatures show that NTU-model has a maximum error

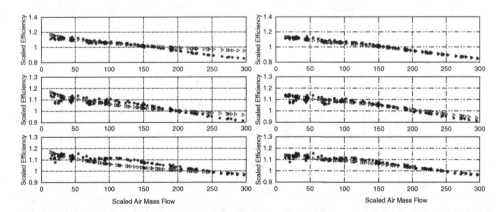

Figure 7.44 Left: Intercooler effectiveness ε for the NTU-model for three fan speeds. Right: Intercooler effectiveness ε from the REG-model for three fan speeds. Measured data is represented by $*$ and the model is represented by $\triangleright$

of 8 K and REG-model 1 K in the exit temperatures. The REG-model thus agrees better with measured data compared to the NTU-model, but it must be remembered that the REG-model has four tuning parameters while the NTU-model has only one. In both cases, the relative error in temperature is rather small since it should be calculated in absolute temperature (Kelvin). Furthermore, it is interesting to note that larger errors in the efficiency can be tolerated for low flows $\dot{m}_{air}$ since the temperature rise over the compressor, from T_{af}, is smaller which finally results in smaller errors in T_{ic}. From this it is also understood that it is important to make the least squares fit to minimize the error in T_{ic} instead of in ε, which also is natural since we want to model the temperature change. Note that the REG-model is linear in the model parameters in both formulations, and a linear least squares method can be used for determining the parameters.

7.13 Throttle Plate Motion

In Section 7.3, the flow through a throttle was modeled and we now turn to the mechanical system that determines the throttle position α, which is a part of the throttle servo system for the engine. Throttle systems have been modeled in, for example, Deur et al. (2004), Eriksson and Nielsen (2000), Scattolini et al. (1997), and Thomasson and Eriksson (2011), and the presentation here is a compilation of these, leading to a control oriented model that can be parameterized from input output data. An example of a throttle servo system is shown in Figure 7.45 and a process model for the electronic throttle body is shown in Figure 7.46.

The system components and models are: a *chopper* converts the control signal u_{th} to electric motor inputs and it is modeled with a constant that can depend on the battery voltage $u_{ch} = K_{ch}(U_{batt}) u_{th}$. An *electric motor* that produces the torque M_a and has a back EMF $u_{emf} = K_v \omega_m$ and an armature system for the current and torque development, that is modeled as a first order system $M_a = \frac{K_a}{\tau_a s+1} u_a$ with time constant τ_a. A set of *gearbox* transfers the motor torque and angular position to the throttle axle and it is modeled using a gear ratio, giving $M_{in} = i_{th} M_a$. The *throttle plate* is mounted on the rigid *throttle shaft*, on which a potentiometer (sometimes two for hardware redundancy) is mounted for measuring the position. There can be an *aerodynamic torque* $M_{air}(\alpha, \dot{m}_{air})$ that depends on the mounting of the throttle plate and on the aerodynamic forces on the plate, see Morris and Dutton (1989), but it can in many cases be neglected or simply be included as a disturbance in the model. There are some plate designs

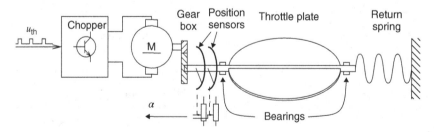

Figure 7.45 Sketch of a throttle with the included components. The control signal is a PWM signal, it is transferred to the motor via a chopper. The motor is connected to the throttle plate shaft, often via a gearbox. Further, there is often a return spring that acts on the main shaft and returns the throttle to a predetermined limp-home position if there is no input. See Figure 7.47 for the limp-home spring

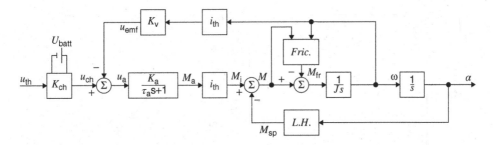

Figure 7.46 Block diagram, showing a process model for the throttle. The torque contributions on the throttle plate are the armature torque, M_a, the friction torque, M_{fr}, and the spring torque, T_s. The armature is modeled with a first order system with gain K_a and time constant τ_a. The chopper and back electromotive torque are modeled with the proportional gains, K_{ch} and K_v, from the control signal and throttle plate angular velocity, u_{th} and ω, respectively

where the plate is made eccentric, to make the throttle self-closing if there is a power electronics failure, and for these an air flow torque model can be needed (Eriksson and Nielsen, 2000). There is furthermore *friction* in the system $M_{fr}(M, \omega, \alpha)$ from the gearbox, motor bearings, and shaft, in the form of both static and dynamic friction. Finally there is often a *return spring* that brings the throttle to a predetermined position (called limp-home position) α_{lh}, if there is no input power $M_{sp}(\alpha)$.

Combining the components above and using Newton's second law for the throttle shaft, gives the following differential equations that govern the throttle shaft position

$$\tau_a \frac{dM_i}{dt} = (i_{th} K_a [K_{ch}(U_{batt}) u_{th} - K_v i_{th} \omega] - M_i) \tag{7.72a}$$

$$J \frac{d\omega}{dt} = M_i - M_{fr}(M, \omega, \alpha) - M_{sp}(\alpha) - M_{air}(\alpha, \dot{m}_{air}) \tag{7.72b}$$

$$\frac{d\alpha}{dt} = \omega \tag{7.72c}$$

Here, the first equation is the armature dynamics and J is the total throttle inertia $J = J_{th} + i_{th}^2 J_m$. The basic structure is now in place and our attention now turns to details in the components.

Limp-Home Spring and Friction Nonlinearities

The limp-home springs and friction are nonlinearities that have a significant influence on the throttle behavior. The result from slow ramps on the input is shown in Figure 7.47, illustrating their influence and significance. The Figure 7.47a and b show the ramp inputs u_{th} and the corresponding output α, while the middle plot shows the same data with u_{th} on y-axis and α on the x-axis. In Figure 7.47c friction is seen in the hysteresis between the up and down ramps, while the limp-home is seen in the discontinuities just below 30% throttle opening.

Starting with friction there are two effects, the static friction M_{fs} and dynamic friction M_{fv}. Here the focus is on the static friction, while the dynamic friction will be discussed later and modeled in (7.75). The *static friction* is seen in Figure 7.47c, where it is the cause of the difference between the up and down ramps. The static friction most frequently modeled is the

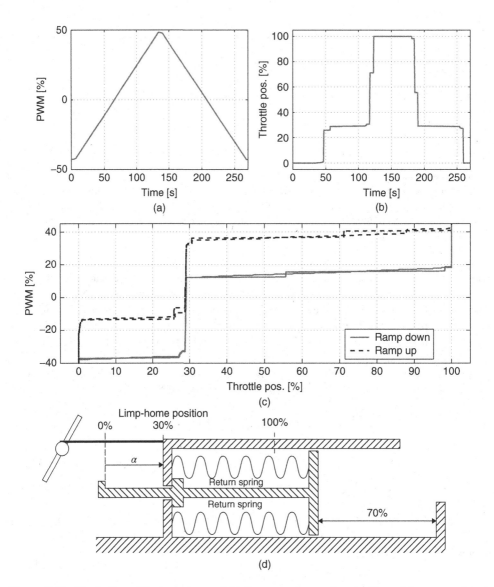

Figure 7.47 Illustration of the importance of friction and return spring. (a) Up and down ramp input with duration 270 s, (b) Resulting throttle position, (c) Input and output relation for two ramps one 270 s from the top two plots and one other with total duration 70 s, (d) Illustration of the two return springs that gives the throttle its resting position, called the limp-home position, at 30% opening

classical Coulomb friction model

$$M_{\mathrm{fs}}(M, \omega) = \begin{cases} M & \text{if } \omega = 0 \text{ and } |M| < M_{\mathrm{c}} \\ M_{\mathrm{c}} sgn(\omega) & \text{otherwise} \end{cases} \qquad (7.73)$$

where the friction torque is equal to the applied torque, M, when $\omega = 0$ and the applied torque is less than the Coulomb friction, M_{c}. Otherwise the friction torque is equal to Coulomb friction in the opposite direction of motion. Several papers have included more complex, dynamic friction models, for example the inclusion of the Stribeck effect. However, in many cases only Coulomb friction is used, which is frequently motivated by the fact that it is simpler and has fewer parameters to determine and foremost that it is often sufficient for throttle servo control design (Deur et al., 2004), (Thomasson and Eriksson, 2011).

The *limp-home* nonlinearity comes from the two spring system, see Figure 7.47d, that pulls the throttle plate towards the limp-home position if there is no input. The spring torque is piecewise linear but the torque differs, depending on whether the throttle plate is inside or outside of the limp-home region. The slope of the $u_{th}(\alpha)$ curve is almost flat above and below the limp-home region, with a very sharp transition between them. There is approximately 30% increase in the control signal from fully closed to fully open throttle, where about 20% is in a narrow region of 0.5–2.0° around the limp-home position. The spring torque can be modeled as a piecewise linear function

$$M_{sp}(\alpha) = \begin{cases} m_{lh}^+ + k^+(\alpha - \alpha_{lh}^+) & \text{if } \alpha > \alpha_{lh}^+ \\ m_{lh}^+(\alpha - \alpha_{lh})/(\alpha_{lh}^+ - \alpha_{lh}) & \text{if } \alpha_{lh} < \alpha \leq \alpha_{lh}^+ \\ m_{lh}^-(\alpha_{lh} - \alpha)/(\alpha_{lh} - \alpha_{lh}^-) & \text{if } \alpha_{lh}^- < \alpha \leq \alpha_{lh} \\ m_{lh}^- - k^-(\alpha_{lh}^- - \alpha) & \text{if } \alpha \leq \alpha_{lh}^- \end{cases} \qquad (7.74)$$

These parameters in the friction and limp-home models can be identified from ramp inputs, for example those such that were shown in Figure 7.47 (see Thomasson and Eriksson (2011) for a discussion on identification).

A third nonlinearity is saturation at the position limits, that is fully open and fully closed throttle positions. This is easily included as saturations in the throttle angle integrator, that is the output of (7.72c) is saturated at 0% and 100% openings.

Simplifying the Throttle Motion Model

In the next step the model (7.72) will be slightly simplified to enable easy parameterization from experimental data. The armature time constant, τ_a, is typically very small, approximately 1 ms (Deur et al., 2004), and this effect is therefore often neglected when the purpose is throttle servo control design. The armature torque can then be divided in to

$$M_a = K_a K_{ch} u_{th} + K_a K_v \omega = M_u + M_{emf}$$

Translating these torques through the throttle gear box enables the throttle motion model to be expressed as four torques that act on the throttle plate shaft. These four are: driving torque from the DC motor, $i_{th} M_u$, the back electromotive torque, $i_{th} M_{emf}$, the spring torque, M_{sp}, and the friction torque, M_{fr}, (that was divided into static M_{fs} from (7.73) and dynamic M_{fv} components).

Further, the dynamic (viscous) friction and electromotive torque are both linear functions in angular velocity, acting in the opposite direction of motion. These effects cannot be separated from each other when studying input output data for the throttle and are therefore lumped into a single torque model that dampens the throttle motion

$$M_{fv}(\omega) + i_{th} M_{emf}(\omega) = K_{fe} \omega \qquad (7.75)$$

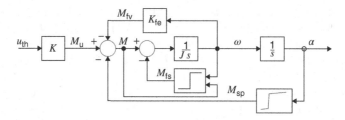

Figure 7.48 A block diagram showing the simplified process model for the throttle motion. The model captures the most important throttle dynamics and is suitable for throttle servo control design

where the back EMF in the motor is expressed as an equivalent torque that acts on the throttle shaft, using the gear ratio of the throttle gear box. Combining (7.76), (7.73), (7.74), and (7.75), and using Newton's second law gives the differential equation for the throttle plate angular velocity. The end result is thus the following throttle model.

Model 7.22 Throttle Motion and Position

$$J\frac{d\omega}{dt} = -K_{fe}\,\omega - M_{sp}(\alpha) - M_{fs}(M, \omega) + Ku_{th} \tag{7.76a}$$

$$\frac{d\alpha}{dt} = \omega \tag{7.76b}$$

where the submodels for the static friction (7.73) and limp-home (7.74) are included. The output angle needs to be saturated at fully open and fully closed positions.

The resulting throttle motion model is also shown as a block diagram in Figure 7.48. This model captures the most important parts of the throttle dynamics and it also forms the basis for the controller structure for throttle servo control in Section 10.3.

7.13.1 Model for Throttle with Throttle Servo

In many cases, the throttle servo is already designed and it can be sufficient to either: measure the throttle angle α directly and use as input to the throttle flow model (7.11), or lump the throttle servo and throttle motion together and represent them as one system, from the throttle reference α_{ref} commanded by the control system to the throttle angle. A linear system often suffices for describing the combined servo and throttle plate movement, capturing the fact that there is a time lag between commanded and achieved angle. A simple model for the throttle system is thus as below.

Model 7.23 Throttle Motion with Servo Included

$$\alpha = H_{th}(s)\,\alpha_{ref}$$

8

Turbocharging Basics and Models

Turbocharged engines have evolved from being exotic to almost a commodity in many vehicles, due to the trend of improved fuel economy by downsizing and supercharging. Control systems and models play an important role in the delivery and handling of the turbo power, so a driver can get a consistent engine and vehicle response in every driving situation. This chapter introduces supercharging devices and layouts in Section 8.1 and later describes models for the compressor and turbine that fit into the MVEM framework.

Turbocharger thermodynamics and generic model structures for compressors and turbines are introduced in Section 8.2. Dimensionless and corrected quantities, used to represent the turbocharger performance, are discussed in Section 8.3 and a refined model structure is introduced. Details of how turbocharger performance is specified and determined are discussed in Section 8.4. Models for the flow and efficiency in compressors and turbines are covered in Sections 8.5 to 8.7, and the interaction with a complete engine system is discussed in Sections 8.8 to 8.10, where two complete MVEM models with turbochargers are given, one for a gasoline and one for a diesel engine.

8.1 Supercharging and Turbocharging Basics

Supercharging is the collective name for several methods that increase the intake air density, that is, methods that charge extra air to the cylinder, and one particular method is called turbocharging. Supercharging is achieved by compressing the gas before the cylinder, and can be implemented by several methods. Different options are available both for the selection of compression method and for the method of driving the compressor, see Figure 8.1.

Mechanical Supercharging
By mechanical supercharging it is meant that a separate compressor, pump, or blower, is driven by the engine's crankshaft and used to compress the air. Mechanical supercharging can give an engine with a high torque curve over the entire engine operating range, this gives good driving performance and is the main advantage. The drawback is that the mechanical energy that drives the compressor is taken directly from the engine shaft, which has a negative effect on the fuel consumption when the compressor is engaged. This method is often used in high performance engines where it is important to have good performance and where a sacrifice in consumption isn't crucial.

Modeling and Control of Engines and Drivelines, First Edition. Lars Eriksson and Lars Nielsen.
© 2014 John Wiley & Sons, Ltd. Published 2014 by John Wiley & Sons, Ltd.
Companion Website: www.wiley.com/go/powertrain

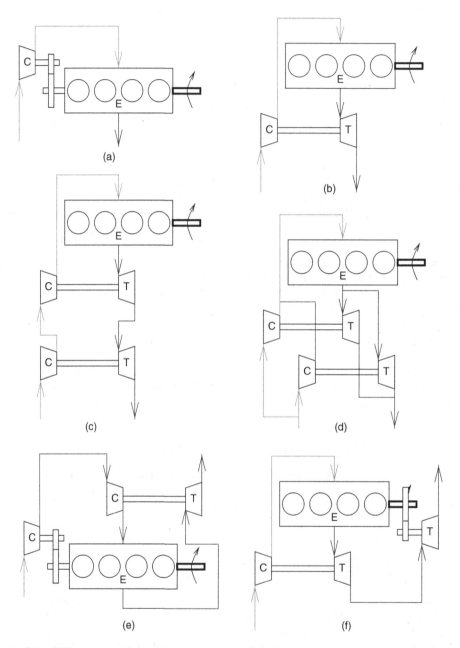

Figure 8.1   Different ways of supercharging an engine. (a) Mechanical supercharging. (b) Turbocharging. (c) Two-stage turbocharging, serial. (d) Two-stage turbocharging, parallel. (e) Engine driven compressor and turbocharger. (f) Turbocharging with turbocompound

An alternative method for driving the compressor is to use an electric motor, instead of the mechanical coupling to the engine shaft. This gives one extra degree of freedom for the compressor since the compressor speed can be selected independently of the engine speed.

Turbocharging

The most common way to supercharge an engine is by using a *turbo* (or *turbocharger*) where a turbine and compressor are mechanically coupled to each other. The combustion engine does not fully use the chemical energy in the fuel, and the turbine utilizes the enthalpy in the exhaust gas to drive the compressor. The turbine recovers some of the energy which is lost in the thermodynamic cycle, that is $\Delta h_{\text{loss}} = c_{\text{p}} (T_4 - T_1)$.

The obvious benefit with a turbo is that it recovers some of the energy lost in the thermodynamic cycle to compress the air. A turbocharged engine has a drop in the torque for low speeds, see Figure 8.2, which is coupled to a region of instability in the compressor, this instability is called *surge*. The lowest speed, for which the turbocharged engine can reach its maximum torque, is a result of the interaction between the turbo and engine. A smaller turbo gives better low speed torque, but as drawback it decreases the maximum engine power. A design trade-off between the low speed torque and maximum power is thus inherent in turbocharged engines. In the mid-speed range the torque curve of a turbocharged engine is shaped by the control system.

Multistage Charging

Another turbocharging method is *two-stage turbocharging*, where two turbine and compressor stages are used. They can be arranged either in series or in parallel, see Figure 8.1. Control valves are used to engage, shut off, or bypass the different compressors and turbines in the turbos. Engines with V-configurations often use the parallel configuration, with one turbo for each bank. A version of the series configuration is the *series sequential* arrangement, that has

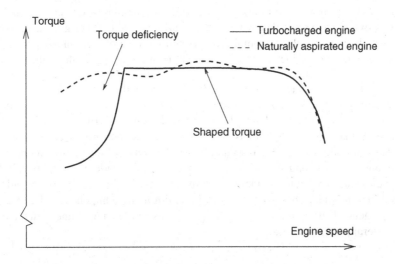

Figure 8.2 A sketch that illustrates the torque curves for two engines with similar performance in torque and power. A turbocharged engine exhibits a drop in torque at low speeds while the naturally aspirated engine gives a higher torque, see also Figure 4.5. The torque curve of a turbocharged engine can be shaped by the control system in the mid speed range to give a flat behavior

two turbos of different size. The second (high pressure) turbo is the smaller one and is used at low engine speeds. At high engine speeds the smaller turbo is bypassed and the larger (low pressure) turbo is used solely. Relating to the design trade-off, shown in the torque curve in Figure 8.2, the smaller (high pressure) turbo is used to increase the low speed torque and thus reduce the low speed torque deficiency of the turbocharged engine, while the larger one is used to complete the power demand from the engine.

Combining a mechanical compressor and turbo, see Figure 8.1, is also used to improve the low speed region and remove the design trade-off for the turbo compressor. Furthermore, power can also be extracted from the exhaust gases through a turbine connected to the crankshaft, this is called *turbocompound*.

Another way is to use *pressure wave supercharging*, which uses a special device, called Comprex. In the device, the enthalpy in the exhaust gas is used to create acoustic waves in a rotating drum. These acoustic waves are used to compress the air in the drum and deliver it to the intake. See Heywood (1988, Chapter 6) and Guzzella et al. (2000), Spring et al. (2007), and Weber et al. (2002) for a more thorough description of the device and its controls.

After this brief description of different supercharging concepts, we turn our attention to the operating principles and modeling of compressors and turbines. The presentation below and principles for the modeling are independent of the configuration used, and thus gives a framework for modeling of turbocharger components in a general setting.

8.2   Turbocharging Basic Principles and Performance

A turbocharger (often called just turbo) consists of a centrifugal compressor and radial turbine, see Figure 8.3 for a picture. Turbines and compressors belong to the category called *turbomachinery*, for which Dixon (1998) gives the following definition:

> We classify as *turbomachines* all those devices in which energy is transferred either to, or from, a continuously flowing fluid by the dynamic action of one or more moving blade rows. The word turbo or turbinis is of Latin origin and implies that which spins or whirls around. To narrow the range of machines we usually look at enclosed machines in which a finite amount of air passes through.

The turbomachine performance has a big impact on engine and vehicle performance, and when selecting a compressor and turbine for an engine it is therefore necessary to study the performance of the turbo and its interaction with the engine. Engine performance has been discussed in the preceding chapters and attention is now on turbocharger performance, which is analyzed and modeled using the laws of thermodynamics, with the aim of developing models that can be used for analysis and control. Turbocharger modeling is considered complex and difficult, this is, however, mostly due to the fact that there are many different ways of formulating the models and that the model becomes very detailed with many components. Here we start by introducing a general structure, conveying the principles, and then continue with the details to make the general principles clear.

8.2.1   Turbochargers in Mean Value Engine Models

Turbochargers fit into the category of flow components and are thus easy to include in the component-based modeling framework that was introduced in Section 7.1.2. The structure of

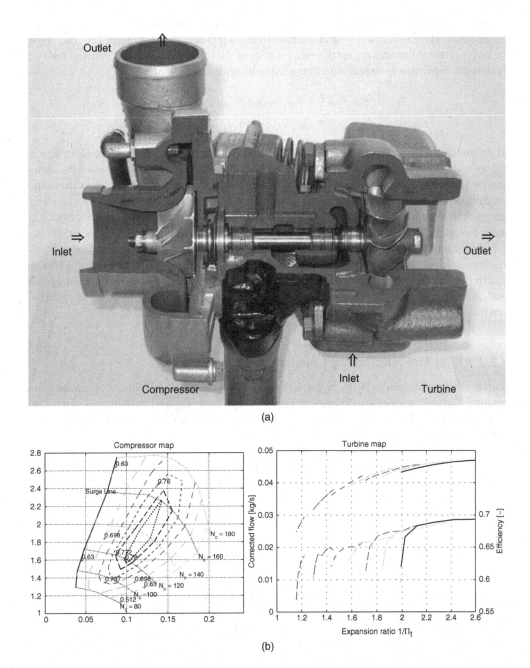

Figure 8.3 (a) Picture of a turbo consisting of a compressor (left) and turbine (right) mounted on a common shaft. In between the compressor and turbine there are channels for cooling and lubricating the turbo and its bearings. (b) Turbocharger maps that are used to specify the performance of compressors (left) and turbines (right). Maps contain information about how pressure ratio, rotational speed, mass flow, and efficiency are connected

a compressor and turbine model in an MVEM architecture has mass flow and temperature components, like a restriction, and in addition to this a power consumption (or generation for the turbine) that influences the rotational dynamics of the turbo shaft. Compressor and turbine models are conveniently expressed using a structure with four functions

$$\dot{m} = f_1(p_{us}, p_{ds}, T_{us}, \omega_{tc}) \qquad \eta = f_2(p_{us}, p_{ds}, T_{us}, \omega_{tc})$$

$$T = f_3(p_{us}, p_{ds}, T_{us}, \omega_{tc}) \qquad \dot{W} = f_4(p_{us}, p_{ds}, T_{us}, \omega_{tc})$$

where an intermediate step, efficiency, is introduced that clarifies the structure and interaction between temperature and power submodels. Functions f_1 and f_2 are extracted from the compressor and turbine maps, see Figure 8.3, where Section 8.3 gives the details of how the maps are determined, interpreted, and used. Functions f_3 and f_4 build upon the thermodynamics and are given in Sections 8.2.2 and 8.2.3 for the compressor and turbine respectively. Sections 8.6 and 8.7 then give details and examples of compressor and turbine models used in the literature.

Turbines and compressors are flow devices that continuously exchange fluid work with mechanical work, and therefore the analysis is made for the power (flow of energy) to and from the devices. Figure 8.4 gives a sketch of turbocharger showing the system boundaries and power transfers used in the thermodynamic analysis. In the analysis that follows it is assumed that the fluid is an ideal gas with c_p constant over the process being studied.

8.2.2   First Law Analysis of Compressor Performance

A compressor takes power $\dot{W}_c$ from the shaft and compresses the fluid element from one temperature T_{01} and pressure p_{01} to another higher temperature T_{02} and pressure p_{02}. First law analysis, Figure 8.4, of a steady flow control volume directly yields the shaft power $\dot{W}_c$ that is used for the compression

$$\dot{W}_c = \dot{m}_c(h_{02} - h_{01}) = [c_p \text{ constant}] = \dot{m}_c\, c_{p,c}\, (T_{02} - T_{01}) \qquad (8.1)$$

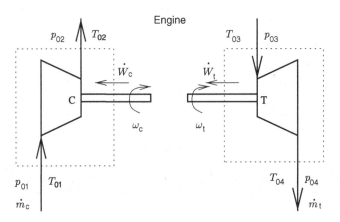

Figure 8.4 Sketch of a turbocharger, showing the pressures and temperatures at the different locations around the device. The dotted lines show control volume boundaries used in a first law analysis of compressor and turbine performance

which is the power that the compression process consumes. To determine how efficient this particular compression process is, it is compared to an ideal process. Compressor efficiency is defined as

$$\eta_c = \frac{\text{Power requirement by an ideal process}}{\text{Actual power consumed}} = \frac{\dot{W}_{c,\text{ideal}}}{\dot{W}_c} \tag{8.2}$$

The process that requires the least amount of energy for compressing a fluid element from p_{01} to p_{02} is the isentropic process which is the ideal process. The connection between pressures and temperatures for an isentropic state change is given by (5.3), which becomes

$$\frac{T_{02,\text{ideal}}}{T_{01}} = \left(\frac{p_{02}}{p_{01}}\right)^{\frac{\gamma-1}{\gamma}}$$

The smallest power required for the compression can thus be expressed as

$$\dot{W}_{c,\text{ideal}} = \dot{m}_c \, \Delta h_{0s} = \dot{m}_c \, c_{p,c} \, T_{01} \left\{ \left(\frac{p_{02}}{p_{01}}\right)^{\frac{\gamma-1}{\gamma}} - 1 \right\} \tag{8.3}$$

which gives the following expression for the compressor efficiency

$$\eta_c = \frac{\dot{W}_{c,\text{ideal}}}{\dot{W}_c} = \frac{\left(\frac{p_{02}}{p_{01}}\right)^{\frac{\gamma-1}{\gamma}} - 1}{\frac{T_{02}}{T_{01}} - 1} \tag{8.4}$$

All quantities occurring in this expression can be determined from measurements in a turbocharger gas stand, which is exactly what is done when the compressor performance is determined experimentally. The procedure and data treatment will be discussed later in Section 8.4. The compressor characteristics are determined from the performance data and used to build models for compressor mass flow $\dot{m}_c = f_{\dot{m},c}(p_{01}, p_{02}, T_{01}, \omega_c)$, and efficiency $\eta_c = f_{\eta,c}(p_{01}, p_{02}, T_{01}, \omega_c)$. With these two models given, the compressor-out temperature can be determined by solving (8.4) for $T_{02} = T_c$. Compressor power consumption can then be determined from (8.1). In summary we get the following generic compressor model.

Model 8.1 Generic Compressor Model

$$\dot{m}_c = f_{\dot{m},c}(p_{01}, p_{02}, T_{01}, \omega_c) \tag{8.5a}$$

$$\eta_c = f_{\eta,c}(p_{01}, p_{02}, T_{01}, \omega_c) \tag{8.5b}$$

$$T_c = T_{01} + \frac{T_{01}}{\eta_c} \left\{ \left(\frac{p_{02}}{p_{01}}\right)^{\frac{\gamma-1}{\gamma}} - 1 \right\} \tag{8.5c}$$

$$\dot{W}_c = \dot{m}_c \, c_{p,c} \, (T_c - T_{01}) \tag{8.5d}$$

This compressor model is formulated so that it fits into the component-based MVEM framework, introduced in Section 7.1.2. It thus determines the mass transfer and temperature of the flowing gas provided surrounding pressures, temperatures, and angular velocity. The power consumption interacts with the turbine power production and as a next step we thus turn to the turbine.

8.2.3 First Law Analysis of Turbine Performance

A turbine takes energy from the fluid by expanding the gases from a higher pressure and temperature to a lower pressure and temperature, and delivers it as power to the shaft. Turbine power is also determined from a first law analysis, Figure 8.4, of a steady flow control volume. The analysis yields the shaft power $\dot{W}_t$ as

$$\dot{W}_t = \dot{m}_t (h_{03} - h_{04}) = [c_p \text{ constant}] = \dot{m}_t c_{p,t} (T_{03} - T_{04}) \tag{8.6}$$

This expansion process does not extract as much energy as an ideal process would, and the ratio defines turbine efficiency

$$\eta_t = \frac{\text{Actual power delivered}}{\text{Power delivered by an ideal process}}$$

Isentropic expansion is the best achievable process. Manipulations, similar to those for the compressor, yield the following expressions for the temperature ratio and work for an ideal process

$$\dot{W}_{t,\text{ideal}} = \dot{m}_t c_{p,t} (T_{03} - T_{04,\text{ideal}}) \quad \text{and} \quad \frac{T_{04,\text{ideal}}}{T_{03}} = \left(\frac{p_{04}}{p_{03}}\right)^{\frac{\gamma-1}{\gamma}}$$

$$\Rightarrow \quad \dot{W}_{t,\text{ideal}} = \dot{m}_t c_{p,t} T_{03} \left\{ 1 - \left(\frac{p_{04}}{p_{03}}\right)^{\frac{\gamma-1}{\gamma}} \right\} \tag{8.7}$$

Which gives the following expression for the turbine efficiency

$$\eta_t = \frac{\dot{W}_t \dot{W}_{t,\text{ideal}}}{=} \frac{1 - \frac{T_{04}}{T_{03}}}{1 - \left(\frac{p_{04}}{p_{03}}\right)^{\frac{\gamma-1}{\gamma}}} \tag{8.8}$$

Here it must be noted that the turbine often operates at high temperatures and there is significant heat transfer from it. This heat transfer is not accounted for in the control volume analysis (see Figure 8.4) and therefore (8.8) overestimates the turbine efficiency. How this is handled will be discussed in detail in Section 8.4.5.

In the same way as for the compressor, the equations for generated power and out flowing temperature of the turbine can be collected and give the following model.

Model 8.2 Generic Turbine Model

$$\dot{m}_t = f_{\dot{m},t}(p_{03}, p_{04}, T_{03}, \omega_t) \tag{8.9a}$$

$$\eta_t = f_{\eta,t}(p_{03}, p_{04}, T_{03}, \omega_t) \tag{8.9b}$$

$$T_t = T_{03} - \eta_t T_{03} \left\{ 1 - \left(\frac{p_{04}}{p_{03}}\right)^{\frac{\gamma-1}{\gamma}} \right\} \tag{8.9c}$$

$$\dot{W}_t = \dot{m}_t c_{p,t} (T_{03} - T_t) \tag{8.9d}$$

8.2.4 Connecting the Turbine and Compressor

Turbine and compressor power are connected through the mechanical shaft where there also are some losses, which are defined as the mechanical efficiency η_m. At steady state the connection between compressor and turbine power is

$$\dot{W}_c = \eta_m \, \dot{W}_t \tag{8.10}$$

During transients there is an imbalance between produced and consumed power which gives an acceleration (or deceleration) of the shaft. This dynamic response is modeled using the Newton's second law

$$\frac{d\omega_{tc}}{dt} = \frac{1}{J_{tc}} \left(\frac{\dot{W}_t}{\omega_{tc}} \eta_m - \frac{\dot{W}_c}{\omega_{tc}} \right) \tag{8.11}$$

where ω_{tc} is the angular velocity and J_{tc} is the inertia. It is of practical importance to consider that this model is singular when $\omega_{tc} = 0$. This can be circumvented in most simulation environments by specifying a lower limit on the turbine speed $\omega_{tc,\,min}$, which is not a big loss of generality since the turbocharger only gives a significant influence when it is rotating at higher speeds.

Another frequently used approach is to exchange the mechanical efficiency and instead add a term for the shaft friction.

Model 8.3 Turbo Shaft Dynamics

$$\frac{d\omega_{tc}}{dt} = \frac{1}{J_{tc}} \left(\frac{\dot{W}_t}{\omega_{tc}} - \frac{\dot{W}_c}{\omega_{tc}} - M_{fric}(\omega_{tc}) \right)$$

where $M_{fric}(\omega_{tc})$ is the friction torque, often modeled as a linear or quadratic function in rotational speed.

Mechanical efficiency is sometimes also included in the turbine power in this model. This formulation has been reported to help stabilize the simulation during transients. The reason is that for high turbo speeds there might be uncertainties in the performance maps and the turbocharger speed can rise substantially moving into this region. Adding a friction term with quadratic speed dependence damps such tendencies.

8.2.5 Intake Air Density Increase

The amount of fresh air charge in a cylinder is proportional to the intake pressure p_{im} and inversely proportional to the temperature, according to

$$m_{a,c} = \eta_{vol} \frac{p_{im} \, V_d}{R \, T_{im}}$$

The compression in the compressor increases both pressure and temperature, and an efficient process has a lower temperature increase for the same pressure increase. The isentropic process (5.3) gives the lowest temperature increase, that is

$$p_2/p_1 = (T_2/T_1)^{\frac{\gamma}{\gamma-1}} > (T_2/T_1)$$

since $\frac{\gamma}{\gamma-1} > 1$. This leads to a question about what conditions for the compression will result in a density increase, and that question is answered in the following example.

Example 8.1 (Density Increase)   Consider the compression achieved with a compressor that has constant efficiency. Provided a pressure increase, what is the limiting efficiency when there no longer is a density increase?

The criterion for no density increase is $p_2/p_1 = T_2/T_1$ inserting this into (8.4) yields

$$\eta_c = \frac{\left(\frac{p_2}{p_1}\right)^{\frac{\gamma-1}{\gamma}} - 1}{\frac{p_2}{p_1} - 1}$$

which is a monotonously decreasing function in p_2/p_1 with $\lim_{x\to 1+} \eta(x) = \frac{\gamma-1}{\gamma}$. For $\gamma = 1.4$ this shows that if the efficiency is greater than $\frac{\gamma-1}{\gamma} \approx 29\%$ there will be a density increase for all pressure ratios.

For the cases of interest for turbocharging, the efficiency is greater than 30% which will result in a density increase.

The Effect of an Intercooler

Charge cooling with a heat exchanger (called *aftercooler* or *intercooler*) after compression, prior to entry to the cylinder, can be used to further increase the intake density. There are several benefits of adding an intercooler, for example decreased knock tendency (Section 6.2.2) in SI engines, lower NO emissions, and so on. In SI engines, the most important benefit of the intercooler is that it decreases the knock tendency and makes it possible to either charge more air or use a higher compression ratio without destroying the engine by knock, see Section 6.2.2. There are also drawbacks, such as it adds a pressure drop to the intake system (see Figure 7.6) and it also needs space in the engine compartment. However, the benefits overcome these drawbacks and an intercooler is installed on more or less all automotive turbocharged engines.

The main benefit is the density increase, which has the positive side effect that it reduces the compressor power requirement, for the same requested engine torque (or air mass to the cylinder). With an intercooler after the compressor there will be less mechanical power required for compression to a desired density. In a turbocharged engine this means a reduced turbine power requirement, which in its turn leads to less engine back pressure and thus less pumping work.

8.3   Dimensional Analysis

Dimensional analysis is a formal procedure where a set of operating variables is reduced to a smaller set of dimensionless groups. These are, for example, used to evaluate turbomachine performance and as aid in the design process. However, the most important usage is that they give valuable insight into the dominating effects and important parameters of operation. This insight is used both when turbocharger performance is determined experimentally and when compressors and turbines are modeled. In fluid mechanics, the set of dimensionless numbers received depends on the assumption of whether or not the fluid is incompressible or compressible, see, for example, Massey (1998). For centrifugal compressors and radial turbines, the compressible numbers are most important. For more information see dedicated books on turbo machinery, such as Dixon (1998), Lewis (1996), and Watson and Janota (1982).

8.3.1 Compressible Fluid Analysis

When analyzing turbomachine performance, we look at how variables like isentropic *stagnation enthalpy* change Δh_{0s}, efficiency η, and *power* $P = \dot{W}$ depend on the operating variables. Let the machine have diameter D and an operating condition defined by the inlet stagnation pressure p_{01}, inlet stagnation temperature T_{01}, mass flow rate $\dot{m}$, and rotational speed N. Fluid properties are ideal gas constant R, viscosity μ, and ratio of specific heats γ. In the analysis the stagnation *inlet density*, $\rho_{01} = p_{01}/R\,T_{01}$, and stagnation *speed of sound* $a_{01} = \sqrt{\gamma\,R\,T_{01}}$ are often used instead of p_{01}, and T_{01}. For the normal operating range of compressors and turbines, the efficiency, flow, and power may be represented by the following functional relations

$$[\eta, \dot{m}, P] = f(\Delta h_{0s}, N, D, \rho_{01}, a_{01}, \gamma, \mu) \tag{8.12}$$

where $f(\cdot)$ is a 3-dimensional vector function (used for compact notation).

A dimensional analysis for a turbomachine, with a compressible fluid and with ρ_{01}, N, and D selected as common factors, yields the following functions and dimensionless groups

$$\left[\eta, \frac{\dot{m}}{\rho_{01}\,N\,D^3}, \frac{P}{\rho_{01}\,N^3\,D^5}\right] = f(\frac{\Delta h_{0s}}{N^2\,D^2}, \frac{N\,D}{a_{01}}, \frac{\rho_{01}\,N\,D^2}{\mu}, \gamma) \tag{8.13a}$$

$$\Phi = \frac{\dot{m}}{\rho_{01}\,N\,D^3}, \quad \hat{P} = \frac{P}{\rho_{01}\,N^3\,D^5}, \quad \Psi = \frac{\Delta h_{0s}}{N^2\,D^2} \tag{8.13b}$$

These are named *flow coefficient*, Φ, *power coefficient*, $\hat{P}$, and *energy transfer coefficient*, Ψ. On the right-hand side the second group is a *blade Mach number*, $Ma = \frac{N\,D}{a_{01}}$, since $N\,D$ is proportional to the blade speed, while the third is a form of *Reynolds number*, $Re = \frac{\rho_{01}\,N\,D^2}{\mu}$. It is worth noting that different definitions of the dimensionless quantities Φ, Ψ, Re, and Ma, can be encountered. For example, the energy transfer coefficient $\Psi = \frac{\Delta h_{0s}}{N^2\,D^2}$ is sometimes defined as $\Psi = \frac{\Delta h_{0s}}{\frac{1}{2}U_2^2}$, where the blade tip speed U_2 is used in the denominator. The three equations in (8.13a) are not fully independent, since the third group can be expressed in terms of the others. This is seen by expanding the power production in the power coefficient, which for a compressor is $P = \dot{m}\,\Delta h_{0s}/\eta$ and yields

$$\hat{P} = \frac{P}{\rho_{01}\,N^3\,D^5} = \frac{\dot{m}\,\Delta h_{0s}/\eta}{\rho_{01}\,N^3\,D^5} = \frac{\dot{m}}{\rho_{01}\,N\,D^3}\,\frac{\Delta h_{0s}}{N^2\,D^2}\,\frac{1}{\eta} = \Phi\,\Psi/\eta$$

Similarly for a turbine, we get $\hat{P} = \Phi\,\Psi\,\eta$. If we assume a perfect gas, the state equation and other thermodynamic relations can be used to eliminate ρ_{01} and a_{01} and give expressions that are expressed in pressure, temperature, and ideal gas constant. Groups of dimensionless numbers can be combined in to new groups and alternative expressions can thus be developed for (8.13a), and the following combinations give a set of variables that are often used

$$\Phi\,Ma = \frac{\dot{m}}{\rho_{01}\,a_{01}\,D^2} = \frac{\dot{m}\,R\,T_{01}}{p_{02}\,\sqrt{\gamma\,R\,T_{01}}\,D^2} = \frac{1}{\sqrt{\gamma}}\,\frac{\dot{m}\sqrt{R\,T_{01}}}{p_{01}\,D^2} \tag{8.14}$$

$$\hat{P}\,Ma^2 = \frac{P}{\rho_{01}\,N\,D^3\,a_{01}^2} = \frac{\dot{m}\,c_p\Delta T}{\rho_{01}\,N\,D^3\,\gamma\,R\,T_{01}} = \frac{\Phi}{\gamma - 1}\frac{\Delta T}{T_{01}} \tag{8.15}$$

$$\frac{\Psi}{(Ma)^2} = \frac{\Delta h_{0s}}{a_{01}^2} = \frac{c_p\,T_{01}\left[\left(\frac{p_{02}}{p_{01}}\right)^{\frac{\gamma-1}{\gamma}} - 1\right]}{\gamma\,R\,T_{01}} = \frac{\left[\left(\frac{p_{02}}{p_{01}}\right)^{\frac{\gamma-1}{\gamma}} - 1\right]}{\gamma - 1} = f\left(\frac{p_{02}}{p_{01}}, \gamma\right) \tag{8.16}$$

Utilizing the fact that γ is dimensionless, enables us to extract a new set of dimensionless numbers and the corresponding relations

$$\left[\eta, \frac{\dot{m}\,\sqrt{R\,T_{01}}}{D^2\,p_{01}}, \frac{\Delta T_0}{T_{01}}\right] = f\left(\frac{p_{02}}{p_{01}}, \frac{N\,D}{\sqrt{R\,T_{01}}}, Re, \gamma\right)$$

The second group on the left side is often used and it is termed the *flow capacity*.

Corrected Parameters

When representing the performance of a particular compressor or turbine, a set of simplifications are made. The influence of the Reynolds number is usually considered small, so it is often disregarded. In a given machine, with given size (D) and fluid, one also often neglects the variations in R, γ, and Re, which gives the following functional expressions

$$\left[\eta, \frac{\dot{m}\,\sqrt{T_{01}}}{p_{01}}, \frac{\Delta T_0}{T_{01}}\right] = f\left(\frac{p_{02}}{p_{01}}, \frac{N}{\sqrt{T_{01}}}\right) \tag{8.17}$$

Note that the independent variables in the last expression are not dimensionless and these quantities are named *corrected mass flow* and *corrected speed*

$$\dot{m}_{co} = \frac{\dot{m}\,\sqrt{T_{01}}}{p_{01}} \qquad \text{and} \qquad N_{co} = \frac{N}{\sqrt{T_{01}}} \tag{8.18}$$

When using these for turbines, they are sometimes called *turbine flow parameter* (TFP) and *turbine speed parameter* (TSP). Furthermore, these quantities are sometimes normalized, with a reference condition specified by T_r and p_r, which gives an alternative definition for the *corrected mass flow* and *speed*, where they are defined as

$$\dot{m}_{co} = \dot{m}\frac{\sqrt{T_{01}/T_r}}{(p_{01}/p_r)} \qquad \text{and} \qquad N_{co} = \frac{N}{\sqrt{T_{01}/T_r}} \tag{8.19}$$

We have now reached the point where we have the expressions for how compressor and turbine data are represented in the maps, compare Figures 8.5 and 8.3.

There are thus two different definitions of the same variable and this implies that performance data received from the manufacturer must be analyzed carefully since some manufacturers use the first representation while others ues the second. Finally, it is worth remarking that this normalization is important to take into account when representing the performance, since it handles changes in inlet conditions, that is changes in temperature and pressure when driving in hot or cold weather or at high or low altitudes. This is especially important for the turbine since it can have significant variations in inlet pressure and temperature when the operating conditions of the engine change.

8.3.2   Model Structure with Corrected Quantities

We know that there exist relations between the performance variables, and dimensional analysis gives us hints about the structure. This knowledge can be used when developing mean value models for turbochargers, for example (8.17) is formulated so that the inputs and outputs match those of the mean value engine models. Utilizing this for extracting flow and efficiency as outputs, while providing speed and pressures as inputs, gives the following structure

$$\dot{m}_{co} = f_1(\frac{p_{02}}{p_{01}}, N_{co}) \quad \text{and} \quad \eta = f_2(\frac{p_{02}}{p_{01}}, N_{co}) \tag{8.20}$$

Based on this structure, the generic models for the compressor flow (8.5a) and efficiency (8.5b) can be expressed using the corrected quantities, that is either (8.18) or (8.19). Using the second version of the corrected quantities gives the following model.

Model 8.4   Compressor Model with Corrected Quantities

$$\dot{m}_c = f_{\dot{m},c}(p_{01}, p_{02}, T_{01}, \omega_c) = \frac{(p_{01}/p_r)}{\sqrt{T_{01}/T_r}} f_1(\frac{p_{02}}{p_{01}}, \frac{N}{\sqrt{T_{01}/T_r}}) \tag{8.21a}$$

$$\eta_c = f_{\eta,c}(p_{01}, p_{02}, T_{01}, \omega_c) = f_2(\frac{p_{02}}{p_{01}}, \frac{N}{\sqrt{T_{01}/T_r}}) \tag{8.21b}$$

Model structures for turbine flow (8.9a) and efficiency (8.9b) can be expressed analogously with the corrected quantities. When compressor and turbine performance are determined, using gas stand test cell measurements, the tests can focus on determining the two functions f_1 and f_2 in (8.21), which has the benefit that two-dimensional functions instead of four-dimensional functions need to be determined.

8.4   Compressor and Turbine Performance Maps

Turbocharger performance is usually represented using maps that connect the speed and pressure over the device to the flow and efficiency, see Figure 8.5 for examples of typical compressor and turbine maps. These maps are graphical representations of (8.20) and it is seen in the figure that corrected quantities are used for speed and mass flow. Performance maps are usually supplied by the manufacturer and they are determined in a flow test bench, called *gas stand*, where flows, pressures, temperatures, and turbocharger speed are measured for several operating conditions. The test stand and the procedure for data acquisition will be outlined in Section 8.4.3. After the data acquisition, quantities like speed and flow are corrected for inlet air density variations to the corrected flow $\dot{m}_{c,co}$ and corrected speed N_{co}. When using the maps it is important to be careful about what version of the corrected quantities are used, and in Figure 8.5 the version (8.19) with reference state is used. Therefore the maps contain information about the reference conditions.

8.4.1   The Basic Compressor Map

Compressor performance is represented by the compressor map, see Figure 8.5a. It has corrected mass flow of the compressor on the x-axis, and pressure ratio $\Pi_c = \frac{p_{02}}{p_{01}}$ over the

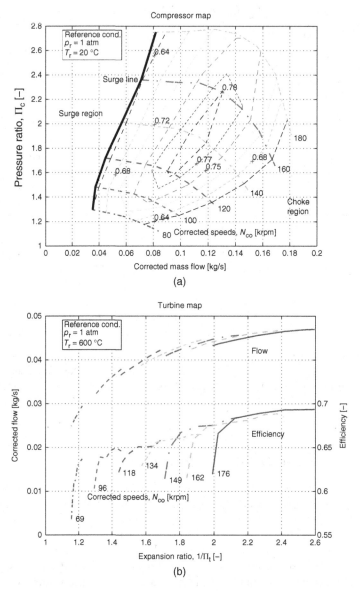

Figure 8.5 (a) Standard compressor map. The speed lines (solid) show how the pressure ratio and mass flow are connected for constant corrected speeds, and the speed is here given in 1000 rpm. The lines of constant efficiency (dashed) show where the compressor is most effective. The surge line (thick solid) shows the stability limit of the compressor, for high flows choking can occur in the compressor. (b) The standard turbine map has two y-axes that show how the expansion ratio $\frac{1}{\Pi_t}$ is connected to turbine mass flow respectively turbine efficiency for different corrected speeds. The solid lines are speed lines

compressor on the y-axis. In the map, lines of constant speed and lines of constant efficiency are plotted. The lines of constant corrected speed are often called *speed lines,* and they describe the pressure and flow characteristics of the compressor. The speed lines are thus a graphical representation of the left equation in (8.20).

The dashed lines in the compressor map in Figure 8.5 are lines of constant efficiency, and thus show how the efficiency varies with corrected mass flow and pressure ratio. The efficiency map is a graphical representation of the equation to the right in (8.20). In the compressor map in the figure, the reference conditions were set to $p_r = 1$ atm and $T_r = 20°C$.

Limits on Compressor Operation

There are four limits on the compressor operation that can be seen in the compressor map in Figure 8.5. The first two are dangerous and can destroy the compressor.

Maximum speed   Rotational speed needs to be limited, since a too high speed can cause mechanical damage due to high centrifugal forces. This limit is the maximum speed line in the upper right part of the compressor map.

Surge   The surge line, to the left of the compressor map, shows the border of a region with unstable operation of the compressor flow. A too high pressure ratio over the compressor causes the flow in the compressor to reverse, which limits the pressure that can be generated by the compressor. The flow reversals can induce imbalanced loading on the compressor and turbo shaft, which can cause the blades to hit the casing or destroy the bearings. More information about surge is given in Section 8.6.5.

Choking   Maximum flow is limited due to choking (sonic conditions) in some components in the compressor. This occurs in the lower right corner of the compressor map and can also occur in the restriction region.

Restriction   When the pressure ratio is below 1, at the bottom of the compressor map, the compressor acts as a flow restriction. This can, for example, occur either early in transients when the turbo speed is low and the engine pumps air out of the intake, or if the compressor is used in a multistage arrangement when an upstream compressor is doing the compression.

8.4.2   The Basic Turbine Map

Turbine performance is often represented by a turbine map, shown in Figure 8.5b. The turbine map has expansion ratio $\frac{1}{\Pi_t} = \frac{p_{03}}{p_{04}}$ on the x-axis, and it has double y-axes, one for the corrected flow and one for the efficiency. Turbine flow is thus specified using the expansion ratio over the device on the x-axis and the flow on the y-axis. Added to this, the variations for different corrected speeds are also shown by a set of lines where each has a constant corrected speed. One can already here note that the speed has only a small influence on the pressure ratio and flow characteristics for a turbine.

The lower lines, in the turbine performance plot, describe the turbine efficiency with the pressure ratio over the device on the x-axis and the efficiency on the y-axis. Turbine performance is presented using corrected quantities where the reference conditions were set to $p_r = 1$ atm, and $T_r = 600°C$ in Figure 8.5. As for the compressor, the turbocharger map can be seen as a graphical representation of the flow and efficiency characteristics (8.20).

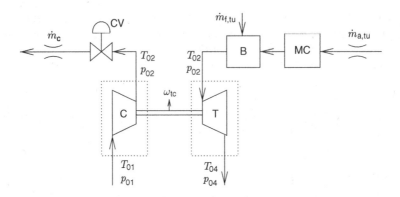

Figure 8.6 Sketch of the setup, used for determining compressor and turbine performance. The mass flows, pressures, and temperatures that are indicated in the sketch are measured. The mechanical compressor, MC, pumps air that drives the turbine. Burner, B, increases the enthalpy and temperature to the turbine, and is positioned after the mechanical compressor. A control valve, CV, after the compressor, controls the compressor pressure p_{02} and flow $\dot{m}_c$

8.4.3 Measurement Procedures for determining Turbo Maps

Compressor and turbine performances are commonly determined together by mounting the turbo in a *gas stand*, see Figure 8.6. The flows to the turbine and compressor are controlled independently, and the quantities shown are measured and used to determine the performance. Three actuators are used to control the operating condition of the turbo: on the compressor side a control valve CV, and on the turbine side a mechanical compressor MC and a burner B. The gas stand is used to determine a sequence of speed lines, and the actuators are used to control the state of the turbocharger. Procedures for measuring and treating the data when determining turbo performance can be found in several standard documents, for example ASME (1997), Chapman and Shultz (2003), *SAE J1826 –Turbocharger Gas Stand Test Code* (1995), and *SAE J922 –Turbocharger Nomenclature and Terminology* (1995). These documents cover, with varying detail, the setup and measurement precautions as well as treatment and presentation of data.

Most maps are measured around the compressor or turbine *design point*, which is the point (or area) where the compressor is designed to give best efficiency. Little data is usually available for low speeds or low pressure ratios, the low efficiency areas, this selection is made since most engine designers are interested in the maximum power or torque from the engine. The measured efficiency approaches zero as the operating conditions approach $\Pi_c = 1, \dot{m}_c = 0$ and the turbocharger approaches zero speed.

Outline of the Procedure

The procedure for determining the compressor's performance can be achieved by considering the compressor map with its speed lines in Figure 8.5. First of all, the speed range of interest is determined and a set of speed lines are set up to be measured. For each speed line the minimum mass flow at the surge line as well as a maximum flow are determined. For the maximum flow there are two options, one is to select the point where the compressor control valve (CV) is fully open and the other is to select the point where the efficiency has dropped below a certain limit (e.g., 58% has been selected by some manufacturers). This line is sometimes called the *choke line*, even though the compressor might still have a margin to choke. With the mass flow limits given, a number of mass flow points are spread out evenly on the speed line between the points.

Turbine inlet temperature is also maintained as constant so that the turbine inlet is close to the reference temperature, which ensures that the corrected speed for turbine is also constant. There are thus three controlled variables, and they are paired in the following way with the three actuators:

- compressor mass flow is controlled by the control valve CV,
- turbo speed is controlled by the mechanical compressor MC, and
- turbine inlet temperature is controlled by the burner B.

The latter two are coupled to each other, but they can be controlled using two cascaded controllers: temperature by the burner in an inner loop and turbocharger speed by the mechanical compressor in an outer loop. Before a measurement can be performed, the system must be stabilized thermally, which can take several minutes and makes the procedure costly in terms of experimental time.

Data Reduction Benefits with Dimensionless Numbers

A major benefit of dimensionless numbers is the reduction of expensive measurements needed for determining turbocharger performance. To be able to use the measured performance we want to know how it changes with, for example, the inlet pressure, such as when driving from sea level to higher altitudes. Many of these dependencies are given by the dimensionless numbers and we thus don't need to make measurements for all possible conditions and combinations. The data reduction is illustrated in the following example.

Example 8.2 (Data Reduction with Dimensional Analysis)   Compare the experimental time needed for doing experiments directly for compressor and turbine models with that of using the corrected quantities.

In the compressor and turbine models, that is (8.5a)–(8.5b) and (8.9a)–(8.9b), there are four dimensions. To determine these functions experimentally, using 7 measurements in each dimension, we would need 2401 measurements in total. If each measurement point takes 15 minutes, due to the wait time for the device to stabilize thermally, then this would take about 600 hours or 15 weeks (with a 40 hours work-week). Using the knowledge that the corrected quantities (8.21) describe the behavior, there are only two dimensions to cover, which needs only 49 points, that is 12 hours.

Besides reducing the number of measurement points, the dimensionless numbers are also useful when modeling compressors and turbines, and are the basis for some of the models developed in Sections 8.6.3 and 8.6.2.

8.4.4   Turbo Performance Calculation Details

Details of how turbocharger performance is calculated from measurement data and the assumptions give insight into how the data should be interpreted and used in the modeling. A control volume analysis of the compressor is the basis for the performance calculations and is shown

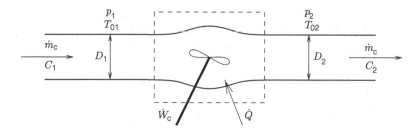

Figure 8.7   Control volume for the open system around a compressor (similar for turbine) that is used to analyze the performance. Pressures and temperatures are measured at inlet and outlet (p_1, T_1, p_2, and T_2) and the diameters of the measurement flanges (D_1 and D_2) are also measured. There is no mass storage and therefore the mass flows in and out are the same $\dot{m}_c$, while the inlet and outlet gas velocities differ and are denoted C_1 and C_2 respectively

in Figure 8.7. The turbine has a similar setting but the direction of the mechanical work is out from the control volume. The sketch shows the quantities that are measured and used in the performance calculations.

A compressor (and turbine) is treated as an open steady flow system, where the influence of potential energy can be neglected. This gives the following expression for the first law

$$\dot{W}_c + \dot{Q} = \dot{m}_c \left((h_2 + \frac{1}{2}C_2^2) - (h_1 + \frac{1}{2}C_1^2) \right)$$

The enthalpy at position i is h_i and the *stagnation* or *total enthalpy* is defined by

$$h_{0i} = h_i + \frac{1}{2}C_i^2$$

which describes the total energy transfer at the inlet. *Stagnation* or *total temperature* as well as *stagnation* or *total pressure* can also be defined. For an ideal gas with constant c_p the following relations are received

$$T_{0i} = T_i + \frac{C_i^2}{2\,c_p} \quad \text{and} \quad p_{0i} = p_i + \frac{\rho_i\,C_i^2}{2} \tag{8.22}$$

where an incompressible fluid is assumed for the pressure. Inlet and outlet total temperatures are measured with probes in the flow, often several around the circumference of the measurement station, and give a reading that is close to the total temperature. Pressures are measured using four or more holes around the circumference of the pipe, that are connected via tubes to one pressure sensor. This gives a measurement of the *static pressure p_i*, where i is one of the four points indicated in Figure 8.6. The assumption of incompressible fluid in (8.22) is often applicable in turbocharger measurements, since the fluid velocity is relatively low at the measurement stations. If there is a need to treat it as a compressible fluid, then the relation $\frac{p_{0i}}{p_i} = \left(\frac{T_{0i}}{T_i} \right)^{\frac{\gamma-1}{1}}$ couples the static and total pressures and temperatures to each other.

Conversion Between Total and Static Quantities

Total temperatures T_{0i} and static pressures p_i are measured and there is thus a need to convert these to static temperatures T_i and total pressures p_{0i}, for example, the total pressure is needed for the efficiency calculations. In the conversion the mass flow $\dot{m}$, cross-section area A (based on the diameters D_i at the measurement stations), velocity $C_i = \frac{\dot{m}_i}{\rho_i A_i}$, and density $\rho_i = \frac{p_i}{R T_i}$ are used. Inserting these into the temperature relation in (8.22) and solving for the static temperature gives

$$T_i = \frac{A^2 p_i^2 c_p}{R^2 \dot{m}_i^2} \left(\sqrt{1 + 2\frac{R^2 \dot{m}_i^2 T_{0i}}{A^2 c_p p_i^2}} - 1 \right) \tag{8.23}$$

where the total temperature T_{0i} and the static pressure p_i are used. Using the pressure relation in (8.22) with the static temperature given by (8.23) and the measured static pressure gives the total pressure

$$p_{0i} = p_i + \frac{R \dot{m}_i^2 T_i}{2 A^2 p_i} \tag{8.24}$$

(8.23) and (8.24) make up a two-step procedure for determining both static temperature and total pressure. If only the total pressure is of interest, then the following equation gives the total pressure directly from the total temperature and static pressure

$$p_{0i} = \left(1 - \frac{c_p}{2 R} \right) p_i + \frac{c_p}{R} \sqrt{p_i^2 + \frac{2 R^2 \dot{m}_i^2 T_{0i}}{c_p A^2}}$$

Total to Total and Total to Static Isentropic Compressor Efficiencies

The compressor efficiency is a measure of how much work an isentropic compression process would require compared to energy the gas gained from the actual compression. The compressor *total to total* isentropic efficiency was derived in Section 8.2.2 and the expression was

$$\eta_{cTT} = \frac{\left(\frac{p_{02}}{p_{01}} \right)^{\frac{\gamma-1}{\gamma}} - 1}{\frac{T_{02}}{T_{01}} - 1} \tag{8.25}$$

which includes the kinetic energy. Another measure of the efficiency is the total to static efficiency. The motive for introducing this efficiency is that it is difficult to recover the kinetic energy (through controlled expansion) so the total pressure can be reached. Therefore, it is often considered more reasonable to use the static pressure after the compressor as a measure of the potential to do work. Compressor *total to static* isentropic efficiency is

$$\eta_{cTS} = \frac{\left(\frac{p_2}{p_{01}} \right)^{\frac{\gamma-1}{\gamma}} - 1}{\frac{T_{02}}{T_{01}} - 1} \tag{8.26}$$

These two efficiencies are compared for two compressors in Figure 8.8, and one can see that there is a difference between them, where a larger flow gives a larger deviation.

It is worth noting that the thermodynamic properties vary with temperature and so does the ratio of specific heats, that is it is a function of temperature $\gamma(T)$ where, for example, at 300 K it has $\gamma = 1.400$ and at 450 K it has $\gamma = 1.392$. Figure 8.8 also shows this influence on the compressor efficiency, by comparing η_{cTT} for one case where inlet temperature is used $\gamma(T_1)$

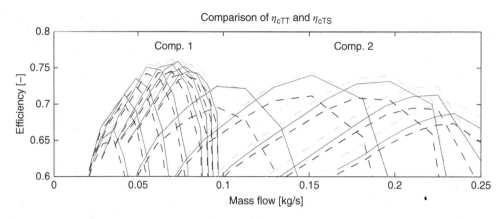

Figure 8.8 Comparison of the different efficiencies η_{cTT} (solid) and η_{cTS} (dashed), for two different compressors. Compressor 2 is bigger but the diameter at the measurement flange is the same, so it has higher speeds which result in a larger difference between total and static pressure. The data shown in dash dotted lines comes from when η_{cTT} is calculated using γ at inlet conditions only

and one case where average temperature is used $\gamma(\frac{T_1+T_2}{2})$. In many cases these variations are neglected in the modeling, but when a detailed analysis is performed attention needs to be paid to these details too.

8.4.5 Heat Transfer and Turbine Efficiency

Heat transfer can be significant in turbochargers, which can have an impact on the analysis of turbocharger performance. For example heat transfer, from the turbine through the housing to the compressor, influences the efficiency calculated from the first law analysis in Figures 8.4 and 8.7 since the analysis neglects the heat transfer effects.

In the compressor, there is often heat transfer from the housing to the gas, which increases the gas temperature in the compressor and thus lowers the calculated compressor efficiency. There can also be heat transfer from the hot gases to the pipes after the compressor, that can lower the temperature and can thus increase the calculated compressor efficiency. It is thus apparent that the heat transfer has an impact on the compressor efficiency that one needs to be aware of, but it is less important in the compressor. However, the coupling between corrected mass flow and pressure ratio has been shown to be fairly insensitive to changing heat transfer conditions, see, for example, Casey and Fesich (2009), Cormerais et al. (2006), Shaaban (2004), and Sirakov and Casey (2011).

In the turbine, the heat transfer is more significant due to its higher gas temperatures, compared to the ambient, housing, and compressor. It therefore produces a significant temperature difference that drives heat transfer, which is lost to the surroundings, the housing with oil and possibly water cooling, as well as the compressor. Turbine efficiency is more influenced than the compressor and special care must therefore be exercised when calculating the turbine performance and the associated efficiency. Using (8.6) directly as a measure of the power generated by the turbine overestimates the power generation, since the heat transfer increases the temperature drop. As a consequence, the calculated efficiency, from (8.8), will be much too high, values reaching over 300% can easily be attained, see Figure 8.9.

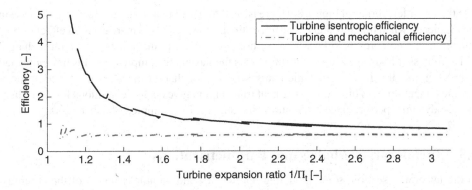

Figure 8.9 Comparison of turbine efficiency calculations with (8.8)–solid line, and (8.27a)–dashed line. Heat transfer from the gas on the turbine side results in high isentropic efficiency, since heat transfer is neglected in (8.8), and the effect increases with lower flows (and expansion ratios)

The core problem with turbine efficiency is that heat transfer is neglected in the first law analysis that is used to determine the turbine power (8.6). Another, better, way to obtain the turbine power is to use the mechanical efficiency (8.10) and the compressor power (8.1). This is then compared to the ideal power from the turbine (8.7) that is determined from the pressure ratio, giving

$$\eta_t = \frac{\dot{W}_c/\eta_m}{\dot{W}_{t,\text{ideal}}}$$

In the next step, the turbine and the turbocharger mechanical efficiency are lumped together. *Total to total* and *total to static* isentropic turbine efficiencies can be defined for the turbine, similar to those of the compressor, giving the following expressions for the lumped process efficiency

$$\eta_{tTT}\,\eta_m = \frac{\dot{m}_c\,c_{p,c}\,(T_{02} - T_{01})}{\dot{m}_t\,c_{p,t}\,T_{03}\left(1 - \left(\frac{p_{04}}{p_{03}}\right)^{\frac{\gamma-1}{\gamma}}\right)} \tag{8.27a}$$

$$\eta_{tTS}\,\eta_m = \frac{\dot{m}_c\,c_{p,c}\,(T_{02} - T_{01})}{\dot{m}_t\,c_{p,t}\,T_{03}\left(1 - \left(\frac{p_4}{p_{03}}\right)^{\frac{\gamma-1}{\gamma}}\right)} \tag{8.27b}$$

Either one of these efficiencies can be reported as the turbine efficiency when the manufacturers provide the data, and one must read the data sheet carefully to see which efficiency is used. Furthermore, one also needs to remember that the turbine efficiency includes the mechanical efficiency.

Figure 8.9 shows a comparison between the turbine efficiency calculations with (8.8) and (8.27a). In the first equation the heat transfer from the gas on the turbine side adds to the energy calculation and results in a high isentropic efficiency, since heat transfer is neglected in (8.8). It is seen in the figure that the effect of heat transfer increases for lower flows (expansion ratios). Using (8.27a) the efficiency is little effected by this heat transfer, which explains why it is the preferred procedure.

In the lower left corner of Figure 8.9 there is, also for (8.27a), a slight trend with an increase in the efficiency when the flow and expansion ratio are lowered. This increase is an effect of heat transfer to the compressor side, where the hot gas on the turbine side heats the turbocharger and compressor housing, which in its turn heats the gas in the compressor. This has motivated research in the direction of adiabatic maps, where turbocharger maps are measured at lower turbine temperatures. This reduces the heat transfer and gives better conditions for describing the aerodynamic performance of the turbocharger, see, for example, Shaaban (2004).

8.5 Turbocharger Models and Parametrizations

In the upcoming section, some modeling approaches and parametrizations of the compressor and turbine will be described. The description of the models will be done at three levels. (1) First a detailed description of the operation of device (compressor or turbine) is given, which allows some submodels for the different processes to be developed. (2) Then a selection of models that have a structure that has been inspired by the physics of the process will be described. (3) Finally, data driven or curve fitting models will be described. In addition, it is also nevessary to address the fact that the measured maps do not cover the whole operating range of the turbocharger, and extrapolation has to be done by the model outside the measured points. Before going into the physics, we start with the most straightforward models that are based purely on measured data and implemented as tables.

8.5.1 Map Interpolation Models

Interpolation of the flow and efficiency maps is a frequently used approach when including turbochargers in simulation models. It is often used since it is easy to apply to data and interpolation routines are straightforward to use. In the MVEM framework, interpolation of data directly fits into the structure of MVEM models, (8.21), where there is one map for corrected flows and one for the efficiency.

When using interpolation there are two things that need to be attended to. The first is to ensure that the interpolating variable data is monotonous so that the interpolation is unique. One example of when this is necessary is seen in the compressor map in Figure 8.5 where the speed line for the highest speed has a slight upward slope when going from the surge line. This means that the pressure ratio does not uniquely determine the corrected flow, and this must be handled. Preprocessing of turbocharger data is often performed, and the data is adjusted slightly so that interpolation becomes unique. The second precaution is to add extrapolation support, since the turbocharger compressor frequently operates outside the regions measured in the maps. The extrapolation is often chosen arbitrarily and linear extrapolation is frequently used, see Kao and Moskwa (1995).

8.6 Compressor Operation and Modeling

In the upcoming section we will describe some different modeling approaches and parametrizations of the turbine and compressor. The first is based on the physics and the others are based on a mix of physics and curve fitting. Models relating the flow to the pressure ratio over the compressor are described in the first section, and then some efficiency models will be described, but first is a short overview of the operation of centrifugal compressors.

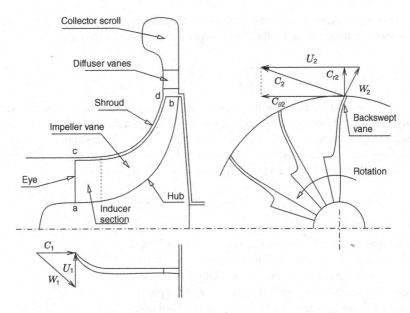

Figure 8.10 Sketch of a compressor with a small backsweep angle. The important parts are named and the velocity triangles at inlet and outlet are sketched. The line c–d is called the shroud and the line a–b is called the hub

Figure 8.10 shows the components in a centrifugal compressor. The gas enters at the eye with velocity C_1, and it has a relative velocity W_1 seen from the rotating compressor wheel. If the blade at the inducer has an angle that perfectly matches W_1 there is no loss of kinetic energy, otherwise there are losses and they are called *incidence losses*. (For the geometry shown in Figure 8.10 the inlet angle $\beta = \arctan(\frac{U_1}{C_1})$, for optimum mass flow, is around 55–$60°$). The impeller transfers mechanical energy to the gas by accelerating it. At the tip it has raised the velocity of the fluid to C_2. At the exit the flow angle can also deviate from the ideal and there the loss is called *slip* and there is an associated *slip angle*. The velocity of the fluid is then decreased through a controlled deceleration, in the diffuser, which converts the kinetic energy to static pressure.

8.6.1 Physical Modeling of a Compressor

Processes that influence compressor work and efficiency will be described in this section. First the minimum amount of work that is required to move a fluid element from one pressure state to another is studied, and then the losses from the compressor shaft to the fluid are discussed.

An adiabatic compression process from a state with pressure p_{01} and T_{01} to another state, with pressure p_{02} consumes the specific energy given by

$$\Delta h_{0s} = c_p \, T_{01} \left[\left(\frac{p_{02}}{p_{01}} \right)^{\frac{\gamma-1}{\gamma}} - 1 \right] \qquad (8.28)$$

If Δh_{0s} is known, for example in terms of compressor operating conditions, such as shaft speed, mass flow rate, friction and so on, then the compressor pressure ratio is given by

$$p_{r,c} = \frac{p_{comp}}{p_{af}} = \left(1 + \frac{\Delta h_{0s}}{c_p T_{01}}\right)^{\frac{\gamma}{\gamma-1}} \tag{8.29}$$

Which gives a connection between the energy and pressure ratio. The rest of this section is thus devoted to describing the specific energy required for isentropic compression, Δh_{0s}.

Δh_{0s} is given by subtracting the losses from the actual specific energy input $h_{0,act}$. The losses are divided into incidence losses, which occur when the in-flowing gas does not have the optimal direction relative the blade, and flow friction losses. That is:

$$\Delta h_{0s} = \Delta h_{0,act} - \Delta h_{loss} \simeq \Delta h_{0,act} - (\Delta h_{fric} + \Delta h_{inc}) \tag{8.30}$$

We start with a calculation of the energy conversion that takes place in the rotor. Figure 8.11 shows a sketch of a rotor spinning with the angular velocity ω. The fluid enters the rotor with the tangential velocity C_{θ_1} and leaves with the tangential velocity C_{θ_2}. The power supplied to the rotor, $\dot{W}$, is given by the product of the torque the rotor exerts on the fluid and the angular velocity, ω, of the rotor. Denoting the torque with M, this becomes.

$$\dot{W} = M\omega$$

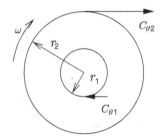

Figure 8.11   Simple rotor that illustrates the entries in the Euler turbine equation

The torque is equal to the time rate of change of the net angular momentum of the fluid. From the figure it is seen that the angular momentum of a mass element, m, entering the rotor is $r_1 m C_{\theta_1}$ and that of a mass m leaving is $r_2 m C_{\theta_2}$. That results in the following

$$M = \frac{d}{dt}(r_2 m C_{\theta_2} - r_1 m C_{\theta_1}) = \dot{m}(r_2 C_{\theta_2} - r_1 C_{\theta_1})$$

where the last equality arise from the fact that for steady flow the velocities C_{θ_1} and C_{θ_2} are independent of time. The expression for the power becomes:

$$\dot{W} = \omega \dot{m}(r_2 C_{\theta_2} - r_1 C_{\theta_1})$$

dividing by $\dot{m}$ gives an expression for the supplied specific energy, $\Delta h_{0,act}$ is obtained:

$$\Delta h_{0,act} = \omega(r_2 C_{\theta_2} - r_1 C_{\theta_1})$$

Suppose that $r_2 C_{\theta_2} \gg r_1 C_{\theta_1}$ so that the latter term can be omitted, then the final expression becomes:

$$\Delta h_{0,act} = \omega r_2 C_{\theta_2} = U_2 C_{\theta_1} \tag{8.31}$$

where ωr_2 has been replaced with the rotor tip speed U_2. This is the general equation for the specific work that has to be supplied to the rotor, the only restriction is the assumption of one-dimensional flow.

Flow Friction Losses

These energy losses can be derived from one-dimensional flow analysis according to Watson and Janota (1982). **Inlet casing loss** may be neglected in small automotive turbochargers with inbord bearings. The air filter will dominate the losses prior to the compressor. If they are significant they can be described by (7.4). **Impeller losses** can occur for several reasons: (1) Surface viscous friction modeled as pipe wall friction. (2) Diffusion and blade loading losses due to boundary layer growth, separation, and mixing. These losses are usually not dominant in automotive turbochargers and are usually lumped together with the surface viscous friction. (3) Blade incidence loss due to off-design angle of the gas at the inducer. (4) Shock losses associated with sonic flow at the inducer entry under high pressure ratio operation. (5) Recirculation losses due to back-flow along the shroud from the impeller tip to the inducer and flow across the blade tip from the pressure to suction side of the vanes due to clearances. (6) Disc friction losses due to shearing of the gas between the back face of the impeller and the adjacent stationary surface. **Diffuser losses** and **collector loss** are also present and the interested reader is referred to Watson and Janota (1982).

To model the friction losses, Δh_{fric}, we assume, as a first approximation, that it is sufficient with a quadratic flow dependence as it is for normal turbulent pipe flow (7.4). The loss due to friction thus becomes.

$$\Delta h_{\text{fric}} = c_1 \dot{m}_c^{c_2} = [\text{choose } c_2 = 2] = c_1 \dot{m}_c^2 \qquad (8.32)$$

Incidence Losses

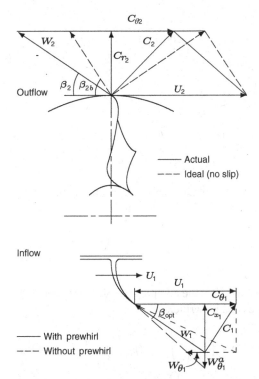

— Actual
--- Ideal (no slip)

— With prewhirl
--- Without prewhirl

Figure 8.12 Velocity triangles for the compressor, showing how the prewhirl and incidence losses changes the velocity components

In the following, a more thorough analysis of the compressor impeller velocity triangles will be made and we will deal with how to characterize the energy losses and the determination of the tangential velocity component, C_{θ_2}, in terms of compressor geometry. Figure 8.12 shows the velocity triangles at the entry (eye) and exit (tip) for the general case of an impeller with a pre-whirl axial inducer and backsweep blades. The notation in the figure is explained below.

The incoming gas arrives at the impeller eye at an absolute velocity C_1 with the axial and tangential components denoted by C_{x_1} and C_{θ_1} respectively. The property U_1 is the tangential component of the blade tip speed and is calculated as:

$$U_1 = r_1 \omega$$

where ω is the angular velocity of the impeller. Since the velocity triangle at the eye hub differs from that at the eye tip, a mean representative value must be considered. This is considered to occur at the radius r_1 that divides the eye into two annuli of equal area.
W_1 is the velocity of the gas relative to the inducer blade. The optimal direction of W_1 would be when this vector has the angle β_{opt} to the tangential velocity C_{θ_1}. In this case the gas does not have to change direction when it reaches the blade. Normally W_1 will deviate from the optimal direction and this will be used later to calculate the incidence losses.

For the case of the exit, we have the equivalent velocity components; C_2 is the absolute velocity, which is divided into a tangential and a radial component C_{θ_2} and C_{r_2}. The tip blade speed is U_2, and W_2 is the speed of the gas as seen from the blade tip. Ideally this velocity component will be directed to the extension of the blade, the angle β_{2b} in the figure, but due to slip the actual direction will differ from β_{2b}.

To calculate the expression for the incidence loss, we use the same assumption as the model for the incidence loss presented in Watson and Janota (1982), this approach is also followed in Müller et al. (1998). The assumption is that the kinetic energy loss is associated with the destruction of the tangential component of W_1, denoted W_{θ_1}, as the fluid adapts to the blade direction; the incidence losses are given by:

$$w_{\text{inc}} = \frac{1}{2} W_{\theta_1}^2 \tag{8.33}$$

Looking at Figure 8.12 and using some geometry gives

$$\frac{W_{\theta_1}^a + C_{x_1}}{U_1 - C_{\theta_1}} = \tan(\beta_{\text{opt}}) \Rightarrow$$

$$\Rightarrow W_{\theta_1}^a = (U_1 - C_{\theta_1}) \tan(\beta_{\text{opt}}) - C_{x_1}$$

furthermore

$$W_{\theta_1} = W_{\theta_1}^a \cot(\beta_{\text{opt}}) = U_1 - C_{\theta_1} - C_{x_1} \cot(\beta_{\text{opt}})$$

hence (8.33) could be formulated as:

$$w_{\text{inc}} = \frac{1}{2}(U_1 - C_{\theta_1} - C_{x_1} \cot(\beta_{\text{opt}}))^2 =$$

$$= \frac{1}{2}(U_1^2 - 2U_1 C_{x_1} \cot(\beta_{\text{opt}}) + C_{x_1}^2 \cot^2(\beta_{\text{opt}}) +$$

$$+ C_{\theta_1}(C_{\theta_1} - 2U_1 + 2C_{x_1} \cot(\beta_{\text{opt}})))$$

Compressors with axial inlet flow have no inlet prewhirl in the ideal case. Assuming that conditions are close to ideal, the inlet prewhirl is set to zero, $C_{\theta_1} = 0$ which implies that $C_{x_1} = C_1$ and C_1 is given from the continuity equation:

$$C_1 = \frac{\dot{m}_c}{\rho_1 A_1} \tag{8.34}$$

where A_1 is the inducer inlet cross-sectional area and ρ_1 is the compressor inlet static density. Using this, the expression for the incidence losses can be written as:

$$w_{\text{inc}} = \frac{1}{2}\left(U_1^2 - 2U_1 \frac{\dot{m}_c}{\rho_1 A_1} \cot(\beta_{\text{opt}}) + \left(\frac{\dot{m}_c}{\rho_1 A_1}\right)^2 \cot^2(\beta_{\text{opt}})\right) \tag{8.35}$$

The actual supplied specific energy given by (8.31) can be rewritten in terms of the radial component of the absolute gas velocity at the tip, C_{r_2}, and the backsweep angle, β_{2b};

$$w_{\text{in,act}} = U_2 C_{\theta_2} = U_2^2\left(1 - \frac{C_{r_2}}{U_2} \cot(\beta_{2b})\right) \tag{8.36}$$

By making the assumption of an impeller with no backsweep, that is $\beta_{2b} = 90°$, this expression reduces to:

$$\Delta h_0 = U_2^2 \tag{8.37}$$

Inserting (8.32), (8.35), and (8.37) into (8.30) and recognizing that U_1 is related to U_2 as D_1 is related to D_2

$$\frac{U_1}{U_2} = \frac{D_1}{D_2}$$

results in the following expression after some algebraic manipulation and regrouping

$$h_{0s} = U_2^2 \left(s_1 \left(\frac{\dot{m}_c}{U_2} \right)^2 + s_2 \left(\frac{\dot{m}_c}{U_2} \right) + s_3 \right) \tag{8.38}$$

where s_i are functions of U_2 and design parameters but in the model they are used as tuning constants. This expression for h_{0s} can be used together with (8.28) to get a description that connects the mass flow, speed, and pressure ratio together.

$$c_p T_{01} \left[\left(\frac{p_{02}}{p_{01}} \right)^{\frac{\gamma-1}{\gamma}} - 1 \right] = U_2^2 \left(s_1 \left(\frac{\dot{m}_c}{U_2} \right)^2 + s_2 \left(\frac{\dot{m}_c}{U_2} \right) + s_3 \right) \tag{8.39}$$

This equation connects the mass flow and pressure ratio to each other and is a second order polynomial in mass flow that can be solved for the mass flow. Furthermore, the compressor efficiency is attained by dividing (8.36) by (8.28). With $\dot{m}_c$ and η_c defined by the equations indicated above, these now fit into the MVEM framework introduced in (8.5a) and (8.5b). There are also other physics based models for centrifugal compressors, see, for example, Gravdahl (1998), Müller et al. (1998), Vigild (2001), Watson and Janota (1982).

8.6.2   Compressor Efficiency Models

Compressor efficiency can be modeled from the physics, which was done implicitly in Section 8.6.1. However, we now turn to models that rely on the dimensionless quantities and use insight from analysis of compressor data. In particular there is a connection between the efficiency η_c and the dimensionless flow parameter

$$\Phi = \frac{\dot{m}_c}{\rho_{01} N D^3} = \frac{\dot{m}_c}{N D^3} \frac{R T_{01}}{p_{01}} \tag{8.40}$$

This connection is shown for two centrifugal compressors of different size in Figure 8.13, where it can be seen that the speed lines are gathered close together. Compare with Figure 8.8 where there is a spread with flow. This indicates that the flow parameter can give a good description of the efficiency. A quadratic function in Φ can be used as a first approximation for the efficiency, when it is sufficient that the model describes the efficiency close to the design speed. Leaving the compressor design point results in a drop in efficiency, and Eriksson (2007) shows that the optimum position changes fairly linearly with speed. To describe the dependence the model needs to include both Φ and speed (or Π_c). A quadratic form in Φ and N_{co} (or Ma) is therefore suggested as a first model for the efficiency.

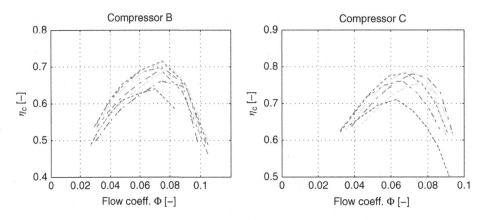

Figure 8.13 Compressor efficiency η_c as function of flow parameter Φ for two different compressors. It can be seen that the flow parameter collects the speed lines close together

---

Model 8.5 Quadratic form Compressor Efficiency

$$\chi(\Phi, N_{co}) = \begin{bmatrix} \Phi - \Phi_{max} \\ N_{co} - N_{co, max} \end{bmatrix} \tag{8.41}$$

$$\eta_c(\chi) = \max(\eta_{c, max} - \chi^T Q_\eta \chi, \, \eta_{c, min}) \tag{8.42}$$

where $Q_\eta \in \Re^{2 \times 2}$ is a symmetric and positive definite matrix. Φ_{max}, $N_{co, max}$, $\eta_{c, max}$, $\eta_{c, min}$ and the elements in Q are tuning parameters.

Note that Q_η must be symmetric to produce a quadratic form and positive definite for it to have a unique maximum for the efficiency. In the model, the efficiency drops when speeds and flows leave the nominal region and a minimum value is therefore $\eta_{min} > 0$ added to (8.42), to ensure that it remains positive.

Variants of this are also often used. Studying the compressor iso-efficiency lines (see e.g., Figure 8.5) it is seen that they look like ellipses and this can be used to describe the compressor map. In Guzzella and Amstutz (1998), a quadratic form in flow and pressure ratio is used to describe the efficiency, i.e. using (8.42) but exchanging (8.41) with

$$\chi(\dot{m}_c, \Pi_c) = [\dot{m}_c - \dot{m}_{c, max} \quad \Pi_c - \Pi_{c, max}]^T \tag{8.43}$$

Looking a little closer at the iso-efficiency lines, compare Figure 8.5, one sees that they are slightly skewed with a changing center. Exchanging Π_c with $\dot{m}_c$ and $\sqrt{\Pi_c - 1}$ gives more elliptic iso-lines, which indicates that the component

$$\chi(\dot{m}_{co}, \Pi_c) = [\dot{m}_{co} - \dot{m}_{co, max} \quad \sqrt{\Pi_c - 1} - \sqrt{\Pi_{c, max} - 1}]^T \tag{8.44}$$

can be used in the quadratic form (8.42), for the compressor efficiency.

Other frequently used modeling options are multiple regression models, for example the following from Sokolov and Glad (1999)

$$\eta_c(\dot{m}_c, N_{tc}) = a_4 + a_5 N_{tc} + a_6 N_{tc}^2 + a_7 \dot{m}_c + a_8 \dot{m}_c^2 + a_9 \dot{m}_c N_{tc} \qquad (8.45)$$

Multiple linear regression models have the advantage that they are easy to tune due to the fact that they can be tuned directly with the normal linear least squares method. Other models might need to be tuned using nonlinear optimization methods for best fit to data.

Remarks

If the compressor will have varying inlet conditions, such as weather and altitude changes, then it is beneficial to use corrected quantities. Most models that don't include such effects can easily be augmented, for example flows and speed in (8.43) and (8.45) can be exchanged for corrected quantities. The efficiency models presented above need the mass flow, or flow parameter, being an output from the compressor model. They thus rely on an accompanying submodel for compressor mass flow, which is the topic of the next section.

8.6.3 Compressor Flow Models

Compressor mass flow is an important output from a compressor model, when integrated in an MVEM. In the models, speed, pressure ratio, and inlet conditions are known and used to calculate the mass flow. One frequently encountered approach is to model the compressor mass flow with the aid of the dimensionless numbers

$$\Psi = \frac{\Delta h_{0s}}{N^2 D^2} = \frac{c_p T_{01}\left(\left(\frac{p_{02}}{p_{01}}\right)^{\frac{\gamma-1}{\gamma}} - 1\right)}{N^2 D^2} \qquad (8.46)$$

$$\Phi = \frac{\dot{m}_c}{\rho_{01} N D^3} = \frac{\dot{m}_c}{N D^3} \frac{R T_{01}}{p_{01}} \qquad (8.47)$$

Dimensional analysis shows that there is a relation between the dimensionless energy Ψ and flow Φ coefficients, see for example Dixon (1998) and Lewis (1996) for a longer discussion. Figure 8.14 shows that the speed lines in the compressor map collapse to almost one single

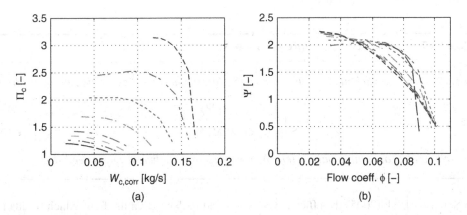

(a)

(b)

Figure 8.14 Dimensional analysis of compressor performance collects the speed lines of the map closer together. (a) Compressor speed lines. (b) Compressible dimensionless numbers

curve in the dimensionless numbers. There is a variation with the operation condition, Mach number (speed), Reynolds number, and so on, as indicated in (8.13).

The following compressor flow model, presented in Jensen et al. (1991), uses a rational function to describe the relation between Φ, Ψ, and Ma. The equations can be formulated such that the compressor mass flow is described by pressure ratio, speed, and compressor intake conditions and the model is defined by the following.

Model 8.6 Jensen et al. Compressor Flow Model

First the energy coefficient (8.46) and the Mach number at the ring orifice of the compressor $Ma = \dfrac{N D}{\sqrt{\gamma R T}}$ are calculated. Then the Mach number polynomials and the quotient for Φ are calculated and the mass flow is finally attained by solving (8.47) for the mass flow

$$k_i(Ma) = k_{i1} + k_{i2}\, Ma + k_{i3}\, Ma^2 \tag{8.48a}$$

$$\Phi = \frac{k_3(Ma)\Psi - k_1(Ma)}{k_2(Ma) + \Psi} \tag{8.48b}$$

$$\dot{m}_c = \Phi\, \rho_{01}\, N D^3 \tag{8.48c}$$

where $k_{i,j}$ are nine tuning parameters that are fitted to the compressor map.

This model is fairly flexible and contains nine parameters that need to be adjusted to map data where a nonlinear least squares problem is solved to find the parameters. Different variants of this approach have been reported, where modifications are made to the expression for the quotient (8.48b) or the orders of the polynomials in (8.48a), see, for example, Jung et al. (2002), Kao and Moskwa (1995), Martin et al. (2009), and Moraal and Kolmanovsky (1999).

Another model, presented in Sorenson et al. (2005), builds on the following observation of experimental data: by dividing the energy coefficient by the efficiency, the data exhibits a linear relation in Φ, this is formulated as

---

Model 8.7 Sorenson et al. Compressor Flow Model

First the energy coefficient is calculated from (8.46) and then the mass flow is determined from

$$\Phi = k_0 - k_1\, \Psi / \eta_c \tag{8.49}$$

$$\dot{m}_c = \Phi\, \rho_{01}\, N D^3 \tag{8.50}$$

where k_1 and k_2 are tuning parameters. A submodel for the efficiency η_c is needed.

In Sorenson et al. (2005) the efficiency was selected to depend on the flow, which results in an implicit function for the flow. This is solved in simulation using a delay of one simulation step, and the flow is fed into the efficiency model one step later.

Another model, presented in Eriksson (2007), builds on the observation that the speed lines in Figure 8.14 are approximated by a quarter of an ellipse. In a first approximation, Ψ and Φ are therefore represented by

$$\left(\frac{\Phi}{k_1}\right)^2 + \left(\frac{\Psi}{k_2}\right)^2 = 1 \qquad (8.51)$$

where k_1 and k_2 are tuning parameters. Formulating the equations such that the compressor flow is described by pressure ratio, speed, and compressor intake conditions gives the following model.

Model 8.8 Basic Ellipse Compressor Flow Model

First the head parameter is calculated from (8.46) and then the mass flow is determined by solving (8.51) for Φ and solving (8.47) for the mass flow in the following way

$$\Phi = k_1 \sqrt{1 - \left(\frac{\min(\max(\Psi, k_2), 0)}{k_2}\right)^2} \qquad (8.52)$$

$$\dot{m}_c = \Phi \, \rho_{01} \, N \, D^3 \qquad (8.53)$$

Some implementation details related to the quarter of the ellipse are included in the first equation. Firstly, if Ψ gets bigger than k_2 then Φ could become imaginary which the max *selector handles. Secondly, if Ψ is smaller than 0, then the flow would decrease which is avoided with the* min *selector.*

The benefits of this model are that it has few parameters (only two) and it is easy to tune and gives a fair description of the nominal operation of the compressor, see Figure 8.16a. A drawbacks is that it does not extrapolate well for pressure ratios below 1 and can thus restrict the flow to the engine for small angular velocities. Extension that handle this has been presented, see for example Leufven (2013) and Leufven Eriksson (2011) where the curvature and center of the ellipse are allowed to change with corrected speed.

Black box models are also frequently encountered, where a generic model structure is used and parameters are fitted to data. Examples of such are multivariate polynomials, neural networks, or other basis functions with tuning parameters, see for example Moraal and Kolmanovsky (1999).

Remark
It is important that the flow models change with inlet conditions, so that they can follow changes in ambient or compressor inlet conditions. It is therefore beneficial if the models for compressor flow use corrected quantities or the dimensionless numbers Ψ and Φ which are selected to scale with inlet conditions.

8.6.4 Compressor Choke

Compressor models for the nominal operating region, where the normal compressor maps are measured, were described in the previous section. Here the attention is on the restriction,

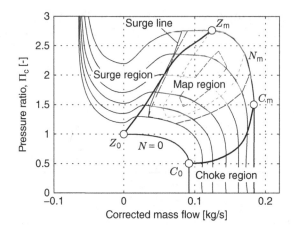

Figure 8.15 Normal compressor map and a sketch of the total operating region. Choke line is marked from C_0 to C_m and zero slope line is marked from Z_0 to Z_m. Speed lines from standstill $N = 0$ to maximum speed N_m

choking and later the surge behavior. These regions are of interest when we are studying dynamic behavior and operation of compressors in multiple stages. For example start up of a compressor or a tip in transient from idling can have pressure ratios over the compressor that are below 1. Figure 8.15 shows a sketch of a compressors full operating region with the normal compressor map, from Figure 8.5 overlaid. The speed lines of the full compressor operation go from standstill, $N = 0$, to full speed N_m. The regions of interest that will be discussed are:

- Normal region: The normal operating region, where the normal map is overlaid.
- Restriction: Operation in the region, $\Pi_c \leq 1$, can occur when another device, engine or another compressor, sucks or pushes gas past the compressor so it acts as a restriction for flow.
- Choking: Conditions where a decrease in pressure ratio does not increase the flow. The border to choke, that is the *choke line*, is marked with the curve from C_0 to C_m, zero speed to maximum speed. This is partly included in the restriction region.
- Surge: Unstable operation with partial of full flow reversal in the compressor.

Compressor Standstill, Restriction, and Choke

Now we turn to the compressor operation at standstill, restriction, and choke. Map data is usually not available in these regions, so this section collects some experimental findings that give a modeler some indications of the expected behavior in these regions. Maximum speed is denoted N_{max} and maximum corrected flow $\dot{m}_{max}$. At *standstill* operation the compressor restricts the flow, in particular as the pressure ratio decreases below unity the mass flow increases until choke is reached, marked with C_0 in Figure 8.15. Experimental data for two compressors, measured in a special gas stand, is presented in Leufven (2013) and it shows that the zero speed line starts at the point $(0, 1)$ and it reaches choke at $(0.5\,\dot{m}_{max}, 0.5)$. The general behavior resembles that of the isentropic restriction (7.10), but with a deviation in curvature between (7.10) and data.

Compressor *restriction* and in particular *choke* is the next topic. A general trend is that the choke border stretches from approximately $(0.5\,\dot{m}_{max}, 0.5)$, to $(\dot{m}_{max}, 1.5)$ and it is a convex curve. The pressure ratio of 1.5 at the maximum speed point C_m is only a rough approximation but it is provided to indicate that compressor choke can occur at fairly high pressure ratios. Some speed lines are observed to bend backwards below choke, but this can often be neglected in the modeling since the compressor operation below choke is very unlikely. For details see Leufven (2013), where five compressors are measured in the restriction and choke region.

A simple model that extends into the restriction region is the following adaptation of the ellipse model. The basic idea is that Ψ approaches a maximum Ψ_{max} at zero flow, see Figure 8.14, thus implicitly describes how the maximum pressure ratio Π_{max} varies with the speed. The other assumption is that the speed lines have an asymptotic point at maximum corrected flow $(\dot{m}_{max}, \Pi_0)$.

Model 8.9 Ellipse Extended to Restriction Region

$$\Pi_{max} = \left(\frac{N^2 D^2 \Psi_{max}}{2 c_p T_{01}} + 1 \right)^{\frac{\gamma}{\gamma-1}} \qquad (8.54a)$$

$$\dot{m}_{c,co} = \dot{m}_{max} \sqrt{ 1 - \left(\frac{\Pi_c - \Pi_0}{\Pi_{max} - \Pi_0} \right)^2 } \qquad (8.54b)$$

where Ψ_{max}, $\dot{m}_{max}$, and Π_0, are model parameters. This flow model now fits into (8.21a) and thus the generic compressor model (8.5a).

This model is compared with compressor data in the right plot of Figure 8.16a.

Model Between Zero Slope and Choke

In the next step, a more advanced model that includes more information about the compressor behavior, will be developed. In most compressor maps, it can be observed that the speed

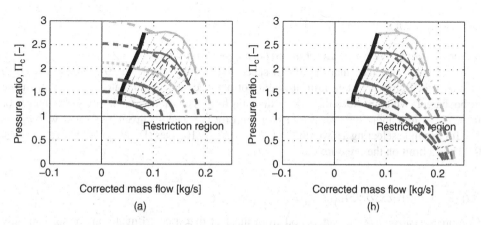

Figure 8.16 Thick dash-dotted lines show model speed lines overlayed on the measured compressor map. (a) The basic ellipse model. (b) Ellipse extended to restriction region

lines have a maximum, that is the point with zero slope. The curve, marked as the Z-line in Figure 8.15, joins the points of zero slope line. It is often used when modeling the compressor behavior, see, for example, Hansen et al. (1981) and Moraal and Kolmanovsky (1999) where the latter names it the zero slope line. Models that describe this behavior have one model or function for the behavior to the right of the Z-line, describing the nominal compressor behavior, and another function to the left, describing surge behavior.

Speed lines have zero slope at the Z-line and infinite slope at choke, that is the C-line. This can be captured by an ellipse, but looking at data it is better if the curvature is allowed to change with speed. For example the higher speeds have speed lines with higher curvature, see, for example Figure 8.16. Using the following modification to the ellipse equation

$$1 = \left(\frac{\Pi_c - \Pi_C}{\Pi_Z - \Pi_C} \right)^{C(N_{co})} + \left(\frac{\dot{m}_{c,co} - \dot{m}_Z}{\dot{m}_C - \dot{m}_Z} \right)^C \tag{8.55}$$

gives a function that allows the center $(\Pi_C, \dot{m}_Z)$ and curvature C to change. A complete model where these are allowed to vary with speed is the following.

Model 8.10 Extended Ellipse for Z to C

$$\Pi_Z(N_{co}) = 1 + c_{1,1} N_{co}^{c_{1,2}} \qquad\qquad \dot{m}_Z(N_{co}) = c_{2,0} + c_{2,1} N_{co} + c_{2,2} N_{co}^2 \tag{8.56a}$$

$$\Pi_C(N_{co}) = c_{3,0} + c_{3,1} N_{co}^{c_{3,2}} \qquad\qquad \dot{m}_C(N_{co}) = c_{4,0} + c_{4,1} N_{co}^{c_{4,2}} \tag{8.56b}$$

$$C(N_{co}) = c_{5,0} + c_{5,1} N_{co} + c_{5,2} N_{co}^2 \tag{8.56c}$$

$$\dot{m}_{c,co} = \dot{m}_Z(N_{co}) + (\dot{m}_C(N_{co}) - \dot{m}_Z(N_{co})) \left(1 - \left(\frac{\Pi_c - \Pi_C(N_{co})}{\Pi_Z(N_{co}) - \Pi_C(N_{co})} \right)^{C(N_{co})} \right)^{\frac{1}{C(N_{co})}} \tag{8.56d}$$

The regions of restriction and choke are studied in detail in Leufven (2013), where this model is applied and further extended. In particular the maximum pressure ratio is shown to be well approximated by the empirical relation (8.56a) with $c_{1,2} = 2.3$. The resulting model is compared to data from an extended compressor map measurement in Figure 8.17, and it gives a good description of the compressor speed line behavior.

A Note on the Lower Speed Limit in Turbocharger Simulation
In Figure 8.17 it is seen that the lowest speed line corresponding to 20 krpm is very close to standstill. Below a certain limit it might not be necessary to simulate the behavior of the turbocharger, which besides the risk of division by 0, is a motive for limiting the integrator in (8.11). A hint about the necessary lower limit for the speed is given by (8.56a) with $c_{1,2} = 2.3$, for example selecting a lower limit of $N_{min} \le N_{max}/10$ gives $(\Pi_c(N_{min}) - 1)/(\Pi_{max} - 1) \le 0.005$ as a limit on the pressure ratio.

8.6.5 Compressor Surge

Compressor *surge* can be understood from the fact that a centrifugal compressor can only generate a pressure increase when there is a flow through it. An increase in pressure ratio over

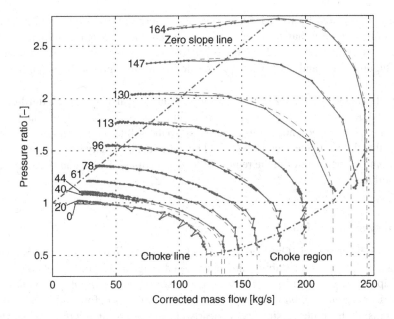

Figure 8.17 Validation of the ellipse model (8.56), dashed lines, that covers the behavior from zero slope line to choke. Measurement data and model from Leufven (2013)

the compressor counteracts the flow, and as the speed lines show a higher pressure ratio this gives a decreased mass flow. If the pressure ratio becomes too high, the compressor cannot maintain the flow necessary for generating the pressure increase and the flow through the compressor breaks down, that is the flow stalls or reverses. In surge, the flow reverses and recovers alternately, which gives an oscillating behavior in the pressures and mass flows. Surge is sometimes also referred to as *compressor pumping*. Furthermore, there are two cases of surge: deep surge and mild surge. In *deep surge* the flow through the compressor is completely reversed until the pressure ratio has decreased sufficiently for the flow to stabilize again. This results in an oscillating mass flow. In *mild surge* the flow oscillates but is not completely reversed.

When entering surge the flow breaks down and there can be an uneven distribution of forces around the compressor wheel, which can give high stress on the compressor blades and turbo bearings. This can in the end cause turbocharger failure, which is why surge is generally avoided. The mechanical vibrations and gas flow pulsations associated with surge give a higher noise and are often a limiting factor when deciding how close to the surge line the compressor is allowed to operate. In this process the engine, compressor, and turbine combination is matched to fulfill the desired surge margin in stationary operation.

Compressor Maps and Surge Modeling

Even though the surge line is marked in the compressor map, it must be pointed out that surge is not a compressor property. Instead it is a system property that depends on both the compressor characteristic and the systems surrounding the compressor, see Greitzer (1981) for the classical axial compressor treatment and Hansen et al. (1981) for an analysis when it is applied to centrifugal compressor. A clear definition of surge does not exist and the surge line is often subjectively determined by the gas stand operator. Therefore, the marked surge line is not a definite limit and surge has to be evaluated in the engine installation. When studying the compressor map, it must be noted that the surge line and the Z-line are not the same. See, for example, Figures 8.15, and 8.17 where the Z-line lies to the right of the surge line for high speeds.

Models often make use of the Z-line, for example Hansen et al. (1981) use a third-order polynomial for the positive flows to the right of the Z-line, a second order polynomial for negative flows, and a third-order polynomial from zero flow to the Z-line. The latter has the parameters selected so that the complete speed-line model becomes continuously differentiable. Note that when the compressor operates in the region with negative mass flow, it can be seen as a poorly designed turbine.

More–Greitzer Surge Model

The More–Greitzer model (Greitzer, 1981) and its variants have become widely used for describing the conditions for stability and operation during surge. The model consists of four model components: *pressure buildup* in the compressor $\hat{p}_c$, *acceleration of mass flow* through the compressor $\dot{m}_c$, *interaction with the pressures* in the volumes before and after the compressor (which mainly becomes p_c since p_{01} is nearly constant), and finally *throttle characteristic* due to its influence on p_c. The model components are illustrated graphically in Figure 8.18, and here the focus will be on the compressor in the dashed box.

An important part is the *compressor pressure buildup*, $\hat{p}$, that depends on mass flow, compressor speed, and inlet conditions. The compressor characteristics have a local minimum at $\Phi = 0$ and a maximum near the surgeline. They are often described using the dimensionless quantities Φ and Ψ, where the classic and most simple models use a third-order polynomial in Φ for Ψ, giving the following model.

Model 8.11   Qubic Pressure Buildup

First the flow parameter Φ is determined using (8.47) which gives Ψ as

$$\Psi = f(\Phi) = c_0 + c_1\Phi + c_2\Phi^2 + c_3\Phi^3$$

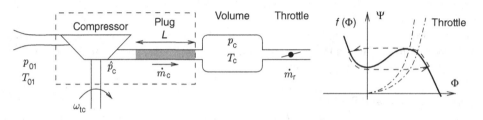

Figure 8.18 Illustration of the components in the More–Greitzer model for compressor surge: compressor pressure buildup, plug with mass, volume and throttle. Solid line: Function for $\Psi = f(\Phi)$, used to describe the compressor pressure buildup $\hat{p}$. Dash dotted: Throttle characteristics for two openings, the one to the right is more open. Dashed – Deep surge cycle

Compressor buildup is given by solving (8.46) for p_{02}

$$\hat{p}_c = f(\dot{m}_c, T_{01}, p_{01}, N) = \left(\frac{1}{2} \frac{\Psi N^2 D^2}{c_p T_{01}} + 1 \right)^{\frac{\gamma}{\gamma-1}} p_{01}$$

Note firstly that this model has the opposite input–output relation compared to the previously introduced models, and secondly that for each flow Φ there is a unique Ψ but not the opposite.

The second part of the model is the *acceleration of mass* in the compressor and surrounding tubes, where the mass is modeled as the gas captured in a pipe of diameter D and length L with a density determined by that in the control volume, that is $m = \rho_c L \frac{\pi D^2}{4}$. The force acting on the mass comes from the pressure difference between that generated by the compressor $\hat{p}_c$ and that in the volume p_c. Newton's second law now gives

$$\rho_c L \frac{\pi D^2}{4} \frac{dV}{dt} = \frac{\pi D^2}{4} (\hat{p}_c - p_c)$$

This equation is now rewritten to describe the acceleration of the mass flow, by noting that the flow velocity in a pipe is $V = \frac{\dot{m}_c}{\rho_c \pi D^2/4}$. Assuming that the density changes slower than the mass flow gives the following differential equation for the mass flow in the compressor.

Model 8.12 Surge Capable Compressor Flow Model

Compressor pressure buildup is described using the compressor characteristic

$$\hat{p}_c = f(\dot{m}_c, T_{01}, p_{01}, N)$$

and compressor mass flow can then be integrated from

$$\frac{d}{dt} \dot{m}_c = \frac{\pi D^2}{4 L} (\hat{p}_c - p_c) \tag{8.57}$$

The mass flow is used in the pressure buildup submodel and is the model output.

To get the full compression system, as in Figure 8.18, the standard models for *control volume* (either (7.27) or (7.24)) and *throttle* (7.11) can be used.

Compressor mass flow is an output (and a state) wherefore the compressor surge model can easily be incorporated as a component in a normal MVEM, since it fits into the generic compressor model. In particular this dynamic model replaces the function (8.5a). A side effect of adding a surge model is that the total model can become stiff, since the state (8.57) contributes with a mode that is fast compared to the turbocharger rotational dynamics. Thus if surge isn't likely to occur in the simulation, or if surge isn't interesting for the problem being studied, then computational time is often gained by omitting surge.

Surge Example

Surge can occur in an SI engine with a throttle when the throttle is closed, see Figure 7.2. A remedy is to add a surge control valve after the compressor, which reduces the pressure after the compressor by leading the mass either directly to the ambient atmosphere or via a return pipe to a position in front of the compressor. The engine control system detects if there is a risk for surge and opens the valve, see Section 10.9.1.

All components are now described and a simple example based on the model above will be used to illustrate the surge phenomenon. The setup is as in Figure 8.18 where the model parameters are set to correspond to those of a normal car engine. Compressor speed is assumed to be constant so that the flow dynamics, which is important for surge, will be clearly visible.

The sequence of events is shown in Figure 8.19 together with the resulting operation in the compressor map. At the starting point, the compressor is operating at a high flow condition with $\dot{m}_c \approx 0.2$ kg/s, this occurs, for example, during an acceleration. At $t = 0.2$ s the throttle is closed, for example at a gear change, which results in a pressure buildup in the volume. When the pressure increases the flow decreases, as they follow along the compressor speed line. At $t = 0.27$ s the forward flow in the compressor can no longer be maintained, due to a too high pressure ratio, and the flow reverses. When the flow is reversed the pressure in the volume decreases since there are now only flows out of the control volume. At $t = 0.28$ s the pressure has decreased sufficiently and forward flow can be reestablished. Now the flow once again increases the pressure and the surge cycle now continues from the beginning.

The surge sequence in the compressor map is shown in Figure 8.19b. The sequence starts down to the right, at the circle. When the throttle is closed the operation follows the speed line, with an increase in pressure and decrease in flow. After a while the operation enters deep surge and the limit cycle of surge is clearly visible. This case shows a deep surge case as the flow is

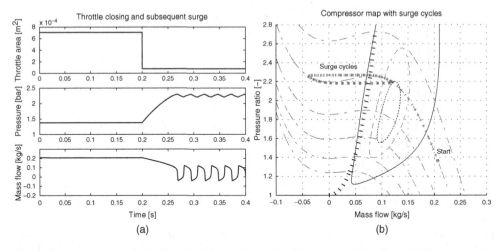

Figure 8.19 (a) A throttle closing, leading to surge. Initially the compressor is operating at a stable point. At $t = 0.2$ s the throttle is closed and the pressure in the system is built up. At $t = 0.27$ s the pressure becomes too high and the flow reverses. When the flow is reversed the pressure decreases, and at $t = 0.28$ s the flow changes direction again and the cycle starts over again. (b) Surge cycles in a compressor map. The thick dotted line shows the border line to where surge occurs. The sequence starts down to the right, at the circle. When the throttle is closed the operation follows the speed line, with an increase in pressure and decrease in flow. After a while the operation enters surge and the limit cycle of surge is clearly visible

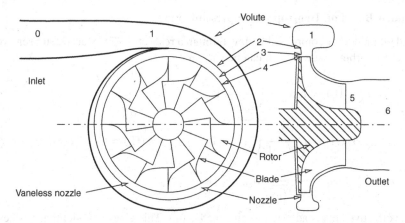

Figure 8.20   Sketch of a radial turbine showing components and the stations that the gas passes. The nozzle in the turbine has no vanes, which is common in many automotive applications. See Figure 8.24 for a turbine with variable vanes

fully reversed. When mild surge occurs the resulting path in the compressor map is a smaller circle close to the surge line.

8.7   Turbine Operation and Modeling

Turbine operation is simply the expansion of a fluid in a controlled manner so that mechanical work can be extracted. To understand the operation of a turbine, we will follow an element through the turbine, see Figure 8.20. Starting at the inlet, 0, the fluid element starts to expand and accelerates through the volute, 1, and through the nozzle, from 2 to 3. The expansion work is thus converted to kinetic energy in the inlet. Then it enters the rotor, 4, and is retarded in the rotor which converts the kinetic energy to mechanical energy. Looking at an element in the rotor, one sees that it is also expanding while it is guided towards the exit and it is thus producing work on the blades. Finally it exits the rotor, 5, and is decelerated in the diffuser as the area expands, 6. With this introduction we will now turn to models for the mass flow and the efficiency.

8.7.1   Turbine Mass Flow

Turbine flow characteristics, Figure 8.5, have a relatively simple appearance and it is fairly easy to get a good description of the turbine flow with simple models. The turbine pressure ratio is defined here as the pressure after divided by the pressure before

$$\Pi_t = \frac{p_{04}}{p_{03}}$$

As mentioned previously, the turbine map is given in terms of the expansion ratio, $1/\Pi_t$ and the turbine corrected mass flow has a strong dependence on pressure ratio and weaker dependence on corrected speed. Therefore, many of the simpler models neglect the speed.

Flow Models Based on Isentropic Compressible Flow

The standard model for compressible flow through a restriction (7.8) has often been proposed as a model for turbine mass flow. It can be written

$$\Pi(\Pi_t) = \max\!\left(\Pi_t, \left(\frac{2}{\gamma+1}\right)^{\frac{\gamma}{\gamma-1}}\right) \tag{8.58a}$$

$$\Psi_0(\Pi) = \sqrt{\frac{2\gamma}{\gamma-1}\left(\Pi^{\frac{2}{\gamma}} - \Pi^{\frac{\gamma+1}{\gamma}}\right)}, \qquad \Psi(\Pi_t) = \Psi_0(\Pi(\Pi_t)) \tag{8.58b}$$

$$\dot{m}_t = \frac{p_{03}}{\sqrt{RT_{03}}} \cdot A_{\mathrm{eff}} \cdot \Psi(\Pi_t) \tag{8.58c}$$

where the effective area is adjusted to give the best fit. This gives a model that reaches choking conditions around $1/\Pi_t \approx 2$, see dash dotted line in Figure 8.21, which is too early. This model does not describe the physics of the flow in a turbine, and many authors have therefore augmented it to fit data, see for example Jensen et al. (1991), Eriksson et al. (2001), Moraal and Kolmanovsky (1999), and Andersson (2005).

In the isentropic restriction there is only one expansion and acceleration from inlet to throat, where the isentropic expansion controls the choking conditions. In a radial turbine, the gases expand in both the stator (inlet, volute, and nozzle) and rotor and this is characterized by the characteristic number, RN, called the degree of reaction. According to Watson and Janota (1982), it is difficult to design a turbine with a degree of reaction far from $RN = 0.5$, which shows that it is possible to increase the flow for expansion ratios up to the order of double the choking limit for a standard throttle. Based on this Eriksson (2007) assumes that $RN = 0.5$, which gives half of the expansion in the stator and the other half in the rotor,

$$\Pi_t = \underbrace{\sqrt{\Pi_t}}_{\text{stator}} \cdot \underbrace{\sqrt{\Pi_t}}_{\text{rotor}}$$

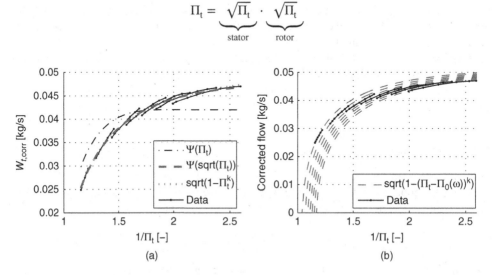

(a)   (b)

Figure 8.21 Comparison between models and data. In both plots the solid lines with dots are measured turbine speed lines. (a) The three models (8.58), (8.59), and (8.60) that do not have a speed dependence speed. (b) Model (8.61) with speed dependence and extrapolation to zero flow

then choking will occur in the stator (or rotor) only when the critical pressure ratio is reached in one of the stages. Based upon this the turbine flow is modeled in the following way.

Model 8.13 Modified Isentropic Restriction for the Turbine

$$\Psi(\sqrt{\Pi_t}) \text{ inserted into (8.58)} \tag{8.59}$$

This model gives an improved agreement with the measured data, see the dashed line in Figure 8.21.

There are also other modifications and a few are summarized here. In Jensen et al. (1991) the model (8.58) is extended with an area A_{eff} that depends on the speed and pressure ratio. The following function is used

$$A_{\text{eff}}(\Pi_t, N_{\text{co}})) = k_1(N_{\text{co}})/\Pi_t + k_2(N_{\text{co}})$$

where the submodels k_i depend on the corrected speed $k_i(N_{\text{co}}) = k_{i,1}N_{\text{co}} + k_{i,2}$. Similarly, Moraal and Kolmanovsky (1999) presents a modification that introduces a pressure dependent term in the Ψ-function, $\dot{m}_{t,\text{co}} = A_{\text{eff}} f(\Pi_t) \Psi(\Pi_t)$ which remedies the too early choke. In Andersson (2005), γ is instead adjusted so that the model gives a good fit to measured data. For the data in Figure 8.21, the best fit is achieved for $\gamma = 4$, which is nonphysical since $\gamma < 1.4$ for exhaust gases. This concludes the modeling where there is a physical motive for the models, and we now turn to curve fitting.

Simple Nonphysical Models

There are simple models that give a good fit to measured data. In Eriksson et al. (2002b) the following two-parameter model is proposed for the corrected turbine flow

---

Model 8.14 Square Root Turbine Flow Model

$$\dot{m}_{t,\text{co}} = k_0 \sqrt{1 - \Pi_t^{k_1}} \tag{8.60}$$

where k_0 and k_1 are tuning parameters, and the latter has been found to lie around $k_1 \approx 2$.

A validation of this model is shown in Figure 8.21 where it is shown to give a good agreement with $k_1 = 2.27$. As is seen in the data, there is a small dependence on speed, which comes from centrifugal forces that oppose the flow direction. If this is needed in the model, it is possible to add a speed dependent change in stagnation pressure ratio Π_0 which gives the following model.

Model 8.15 Square Root Turbine Flow Model with Speed

$$\dot{m}_{t,co} = k_0 \sqrt{1 - (\Pi_t - \Pi_0(N_{co}))^{k_1}} \tag{8.61}$$

where k_0 and k_1 are tuning parameters and the latter has been found to lie around $k_1 \approx 2$.

A comparison between the model and data is shown in Figure 8.21, where it seen that the stagnation pressure ratio (pressure ratio at zero flow) moves with the speed.

8.7.2 Turbine Efficiency

The *blade speed ratio* (BSR) is often used to describe the efficiency of the turbine, and a second- or third-order polynomial in BSR is usually used for the efficiency. The BSR is defined as follows (Watson and Janota, 1982)

$$BSR = \frac{\omega_t\, r_t}{\sqrt{2\, c_p\, T_{03}(1 - \Pi_t^{\frac{\gamma_e-1}{\gamma_e}})}}$$

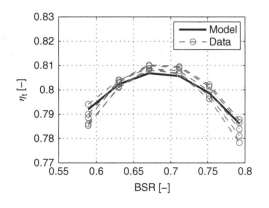

Figure 8.22   BSR and turbine efficiency

The blade speed ratio was originally developed for the impulse turbine where the speed of the fluid is compared to the tip speed of the turbine. If the tip speed is very high, then the turbine wheel moves too fast in comparison with the fluid that enters the turbine wheel. On the other hand, if the turbine speed is too low it can destroy the incoming kinetic energy. In between there is an optimum and this usually occurs for $BSR_{opt} \in [0.6, 0.7]$. A frequently used function for the efficiency is a quadratic function in BSR, see the model below and Figure 8.22.

Model 8.16 Turbine Efficiency with BSR

$$\eta_t(BSR) = \eta_{t,\max} \cdot \left\{ 1 - \left(\frac{BSR - BSR_{opt}}{BSR_{opt}} \right)^2 \right\} \tag{8.62}$$

where $\eta_{t,\max}$ and BSR_{opt} are tuning parameters.

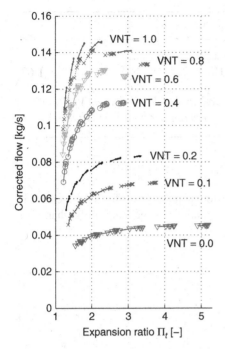

Figure 8.23 Flow characteristics of a VGT turbine, the different vane positions change the flow characteristics of the turbine. (Data from Honeywell)

The turbine inlet can experience significant pulsations from the individual cylinders, while the mean value models treat the average values. To capture the beneficial effects that the pulsations have on the turbine there are sometimes pulse compensation maps added to the efficiency.

8.7.3 Variable Geometry Turbine

Variable geometry turbines (VGT) or variable nozzle turbines (VNT) are used frequently in diesel (CI) engines since they provide the possibility of controlling the turbine while maintaining good efficiency. They would also be beneficial to use on gasoline (SI) engines, but are not utilized much as these have higher exhaust temperatures which the VGT cannot withstand.

In a VGT turbine, they have designed the nozzle so that the effective area of the turbine can be changed using vanes that can be rotated, see Figure 8.24. Changing the effective area also changes the mass flow that can go through the device at a given pressure ratio. To determine the turbine's flow, characteristic measurements are performed for different settings of the nozzle angles, see Figure 8.23. This change in flow characteristic can be modeled by extending the models above with a function (instead of a constant) for the effective area, $A_{\text{eff}}(u_{\text{vgt}})$ where u_{vgt} is the position of the vanes in the nozzle.

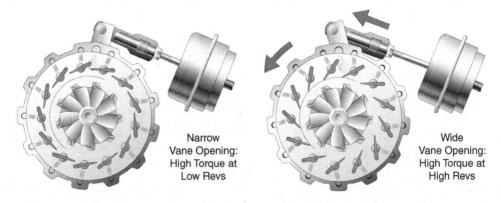

Figure 8.24 Geometry of a variable geometry turbine. The vanes in the turbine nozzle are rotated to guide the flow and change the flow area. Reproduced from Volvo, with permission

---

Model 8.17   Turbine Flow Model with VGT and Speed

$$\dot{m}_{\mathrm{t,co}} = k_0(u_{\mathrm{vgt}}, N_{\mathrm{co}})\sqrt{1 - (\Pi_{\mathrm{t}} - \Pi_0(u_{\mathrm{vgt}}, N_{\mathrm{co}}))^{k_1(u_{\mathrm{vgt}})}} \qquad (8.63)$$

where $k_0(u_{\mathrm{vgt}}, N_{\mathrm{co}})$, $\Pi_0(u_{\mathrm{vgt}}, N_{\mathrm{co}})$ and $k_1(u_{\mathrm{vgt}})$ are submodels that depend on the VGT position and corrected speed, and they can be selected to be polynomials or power functions.

8.8   Transient Response and Turbo Lag

The time lag from when the driver hits the accelerator to when the driver feels the increase in acceleration (i.e., engine torque) is called the turbo lag, and it is larger for a turbo than for a naturally aspirated engine.

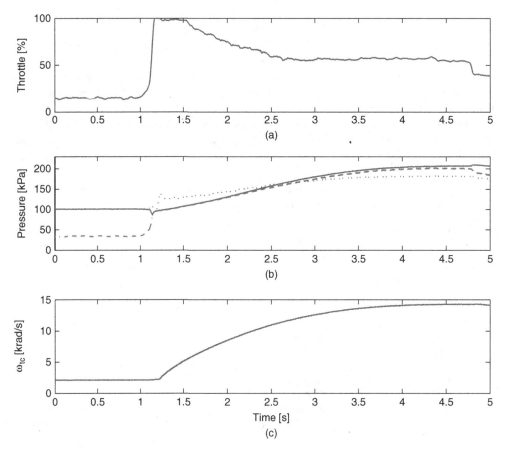

Figure 8.25   A measured tip-in load transient for a turbocharged SI engine. In the transient the engine speed is constant at $N = 2000$ RPM. (a) throttle angle. (b) Intake manifold pressure p_{im} (dash-dotted), pressure after intercooler p_{ic} (solid), and exhaust manifold pressure p_{em} (dotted). (c) Turbocharger speed ω_{tc}

A load transient that exposes the lag for a turbocharged engine is shown in Figure 8.25. During the transient the engine speed is maintained constant at $N = 2000$ RPM and the change in throttle angle is commanded at $t = 1$ s. In the middle plot there are two distinct dynamics visible; a first fast dynamic from $t = 1$ to $t = 1.15$ s and a second slower continuing from the start to $t = 4$ s. In the first phase, the dynamics is dominated by throttle movement dynamics, filling of the intake manifold, and emptying of the volumes in and around the intercooler. These transients in the first phase have a response time similar to those of a naturally aspirated engine. The second dynamics is dominated by the turbocharger speed buildup (shown in the bottom plot), since the mass flow through the compressor depends strongly on the turbocharger speed. Therefore, the turbo lag is mainly due to the slower speed dynamics of the turbocharger.

One way to reduce the time lag is to control the turbo such that the turbo always has a relatively high speed, and then reduce the flow into the intake manifold and the manifold pressure using a throttle. This improves the driveability by reducing the turbo lag, but it also increases the fuel consumption since it increases the exhaust back pressure. Eriksson et al. (2002a) quantified the fuel consumption and transient response trade-off, for a turbocharged SI engine, and show that a transient optimized strategy has a response time that is 0.5 s faster than the fuel consumption optimized, and that the penalty in fuel consumption is about 2% for normal driving conditions. The penalty in fuel consumption from the response time optimized strategy can be higher than 10% at high engine speed and low load conditions.

It is also worth pointing out that $p_{im} > p_{em}$ for $t > 3$ s in Figure 8.25, which means that the gas exchange process produces work on the piston. This favorable condition can occur in high load/low speed operating points. However, in most cases $p_{im} < p_{em}$ and the gas exchange process consumes work, which is the normal case for naturally aspirated engines.

8.9   Example – Turbocharged SI Engine

Now we have finally covered all the components that are necessary for modeling a complete turbocharged engine, and we will compile two examples in the following sections. The first is an SI engine and the other a CI engine with EGR and VGT.

The SI engine model is implemented in Simulink, and the top level diagram is shown in Figure 8.26. A component-based modeling approach has been applied with control volumes in series with restrictions and flow generators (compressor, turbine, and engine). The model consists of:

- Six control volumes for the tubes and other volumes present in the engine, all modeled by (7.27). The components are volumes between air filter–compressor, compressor–intercooler, intercooler–throttle, as well as intake manifold, exhaust manifold, and finally the exhaust pipe between turbine and catalyst.
- Three restrictions for turbulent incompressible flow (7.6). These components are air filter, intercooler, and a lumped model for the pressure drop in the exhaust system, like catalyst and muffler.
- Two compressible isentropic restrictions, (7.8), one for the throttle and one for the wastegate (included in the turbine block).
- The compressor model is based on the dimensionless numbers for the flow, and the efficiency is modeled using a combination of (8.42) and (8.44).
- The intercooler temperature drop is modeled using a regression polynomial similar to (7.70)
- The turbine is modeled using (8.60) for the flow and (8.62) for the efficiency.

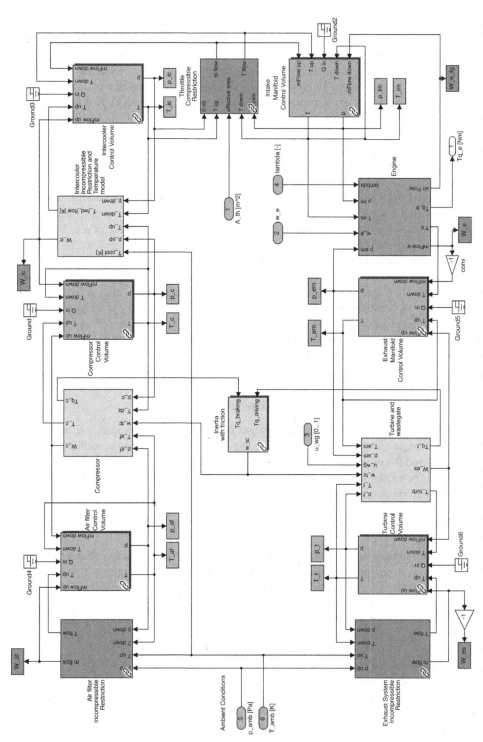

Figure 8.26 Simulink diagram of a turbocharged SI engine. The engine is modeled in a component-based fashion with control volumes in series with restrictions or flow

- The engine flow is parametrized using the volumetric efficiency η_{vol} and the torque from the engine is given by the model described in Section 7.9.

8.10   Example – Turbocharged Diesel Engine

The second example is a model for a turbocharged diesel engine with VGT and EGR. A sketch of the engine configuration is given in Figure 8.27 together with the corresponding implementation in Simulink. A picture of the engine is shown in Figure 8.28 and the full derivation is given in Wahlström and Eriksson (2011a). The CI engine model follows the same modeling principles as the turbocharged SI engine above, and also uses similar model components as the SI engine. To highlight a few points:

- The compressor model is the same as for the SI, it is based upon the dimensionless numbers, for the flow, and the approach illustrated by combining (8.42) and (8.44).
- The turbine is modeled using (8.60) for the flow and (8.62) for the efficiency.
- The engine flow is based on the volumetric efficiency η_{vol}.
- The EGR cooler and intercooler are modeled as perfect coolers so that the outlet temperature is the same as the coolant medium.

Model 8.18   Diesel Engine with EGR VGT

The model is focused on the gas flows, see Figure 8.27, and has five states: intake and exhaust manifold pressures (p_{im} and p_{em}), oxygen mass fraction in the intake and exhaust manifold (X_{Oim} and X_{Oem}), and turbocharger speed (ω_t). These states are collected in a state vector x

$$x = (p_{im} \quad p_{em} \quad X_{Oim} \quad X_{Oem} \quad \omega_t)^T \tag{8.64}$$

The resulting model is expressed in state space form as

$$\dot{x} = f(x, u, N) \tag{8.65}$$

where the engine speed N is an input to the model, and u is the control input vector

$$u = (u_\delta \quad \tilde{u}_{egr} \quad \tilde{u}_{vgt})^T \tag{8.66}$$

which contains mass of injected fuel u_δ, EGR-valve position $\tilde{u}_{egr}$, and VGT actuator position $\tilde{u}_{vgt}$. The EGR-valve is closed when $\tilde{u}_{egr} = 0\%$ and open when $\tilde{u}_{egr} = 100\%$. The VGT is closed when $\tilde{u}_{vgt} = 0\%$ and open when $\tilde{u}_{vgt} = 100\%$.

Manifolds and Turbo Speed

$$\frac{d}{dt} p_{im} = \frac{R_a T_{im}}{V_{im}} (\dot{m}_c + \dot{m}_{egr} - \dot{m}_{ei}), \qquad \frac{d}{dt} p_{em} = \frac{R_e T_{em}}{V_{em}} (\dot{m}_{eo} - \dot{m}_t - \dot{m}_{egr}) \tag{8.67}$$

$$\frac{d}{dt} X_{Oim} = \frac{R_a T_{im}}{p_{im} V_{im}} ((X_{Oem} - X_{Oim}) \dot{m}_{egr} + (X_{Oc} - X_{Oim}) \dot{m}_c) \tag{8.68}$$

$$\frac{d}{dt} X_{Oem} = \frac{R_e T_{em}}{p_{em} V_{em}} (X_{Oe} - X_{Oem}) \dot{m}_{eo}, \qquad \frac{d}{dt} \omega_t = \frac{W_t \eta_m - W_c}{J_t \omega_t} \tag{8.69}$$

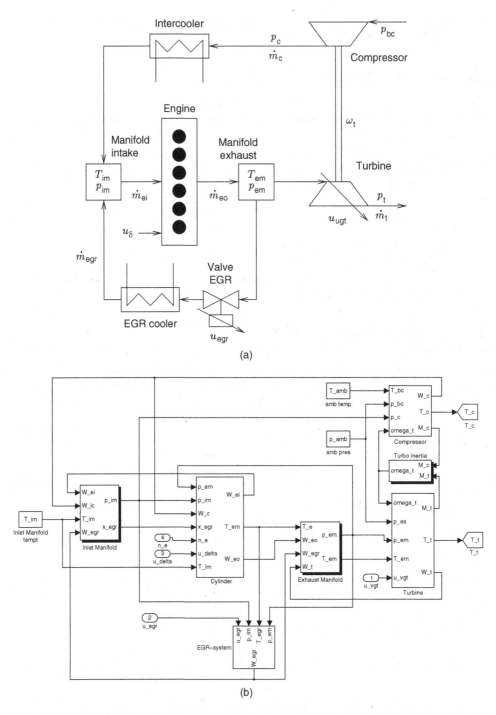

Figure 8.27 (a) Sketch showing the components that are modeled in a turbocharged CI engine with VGT and EGR. (b) A Simulink diagram for a turbocharged diesel engine with VGT and EGR that is implemented in Simulink. (Model from Wahlström and Eriksson (2011))

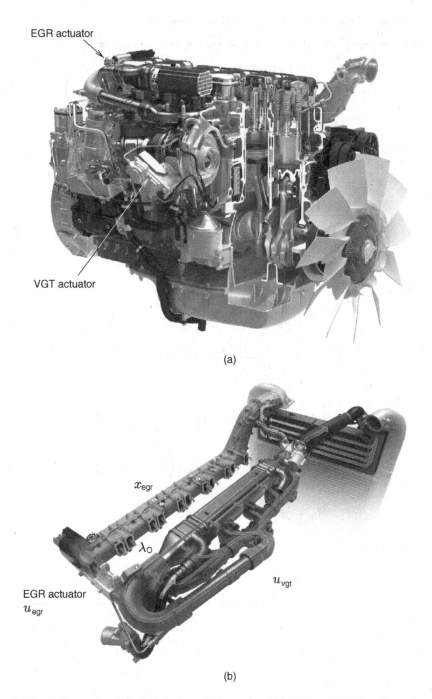

(a)

(b)

Figure 8.28 (a) Illustration of the Scania six-cylinder engine with EGR and VGT that the engine model describes. (b): Illustration of the EGR-system that is modeled in Model 8.18. The performance variables oxygen/fuel ratio λ_O and EGR-fraction x_{egr} used in the diesel EGR VGT control case study in Section 11.5 are also highlighted. Reproduced with permission, images copyright Scania CV AB

Gas Flows into and Out of Cylinders and EGR-System

$$\dot{m}_{ei} = \frac{p_{im} \, N \, V_d}{120 \, R_a \, T_{im}} \eta_{vol}(p_{im}, N), \quad \dot{m}_F = \frac{10^{-6}}{120} \, u_\delta \, N \, n_{cyl}, \quad \dot{m}_{eo} = \dot{m}_F + \dot{m}_{ei} \tag{8.70}$$

$$X_{Oe} = \frac{W_{ei} \, X_{Oim} - W_f \, (O/F)_s}{W_{eo}}, \quad T_{em} = T_{em}\left(\frac{p_{em}}{p_{im}}, \dot{m}_F, \dot{m}_{eo}\right) \tag{8.71}$$

$$\dot{m}_{egr} = \frac{A_{egr}(\tilde{u}_{egr}) \, p_{em} \, \Psi_{egr}\left(\frac{p_{im}}{p_{em}}\right)}{\sqrt{T_{em} \, R_e}} \tag{8.72}$$

Turbine

$$\frac{\dot{m}_t \, \sqrt{T_{em}}}{p_{em}} = A_{vgtmax} \, f_{\Pi t}(\Pi_t) f_{vgt}(\tilde{u}_{vgt}), \quad \Pi_t = \frac{p_{amb}}{p_{em}} \tag{8.73}$$

$$\dot{W}_t \, \eta_m = \eta_{tm}(\omega_t, T_{em}, \Pi_t) \, \dot{m}_t \, c_{pe} \, T_{em}(1 - \Pi_t^{1-1/\gamma_e}) \tag{8.74}$$

Compressor

$$\dot{m}_c = \frac{p_{amb} \, \pi \, R_c^3 \, \omega_t}{R_a \, T_{amb}} \Phi_c(\omega_t, \Pi_c), \quad \Pi_c = \frac{p_{im}}{p_{amb}} \tag{8.75}$$

$$\dot{W}_c = \frac{\dot{m}_c \, c_{pa} \, T_{amb}}{\eta_c(\dot{m}_c, \Pi_c)} (\Pi_c^{1-1/\gamma_a} - 1) \tag{8.76}$$

Engine Torque

$$Me = M_{i,g} - M_{i,p} - M_{fr}, \quad M_{i,p} = \frac{V_d}{4 \, \pi}(p_{em} - p_{im}) \tag{8.77}$$

$$M_{i,g} = \frac{1}{4 \, \pi} \, u_\delta \, 10^{-6} \, n_{cyl} \, q_{HV} \, \eta_{igch} \left(1 - \frac{1}{r_c^{\gamma_{cyl}-1}}\right) \tag{8.78}$$

$$M_{fr} = \frac{V_d}{4 \, \pi} 10^5 \, (c_{fric1} \, N^2 + c_{fric2} \, N + c_{fric3}) \tag{8.79}$$

Performance Variables

$$x_{egr} = \frac{\dot{m}_{egr}}{\dot{m}_c + \dot{m}_{egr}}, \quad \lambda_O = \frac{\dot{m}_{ei} \, X_{Oim}}{\dot{m}_F \, (O/F)_s}, \quad N_t = \omega_t \frac{30}{\pi} \tag{8.80}$$

The model focuses on the gas flows and therefore the inputs are the actuator positions, ($\tilde{u}_{egr}$ and $\tilde{u}_{vgt}$). Actuator dynamics is also important but depends on the current hardware and is omitted above in the gas flow model. For the engine used for modeling and validation, shown in Figure 8.28, the actuator dynamics for EGR and VGT had three additional states and two time delays. A detailed description and derivation of the complete model is given in Wahlström and Eriksson (2011a), together with a tuning methodology and a validation against test cell measurements. The complete model is also available for download from `www.fs.isy.liu.se/Software/`.

9

Engine Management Systems – An Introduction

Engine control systems play an important role in achieving legislated emission limits while at the same time providing a good driving experience for the driver. This is achieved in a dedicated robust controller hardware that goes under several names: Engine Management System (EMS), Engine Control Unit (ECU), or Powertrain Control Module (PCM). The hardware has limited memory and computational power which restricts the functionality that can be implemented, but as with all computers there is a steady increase in performance with each generation. This chapter outlines some general aspects of the EMS hardware and software in Section 9.1 and the basic functionality from sensors and virtual sensors to actuation in Section 9.2. Parameter calibration and storage functionality, in the form of maps, is available for control and calibration engineers and this is touched upon in Section 9.3.

9.1 Engine Management System (EMS)

When engine modeling was introduced, Figure 7.1 gave an engine oriented sketch, with the most important sensors and actuators on the engine. Figure 9.1 looks at it from an EMS-centered point of view with examples of sensors, actuators and internal components in the EMS highlighted. The basic function of the EMS is to process sensor values and deliver control outputs to the actuators in real time, while simultaneously providing hardware protection and monitoring both hardware and software for safety critical engine control functions.

It is already worth pointing out here that safety, monitoring, and diagnosis are crucial components in the EMS and make up a significant portion of the software, for example in Bosch Motronic EMS about half the computing power and memory is dedicated to diagnostic tasks (Dietsche, 2011). These topics will be treated in more detail in Chapter 16.

An EMS can also receive inputs from other sources and send signals to other systems than the engine. The EMS may, for example, for interface directly with an electronically-controlled automatic transmission or communicate with a transmission control unit. In-vehicle networks are used for communication and enable information exchange between different control modules. These often utilize the Controller Area Network (CAN), see lower left of Figure 9.1, bus which is a field-bus that is widely adopted by the automotive industry.

Modeling and Control of Engines and Drivelines, First Edition. Lars Eriksson and Lars Nielsen.
© 2014 John Wiley & Sons, Ltd. Published 2014 by John Wiley & Sons, Ltd.
Companion Website: www.wiley.com/go/powertrain

9.1.1 EMS Building Blocks

An EMS consists of several hardware components, for example one or more microcontrollers (CPU), memory (RAM, Flash, EEPROM), power circuits, communication system (CAN, FlexRay, TCP/IP), sensor input, output drivers, I/O circuitry, status indicators, and diagnostic systems. These are needed to process the information from inputs to outputs. Starting from the inputs, Figure 9.1 shows that there are different types of inputs:

- Analog sensor values, read by analog to digital (A/D) converters.
- Digital inputs from on/off signals, pulses that mark engine events or pulse width modulated (PWM) signals for example an air mass flow sensor. These are either sampled on digital I/O pins or connected so they can generate interrupts in the microcontroller.
- Bi-directional data transfer to and from other units using, for example, CAN bus or FlexRay.

Internally in the EMS there are one or more CPUs that do the majority of the calculations, the program and data is stored in the flash EPROM, enabling the new functions to be downloaded. The program code and data can also be encrypted to protect it from being modified by an unauthorized part. In the EMS there is often also a Timer Unit (TU) that generates wave forms and supports crank angle referenced signals, see Section 9.1.2.

Actuator Hardware Drivers

An EMS also has output drivers for various types of actuators, with different power requirements, converting the low level CPU outputs to high power EMS outputs. The capacity

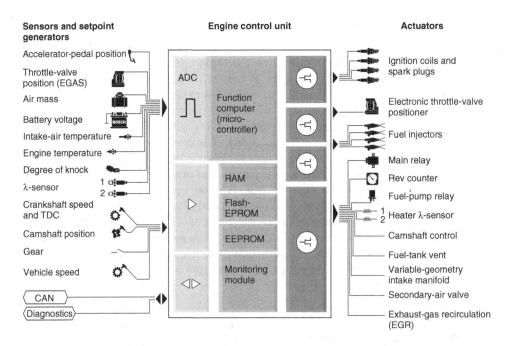

Figure 9.1 Overview of an engine management system for a gasoline engine. *Source:* Bosch Automotive Handbook 8e, Figure 2, Page 535. Reproduced with permission from Robert Bosch GmbH

of the driver stages on-board the EMS depend on the actuator type. Some actuators require less current, like the main relay, heaters for the λ sensors, and so on, while others require more, like the ignition coils, fuel injectors, and the electronic throttle valve motor.

There is also a trend with *intelligent actuators*. For example, the VGT actuator in the heavy duty diesel engine, used in Wahlström and Eriksson (2011b), has a separate control module with servo and integrated power electronics for the actuator. Set-point commands are sent from the EMS to the VGT module, via the CAN bus, which in its turn returns the current position to the EMS.

9.1.2 System for Crank and Time-Based Events

There are events in the EMS that are executed with a fixed frequency, for example sampling of input signals and throttle servo controller, and events that are executed synchronously with the crankshaft rotation, such as ignition and injection. The first are called *time based events* and the latter *crank angle-based events*. In addition, there are also asynchronous events, for example when the driver presses a button. To handle the time-based events there is a scheduler that executes different tasks periodically at a given frequency. Different frequencies, or time bases, are used for different functions depending on their need, where the fastest are executed at several 100 Hz or even higher, while the slowest are at about 1 Hz or even slower.

For crank angle-based events, the engine position is monitored with the crankshaft sensor, shown as *Crankshaft speed and TDC* in Figure 9.1, which has 60 slots (or cogs) whereof 2 are removed, giving it the name 58X wheel. The two missing slots give a reference for TDC positioning. However, they cannot give the engine position completely since an engine cycle is two crank revolutions. Therefore the cam, that rotates at half the crank speed, is used. The sensor marked *Camshaft position* in Figure 9.1 is used to discern between a TDC near combustion or a TDC in the gas exchange phase. The cam phase sensor can also be used to measure the cam phasing if the engine is equipped with a variable cam timing system.

Crank angle events are often managed by a *Timer Unit* (TU) in the EMS, see Figure 9.2. The TU is a co-processor that combines special hardware and low-level microcode to make high-resolution timing measurements and to generate high-resolution pulse signals. It manages the critical timing associated with the EMS I/O and removes a very computation-intensive task from the CPU. There is a bi-directional communication between the CPU and TU, where the CPU can read and write information in the TU and the TU can trigger interrupts in the CPU.

TU tasks are organized in channels where a channel can, for example, be se tup to provide the pulses to a fuel injector, an ignition coil, or a PWM signal. The TU channels process

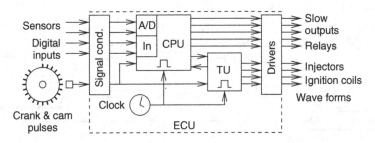

Figure 9.2 A Timer Unit (TU) in the EMS manages the timing of critical pulses and offloads timing of critical tasks from the CPU. It also generates wave form outputs, like injection and ignition pulses, with high precision in timing

cog and timer events and generate I/O signals and interrupts, that is they keep track of the engine position and generate injection and ignition events at the right crank angle positions. In particular, the 58X wheel has a resolution of 6 degrees, but there is a need for controlling the ignition and injection timing with higher accuracy. To meet that requirement, the TU channels can lock on the crankshaft pulses and use the internal timer to increase the resolution, so that ignition and injection pulses can be generated with a resolution of fractions of a degree.

9.2   Basic Functionality and Software Structure

Figure 9.3 shows a calculation sequence used in a rapid prototyping EMS (Backman, 2011) that implements the torque based structure. In the top left corner there is a time scheduler block that generates trigger signals for the underlying blocks, giving execution frequencies of 100 Hz (10 ms) and 400 Hz (2p5 ms) in the prototyping system.

To the left in Figure 9.3 the sensor values enter, together with a selection of the current control outputs. The block *SensorCalc* generates all relevant signals for the EMS, for example raw sensor values, filtered sensor values, as well as non-measured variables that can be calculated from inputs and other measured outputs. Calculated values are often called *virtual sensors* since they produce signals that aren't available as physical sensors, but can be generated using, for example, observers. In the block *Limits* the available information is used to determine the limits on engine torque or exhaust temperature, for example.

9.2.1   Torque Based Structure

Sensor values are combined in the block *TorqueReq* that calculates the requested torque based on the sensor values. This block implements all the functions outlined in the torque based

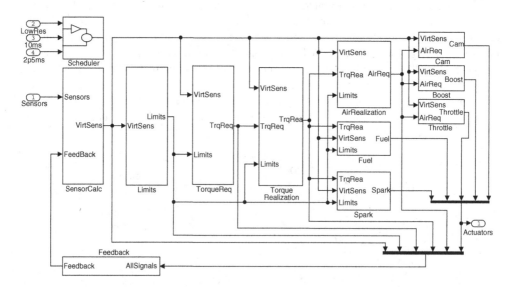

Figure 9.3   Structure of the code of a prototype EMS with torque based structure. Interpretation of driver inputs (and other inputs) into a torque demand is a major part of the powertrain related code

structure of Figures 3.9 to 3.11, generating the outputs' short- and long-term torque. The block includes driver interpretation, vehicle control, and driveline control, and their active limits. The short- and long-term torque are propagated and split in *TorqueRealization,* where essentially the air path (and fuel path) is used to fulfill the long-term demand and the ignition is used to generate the gap between the long- and short-term torque, while the fuel system supports torque interventions which require fuel cut.

Finally, the torque realization (or arbitration) is done by the respective subsystems: air, fuel, and ignition. Here the air system is further divided into the throttle (intake manifold), boosting (turbo), and variable valve actuation (cylinder gas exchange and pumping).

9.2.2   Special Modes and Events

There are also functions and controllers that are executed on demand, for example depending on the operating mode of the engine. Examples are cold start to heat up the catalyst, DPF generation in diesels, low NO_x mode, and high NO_x with SCR generation. Execution of some functions are controlled by the state of the engine and the vehicle, for example closed loop λ-control can only run when the λ-sensor is heated up and operational. Another example is that the diagnosis system might have detected a malfunctioning sensor and then a controller might be disabled, see Section 16.1.3 for a discussion. All in all, a significant engineering job lies in the development and validation of modes, functions, and interactions.

9.2.3   Automatic Code Generation and Information Exchange

Much of the control development, also called *function development*, is done in simulation environments. To increase quality and avoid manual errors there is a drive to generate the EMS code automatically from the simulation environments. This both speeds up the development process and avoids introducing errors in intermediate implementation steps. For example, the controller in Figure 9.3 is compiled and run in a rapid prototyping EMS. A large part of the code in the control loops is generated automatically from the simulation environments, and hardware in the loop simulators are used to validate functionality.

There are also other initiatives for shortening the development time and improving the quality of the generation code and total software. One effort is directed to enabling exchange of code between OEMs and suppliers so that code from a supplier can easily be integrated in different platforms. An example of an initiative is AUTOSAR (AUTomotive Open System ARchitec-ture), which is a development partnership of major automotive industry OEMs and suppliers, as well as tool and software vendors worldwide. The goal of the partnership is to establish a global standard for common software architecture, application interfaces, and methodology of embedded software for vehicle electronics.

9.3   Calibration and Parameter Representation

When the controllers have been designed and implemented there is still the step of calibration and tuning of the controller, for production release. This is an important part of the development process, where performance optimization and trade-offs are fine-tuned before the production. It is necessary because, at the time when the controller is developed, software and hardware, like engine, actuators, and so on, are not finished. Calibration of controller parameters and set points in the final stage for production is a time-onsuming task as there can be about 30 000 parameters that need to be set in an EMS before production. Much of the calibration can be automatized with the support of automatic calibration tools that can run an engine day and night to parametrize the control system. Even though it can be automated, it is time and resource consuming, so designing a controller that is easy to calibrate is thus an important engineering task. Calibration data can be represented as scalar values or as matrices in lookup tables, often referred to as engine maps.

9.3.1 Engine Maps

Maps or lookup tables are convenient for representing nonlinear model components when measurement data is readily available. Some components have a nonlinear dependence on their inputs and a function that describes their behavior might be difficult to derive analytically. If measurement data is available for these it is straightforward to use a map instead of a functional representation. To illustrate the use of maps two examples are given below, the first is a one-dimensional (1D) lookup table for a sensor characteristics and the second illustrates how two-dimensional (2D) lookup tables are used to describe the engine's air mass flow and volumetric efficiency.

1D Map Example: Sensor Characteristics

One example of a 1D map is the translation of the measured λ-sensor voltage U to the actual λ. This translation is used both in engine test cells and in the engine control system, where the sensor delivers a sensor voltage that has to be translated to λ. The set of calibration data, shown to the left in Figure 7.21, is the basis for the map and it specifies the measured values of U in V and λ. This table is the basis for the 1D map $y = f(x)$ where the x-value is the sensor voltage and the y-value gives the corresponding λ. One can also graphically represent the values in a table

U	x_1	x_2	x_3	x_4	$\ldots$	x_n
λ	y_1	y_2	y_3	y_4	$\ldots$	y_n

Between data points, marked with circles in Figure 7.21, the values have to be interpolated. This is usually done through linear interpolation. In engine control systems the grid (data points) for the x-values is often equidistant so that it is easy to determine in what interval the input lies. When the x-data is equidistant, with $x_{i+1} - x_i = \Delta x$ for all i, then the start index j of the interval $[x_j, x_{j+1}]$ that the input x belongs to is determined through

$$j = \left\lfloor \frac{x - x_1}{\Delta x} \right\rfloor + 1 \tag{9.1}$$

where $\lfloor x \rfloor$ means truncation to the nearest smaller integer. With the index given the output y is easily calculated through

$$y = \frac{x - x_j}{\Delta x} y_{j+1} + \frac{x_{j+1} - x}{\Delta x} y_j$$

where x_j, x_{j+1}, y_j, and y_{j+1} are the values at the grid around the interval. If the data isn't equidistant the grid points have to be searched to ensure that $x \in [x_j, x_{j+1}]$ which becomes more time-consuming.

It is also noteworthy that the sensor data in Figure 7.21 can be used to describe both the function $y = f(x)$ and its inverse $x = f^{-1}(y)$ using the same map data, since the function is invertible. However, if λ is used as input the grid for the input is no longer equidistant and (9.1) cannot be used to determine the position of x in relation to the grid points. In this case a search has to be performed.

2D Map Example: Engine Air Flow

Maps can be used to represent the measured values directly, like the mass flow, or they can represent quantities that depend on other values, like the volumetric efficiency. For example $\dot{m}_{ac} = f(N, p_{im})$ can be measured over the engine's operating range by running stationary tests where the engine is controlled to a specified set of points (a grid) determined by, for example, engine speed and intake manifold pressure. When stationary conditions are reached at the operating point, the signals required to calculate the quantity (like, for example, N, p_{im}, $\dot{m}$, and T_{im}) are measured and stored.

A two-dimensional lookup table or map M is represented by the parameter data, z, stored in an $n \times m$ matrix with input variables x and y stored in $n \times 1$ respectively $m \times 1$ matrices, as shown below

	y_1	y_2	$\cdots$	y_m
x_1	z_{11}	z_{12}	$\cdots$	z_{1m}
x_2	z_{21}	z_{22}	$\cdots$	z_{2m}
$\vdots$	$\vdots$	$\vdots$	$\ddots$	$\vdots$
x_n	z_{n1}	z_{n2}	$\cdots$	z_{nm}

The table represents measurements, z_{ij} in the points, (x_j, y_i), and together with a well-behaved interpolation scheme $I(M, x, y)$ the table gives a two-dimensional continuous function $z(x, y) = I(M, x, y)$. Equidistant data is frequently used for the 2D maps together with bilinear interpolation, which is the same as applying linear interpolation in both variables.

Several static measurements have been performed on a naturally aspirated SAAB 2.3-liter engine which has produced values for the lookup tables. A graphical representation of the air mass flow and volumetric efficiency are shown in Figure 9.4. Figure 9.4a shows the measured values directly, while Figure 9.4b shows the map of the volumetric efficiency, which is calculated from the measured variables (N, p_{im}, $\dot{m}$, and T_{im}) using (4.9).

The volumetric efficiency depends on both the engine speed and intake manifold pressure. It is interesting to note that the dependence on manifold pressure agrees well with that exhibited by the theoretical considerations of the expansion of residual gases considered in Section 5.2.4 and Figure 5.8. The fluctuations in engine speed are due to resonance effects in the intake manifold runners. At some engine speeds the resonance gives a positive effect, which gives more air in the cylinder, and at other speeds the resonance gives a negative effect.

It is also possible to use maps of higher dimensions but it might not be preferable, since a higher dimension map requires more memory. The data in the map grows exponentially with the number of dimensions.

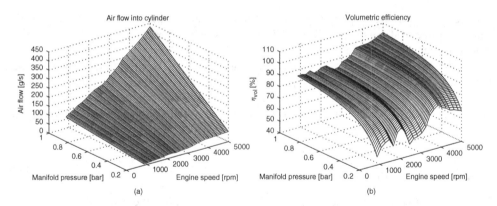

Figure 9.4   Graphic representations of the parameters stored in 2D maps. The air mass flow (a) and the calculated volumetric efficiency (b) are used as examples

9.3.2   Model-Based Development

Traditionally, maps were used extensively in the control system, but with more complex engine hardware this resulted in control systems that were difficult and time-consuming to calibrate, due to the interdependence between variables and quantities. This led to a drive towards model-based methods where models are used more extensively in the development and as a basis for the resulting controller code, by effectively describing the interconnection between parameters and engine quantities. Today, model-based development is helping to reduce the maps and ossorting out the dependencies between the parameters and variables in the system, and also giving good starting points for the calibration. When the controllers are implemented there can still be parameters that are represented by maps and need to be calibrated, for example the volumetric efficiency or operating point dependent set points for controllers. In summary, the systematic use of models and software tool-chains for control and calibration are important for an efficient development of engine control functions.

10

Basic Control of SI Engines

An engine management system (EMS) must meet and compromise between several goals, such as: produce good driveability, maximize the engine performance, minimize the fuel consumption, and give low emissions. These are challenging in themselves, but there are even further challenges. A fact that illustrates this, is that important performance variables aren't measured in a production engine. Recall Figure 4.1, that shows that the engine produces emissions and power as output, and note that neither of these are measured so that we cannot control them directly. It is therefore nevessary to infer the performance from other variables, and Figure 10.1 illustrates the situation where available sensor signals are used to infer the performance outputs. To a large extent, this is where understanding the system and the models, that were built in previous chapters, plays an important role.

The situation described above is standard within automatic control, and the general vocabulary and notation is as follows

- **Performance output, z,** or performance variable, is the variable desired to control, in the engine example it is, for example, torque or emission.
- **Measured output, y,** are the sensor signals available.
- **Control signal, u,** is the actuator output used to achieve the control task.
- **Disturbance, d, v, e,** ... is an unknown and uncontrolled input acting on the system.

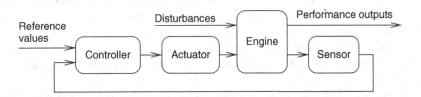

Figure 10.1 Control with performance variables that are not necessarily the same as those measured

Already in Section 3.3, the principles were used, since the whole control structure is built around torque even though it is not measured. The principles will also be used in chapters on driveline control, see Sections 15.2.2–15.2.4, and Figures 15.13 and 15.28. The focus in this chapter is on engine control, and when performance variables cannot be measured this leads to interesting combinations of model-based feedforward

Modeling and Control of Engines and Drivelines, First Edition. Lars Eriksson and Lars Nielsen.
© 2014 John Wiley & Sons, Ltd. Published 2014 by John Wiley & Sons, Ltd.
Companion Website: www.wiley.com/go/powertrain

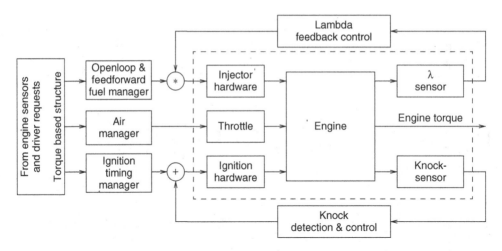

Figure 10.2 Schematic showing three important control loops in an SI engine. The engine hardware with actuators and sensors is inside the dashed box while the control is outside. The topmost is the *(A/F) ratio control* (or λ control) and the bottommost is the *ignition control* loop. The engine torque control path, which essentially controls the air, is shown in the middle

and feedback. Careful feedforward improves transient response, and further, if done right, in the case of sensor failures the engine can operate tolerably in an open loop.

10.1   Three Basic SI Engine Controllers

Given the perspective from the previous page, we now start our journey into the area of engine control. An engine management system for SI engines contains several controllers, and three of the most important ones are shown in Figure 10.2.

These are air, air/fuel ratio (λ), and ignition controllers. The *air manager* is the main path for torque control in an SI engine. It controls the amount of air that is fed to the engine using the throttle (and turbo). There is no feedback from the torque but several feedback loops are used internally, like throttle servo and intake manifold pressure control. Section 10.2 will describe the torque and air path control in more detail and later Section 10.9 will add boost pressure control for turbo engines. Torque also depends on the ignition timing, and their interaction and utilization is discussed in Section 10.8.

The main purpose of the *fuel manager* or *air/fuel ratio controller* is emission reduction. To achieve good emission reduction it is necessary to use both feedback and feedforward. An open loop control is not sufficiently accurate, due to model uncertainties, and feedback is therefore necessary for maintaining λ in the λ-window, that is in the narrow band around λ = 1 where the catalyst is most efficient (recall Figure 6.17). Feedforward is necessary for good controller performance in transients where it determines the basic amount of fuel to inject. Section 10.4 will describe the A/F ratio control, also called λ-control, in more detail.

The main purpose of the *ignition manager* is to get good fuel economy while avoiding knock. At low loads the goal is to run the engine with the most fuel efficient ignition, while at high loads engine knock can pose a limit on the ignition timing and it has to be set later. Different fuels and different ambient conditions have different requirements, motivating the knock feedback loop. Section 10.6 will describe ignition and knock control.

10.1.1 Production System Example

An example of an engine management system for a direct injected SI engine, Bosch DI-Motronic, is shown in Figure 10.3. It integrates the engine management functions outlined in Chapter 9, with torque, air, fuel, and ignition control, as well as emission reduction and diagnostic functions. The functionality that some of the sensors and actuators provide is explained below.

The torque manager uses the accelerator pedal position (25) as input to control the throttle (7), internally the intake manifold pressure (8) and engine speed sensor (19) are also used. The integrated air temperature and air mass flow sensor (6), and the λ-sensor (12) are used together with the manifold pressure sensor (8) in the air manager and λ-controller. The coolant temperature sensor (16) influences both the fuel injection and the spark advance, mainly during start up. The fuel injector (10) injects fuel directly into the cylinders. The fuel pressure sensor (9) and associated regulator has as its main objective to give a well-defined pressure

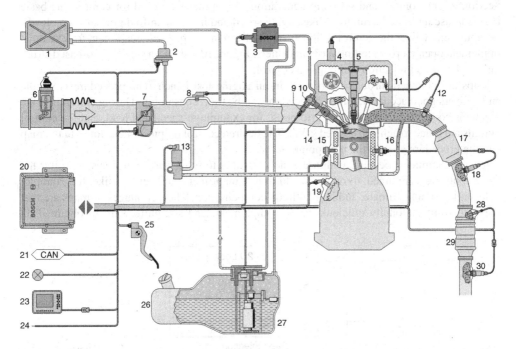

Figure 10.3 Overview of sensors, actuators, and supporting systems for a direct injected gasoline engine. The sensors and actuators are: 1. Carbon canister for capturing evaporating fuel. 2. Canister-purge valve. 3. High pressure fuel pump. 4. Cam phase controller. 5. Spark plug. 6. Hot film air mass flow sensor with integrated temperature sensor. 7. Throttle actuator with angle sensor. 8. Intake manifold pressure sensor. 9. Fuel pressure sensor. 10. High-pressure fuel rail. 11. Cam phase and position sensor. 12. λ sensor. 13. Exhaust gas recirculation (EGR) valve. 14. High-pressure injector. 15. Knock sensor. 16. Engine temperature sensor. 17. Primary catalytic converter (three-way catalyst). 18. λ sensor (optional). 19. Crankshaft position and TDC sensor. 20. Engine management system (EMS). 21. CAN bus interface. 22. Fault lamp. 23. Diagnosis interface. 24. Interface to immobilizer unit. 25. Accelerator pedal module. 26. Fuel tank. 27. In-tank unit with electric fuel pump, fuel filter, and fuel pressure regulator. 28. Exhaust gas temperature sensor. 29. Main catalytic converter (NO_x accumulator plus three-way catalyst). 30. λ sensor. *Source:* Bosch Automotive Handbook 8e, Figure 5, page 541. Reproduced with permission from Robert Bosch GmbH

over the fuel injector (10), so that the amount of fuel injected into the intake will only depend on the injection time and not on the load or intake manifold pressure. The ignition system has coils at the spark plug (5) and a knock sensor (15) that detects knock, so spark advance can be retarded to avoid noise and engine damage. Sensors for crank speed and position (19) and cam phase and position (11) are important for the timing of fuel injection and ignition, and these were discussed in Section 9.1.2. A cam phase actuator (4) controls the intake and exhaust valve timings. The carbon canister (1) prevents fuel vapors in the tank (26) from entering the atmosphere. The carbon canister is emptied by the fuel canister-purge valve (2) where air from the atmosphere is drawn through the activated carbon into the intake manifold by its vacuum.

10.1.2 Basic Control Using Maps

Lookup tables or *maps* are often used in the control system for parameter representation (c.f. Section 9.3.1), control, and set-point generation. When maps are used for control, the basic idea is to use measured quantities, like engine speed and intake manifold pressure (i.e., load), as inputs and deliver the control action as output from the map. These maps can, for example, implement open loop or feedforward controllers. Figure 10.4 shows a simple map-based structure for basic fuel injection and ignition control.

Maps are obtained through calibration in an engine test bench. The procedure is simple and delivers good accuracy, but the procedure can be time-consuming. During calibration the engine is controlled to the desired grid point, for example in speed and load, and the control outputs are tuned to achieve desired goals. The desired control output is then stored in the map and the procedure goes to the next point.

A very common procedure within engine control is to refine the control system, after the basic controllers are tuned, through the addition of maps that handle effects like, for example: cold start, hot/cold climate, and high altitude compensation. The advantages of maps are that they are computationally efficient, easy to implement, and straightforward to obtain if an

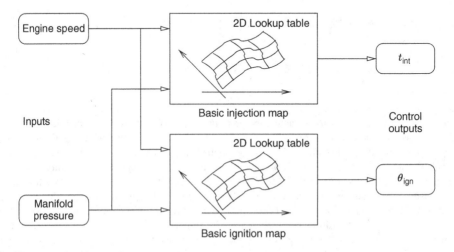

Figure 10.4 Illustration of how maps can be used to achieve the basic control of injection t_{inj} and ignition angle θ_{ign} in a control system. Speed and load (i.e., manifold pressure) inputs are delivered to the maps and the control action is given as output, for example $t_{inj} = f(N, p_{im})$

engine is available. The major disadvantage is that they don't offer extrapolation; when an engine is redesigned the maps might have to be recalibrated which is a time-consuming process. Using maps is sometimes called map-based control.

Another paradigm is model-based control where models are used to describe the functions and the complex interactions between inputs and outputs. Model-based methods are currently being implemented to replace and reduce maps, since they offer the benefit of extrapolation and can reduce the calibration effort. In the end, models can on the other hand be implemented as maps in the final control implementation. For example, if a function is too complex to be executed on the controller hardware, then the lookup table can be generated from the model offline, which in the end can save calibration and execution time.

10.1.3   Torque, Air Charge, and Pressure Control

The torque-based structure was introduced in Section 3.3. and further refined in Section 9.2. It is used to give a systematic framework for handling the drivers' request and other torque influencing functions at vehicle and powertrain level. At the engine level it specifies the desired engine torque $M_{e,ref}$, which is the reference value for the engine torque. We now turn to how this can be achieved and as aids we have the engine torque models in Section 7.9.

10.1.4   Pressure Set Point from Simple Torque Model

As a first step we consider an SI engine that operates at $\lambda = 1$ (and possibly fuel rich), implying that the air determines the energy that can be released in the combustion. For such conditions the torque generation can be described by Model 7.17, that gives a simple relation between engine torque and intake manifold pressure. With the torque $M_{e,ref}$ specified, then (7.53) and (7.54) can be solved for the pressure, giving

$$p_{im,ref} = \frac{1}{C_{p2}} \left(\frac{2\pi n_r}{V_D} M_{e,ref} + C_{p1} \right) \tag{10.1}$$

This gives a reference value for the intake manifold pressure which can be controlled with the help of the throttle servo. The result is a cascade controller structure for the torque with torque to pressure calculations, pressure controller, and throttle servo. Furthermore, there is often an intake manifold pressure sensor available, so that it is possible to have closed loop control of the pressure. The resulting controller structure is depicted in Figure 10.5. A cascade solution is viable since the throttle servo dynamics is faster than the intake manifold pressure dynamics. The throttle servo is a supporting system and its design will be discussed in more

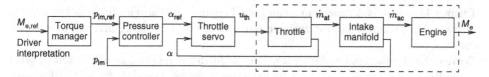

Figure 10.5 Illustration of the cascaded controllers in the torque-based structure. Separation of time scales between the systems is used for separating the designs, where the throttle servo is a fast inner loop and the intake pressure is a slower outer loop

detail in Section 10.3. From the figure we can once again note that there is no feedback from the torque but it uses several feedback controllers internally.

10.1.5   Set Points from Full Torque Model

In the previous controller the simple torque model (7.53) was used, which provided a straight-forward way to calculate set points for the pressure. In the next step we turn to set point generation with the more detailed MVEM for work and torque, described in Section 7.9. Before proceeding, we note that the basic structure with cascade control in Figure 10.5 is the same, but the calculations in the "Torque manager" block are refined.

The torque model (7.55) consisted of the following three components

$$M_e \, n_r \, 2 \, \pi = W_e = W_{i,g} - W_{i,p} - W_{fr}$$

With the submodels for gross indicated work, $W_{i,g}$, pumping work $W_{i,p}$, and friction work W_{fr} inserted we get the following expression

$$M_e \, n_r \, 2 \, \pi = m_f \, q_{LHV} \left(1 - r_c^{\gamma - 1} \right) \eta_{ig} \, \eta_\lambda \, \eta_{ign} - V_D \, (p_{em} - p_{im}) - V_D \, \text{FMEP}(N) \qquad (10.2)$$

Furthermore, fuel mass, air mass, and pressure are related to each other via the air fuel ratio and volumetric efficiency as follows

$$m_f = \frac{m_a}{\lambda \, (A/F)_s} = \frac{\eta_{vol} \, V_D \, p_{im}}{\lambda \, (A/F)_s \, R \, T_{im}} \qquad (10.3)$$

This expression can be inserted into (10.2), which gives an expression for the torque in terms of the pressure. It is worth noting that if all components in (10.2) were constants then the resulting function would be affine in intake manifold pressure, just as (7.53). In fact, this is one of the reasons why the affine model (7.53) gives a good description of the engine torque.

Based on the models above, we can develop three ways of calculating the set points from a given torque reference $M_{e,ref}$ and feed to engine control loops:

- Combine 10.2 and 10.3 and solve for p_{im}. This is useful for stoichiometric, and fuel rich, SI engines and the expression is given below. It is also useful for boost pressure control of both SI and CI engines.

$$p_{im,ref} = \frac{M_{e,ref} \, n_r \, 2 \, \pi + V_D \, (p_{em} + \text{FMEP}(N))}{V_D + q_{LHV} \left(1 - r_c^{\gamma - 1} \right) \eta_{ig} \, \eta_\lambda \, \eta_{ign} \, \frac{\eta_{vol} \, V_D}{\lambda \, (A/F)_s \, R \, T_{im}}} \qquad (10.4)$$

- Solve (10.2) for m_f. This is the main path for CI engines and will be discussed more in Chapter 11, which is dedicated to control of CI engines. It is also useful in SI engines that operate with excessive air, for example direct injected SI engines operating in stratified mode.
- Solve (10.2) for the ignition efficiency. This is suitable if fuel and air are locked by other controllers. For example, if a short-term reduction in torque is needed, this will be discussed in Section 10.6.3.

Here we will use the first approach, that fits into Figure 10.5, and later return to the others.

10.1.6   Pressure Control

Our attention now turns to pressure control, where the design will also be used to exemplify how additional demands, beside control performance, can play an important role in the selection of control structure and design. Pressure control is a link in the chain of torque management that is safety critical for the vehicle, where unintended torque is not allowed to occur and supervision of the chain is needed. See also Section 16.1.1 on Functional safety–unintended torque, for a longer discussion. Furthermore, if the pressure sensor breaks then it should, for dependability reasons, still be possible to drive the vehicle though perhaps with reduced performance. These additional demands on the controller influence its structure and design.

Feedforward and Feedback Control in Combination

Pressure control is achieved with feedforward and feedback controllers in combination, see Figure 10.6. These have the following functionality:

- The feedback controller is used for performance and robustness, ensuring that the desired pressure is attained and thus reducing the effects of model uncertainties and unmeasured disturbances.
- If a fault occurs such that the feedback controller cannot be executed safely, for example if the p_{im}-sensor breaks, then the feedback controller is shutoff and the feedforward controller is run alone. Therefore, the feedforward controller is designed to give the correct stationary control input α_{ref} provided a $p_{im,ref}$ and thus enables a fallback strategy.

In control design the focus is often on controller performance, where the feedback loop is designed first and if the response to a set point change isn't fast enough a feedforward is added to improve the performance. However, due to the dependability demands here, the feedforward is designed first so that the engine can be run without pressure feedback, this is often called open loop pressure control. It thereby provides a fallback strategy and when operation is in open loop pressure control, the focus is mainly on reaching the right stationary values for pressures and throttle angles.

The *feedforward* controller can be implemented using either maps, as outlined in Section 10.1.2, or models. In the first case, the engine is run at the reference pressure and the corresponding throttle angle is stored as z values in the map, and the reference pressure x and engine speed y are stored as grid points for the map. The model-based way to determine the feedforward throttle angle by using MVEM components will be shown in the example below.

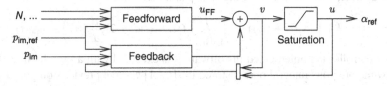

Figure 10.6   Block diagram of the intake manifold pressure controller, showing a more detailed view than Figure 10.5. Additional inputs, indicated at the engine speed (N), illustrate that more inputs from the engine are often used in the calculations. Furthermore, the pressure controller relies on the throttle that can saturate, and this must also be handled by the closed loop controller

The *feedback* is designed and tuned after the feedforward is finished. To achieve good performance it is often gain scheduled with the engine speed, since the dynamics of the intake manifold is speed dependent. Furthermore, the feedback controller is also designed and tuned to give good dynamic performance. See, for example, the response of the turbocharged engine in Figure 8.25, where the throttle makes a significant overshoot to the maximum and then settles at stationary value. With only feedforward, the throttle goes to the stationary value directly, giving the engine a slower torque response. Figure 8.25 also shows that the throttle actuator can saturate, and it is therefore important to account for integrator windup in the feedback controller. This will be discussed for the PID controller below.

Example of Feedforward and Feedback Pressure Control
Based on the discussion above, the following two examples are used to illustrate the combined feedforward and feedback control. First the model-based feedforward generates α_{FF} and then a generic PID is used in the feedback.

Example 10.1 (Model-based feedforward pressure controller)   The engine air mass flow model (7.13), based on the volumetric efficiency (7.15), gives the desired engine air mass flow, as a function of intake pressure and engine speed, as well as other quantities. Stationary, it is the same as the throttle flow. With this flow the throttle area and angle can be determined, using the throttle model (7.11a) and area model (7.12). All in all we get the following calculation scheme

$$\dot{m}_{ac} = f_1(\eta_{vol}(p_{im}, N), p_{im}, N, T_{im}) \tag{10.5a}$$

$$A_{ref} = f_2(\dot{m}_{ac}, p_{im}, T_{im}, p_{th,in}) \tag{10.5b}$$

$$\alpha_{FF} = f_3(A_{ref}) \tag{10.5c}$$

Here (10.5b) uses (7.11c), which states that the temperature is unchanged over the throttle, and therefore only T_{im} is needed in the calculations.

To exemplify the implementation details in the feedback loop, a generic PID will be used, where the feedforward term is included (i.e., $u_{FF} = \alpha_{FF}$). A PID-controller can be implemented in discrete time and achieve anti-windup through tracking in the following way. The interested reader is referred to, for example, Åström and Hägglund (1995) for a deeper discussion on PID-implementation and anti-windup.

Example 10.2 (PID with anti-windup by tracking)   A formulation of the PID-controller in continuous time, with a feedforward term u_{FF} and derivative on only y, is

$$u(t) = K_P\left(\left(y_r(t) - y(t)\right) + \frac{1}{T_I}\int_0^t y_r(t) - y(t)dt + T_D\frac{d}{dt}y(t)\right) + u_{FF}(t)$$

When the controller is implemented in software it is executed with a sample time, T_s. In the code below the previous integral part is stored in $I[i-1]$ and the previous output in $y[i-1]$.

$$I_i = I[i-1] + \frac{K_P}{T_I}T_s(y_r - y) \tag{10.6a}$$

$$v = K_P(y_r - y) + I_i + \frac{K_P T_D}{T_s}(y - y[i-1]) + u_{FF} \tag{10.6b}$$

$$u = \min(\max(u_{\min}, v), u_{\max}) \qquad (10.6c)$$

$$I[i] = I_i + \frac{T_s}{T_t}(u - v) \qquad (10.6d)$$

The discrete time PID is represented by the first two lines, with v being a temporary control output. On row 3 the control output is saturated to the limits. Finally, row 4 has the tracking with time constant, T_t, which adjusts the integral if the control is saturated, and the current integral value is saved in $I[i]$. A guideline for tuning the tracking time to be constant is to start with $T_t \approx T_I$.

This example of a PID controller is generic, but it can be put into the context of closed loop intake manifold pressure control. The link between the PID and the closed loop controller is in Figure 1.6 represented by the components u_{FF}, v and u that are used in the PID-code above. All in all, the two examples above illustrate how open loop and closed loop pressure control can be achieved, combined, and integrated, as well as how the effect of integrator windup can be handled. An example where pressure control is used together with ignition control will be given in Section 10.8.

Feedback Tuning and a Note on Intake Dynamics
The tuning of the feedback can be done using standard methods for PID controllers. In addition, it can be beneficial to include gain scheduling of the controller parameters, since the time constant of the system depends on engine speed. A simplified system with constant volumetric efficiency and with throttle flow as input gives insight into the pressure system dynamics.

Example 10.3 (Time constant of intake manifold with flow control)   Study the isothermal model (7.24) and assume that the engine speed, N, and volumetric efficiency, η_{vol}, in (4.9) are constant. Further assume that we can do exact feedback linearization of the throttle flow, that is invert (7.11), so that $u = \dot{m}_{at}$. Then (7.24) can be written

$$\frac{dp_{im}}{dt} = \frac{R\,T_{im}}{V_{im}}\left(u - \frac{\eta_{vol}\,V_D\,N}{R\,T_{im}\,n_r}p_{im}\right) = -\frac{\eta_{vol}\,V_D\,N}{V_{im}\,n_r}p_{im} + \frac{R\,T_{im}}{V_{im}}u$$

which shows that the time constant for the intake manifold pressure is $\tau_{im} = \frac{V_{im}\,n_r}{\eta_{vol}\,V_D\,N}$, and that it depends on engine speed.

10.2   Throttle Servo

The throttle servo is an important component in the torque-based structure and intake manifold pressure control. Before proceeding, it is worth noting that the pressure controller uses α_{ref}, but one could think that it would also be possible to use u_{th} directly, skipping the throttle servo. However, this gives significant problems due to the friction and poor open loop behavior of the throttle actuator. This is understood by studying Figure 7.47, where it is seen that a u_{th} with 20% PWM can give a throttle position anywhere between 29% and 100%, which is not accurate enough. The pressure controller is therefore designed so that it relies on the throttle servo. From a safety and dependability standpoint, it is also worth mentioning that if the throttle servo malfunctions then the throttle returns to the limp-home position, which is designed to give a safe state for the engine.

10.2.1   Throttle Control Based on Exact Linearization

A description of the throttle plate motion was developed in Model 7.22 and will be used in the development of the throttle servo. The approach described below is based on exact linearization and comes from Thomasson and Eriksson (2011), that to a large extent builds upon Deur et al. (2004). The throttle contains two nonlinearities, limp-home spring and friction, that have significant influence on the throttle behavior, see Figure 7.47. The model (7.76) can be linearized using two nonlinear compensators, that modify the control signal according to

$$u_{\text{th}} = \frac{M_{\text{sp}}(\alpha)}{K} + \frac{M_{\text{fs}}(M, \omega)}{K} + \tilde{u}_{\text{th}} \tag{10.7}$$

This achieves exact linearization of (7.76), but (10.7) is difficult to implement in practice, for example because there is noise in α and since M and ω aren't measured. Therefore these are modified in the following way. Using (7.74) exactly counters the *limp-home torque*, $M_{\text{sp}}(\alpha)$. However, close to the limp-home position, small variations in the measured output could then result in severe chattering of the control signal due to the large slope of the curve in this region. To overcome this, the commanded throttle reference is used as input $M_{\text{sp}}(\alpha_{\text{ref}})$.

Using (7.73) the *friction* could be compensated $M_{\text{fs}}(M, \omega)$, which is an ideal relay that switches at $\omega = 0$, but needs the angular velocity. Further, if the throttle plate is moving away from the reference value, then the friction compensator would add to that motion. The friction compensator is therefore based directly on the tracking error, acting in the direction that reduces the tracking error. An ideal relay is sensitive to noise around $e_\alpha = 0$, which can cause undesirable oscillations around the reference value, to remedy this a small dead zone around $e_\alpha = 0$ is introduced. In summary, limp-home and friction compensations modify the control signal as follows

$$u_{\text{th}} = \frac{M_{\text{sp}}(\alpha_{\text{ref}})}{K} + \frac{M_{\text{f}}(e_\alpha)}{K} + \tilde{u}_{\text{th}} \tag{10.8}$$

Linear Throttle Model

The exact linearization of the throttle model gives the following equation

$$J \frac{d^2 \alpha}{dt^2} = -K_{\text{fe}} \frac{d\alpha}{dt} + K \tilde{u}_{\text{th}}$$

which has the following open loop transfer function

$$G_{\text{th}}(s) = \frac{K}{J s^2 + K_{\text{fe}} s} = \frac{K/K_{\text{fe}}}{s \left(\frac{J}{K_{\text{fe}}} s + 1 \right)} = \frac{K_0}{s (s T_0 + 1)} \tag{10.9}$$

In many cases K, J, and K_{fe} are unknown, but system identification experiments can be performed. Here step response experiments were made for the throttle in Figure 7.47, and $K_0 \approx 20$ and $T_0 \approx 20$ ms were identified.

Design of PD and I Parts

As an example of a control design method, a pole placement method is used, see (Rivera et al., 1986), (Åström and Hägglund, 2006) for more details. Here we specify the desired closed loop system to be a first-order system, since we want to avoid overshoots. The desired closed loop behavior is

$$G_{des}(s) = \frac{1}{s\,\Lambda + 1} \qquad (10.10)$$

where the time constant Λ is used for fine tuning of closed loop response. For the system (10.9) it is known that a PD-controller $u(t) = K_P\,(e(t) + T_D\dot{e}(t))$ can generate a first-order system in closed loop. With the PD controller the closed loop system becomes

$$G_{cl}(s) = \frac{K_P\,(1 + s\,T_D)\,\dfrac{K_0}{s\,(s\,T_0+1)}}{1 + K_P\,(1 + T_D\,s)\,\dfrac{K_0}{s\,(s\,T_0+1)}} = \frac{1}{s\,\dfrac{1}{K_P\,K_0}\dfrac{s\,T_0+1}{s\,T_D+1} + 1} \qquad (10.11)$$

Matching the terms between (10.11) and (10.10) gives the PD controller parameters $T_D = T_0$ and $K_P = \frac{1}{K_0\,\Lambda}$. Here Λ is used to tune the controller where one parameter gives the values on P and D parts using the model. This gives a good starting point for the controller calibration. The total controller is shown in Figure 10.7 where the linear system is controlled by a PID-controller.

Due to model uncertainties and disturbances, an I-part with anti-windup is added to the PD controller. Gain scheduling of the I-part on the control error is also applied for the fine adjustment of the position. This is motivated for when there is a small control error and the controller still needs to make large actions to overcome the friction. It is therefore designed with gain scheduling on the control error, and the interested reader is referred to Thomasson and Eriksson (2011) for more details.

Throttle Servo – Performance Evaluation for the Example

Controller performance is demonstrated for two steps, large and small, in Figure 10.8. The large step shows the maximum performance of the throttle servo, where the control signal saturates during 40 ms and the controller settles after 0.1 s without overshoot. The small step, from 28.4 to 29.4%, is performed around the limp-home position at 28.8%. It shows significant control action while it works against limp-home and friction to achieve the small change in position. One can also see the A/D quantization in the measured position. In general, the throttle servo achieves good performance. Throttle control designs is a well-studied area and more examples can be found in, for example, Scattolini et al. (1997), Eriksson and Nielsen (2000), Canudas de Wit et al. (2001), Vašak et al. (2006), Pavković et al. (2006).

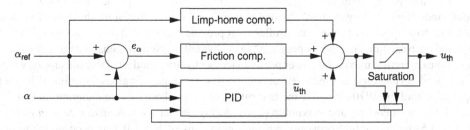

Figure 10.7 The structure of throttle controller with the nonlinear effects of limp-home and friction compensation. The feedback controller is implemented as a PID controller with anti-windup due to the possibility of the control signal saturating

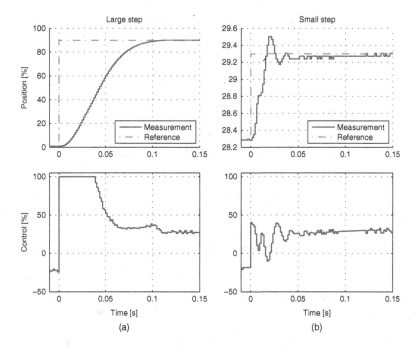

Figure 10.8 Step responses of the throttle with throttle servo. (a) Large steps from 0 to 90% open. (b) Small steps around the limp-home position (at 28.8%). Showing the resulting control design

Throttles Without Limp-home

Most throttles have a limp-home at a safe position, and here the safe position was at 28.8%. There are some throttles that don't have the limp-home in the middle region. One example is the compressor by-pass valve, used on the engine in Figure 12.7, that has its limp-home position at fully open. Another possibility is that it can have its limp-home position at fully closed, as the one in Eriksson and Nielsen (2000) does. Anyway, the method for control design, described above, can still be applied but naturally modified so that it does not have the limp-home compensation.

10.3   Fuel Management and λ Control

Air and fuel management are coupled to each other, and we adopt the view that the air is commanded by the air system and the fuel management is used to control the air-to-fuel ratio (λ). The basic function of the λ controller is to provide a combustible air and fuel mixture. Beyond the basic function, the λ operation can be divided into four regions: with a *rich mixture* $\lambda < 1$ the engine has maximum power per displacement volume and available air supply, which gives good acceleration but reduced fuel conversion efficiency. Engines were tuned for rich operation until the 1970s but after the first oil crisis are used only for cold engines and at full load for catalyst cooling and maximum power during acceleration. A *moderate lean mixture* $\lambda \in [1; 1.5]$ has good fuel economy but high emissions of NO_x. It was used after the first oil crisis until the beginning of the 1980s, when the emission legislations were made stricter. This strategy was used especially at medium loads. *lean mixture* $\lambda > 1.5$ gives high efficiency and reduces the NO_x emissions. It is used in lean A-burn engines at low and medium loads.

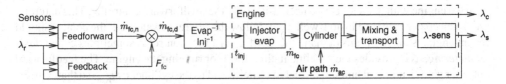

Figure 10.9 Air/fuel ratio controller and the system being controlled, showing a simplified view that includes the important components in the fuel path of the engine

A *stoichiometric mixture* $\lambda = 1$ is nowadays the main mode of operation, used in engines with three-way catalysts (TWC) and represents the state of the art in emission reduction. In the control design below the focus is on control for the TWC so that it can achieve good emission reduction.

10.3.1 Feedforward and Feedback λ Control Structure

In the following section the basic functions of the fuel management system for achieving a desired λ are described. A block diagram of the basic structure for an λ ratio controller is shown in Figure 10.9. It separates the engine and controller and highlights important components in the engine that influence the control design, in particular the injector and fuel evaporation dynamics, mixing into the cylinder, and transport/mixing in the exhaust to the sensor. The controller has four components and they are motivated by:

- Feedback, F_{fc}. In SI engines with three-way catalysts it is important that λ is maintained in a narrow band around $\lambda = 1$, recall Figure 6.17. This requirement on precision can only be achieved with feedback control, using a λ-sensor, since the model errors in $\dot{m}_{ac}$ and variations in $(A/F)_s$ are larger than the acceptable range for λ.
- Feedforward, $\dot{m}_{fc,n}$. Due to the dynamics and foremost the transport delay from the fuel injection to the λ-sensor, there is a need to have a feedforward loop.
- Multiplication. The nominal fuel flow to the cylinders $\dot{m}_{fc,n}$ is given by the feedforward and the feedback loops generates a correction factor F_{fc} that adjusts the fueling. A multiplication is used since the system is controlling the (A/F) quotient. The total controller output is the desired fuel flow to the cylinders

$$\dot{m}_{fc,d} = \dot{m}_{fc,n} \, F_{fc}$$

- Injector and fuel dynamics compensation. These are introduced to simplify the feedforward and feedback control designs, by separately handling the injector characteristics and fuel dynamics.

These will be discussed in more detail in the coming sections, starting with feedforward control then feedback control, and finally injector and evaporation compensation.

10.3.2 Feedforward λ Control with Basic Fuel Metering

There are similar demands on the feedforward fuel control as on the pressure feedforward discussed in Section 10.2.3. In particular, if the feedback is disabled then it should be possible to run the engine with open loop λ control, using only the feedforward controller. This naturally gives reduced performance in how well λ can follow the reference value, but the control should prevent engine stall. With this demand the feedforward fuel control is sometimes called *basic fuel metering*, as it provides the baseline fueling level for running the engine. The feedforward controller gives a nominal amount of fuel to be delivered to the cylinders, $\dot{m}_{fc,n}$, so that it meets the air delivery. Feedforward control can be seen as consisting of two parts. One is equivalent to estimating the amount of air that enters the cylinder, and the other accounts for the fuel property $(A/F)_s$ and desired mixture strength, λ_r.

There are two different methods for determining the basic fueling level, they are called *mass air flow* and *speed density* principles. Both determine an air mass flow and then determine how much fuel to inject based on the air mass flow and the desired λ.

The *mass air flow principle* uses the measured air mass flow $\dot{m}_{at}$ as an estimate of the air flow that will enter the cylinder. It is most suitable for central injection systems where the fuel is injected close to the throttle where the air flow is measured. The basic equation for determining the desired fuel flow then becomes

$$\dot{m}_{fc,n} = \dot{m}_{at} \, \frac{1}{\lambda_r \, (A/F)_s}$$

With port and direct injected engines this approach is of less importance.

The *speed density principle* is based on the volumetric efficiency (4.9) and subsequent Model 7.6, which gives the air mass flow into the cylinder $\dot{m}_{ac}$ provided the engine displacement volume, intake density, and engine speed. The intake manifold density is determined from the pressure and temperature in the intake manifold using the ideal gas law. Based on the Model 7.6 the fuel flow can be determined by

$$\dot{m}_{fc,n} = \eta_{vol}(p_{im}, N, \dots) \, \frac{p_{im} \, V_D \, N}{R \, T_{im} \, n_r} \, \frac{1}{\lambda_r \, (A/F)_s}$$

This strategy is well-suited for multi-point bank injection, sequential fuel injection, and direct injection since they inject the fuel close to the cylinders. At steady state conditions (without measurement errors or model errors in $\eta_{vol}(p_{im}, N)$) the two principles give the same value, since $\dot{m}_{ac} = \dot{m}_{at}$ at steady state.

Transients and Prediction for Performance

Looking more into the details of the feedforward controller, there is also an issue with causality. The fuel injection is timed so that the injection stops when the inlet valve opens (for good evaporation, see Section 6.1.1). From the instant when the injection duration, t_{inj}, is calculated, to when the fuel injection is commanded, there is some time for the manifold pressure or engine speed to change, if the engine is in a transient. This can result in a change in $\dot{m}_{ac}(N, p_{im})$. To improve the performance further the fuel can be metered based on a prediction of the future air mass-flow and an example is shown in Andersson (2005, Ch. 9).

10.3.3 Feedback λ Control

Recall that the feedforward loop is essentially composed by air estimation and multiplication of the fuel property $(A/F)_s$ and reference value λ_r. Feedback is used to mitigate the results of model errors in air estimation and the unknown fuel property $(A/F)_s$ in order to ensure that λ is maintained in a narrow band around $\lambda = 1$ for the TWC. A quotient is controlled in the feedback loop, which makes it feasible to use a multiplication for the correction in the total controller. The product can be formed as follows

$$\dot{m}_{fc,d} = \dot{m}_{fc,n} \, F_{fc} = \dot{m}_{fc,n} \, (1 + \Delta_{fc}) \qquad (10.12)$$

where the last step is used to illustrate that the feedback controller makes adjustments, Δ_{fc}, around the nominal value of 1.

The feedback control problems to solve depend highly on the λ-sensor used, that is switch-type or wide-range sensor. The switch-type sensor gives a special type of solution and will be discussed in more detail below. With a wide-range sensor the problem becomes simpler and it improves the control performance. The feedback design has to cope with the effects of transport delay, mixing dynamics, and sensor dynamics. Due to the delay. Smith-predictors, or similar controllers with dead time compensation, are often considered for the design, see Jankovic and Magner (2011). Furthermore, since there are uncertainties in the parameters, robust controllers like H_∞ based designs can also be considered. There is a substantial amount of literature on the subject and the interested reader is referred to Chang et al. (1995), Guzzella and Onder (2009), Muske et al. (2008), Roduner et al. (1997), and Yildiz et al. (2008).

Feedback Control with Switch-Type λ-sensor

We now turn to feedback control with the switch-type or EGO sensor, recall Figure 7.20 for its relay characteristics. A sketch of the system with the controller is shown in Figure 10.10, where the system is simplified to a time delay system. Signal conditioning is performed on the EGO sensor output, thresholding with U_{ref} so that the controller gets a clean relay signal as input. From control theory we know that relay feedback of a dynamic system will inherently give an oscillation. Its frequency and amplitude depend on the system properties, here time delay and controller parameters. The following example illustrates the function and oscillating signals in a EGO feedback system as well as the benefit of feedback.

Example 10.4 (Illustration of Relay Feedback and Limit Cycle)   An engine has a transport delay of $\tau_d = 0.2$ s from injection to EGO sensor. The feedforward loop, giving $\dot{m}_{fc,n}$, is tuned for normal gasoline with $(A/F)_s = 14.6$ but the engine is running on the reference fuel isooctane with $(A/F)_s = 15.1$. The feedback controller is an I-controller,

$$F_{fc} = (1 + \Delta_{fc}) \qquad \text{with} \qquad \Delta_{fc} = K_I \int e_\lambda dt$$

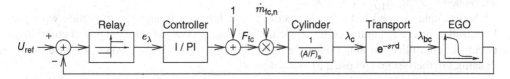

Figure 10.10 Block diagram showing the structure of I and PI control of a time delay system with relay feedback. The voltage reference U_{ref} sets the switching voltage, around 0.5 to 0.6 V, and the relay on the input side conditions the signal e_λ to switch from -1 to $+1$ for the PI controller

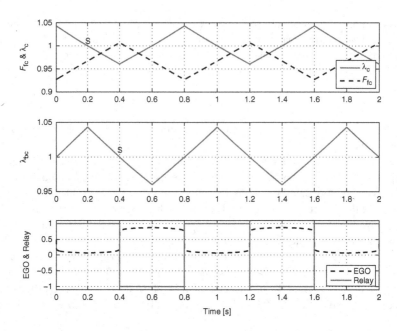

Figure 10.11 Illustration of a limit cycle from relay feedback, as shown in Figure 10.10. The system is a pure time delay controlled with an I-controller, giving the limit cycle a period time of $T = 4\,\tau_d$ and amplitude of oscillation $\tau_d\,K_I$

which is the simplest controller that gives zero stationary error. It has $K_I = 0.2$. The signals from a simulation of the resulting closed loop system, that gives a limit cycle oscillation, are shown in Figure 10.11.

In the top plot λ_c in the cylinder (solid) and the correction factor from the feedback F_{fc} (dashed) are shown. It is seen that on average $\lambda_c = 1$ and $F_{fc} = \frac{14.6}{15.1} \approx 0.97$, which shows that the feedback loop manages to correct λ for the error in fuel. Going into more detail on the signals, we see that e_λ has relay characteristics and is constant in periods, and thus produces ramps in the integrator and F_{fc}. With ramps up and down in the fueling we get ramps down and up in λ_c. The cornerstones for understanding the limit cycle are the switching and the time delay between λ_c and λ_{bc}. In the cylinder, λ_c switches at the point marked S in the top plot, the gas is transported to the exhaust and λ_{bc} switches $\tau_d = 0.2$ s later at S in the middle plot. At the switching of λ_{bc} the EGO also switches. The total oscillation with ramps up and down gets a period time of $T = 4\,\tau_d$. Further, the amplitude of the oscillation is the integration of e_λ over a time delay giving, $K_I \int_0^{\tau_d} 1\,dt = K_I\,\tau_d$.

The example above illustrated the cause of the limit cycle and its properties with an I-controller, where τ_d controlled the cycle time and the product of time delay and controller parameter controlled the amplitude. The following examples illustrate when we have more dynamics in the system and use a PI-controller.

Example 10.5 (Illustration of PI Feedback Tuning) An engine has transport delay $\tau_d = 0.2$ and wall-wetting parameters $X = 0.5$, $\tau_{fp} = 0.2$. The feedforward has the right $(A/F)_s$. A PI

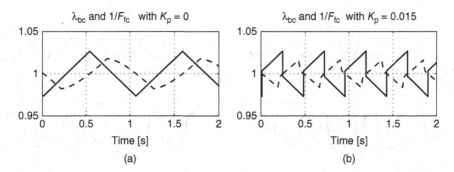

Figure 10.12 Limit cycle in the amount of injected fuel (solid) and in $\lambda_{\rm bc}$ (dashed). (a) I controller. (b) PI controller. To expose the time delay and dynamics, the control signal is plotted for $\frac{1}{F_{\rm fc}}$, which makes it easier to make the comparison with $\lambda_{\rm bc}$

controller is used and parameterized as

$$\Delta_{\rm fc} = K_{\rm P}\, e_\lambda + K_{\rm I} \int e_\lambda\, dt$$

Further, two different settings of the PI-parameters are used.

$$\begin{cases} K_{\rm I} = 0.1 \\ K_{\rm P} = 0 \end{cases} \quad \text{and} \quad \begin{cases} K_{\rm I} = 0.1 \\ K_{\rm P} = 0.015 \end{cases}$$

Simulation results for the two PI controller tunings are shown in Figure 10.12.

The behavior for the I-controller is essentially the same as Figure 10.11, where the difference comes from the additional wall-wetting dynamics. The limit cycle time for $K_{\rm P} = 0$ is about 1 s, which is slightly larger than $4\,\tau_{\rm d}$ that a pure time delay system gives. Decreasing $K_{\rm I}$ decreased the amplitude when comparing Figures 10.11 and 10.12, but the lengthened period makes the amplitude slightly larger than $\tau_{\rm d}\, K_{\rm I}$. Adding a P-part gives a jump back when e_λ switches, as seen in Figure 10.12b. The resulting limit cycle time for $K_{\rm P} = 0.015$ is close to 0.5 s (i.e., $\sim 2\tau_{\rm d}$). If the system contained a pure time delay the criterion for doubling the frequency, compared to an I-controller, is $K_{\rm P} = \frac{\tau_{\rm d}\, K_{\rm I}}{2}$. The amplitude is tuned by $K_{\rm I}$ and then $K_{\rm P}$ can be set to give the desired jump back size to increase the frequency. This gives some insight into the basic tuning of the PI based feedback controller.

Two engine experiments with I and PI controllers, performed in Berggren and Perkovic (1996), are shown in Figure 10.13. The same general behavior, with ramps and switches, as the simulations in Figure 10.12 is seen. For a normal engine with EGO feedback control, the limit cycle frequency at idle is in the order of 1–2 Hz and with increased speed the frequency increases as the time delay goes down.

Benefits of Feedforward
The examples above illustrate the benefit of feedback and we now turn to an illustration of how the feedforward helps maintain λ near 1 during transients. A simulation is shown in Figure 10.14, where a pure feedback controller and a combined feedforward and feedback

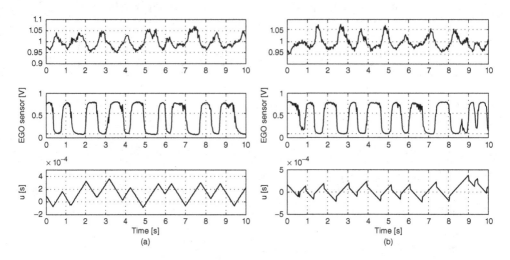

Figure 10.13   An engine with EGO feedback, run in steady state with I (a) and PI (b) control

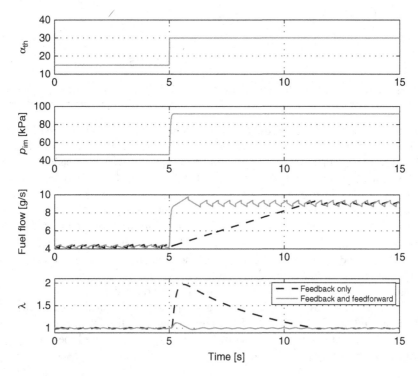

Figure 10.14   Tip-in transient with and without feedforward control of λ. With only feedback it is not possible react to the tip-in transient, and there is a significant control error due to transport delay and limited information from a discrete sensor. Feedforward reduces the control error significantly

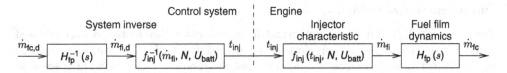

Figure 10.15   Illustration of how the compensation for the fuel injector and fuel dynamics systems use the inverse of their respective subsystems

controller are compared. It is seen that the pure feedback controller reaches $\lambda \approx 2$, which means that the engine goes beyond the lean limit and will misfire. Feedforward reduces the control error in λ significantly. There are still some small errors in the transient caused by wall-wetting dynamics, that are not yet compensated for, so this is the topic of next section.

10.3.4   Fuel Dynamics and Injector Compensation

The purpose of the fuel injection system is to deliver a desired fuel flow into the cylinder, $\dot{m}_{fc,d}$. The right side of Figure 10.15 shows the fuel injector and fuel dynamics that need to be accounted for in the control of the A/F ratio. Also shown in the figure is an illustration of how these are compensated for, using the inverses of their respective models. We first turn to the fuel dynamics, which is the behavior from the injector to the cylinder that was modeled and described in Section 7.7.3. For *direct injected engines* there is no wall-wetting dynamics so there is no need to do compensation, that is $\dot{m}_{fi,d} = \dot{m}_{fc,d}$.

For *port or manifold fueled engines* the wall-wetting dynamics can be significant, especially for cold engines. A compensation can be developed based on Model 7.12. If we assume that the τ_{fp} and X parameters are constant, then the model becomes a linear time invariant system and has the following transfer function

$$H_{fp}(s) = \frac{s\,\tau_{fp}(1-X)+1}{s\,\tau_{fp}+1} = (1-X) + \frac{X}{s\,\tau_{fp}+1}$$

This transfer function has relative degree 0, and as the right hand side shows this comes from the presence of the direct term $(1-X)$. The model can thus be inverted without causality problems and the inverse can be used in the compensation directly

$$\dot{m}_{fi,d} = H_{fp}^{-1}(s)\,\dot{m}_{fc,d}$$

The resulting compensation is illustrated in Figure 10.15. This is only applicable if the wall-wetting model is causally invertible, if not an approximate inverse can be used. An example of how the wall-wetting effect can be compensatedf, using an observer, allowing for varying parameters and non invertible systems, will be illustrated later in Section 10.4.5. The transient compensation is mostly performed in open loop and the fuel film compensator parameters are determined by measurement and calibration, but adaptive fuel-film compensators have also been proposed, see for example Moraal (1995), Moraal et al. (2000), Muske et al. (2008).

Even though wall-wetting dynamics might not be present there can be other fuel system dynamics that have similar effects like, for example, the fuel rail pressure. With a model, they can also be compensated for in similar ways as the wall-wetting effects.

Fuel Injector Characteristics

The purpose of the fuel injection system and the fuel injector controller is to deliver a desired fuel flow to the engine, $\dot{m}_{fi,d}$. Fuel injection systems and their basic operation were described in Section 6.1.1, while models for injectors and the resulting fuel flow were covered in Section 7.7.2. In the control systems the fueling (and also air charge) is often expressed in mass per stroke or mass per cylinder per stroke, telling us how much fuel to inject for each combustion event. The mass that each injector should inject can be calculated from the desired fuel mass flow through

$$m_{fi,inj} = \frac{\dot{m}_{fi,d}\, n_r}{n_{cyl}\, N} \tag{10.13}$$

With the fuel mass given, the injector characteristic (7.28) (or (7.30) for a direct injected engine) can be inverted to give the fuel injector opening time. For a port fuel injected (PFI) engine the desired injector opening time becomes

$$t_{inj} = \frac{m_{fi,inj}}{C_0 \sqrt{\Delta p_{inj}\, \rho_{fuel}}} + t_0(U_{batt}) = [\text{PFI engine}] = C_1\, m_{fi,inj} + t_0(U_{batt}) \tag{10.14}$$

where C_0 is the fuel injector constant that is lumped into C_1, together with pressure and density that are fairly constant for PFI engines.

Rail Pressure Control and Compensation

With advanced injection systems there is also the possibility of varying the fuel pressure in the injection rail, see Section 7.7.2, so that the injection can be optimized for different conditions. This gives two things to handle in the control system:

- Control of the fuel pressure using the fuel pump (no 3 in Figure 10.3) and fuel rail pressure sensor (no 9 in Figure 10.3). This is a regulator problem where feedback from the pressure sensor is used to control the pump flow. The feedback can be complemented with feedforward from the injected fuel flow, to improve the performance of the rail pressure control in load changes.
- The injector opening time, t_{inj}, needs to be adjusted when the fuel rail pressure varies, so that the right amount of fuel is delivered. For moderate fuel injection pressures (10.14) can be used, while high injection pressure systems need a more advanced model, like (7.30).

Beside pressure variations there are density variations, due to fuel temperature variations. This effect can also be included in the injector opening time calculation to increase accuracy.

The fuel injector controller essentially inverts the fuel injector characteristics, see Figure 10.9. The main reason for having it is that it removes the injector control from λ feedback and feedforward controllers that can then concentrate on the flows of fuel, instead of working with injector timings and compensations, making the control design a little easier.

10.3.5   Observer Based λ Control and Adaption

An *observer*, based on the MVEM in Chapter 7, will be used to give an example of a model-based λ-controller. The observer is constructed so that it will estimate the air flow into the cylinder, the fuel puddle mass, and so on. Before starting we recall the general equations for an observer for a nonlinear system. The starting point is a model

$$\begin{cases} \dot{\mathbf{x}} = \mathbf{f}(\mathbf{x}, \mathbf{u}) \\ \mathbf{y} = \mathbf{g}(\mathbf{x}, \mathbf{u}) \end{cases}$$

where $\mathbf{x}$ are the states, $\mathbf{u}$ the inputs, and $\mathbf{y}$ the outputs. A nonlinear observer for such a system can be expressed as

$$\begin{cases} \dot{\hat{\mathbf{x}}} = \mathbf{f}(\hat{\mathbf{x}}, \mathbf{u}) + K(\mathbf{y} - \hat{\mathbf{y}}) \\ \hat{\mathbf{y}} = \mathbf{g}(\hat{\mathbf{x}}, \mathbf{u}) \end{cases}$$

where K is the observer gain. The model can also be used to produce unmeasured signals, that is it acts as *virtual sensor*, that the control system can use. The upcoming Example 10.6 will show how $\dot{m}_{ac}$ and m_{fp} can be observed, based on the models in Chapter 7.

Fuel Control and Adaptation

An observer-based λ controller will be developed in the example below. It uses the observer for the feedforward control while the feedback loop updates and corrects the nominal $(A/F)_s$ ratio. During steady state conditions, the closed loop λ controller can be used to update the data and maps that are used for feedforward compensations in order to reduce the impact of aging or changing environmental conditions.

Example 10.6 (Fuel control with $\dot{m}_{ac}$ and m_{fp} predictors)   In modeling measured signals, engine speed, intake manifold temperature, and throttle angle are inputs. Ambient pressure p_{amb} and temperature T_{amb} are assumed to be constant. Furthermore, the fuel injector characteristic is assumed to be compensated for, as in (10.14), so that the controller commands injected fuel mass flow.

Model: In the first step the model equations, that describe the system, are collected. The manifold pressure is modeled in (7.24), using (7.13) for cylinder air mass flow, $\dot{m}_{ac}$, and (7.11) for throttle air mass flow, $\dot{m}_{at}$. Combining it with the fuel dynamics model (7.31), gives the following nonlinear state space model.

$$\begin{cases} \frac{dp_{im}}{dt} & = \frac{R\,T_{im}}{V_{im}}(\dot{m}_{at}(\alpha, p_{amb}, p_{im}, T_{amb}) - \dot{m}_{ac}(N, p_{im}, T_{im})) & \textit{States} \\[2mm] \frac{dm_{fp}}{dt} & = X\dot{m}_{fi} - \frac{1}{\tau_{fp}}m_{fp} \\[2mm] \dot{m}_{at}(\alpha, p_{amb}, p_{im}, T_{amb}) & = \frac{p_{amb}}{\sqrt{R\,T_{amb}}}Q_{th}(\alpha)\Psi(\Pi(\frac{p_{im}}{p_{amb}})) & \textit{Measured outputs} \\[2mm] p_{im} & = p_{im} \\[2mm] \dot{m}_{ac}(N, p_{im}, T_{im}) & = \eta_{vol}(N, p_{im})\frac{V_D\,N\,p_{im}}{n_r\,R\,T_{im}} & \textit{Non-measured outputs} \\[2mm] \dot{m}_{fp,c} & = \frac{1}{\tau_{fp}}m_{fp} \end{cases}$$

Observer: The first two outputs are measurable and can be used in the observer for updating the states. The two other outputs will be used in the controller as they describe the air mass flow

$\hat{m}_{ac}$, and the contribution of fuel flow from the wall-wetting effect, $\hat{m}_{fp,c}$ to $\hat{m}_{fc}$. The observer, with the feedback gain matrix K, then becomes:

$$\begin{cases} \dfrac{d\hat{p}_{im}}{dt} = \dfrac{R\,T_{im}}{V_{im}}(\hat{m}_{at} - \hat{m}_{ac}) + K_{11}(\dot{m}_{at} - \hat{m}_{at}) + K_{12}(p_{im} - \hat{p}_{im}) \\[2mm] \dfrac{d\hat{m}_{fp}}{dt} = X\dot{m}_{fi} - \dfrac{1}{\tau_{fp}}\hat{m}_{fp} + K_{21}(\dot{m}_{at} - \hat{m}_{at}) + K_{22}(p_{im} - \hat{p}_{im}) \\[2mm] \hat{m}_{at} = \dfrac{p_{amb}}{\sqrt{R\,T_{amb}}}Q_{th}(\alpha)\Psi(\dfrac{\hat{p}_{im}}{p_{amb}}) \\[2mm] \hat{p}_{im} = \hat{p}_{im} \\[2mm] \hat{m}_{ac} = \eta_{vol}(N,\hat{p}_{im})\dfrac{V_D\,N\,\hat{p}_{im}}{n_r\,R\,T_{im}} \\[2mm] \hat{m}_{fp,c} = \dfrac{1}{\tau_{fp}}m_{fp} \end{cases}$$

It is worth commenting that m_{fp} isn't observable, but the mode is stable which makes the system detectable. To get an observable system one could add models for air/fuel ratio including transport, mixing, and λ-sensor output.

Controller: The fuel that should enter the cylinder can be determined by (7.35), provided the air mass flow estimate, $\hat{m}_{ac}$ and $(A/F)_s$ we get the fuel

$$\hat{m}_{fc} = \hat{m}_{ac}\frac{1}{(A/F)_s} \tag{10.15}$$

and the amount of fuel to inject can be calculated from the fuel dynamics model (7.32),

$$\dot{m}_{fi} = \frac{1}{1-X}(\hat{m}_{fc} - \frac{1}{\tau_{fp}}m_{fp}) = \frac{1}{1-X}(\hat{m}_{fc} - \hat{m}_{fp,c})$$

Feedback: A PI controller that measures the exhaust λ ensures that λ is correct during steady state. It can be an adaptive element that estimates

$$\frac{1}{(A/F)_s} = \frac{1}{(A/F)_{s,n}}\underbrace{(1 + PI(e_\lambda))}_{\text{feedback contr.}}$$

in (10.15). Here, e_λ is the error in λ and $(A/F)_{s,n}$ is a nominal $(A/F)_s$ (for example, that which was used during calibration). The deviation from nominal is caught by the feedback loop that has a swing around 1.

The resulting controller is displayed in Figure 10.16.

Extending the Feedback Adaptation to Maps

In the example above we made the interpretation that one parameter $(A/F)_s$ was updated by the feedback loop. With an augmented system model and feedback from appropriate sensors an observer can also be designed to update entire maps. When updating maps one needs to look into observability and other properties and the interested reader is referred to Höckerdal et al. (2011) for details and an example of one-dimensional map updating, and Höckerdal et al. (2012) for a two-dimensional map update for the volumetric efficiency, η_{vol}.

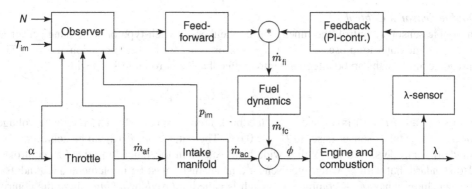

Figure 10.16 Block diagram illustrating engine system, observer, feedforward loop, and feedback λ controller

Observer Feedback Choices

Sensors and their placement influence the output of the observer, even though the engine system is essentially the same. Sensor dynamics can be included in the model, for example if they have time constants that are slower or are in the same order as the system. Examples of different sensor configurations are the choice of sensor for air mass flow $\dot{m}_{at}$, or intake manifold pressure p_{im}. With these two sensors we get three different configurations,

- Only air mass flow sensor, and no pressure sensor.
- Only intake manifold pressure sensor.
- Both air mass flow sensor and pressure sensor.

The impact of these choices on the system is that the number of outputs change and that the observer gain K is tuned in different ways, this is investigated more closely in Jensen et al. (1997).

10.3.6   Dual and Triple Sensor λ Control

A mid-brick or post-catalyst λ sensor is often required for catalyst diagnosis, see Section 16.5.1, and when it is available it can be used for control to further improve the catalyst efficiency. One reason for improvement comes from the fact that the switch-type sensor has a slight difference in dynamics between rich and lean steps which can provide a slight offset in the control. Another is that a pre-catalyst sensor is exposed to high temperatures and variations in untreated gas composition which change the λ to voltage characteristics, this occurs for both the switch-type and wide-band sensors. A post-catalyst sensor is less exposed to these variations and also adds information about the catalyst state, for example oxygen storage. A single feedback loop with only a post-catalyst sensor is therefore tempting, but it is not enough since the additional delay and catalyst dynamics give a too low bandwidth in the feedback loop. Therefore, cascade control is used with a faster inner loop and slower outer loop. According to Dietsche (2011) there are also *triple sensor* configurations, as shown in Figure 10.3, used for reaching SULEV requirements (see Section 2.8) where the third sensor is used in an extremely slow outer loop.

Double Sensor λ Control

In double sensor λ control, the inner loop uses either a switch-type or wide-band sensor at λ_{bc}, while the outer loop uses a switch type sensor at λ_{ac} for highest precision at $\lambda = 1$. The cascade control with can be integrated into the feedback controller (10.12) with

$$\Delta_{fc} = FB_i(\lambda_{bc}, FB_o(\lambda_{ac}, U_{ref}))$$

where FB_i are FB_o are inner and outer feedback loops respectively. The reference voltage for the post-catalyst sensor is set to $U_{ref} \approx 0.6$ V. The outer loop, FB_o, can be achieved with standard controllers, like PI. With a wide-band sensor in the inner loop the FB_i control is also standard, but with a switch-type sensor a mechanism must be implemented that adjusts the switching behavior. A common approach is called *delayed switching* where the control is delayed for a time t_d after a switch in λ_{bc} has been detected, see Figure 10.17. In delayed switching, the time t_d, during which the inner loop control output is held constant, is controlled by the outer loop

$$t_d = FB_o(\lambda_{ac}, U_{ref})$$

Delayed switching can adjust the average lambda from $\lambda = 1.0$ within limits dictated by the I part. Figure 10.17 shows an example where the average λ is controlled to 1.01 and 0.99. In the example the outer loop, FB_o, is a slow I-controller and the inner loop FB_i uses delayed switching. In the inner loop, shown in Figure 10.10, the delayed switching mechanism is introduced in the PI controller after the relay that provides the switching information.

10.4   Other Factors that Influence λ Control

In the previous section on λ control, the focus was on normal engine operation where the emission reduction with the TWC is the key factor. For smooth, reliable, and efficient engine operation over the entire operating region there are additional requirements that influence the

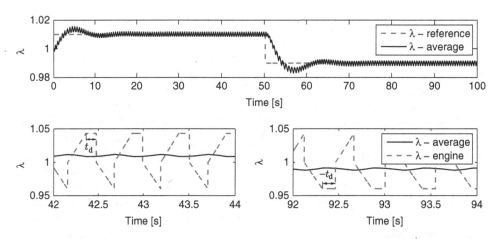

Figure 10.17   Example where delayed switching on the inner loop is used to modifying the average λ. When a switch in the inner loop λ is detected, the value of the control is held for t_d s

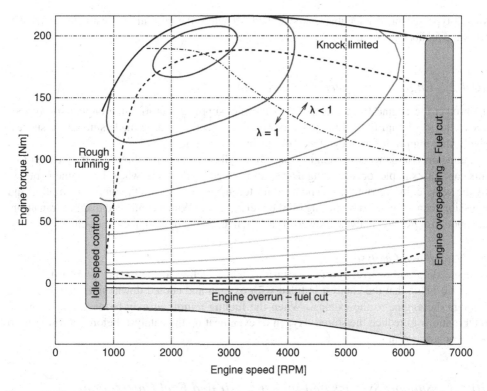

Figure 10.18   Sketch of an engine map, illustrating that there are control strategies and trade offs that are active in different operating regions. The dash dotted line shows the transition between fuel enrichment and $\lambda = 1$, and the dashed shows where optimal ignition is abandoned

λ control. These requirements impose trade offs that can change either the control mode or reference value for λ, see Figure 10.18, and these will be discussed in the sections below.

10.4.1   Full Load Enrichment

Fuel enrichment has the main effect of cooling the cylinder charge, through fuel evaporation, which reduces the temperature at IVC. As a result the air density and air mass flow to the cylinders also increase, see, for example, (7.15). Furthermore, it adds more mass to the cylinders which all in all reduces the temperature after the combustion. Full load enrichment is used to achieve three goals: firstly and most importantly, protecting the catalyst (and turbine on turbocharged engines) from thermal damage through reduced exhaust temperature. Secondly, reducing the knock tendency of the engine through decreased temperature at IVC and the subsequent lower temperature of the unburned gas during the combustion process. Thirdly, maximum power and maximum acceleration is attained if the engine is run slightly rich, as more energy is available due to the increased density.

The transition to a rich mixture is made smooth by a gradual increase in fuel as the engine load (or power) gets closer to maximum. The outer loop for λ control is switched off, since the catalyst and emission control are abandoned in favor of protection. If the inner loop has a wide-band sensor, then it is used with reference values below 1. If the inner loop has a switch-type sensor, the feedback is switched off while the current best estimate of $(A/F)_s$ from the feedback is used.

10.4.2 Engine Overspeed and Overrun

To protect the engine from *overspeeding* the fuel supply is cutoff if the engine speed goes above the maximum limit. Engine overspeeding protection can be made smooth by successively shutting off different cylinders, or closing the throttle.

Engine *overrun* is when the engine runs above idle speed but the throttle is in idle position, this can, for example, occur during deceleration or going downhill when the engine is pulled by the vehicle. In overrun there is no need for torque generation and the fuel consumption and emissions can be reduced by cutting off the fuel injection. With a controlled overrun transition, a smooth progression between power-off and active engine operation can be achieved.

Fuel Cutoff and Catalyst

Overspeeding and overrun are two occasions when fueling to the engine is cut off. A side effect of this is that the catalyst is filled with oxygen and cooled down, since only air goes through the engine to the exhaust. When the fueling returns, it may be necessary to make interventions to reduce the stored oxygen or even heat up the catalyst, before returning to to $\lambda = 1$ control again.

10.4.3 Support Systems that Influence Air and Fuel Calculation

There some support systems that influence the air and fuel calculations, and Figure 10.19 shows three examples: EGR control, carbon canister purge, and crank case ventilation, that will be discussed below.

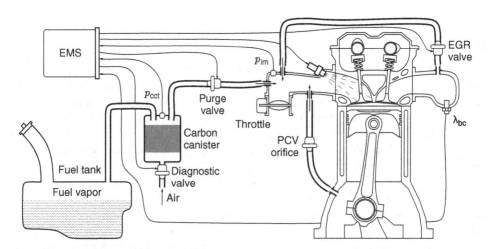

Figure 10.19 Example of disturbances to the air and fuel metering caused by additional flows of air and fuel into the intake manifold

EGR Control

In exhaust gas recirculation (EGR) burned gases are extracted from the exhaust system, returned to the intake system, and mixed with the air–fuel mixture. There is always some degree of internal EGR, see Section 5.2.4 and Example 5.1, which is due to the exhaust and induction processes and the valve overlap. The reasons for using EGR are that it reduces the engine-out emissions of NO_x, and also the part load fuel consumption. The already combusted gas reduces the peak combustion temperature, which in turn can reduce the NO_x formation up to 60%. Part load efficiency is improved since the addition of EGR increases the intake manifold pressure, which reduces the pumping work for the same amount of air and fuel. Cooled EGR can also be used to reduce the knock tolerance. There is a limit, called the EGR tolerance, that specifies how much EGR the engine can tolerate before misfire is generated. The practical upper limit on EGR ratio is therefore imposed by the increase HC emissions and fuel consumption, near the misfire limit, in combination with rougher engine operation.

The EGR is controlled by an electro-pneumatic actuator connected to the EMS, see Figure 10.19. The EGR is shut off during idling, and when no significant NO_x emissions are produced in any case and there are high amounts of internal EGR. In part load operation EGR is usually applied, as it is here that its potential is greatest. The unfavorable pressure conditions at wide open throttle pose limitations on the EGR ratio since there is not enough pressure difference between the exhaust and intake systems to produce EGR flow. Furthermore, at maximum torque, the amount of air that can be inducted is a limiting factor, and EGR is shut of since it competes with air for the space in the cylinder. Two practical issues that introduce margins for EGR rates and calibration are: (i) clogging of valves and plumbing because of exhaust-gas deposits, (ii) aging of the engine reduces its EGR tolerance.

When considering λ control it is important to account for the fact that EGR adds additional gases to the intake and cylinder. Therefore, it is not possible to use the volumetric efficiency to calculate the air mass flow, instead the total mass flow has to be divided into air and EGR mass flow, as mentioned in Section 7.4.1.

Evaporate Emission Control – Purge

As fuel evaporates in the fuel tank, hydrocarbons can be discharged into the atmosphere. This effect increases with increased fuel temperature. The legal requirements regarding evaporative emissions can be satisfied with a carbon canister which stores the hydrocarbons from the fuel tank and ventilates it by drawing air through the canister into the intake system, see Figure 10.19. In normal operation the diagnostic valve, discussed in Section 16.5.3, is open and the fuel vapors are ventilated exclusively via the carbon canister. Since the canister has only a limited capacity it must be continuously regenerated (purged) by drawing fresh air through it into the intake system.

When the engine is running, air is drawn through the diagnostic valve and carbon canister into the engine intake. As a side effect it produces an increase in fuel flow from the purge system that the example below will illustrate.

Example 10.7 ((A/F) changes from the purge system)   An engine runs with isooctane fuel at $\lambda = 1$. Some fuel has evaporated and is stored in the carbon canister. When the canister purge valve is opened, 1% of the intake-gas volume-flow is replaced by gaseous fuel that enters from the purge system. How does this effect λ?

The $\lambda = 1$ condition before the opening of the valve yields

$$\frac{\dot{m}_{ac}}{\dot{m}_{fc} \, (A/F)_s} = 1 \; \Rightarrow \; \dot{m}_{ac} = \dot{m}_{fc} \, (A/F)_s$$

For isooctane the stoichiometric mixture strength is $(A/F)_s = 15.1$ and the mass ratio

$$\frac{m_f}{m_a} = 4.773 \frac{8 \cdot 12.001 + 18 \cdot 1.008}{32 + 3.773 \cdot 28.16} = 3.94$$

The additional fuel-mass flow is thus $\Delta \dot{m}_{fc} = 0.01 \, \dot{m}_{ac} \, \frac{m_f}{m_a} = 0.0394 \, \dot{m}_{ac}$. With the increase in fuel flow and decrease in air flow the new λ becomes

$$\lambda_{purge} = \frac{0.99 \, \dot{m}_{ac}}{\dot{m}_{fc} + \Delta \dot{m}_{fc}} \frac{1}{(A/F)_s} = \frac{\dot{m}_{ac}}{\dot{m}_{fc} \, (A/F)_s} \frac{0.99}{1 + 15.1 \cdot 0.0394} = 0.62$$

Under these, somewhat extreme, conditions λ decreases by as much as 38%.

Since the canister-purge system produces significant changes in λ it is required that the vapors from such a system must be controlled, both for maintaining exhaust emissions within the desired limits as well as ensuring good driveability. Under certain conditions the canister purge system is either switched off (idle) or it remains ineffective (insufficient vacuum at full-load conditions).

Crank Case Ventilation

Positive crank case ventilation (PCV) is a measure taken to reduce the hydrocarbon emissions from piston blow-by. Fuel that enter the crank case of the engine can later escape to the atmosphere, for example, through holes or leakages in the seal of the engine. Therefore, it is required that the gases that escape to the crank case are led back into the intake and later combusted. This adds a small but additional flow of both air and fuel to the intake, but it is small to negligible since it has already been measured by the air flow meter and fuel injector.

10.4.4   Cold Start Enrichment

At cold start the engine needs to be fed with more fuel, compared to a fully warmed up engine operating at the same speed and load. One reason is that for good combustion quality the fuel must be vaporized before the combustion. At cold conditions the fuel does not evaporate completely, and in order to produce a combustible mixture the amount of injected fuel is increased. Another effect is that the valves and walls don't heat the charge as much, which gives a higher volumetric efficiency which also requires additional fuel.

10.4.5   Individual Cylinder λ-control

Some techniques have been proposed for determining the (A/F) ratio for the individual cylinders using only one λ sensor after the confluence point in the exhaust. The techniques can be utilized for controlling the fuel and minimizing the maldistribution of λ between the cylinders. Maldistribution of λ between the cylinders can occur for several reasons, for example: intake system design, production tolerances in injectors or inlet valves, and aging and wear of injectors or inlet valves. Individual cylinder control is beyond the scope of the book and the interested reader is referred to Cavina et al. (2010), Grizzle (1991), Kainz and Smith (1999), and Shiao and Moskwa (1996).

10.5 Ignition Control

We now turn to the ignition timing, which is the third of an SI engine's main loops. It is important since it affects the major outputs of the engine such as: torque, efficiency, emissions, and engine knock. The basic function is to initiate a flame kernel that can develop into a turbulent flame and propagate through the combustion chamber. *Ignition control* means control of both the energy and timing of the spark, where the energy is used to ensure that combustion starts, while the timing is used to position the combustion relative to the crank motion. Spark energy control goes under the name *dwell time control*, while spark timing control is also called *spark advance control*. The term *spark advance* refers to the crank angle position before TDC (BTDC), where the spark discharge occurs, that is a spark advance of 20° corresponds to an ignition timing of 20° BTDC. As indicated in Figure 10.2, the ignition controller is

$$\theta_{ign} = \theta_{ol} + \Delta\theta_{fb}$$

which consists of an open loop controller, for maximum efficiency, and a feedback loop, for knock protection.

Maximum Brake Torque Control – Open Loop Control
The ignition timing that gives maximum torque in a dynamometer (brake) or vehicle is called *maximum brake torque* (MBT) timing. Timings that are advanced (more spark advance) or retarded (less spark advance) from MBT timing result in a lower output torque, see Figure 7.29. For an engine, MBT timing normally lies somewhere between 15 and 40° and varies with the operating condition. The MBT timing depends in a complex way on engine design, operating condition, and the properties of the fuel, air, and burned gas mixture. The main purpose of the open loop controller, during normal operation, is to fulfill

$$\theta_{ol} = \theta_{ign,opt}(\omega_e, m_f, \lambda, x_r)$$

in (7.58), igniting the mixture at the optimum position, independent of the operating condition. Optimum spark advance depends mostly on intake manifold pressure, p_{im}, and engine speed, N, and then various compensation factors are employed to account for variations in, for example, EGR, inlet air temperature, and coolant temperature. Spark advance controllers are calibrated open-loop systems, see the ignition timing manager block in Figure 10.2.

There are operating condition a where the optimal ignition timing is abandoned, see Figure 10.18. For example, near maximum torque the knocking can be limiting and the optimal ignition timing has to be abandoned. Another case is at low speeds where the engine can induce vibrations in the driveline which can be reduced with later ignition timing.

Empirical Rules for Calibration and Feedback Control

There are empirical rules that relate the cylinder pressure trace and mass fraction burned profile to MBT timing, and two examples are:

- MBT timing results in a position for the pressure peak (PPP) that occurs at 16° independent of the engine speed and load, see, for example, Hubbard et al. (1976). This is illustrated in Figure 10.20. Actually, the optimum PPP varies with engine design, but is fairly constant for an engine and lies between 12° and 20° ATDC.
- MBT timing results in the fact that the 50% mass-fraction burned curve occurs at 9° after TDC, see, for example, Matekunas (1984), Sellnau and Matekunas (2000). The exact location can vary with engine and design, but the method is popular and it is often referred to as MFB50.

These empirical rules relate the result of the spark advance to the efficiency, and can thus be used during calibration or in a feedback controller. But they also require extra sensors, like cylinder pressure. Feedback control of the ignition timing that use direct in-cylinder measurements of the combustion for optimizing engine efficiency has so far only been seen on the

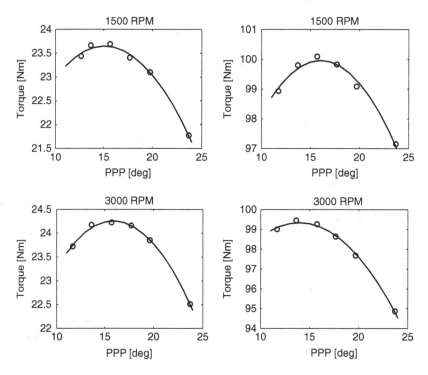

Figure 10.20   Peak pressure position (PPP) compared to the maximum output torque, over a wide range of engine operation conditions. For all operating conditions, the maximum torque gives a PPP that lies around 15–16° ATDC. In each plot, only the spark timing is changed while the injected fuel, engine speed, and throttle angle are held constant. Each circle is computed as a mean value from 200 consecutive cycles in the same operating point

research bench, here exemplified by some early works are Eriksson et al. (1997), Hubbard et al. (1976), Powell (1993), and Sellnau and Matekunas (2000). With the drive for improved engine efficiency, the current practice with open loop strategies might in the future be replaced by more advanced feedback strategies.

10.5.1   Knock Control – Feedback Control

Engine knock was discussed in Section 6.2.2, where it was pointed out that knocking gives a noise that can be a nuisance for the driver and if it is left unattended it can also lead to engine failure. Knock tendency depends largely on the end gas temperature, and Figure 10.21 shows a simulation of the end gas temperature, assuming ideal gas and adiabatic compression of the end gases, for different ignition angles. The figure shows that the end gas temperature can be controlled by the spark advance, and that a later ignition timing gives a lower end gas temperature. Knock control uses this effect and aims at controlling the engine as close to the knocking limit as possible, allowing a few knocking cycles to occur but preventing knock from occurring too often.

Historically, knock was avoided by having a conservative spark advance calibration, selected so that knocking would not occur under any driving or environmental conditions. This meant that at high loads the spark advance is retarded (later ignition) from the optimum, which results in a conservative schedule. Current practice is to control knock with a feedback system that relies on a measurement of knock intensity. When a knock is detected, the spark advance is retarded, which decreases the end-gas temperature and the knock tendency. Knock control is performed for individual cylinders and the knock controller overrides other spark advance control functions.

Knock Sensing
There are several ways to sense the combustion knock, and Figure 10.22 shows three sensor examples: knock senor, cylinder pressure, and ion sensing. For these three the corresponding raw signals are shown in Figure 10.23. Knock is easily seen in the cylinder pressure

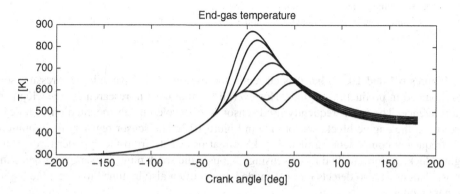

Figure 10.21   The mechanism for controlling knock is by controlling the end-gas temperature. A later ignition timing gives a lower end-gas temperature and thus a lower risk of knock

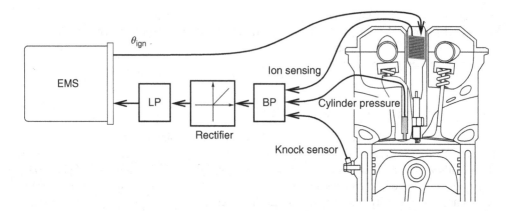

Figure 10.22 Knock is controlled by the ignition timing. Three examples of possible knock sensors: ion sensing, cylinder pressure, and knock sensor (accelerometer). Knock detection is by band pass (BP) filtering the resonant modes, half wave rectifying, and low pass filtering the signal. The EMS samples the signal in a window where knock is liable to occur, that is, from around TDC to approximately 40 degrees after TDC, the maximum value gives a measure of the knock energy

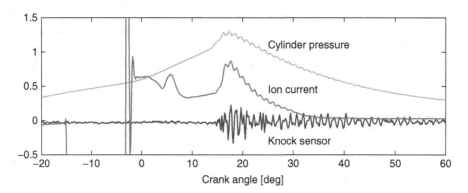

Figure 10.23 Three signals for the same knocking cycle and cylinder. The spark ignites the mixture at 15° BTDC and an auto ignition occurs around 14° ATDC. The ringing associated with the knock is clearly visible in all three signals

(see Figures 6.4 and 10.23) but it is currently too expensive to have cylinder pressure sensors mounted in production engines. However, it is often used in research and development as a reference. The most frequently used sensor is an accelerometer mounted on a strategic position on the engine block, sensor (15) in Figure 10.3. The sensor registers the vibrations with frequency components around 5–8 kHz, that propagate from the cylinder through the engine block. A third method is ion sensing, that uses the spark plug as a sensor for detecting the oscillations, it also detects the first mode but can often also be tuned to detect the higher resonant modes.

All three signals, shown in Figure 10.23, show the 5–8 kHz oscillation that is associated with knock. Figure 10.22 shows the components in a knock detection system where filters implemented in hardware are used, with more powerful CPU more of the detection is moved into

the EMS hardware and software. Independently of the detection implementation, a measure
of the knocking energy E_k is produced and this information is used for feedback.

Knock Feedback Control

The controller acts on each cycle and the independent variable is the cycle index i, it also
operates on all cylinders. If the knock energy at cycle i is larger than a threshold $E_k \geq E_{Thres}$,
then it is said that knock is detected and this is indicated by $D_k[i]$. To protect the engine, when
knock is detected, the feedback controller ignites later (retarded ignition), and this is done
with a retard gain K_{re}. On the other hand, if no knock is detected the ignition can be advanced
towards optimum, with an advance gain K_{ad}. This gives the following scheme for the feedback
controller

$$\Delta\theta_{fb}[i+1] = \max(\Delta\theta_{fb}[i] - K_{ad} + K_{re}\, D_k[i], 0) \qquad (10.16)$$

where the max selector ensures that the ignition is not advanced past the nominal ignition
timing, which during normal operation is near MBT timing. Figure 10.24a illustrates the action
of the controller when there are knock events (dashed).

Figure 10.24b shows the resulting spark advance for a five-cylinder engine, controlled by
the feedback controller. The circles mark the knock events and the pattern from the left plot
is easily recognized. There is a spread between the cylinders due to engine wear from varying
oil and soot loading in the cylinders.

In the controller, there are two tuning parameters: the advance and retard gains, K_{ad} respectively K_{re}. These are tuned to give a certain knock probability which is p. If we assuming steady
state operation in average, the advances and retards will cancel out. Knock probability p and
controller gains are then related to each other according to $K_{re}\, p = K_{ad}\,(1-p)$ which can be
re-arranged to

$$p = \frac{K_{ad}}{K_{re} + K_{ad}} \qquad K_{ad} = \frac{p}{1-p}\, K_{re} \qquad (10.17)$$

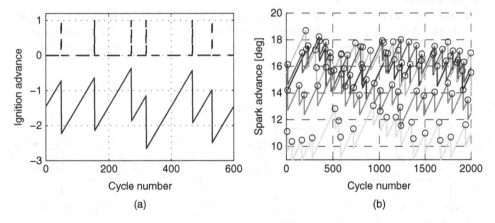

Figure 10.24 Knock feedback control. (a) Illustration of knock control with knock detection, $D_k[i]$,
(dashed), and the advance and retard for knock events (solid), where the advance has the opposite sign as
the ignition angle, that is $-\Delta\theta_{fb}$. (b) Closed loop spark advance for a supercharged five cylinder engine
running at $N = 1500$ rpm, $p_{im} = 120$ kPa, and $r_c = 12$. Circles indicate the knock events

In Figure 10.24, from Thomasson et al. (2013), the controller retard gain was set at $K_{re} = 1.5$ degrees (typical for a production system), and the increment gain for a $p = 1\%$ target knock rate then follows from 10.17, giving $K_{ad} = 1.5/99$. Counting knock instances, one sees that there are about 5 knocks for 500 cycles, which agrees with the 1% set point.

Stochastic and Adaptive Ignition Control

The knock controller (10.16) continuously searches for the knock limit, which results in continuous change in control action. To remedy this, a stochastic knock control approach has been proposed, that uses the likelihood ratio for controlling the knock probability and changes the control only if it is needed. As a result there is less control action for the same knock ratio and performance, see for example, Peyton Jones et al. (2013), Thomasson et al. (2013).

The controllers above react and adjust to knock, which might limit the control performance in transients. For example, knock depends on the fuel, and if a low octane fuel is used the knock controller would need to continuously adjust to the varying knock demands. A faster knock control action in transients can be achieved by storing a low pass filtered knock control action in a map for each operating point. When the engine returns to the operating point, the retard is read from the stored map and ends up in the correct point, see, for example, Kiencke and Nielsen (2005).

10.5.2   Ignition Energy – Dwell Time Control

There are basically two different types of electrical ignition systems: capacitive discharge and inductive discharge systems, see Figure 10.25. The inductive is most common since it has greater efficiency than the capacitive discharge system. In a capacitive discharge system, energy is first stored in a capacitor, and at the time of ignition the capacitor discharges its energy through the primary circuit of the coil, which induces a high voltage at the secondary circuit. In an inductive discharge system, a current is led through the primary circuit store energy in the coil, and at the ignition the primary current is cut off which induces a high voltage in the secondary circuit.

The high voltage peak in the secondary circuit gives a spark at the spark plug. The dwell time t_{dw} is the time before θ_{ign} that the ignition system charges the capacitor or the primary winding of the ignition coil, increasing its energy. The energy stored in a capacitor with capacitance C

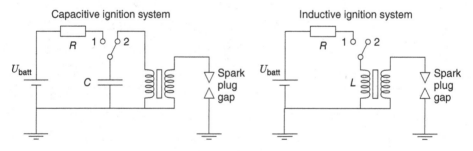

Figure 10.25   Sketch of inductive and capacitive ignition systems, with ignition coils and spark plug. When the switch is moved from position 2 to 1 the energy storage starts, as charge in capacitive system and as a magnetic field in the inductive system. At θ_{ign} the switch is moved from 1 to 2, which produces a high voltage at the secondary winding and a spark at the spark plug

and coil with inductance L are given by the expressions below where there is a resistance R in the circuits.

	Energy	Voltage or current
Capacitor	$W = \dfrac{C\,U_C(t_{dw})^2}{2}$	$U_C(t_{dw}) = U_{batt}\left(1 - e^{-t_{dw}\frac{1}{RC}}\right)$
Coil	$W = \dfrac{L\,I_L(t_{dw})^2}{2}$	$I_L(t_{dw}) = \dfrac{U_{batt}}{R}\left(1 - e^{-t_{dw}\frac{R}{L}}\right)$

Dwell time control is implemented as an open loop controller based on engine speed, load, and dilution, and it is calibrated during the engine calibration procedure.

10.5.3   Long-term Torque, Short-term Torque, and Torque Reserve

As mentioned above, the main purpose of ignition control is high efficiency, but there are some occasions when this is abandoned. In Section 3.3.6 the three torques *long-term*, *available*, and *short-term* torque were introduced, to enable flexible control of the powertrain torques. Long-term torque, $M_{e,l}$, is used to prepare the engine so that it will be able to produce the desired torque; short-term torque, $M_{e,s}$, is used for shorter interventions or other control actions, while available torque, $M_{e,a}$, is the maximum torque that is currently available. Looking at the engine system dynamics, we see that the fuel and ignition have relatively short time constants, while the intake manifold (and turbo dynamics) have longer time constants. Therefore, the air path is suitable for long-term torque, controlling pressures and gas flows to the engine, while the ignition and injection are suitable for short-term interventions. Fuel cutoff can also be used if engine braking is needed, while ignition timing can be used for more smooth torque reduction.

To achieve a desired torque with the ignition timing, it is possible to use the calculation scheme from Section 10.2.2, where the torque model (10.2) is solved for the ignition efficiency and then the ignition efficiency model (7.58) is solved for the ignition. An alternative, which has the advantage that submodels for friction and pumping losses are not needed in the calculations, is to use the notion of *torque reserve*, ΔM_a. It was defined as the gap between the available and short-term torques, $\Delta M_a = M_{e,a} - M_{e,s}$. Using (10.2) we can express it as follows

$$\Delta M_a = M_{e,a} - M_{e,s} = \frac{m_f\, q_{LHV}\, \eta_{ig}\, \eta_\lambda}{n_r\, 2\,\pi\,(1 - r_c^{\gamma-1})} \underbrace{[\eta_{ign,a}(0) - \eta_{ign,s}(\Delta\theta_{ign})]}_{=1} \tag{10.18}$$

This can be solved for $\eta_{ign,s}$ and the ignition efficiency model (7.58) can then be used to solve for $\Delta\theta_{ign}$ (using only positive values) that is added to the optimal ignition timing and gives the resulting ignition actuation

$$\theta_{ign} = \theta_{ign,opt} + \Delta\theta_{ign}$$

An illustration of how this can be used for control will be illustrated in Example 10.9 in Section 10.8.

Torque Reserve, Efficiency, and Catalyst Heating

When ignition efficiency is used to implement the torque reserve, it also affects the temperature out of the cylinder, see Section 5.2.2, where lower efficiency means more q_{loss} which in turn means higher T_4 in the cycle. During shorter interventions, there is no significant effect on the temperatures of the exhaust component's exhaust, due to their thermal inertia. With longer interventions, a torque reserve and knock control heats up the catalyst, which might need to be considered at high load conditions. However, in some cases it can be beneficial, for example, during cold start a torque reserve (reduced engine efficiency) can be used to shorten the catalyst light-off time and thus help to reduce the cold start emissions.

10.6   Idle Speed Control

Idle speed control is a regulator problem where the engine speed is controlled so that the engine is not allowed to drop too low so the engine stalls. It is an important function that influences the feeling and acceptance of the vehicle. For example, a poorly designed idle speed controller, with variations in engine speed or occasionally failing to prevent engine stall, is a nuisance for the driver. There are several items to consider for the idle speed controller:

- Time delay from control action to power stroke, see Section 7.9.4, limits the bandwidth in the feedback loop.
- Use of two actuators to achieve the control action; fast but limited ignition timing and slower but larger torque from the air manager. This can be integrated through torque based structure with the long- and short-term torque.
- Care must be taken to handle the mode switching, for example, entering and leaving idle speed control mode should be smooth for the driver.
- Limited torque actuation, for safety reasons the idle speed control is not allowed to use the full torque actuation, for example, if the driver's foot slips off the clutch pedal by mistake the engine should stall to prevent vehicle runaway.
- Reject disturbances from auxiliary devices that require relatively large torques from the engine, such as: generator, air conditioner, automatic transmission put into drive, or power-steering servo pump. Feedforward from measurable disturbances is used improve the performance, for example, the engine controller can get a signal from the air condition controller before it switches on and prepares proper actions.

Several control designs have been proposed in the literature where much of the attention is on the time delay and dynamics of the system and the separation between the controllers. The following example illustrates how the idle speed controller can be structured to utilize short- and long-term torque.

Example 10.8 A PID controller for the idle speed reference speed uses the short-term torque, $M_{e,s}$, as actuator. Based on this, the long-term torque, $\Delta M_{e,l}$, is modified so that a long-term torque reserve, M_l, is added to the short-term torque. This torque is then used as reference for the air management (10.4). The controller then becomes

$$M_{e,s} = PID(N_{ref} - N) \qquad\qquad (10.19a)$$

$$M_{e,l} = M_{e,s} + \Delta M_l \qquad\qquad (10.19b)$$

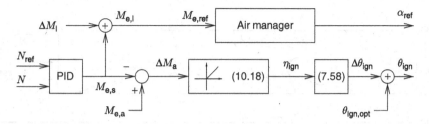

Figure 10.26 Example of the idle speed controller using the ignition angle as a fast actuator and the air system as slower actuator with a long-term torqe reserve as difference

where the PID is implemented with anti-windup and bump-less transfer features to facilitate smooth mode switching when entering and leaving idle speed. An illustration of this idle speed structure is shown in Figure 10.26, where a saturation is added to the torque reserve ΔM_a to avoid negative values, since the ignition controller might demand more torque than what is available.

In the context of the complete engine controller, these two represent the two lower loops in Figure 10.2, and they will be used in the example in the next section.

10.7 Torque Management and Idle Speed Control

The following example illustrates how the torque reserve is used during idle and also how the fuel cutoff and is used during gear changes and engine braking. In the example, a driver follows a portion of the speed profile of the European driving cycle that includes idle, acceleration, gear change, deceleration and stop going back to idle.

Example 10.9 (Vehicle control with torque management and torque reserve) Engine speed during idle is controlled by a feedback loop on the engine speed using the torque manager as actuator with air manager and spark advance control, to control the long- and short-term torques. A long-term torque reserve of $\Delta M_1 = 10$ Nm is used during idle. A portion of the European drive cycle, see Figure 2.22, is used to exemplify the control functions.

The resulting speeds and control actions during launch, gear change, driving, and deceleration are shown in Figure 10.27. The vehicle speed profile is shown in the top plot, together with the engine speed. At 49 s the driver starts closing the clutch to accelerate the vehicle and the idle speed controller is used during the initial take off. One can clearly see the torque reserve in the torque plots and how it is actuated using the ignition timing. At the instant when the clutch is closed, the engine speed drops 20 RPM below the set point and the ignition timing is actuated to quickly generate extra torque and push the idle speed back to the set point. When the clutch is locked, at 51 s, the driver accelerates the vehicle with the accelerator pedal and the torque reserve is set to 0 so that the ignition is at optimum during driving. At 54 s there is a gear change where the torque is reduced by retarded ignition and the engine also enters fuel cutoff. After the gear change, there is acceleration and driving until the deceleration phase starts, where the torque demand goes down and fuel cutoff is used during the deceleration. Finally, when the engine speed reaches idle, the idle speed controller is activated.

Idle speed is frequently encountered in urban driving and it is therefore desirable to reduce the idle fuel consumption which follows the engine speed set point. It is desirable to have as low

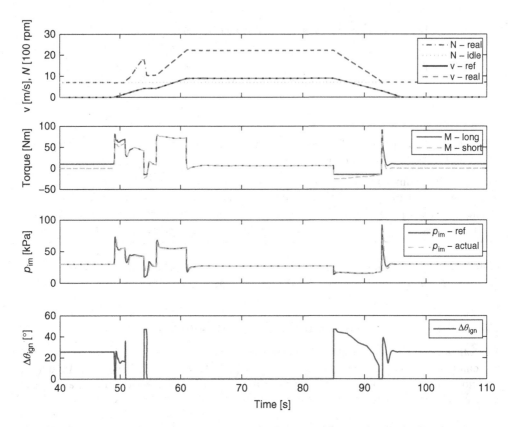

Figure 10.27 A section of the European drive cycle illustrating idle speed control with torque reserve and gear changes, as well as engine braking with fuel cutoff

engine speed as possible, but a low speed gives a longer time delay from control action to torque development, which makes the control problem more challenging. This leads to the trade-off between the risk of stall and engine fuel consumption, which is also present in the trade-off with the torque reserve. In addition, the idle speed set point can be increased at occasions, for example at cold start, to shorten the catalyst warm up time.

10.8 Turbo Control

Compared to a naturally aspirated SI engine, the turbocharged SI engine has two additional actuators and control loops. One is *anti-surge control*, where the goal is to avoid compressor surge. The other is *boost control* that supports the torque control and uses the wastegate actuator and a boost pressure sensor p_{ic}. Figure 10.28 shows an example of the surge control valve with a pneumatic wastegate actuator system.

10.8.1 Compressor Anti-surge Control

Surge is a compressor flow instability that can cause noise and potentially turbocharger failure, recall Sections 8.4.1 and 8.6.5. The turbine and compressor are matched so that surge should

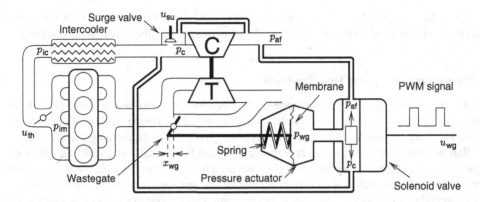

Figure 10.28 Illustration of an anti-surge valve and a pneumatic wastegate actuator system. The wastegate actuator position x_{wg} is controlled by the solenoid valve that uses the boost pressure p_c to generate the pressure on the membrane and the pneumatic actuator force. There are also other actuator arrangements that use, for example, the brake system vacuum

not occur during normal operation. However, as the example of a surge cycle in Section 8.6.5 shows, surge can occur if the throttle is closed rapidly, which causes the pressure ratio over the compressor to rise and forces the compressor into surge.

Surge can be avoided by reducing the pressure in the volume after the compressor. A control valve either leads the gas back to before the compressor, as in Figure 10.28, or releases the compressed gas out to the atmosphere. An advantage with the configuration in the figure is that the mass in the system is conserved and the mass flow sensor can be utilized to determine the mass in the system. The anti-surge valve (sometimes also called surge, bypass, blow-off, or dump valve) is currently a binary valve that is either open or closed. The anti-surge valve is controlled by the EMS, which opens the surge control valve if it detects throttle closing and a possibility of high pressure ratios and low or reversed flows.

In V-engines with dual turbochargers that feed the same intake system, an instability phenomenon called co-surge can occur. In co-surge the compressors alternate into and out of surge and this can occur during normal boosted operation if the compressors are disturbed and the individual speeds start to deviate from each other. In such engines, it is necessary to monitor and control co-surge.

10.8.2 Boost Pressure Control

The wastegate is primarily used to control the boost pressure, that is the pressure after the compressor or intercooler, see Figure 10.28 for an illustration. The wastegate allows some of the gases to bypass the turbine and reduces the turbine power and thus the compressor power. The controller is foremost tuned to give the desired torque curve. Its basic functionality is to open the wastegate when the pressure after the compressor goes over a certain limit, and this can be achieved using mechanical designs. A problem with mechanical design of the wastegate controller is that the wastegate opening is only determined by the wastegate spring stiffness and pressure differences.

Boost pressure control in modern engines is therefore integrated into the torque-based structure, and it uses a boost pressure sensor, often the pressure before the throttle p_{ic}, for feedback control. An example of a boost pressure controller will be given in Section 10.9.3, but before going into the specifics we will look at a trade-off between boost pressure and throttle control.

Efficiency and Response Time Trade-off

In boost pressure control, there is a trade-off between fuel economy and response time. Boost pressure control is about delivering air to the engine, which is also the purpose of throttle control. There are two actuators for the same purpose, giving the controller a degree of freedom. It thus becomes interesting to see how the air can be delivered most efficiently and what the trade-offs are. The traditional mechanical boost pressure controller, that gives the *best transient response*, closes the wastegate as much as possible and only opens it when the boost pressure has reached its maximum $p_{ic,max}$. The controller thus tries to control the boost pressure to the maximum according to

$$u_{wg} = PID(p_{ic} - p_{ic,max})$$

This strategy strives to *always close* the wastegate as much as possible, while the air flow to the engine is controlled by the throttle.

In Eriksson et al. (2002a), a *fuel optimality statement* was derived for an SI engine, showing that the most fuel efficient operation is to reduce the back pressure and thus open the wastegate as much as possible, provided that sufficient air is delivered. This controller can be seen as *always opening* the wastegate as much as possible. In low loads, the wastegate is open and the load is controlled by the throttle, when the throttle reaches fully open the wastegate takes over and controls the intake pressure.

$$u_{wg} = \begin{cases} 0 & \text{if} \quad u_{th} < 1 \\ PID(p_{im} - p_{im,ref}) & \text{if} \quad u_{th} = 1 \end{cases}$$

Quantifying Gain in Fuel Economy and Loss in Response Time

An example with an engine that has maximum boost of $p_{im} = 2$ bar is used in Eriksson et al. (2002) to study the gains in fuel economy and losses in response time between the different controllers. The gains in fuel economy over the engine operating region are shown in Figure 10.29. It is seen that in the normal driving region there is a fuel consumption gain of about 1.5–4% when running the fuel optimal controller. Experimental data from an engine test cell showed 1.9% improvement in fuel consumption in a low load and low speed region. This agrees well with the results in Figure 10.29.

The next comparison is to see the impact that it has on the transient response. For each point in the engine map, the transient response is defined as the time it takes to reach 90% of full torque while maintaining constant speed. The resulting time for the time optimal controller is shown in Figure 10.30a, while the additional time it takes for the fuel optimal controller is shown in Figure 10.30b. It is seen that the fuel optimal controller is about 1 s slower for the worst case when compared to the transient optimal.

Trade-off Using Throttle Δp_{th} Controller

A straightforward way to trade-off between efficiency and transient response is to provide a reference value for the boost pressure controller $p_{ic,ref}$ that uses the intake manifold reference

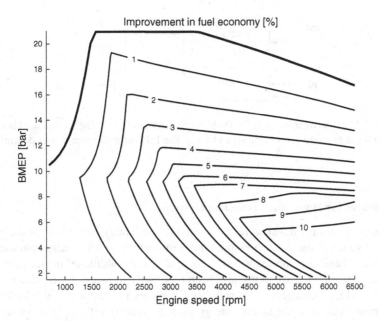

Figure 10.29 Gains in fuel economy for the fuel optimal boost control strategy compared to the time optimal

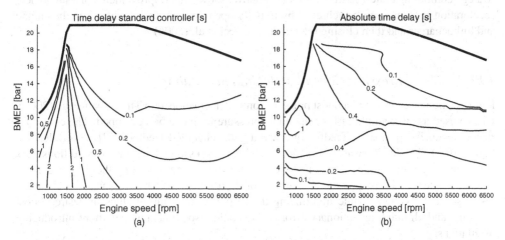

Figure 10.30 Loss in response time between time optimal and fuel optimal boost control strategies. (a) Response time of time optimal strategy. (b) Additional time for the fuel optimal strategy

$p_{\text{im,ref}}$ from the air manager adds a pressure margin Δp_{th} over the throttle, that is

$$p_{\text{ic,ref}} = p_{\text{im,ref}} + \Delta p_{\text{th}}$$

This provides the throttle with an air reserve that gives the driver the feeling of a fast response when changing the accelerator pedal while driving. It can be used to adjust the trade-off between boost and response time. The margin can also be changed dynamically to facilitate an ECO-driving mode, low Δp_{th}, or sports driving mode, high Δp_{th}.

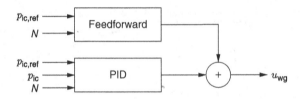

Figure 10.31   The controller structure for the boost pressure controller. The feedforward is a static map from desired boost pressure and engine speed to u_{wg}. The PID controller parameters depend on engine speed

Other Measures for Improving Transient Response

The generic problems for transient response are the availability to boost pressure and that it takes time to generate the boost pressure with a turbocharger. Alternative solutions for this are to use a mechanical compressor instead or add it as an additional device to support the turbocharger. Several other solutions have been proposed and investigated, where one path is to use electrified turbochargers or hybrids, see the survey in Eriksson et al. (2012). Another path is to convert the vehicle kinetic energy in decelerations with variable valve actuation systems and store it as pressurized air in a high pressure tank. This air at high pressure can then be used to boost the engine during accelerations, see Dönitz et al. (2009) for an example. A strategy combining clutch control and boost control, shows that improvements in total vehicle acceleration response can be achieved by initially opening the clutch to speed up the engine and turbocharger and then closing the clutch, see Frei et al. (2006).

10.8.3   Boost Pressure Control with Gain Scheduling

In this section an example of a boost pressure controller, developed in Thomasson et al. (2009), will be used as an example of how the boost pressure loop can be developed. The controller structure consists of a static feedforward and a gain scheduled feedback PID controller, see Figure 10.31. The feedforward gives the desired boost pressure at stationary conditions. It is determined by running the engine at stationary conditions and recording the duty cycle u_{wg} needed for desired boost pressures. The task of the PID controller is to shape the dynamic response of the system, while minimizing the response time during steps in desired boost pressure, and eliminating stationary error. This should also be achieved without introducing oscillations.

PID Tuning Method

From an industrial perspective, it is desirable for a tuning method to be simple, fast, and easy to automate. To tune the PID controller, some experiments for gathering process knowledge are needed. The complete MVEM from Section 8.9, together with the pneumatic wastegate actuator model, are used with experimental data to gain insight into the system dynamics. Experimental results indicate that at least a second-order system is needed to describe the system behavior for steps in input signal. Several simulations with the MVEM and wastegate model were performed for different speed and load conditions, and they confirmed that a second-order behavior gives a good approximation of the system dynamics. However, there is a significant variation of the system dynamics with engine speed.

Step responses in u_{wg} are suggested for model identification experiments, and the step responses are calculated for several engine speeds in the range of interest. For each speed we have the following process model

$$G(s) = \frac{K(N)}{T^2(N) s^2 + 2 T(N) \zeta(N) s + 1} \quad \text{with } \zeta \leq 1 \tag{10.20}$$

where K is given by the static gain from wastegate to boost pressure p_b: $K = \frac{\Delta p_b}{\Delta wg_{dc}}$, ζ is a function of the pressure overshoot p_{os} ($\zeta = f\left(\frac{p_{os}}{\Delta p_b}\right)$), and T scales the step response time.

Figure 10.32 shows step responses from measured data and the fitted process model for two engine speeds. It is seen that there is a significant variation in dynamics between speeds and that the second-order model, with different parameters for different engine speeds, gives a good description of the pressure behavior.

The suggested parameter tuning is based on the IMC-framework for controller design or Q-parametrization (Garcia and Morari, 1982). In general, the controller $F(s)$ is not a PID controller, but for many simple process models it is (Rivera et al., 1986). In this case, when $G(s)$ is a second-order system, and with the closed loop reference model chosen to be $G_d(s) = 1/(\Lambda s + 1)$, we get a PID controller as a result. This choice of $G_d(s)$ as a first-order system is motivated by the desire to suppress disturbances without introducing oscillations. The parameter Λ is a tuning parameter that can be interpreted as the time constant for how fast the controller will react to a control error. All in all this methodology gives

$$K_P(N) = \frac{2 T(N) \zeta(N)}{\Lambda K(N)} \quad K_I(N) = \frac{1}{\Lambda K(N)} \quad K_D(N) = \frac{T(N)^2}{\Lambda K(N)} \tag{10.21}$$

Since the parameters in the second-order model are engine speed dependent, the PID parameters will also depend on the engine speed. The choice of the tuning parameter Λ and how it affects the controller will be discussed in the results section.

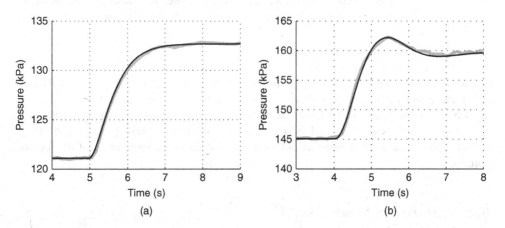

(a)   (b)

Figure 10.32 Measured boost pressure response in a test car to steps in u_{wg} (grey thick line) together with the adapted process model (black thin line) for two different engine speeds. The adapted model shows a very good fit to measured data. For 5500 rpm the pressure peak is slightly sharper compared to the adapted model, but the difference is very small

PID Implementation Aspects

A direct implementation of the ideal PID controller, with an unfiltered derivative, is not appropriate due to high frequency measurement noise. The D part of the controller has been filtered with a low-pass filter with a cutoff frequency of 20 rad/s.

The low pass filtering of the derivative introduces another problem if steps in the reference signal occur. If the derivative acts on the control error, $e = r - y$, and this results in large transients for sudden changes in reference value. For an unfiltered signal this would only be one sample, but when the derivative is filtered it can sustain for several samples. One solution, described in Åström and Hägglund (2006), is to let the derivative act on $e_d = \beta r - y$ where β is chosen between 0 and 1. Here $\beta = 0$ is chosen because the derivative's main task is to reduce the pressure increase after a step in desired boost pressure, when the reference value is reached. For slower and smother changes in boost pressure reference, where a derivative acting on the $e = r - y$ would be preferred, the derivative action is fairly small.

If the integrator is engaged when the control signal saturates, then it can cause undesirably large overshoots due to integrator windup. Therefore, conditional integration is used for anti-windup.

Controller Tuning and Results

The response in boost pressure to a step in reference value should be as fast as possible with a small overshoot (around 5 kPa) and no oscillations. The small overshoot is desirable since it is better to have a small excess than a shortage in engine torque.

The tuning parameter Λ should be chosen so that the transient behavior described above is achieved. If a too small value for Λ is used, the controller will be too aggressive, and introduce oscillations during transients. With a too large value for Λ, disturbances will not be suppressed fast enough and the transient response will be slow. A value of around $\Lambda = 2$ has proved to be a good starting point for calibration. Engine test cell experiments have shown that engines with comparatively bigger turbos, and thus slower and smoother responses, can tolerate a smaller value of Λ. This may also be needed to achieve the desired closed loop response. Furthermore, a system with fast step response and larger overshoots needs larger Λ-values.

Controller Evaluation in a Test Vehicle

The controller is evaluated in a test vehicle on a vehicle dynamometer. Figure 10.33 shows responses for the closed loop system for two engine speeds. Both figures show that the controller successfully achieves the desired performance, that is a fast transient response with a small overshoot in boost pressure, not exceeding 5 kPa, without any significant oscillations. More experimental results, as well as details of the simulation investigation, can be found in Thomasson et al., (2009), together with a discussion about the importance of the feedforward loop.

10.8.4   Turbo and Knock Control

With the EMS, boosting and engine performance can be matched to the engine and fuel. Changing the fuel also changes the engine knock tendency, which can be handled by an electrically controlled valve, Figure 10.28. For example, if there is a tendency for the engine to knock even though the ignition has been significantly retarded by the knock controller, then it is

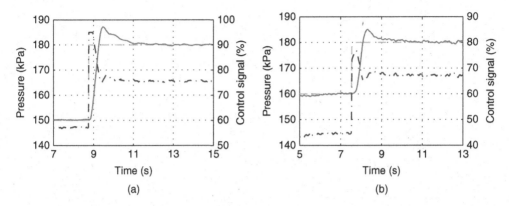

Figure 10.33 Boost pressure step responses for (a) 2500 RPM and (b) 4500 RPM measured in a car. The overshoot is close to 5 kPa

an indication of a fuel with a low octane number. In such a case, the boost pressure can be reduced to recover some ignition timing. The current fuel and engine conditions can be better matched and higher efficiencies and powers can be reached without risk of destroy the engine. A high compression ratio can be selected without the need for a high octane requirement for the engine. This is the case in some turbocharged ethanol engines that have a higher power output when run on ethanol compared to normal gasoline.

10.9   Dependability and Graceful Degradation

As discussed above, a demand on engines and vehicles is that the system should not shutdown completely if a non critical sensor or function malfunctions. When a component malfunctions, the system is allowed to degrade in performance but not to stop working. This is called *graceful degradation*, and is important to consider in the design. The following examples illustrate some possibilities worth considering in the design:

- Exchanging controllers. The pressure control example given above where the feedback controller is shutoff in case of a sensor failure.
- Replacing sensors with calculations. When a sensor fails it can often receive a replacement value from observers or *virtual sensors*. The replacement value is then calculated or estimated with the aid of a model and one or more other sensors.
- Using alternative models. For example, if sensors or subsystems fail the models can be replaced. For example, if the torque calculation (10.4) cannot be used for calculating the torque, then it is possible to fall back on the simpler torque calculation in (10.1).

An advantage with using just an open loop controller, is that if the open loop controller and the system are both stable, then the total system with controller and system will also be stable. With feedback, a stable controller for a stable system might make the closed loop system unstable, which makes the design and precautions more intricate. Falling back to open loop controllers can thus be preferable in some situations.

These aspects will be further discussed in Chapter 16, see especially Section 16.1.4, Accommodation of fault situations.

11

Basic Control of Diesel Engines

Diesel engines have historically been durable and economic, but have been considered slow. This has very much changed, as there has been a dramatic development in diesel engine technology both for personal cars and heavy trucks. Diesel engines for personal cars now combine both comfort and driveability, and for heavy trucks it has been possible to comply with strict environmental requirements while still offering low fuel cost and reliable transportation. There are several components in this technological development, but it is important to realize that computerized **control** is an enabling technology to achieve the functionality and performance.

The scope of this chapter is to introduce the concepts and technologies of basic diesel engine control. An overview of the diesel engine system, its operation, and control goals is given in Section 11.1. Section 11.2 deals with torque control. Even though torque is the fundamental control objective, there are several other objectives related to environment protection as described in Section 11.3. The two main systems that handle these challenges are fuel injection control, described in Section 11.4, and gas flow control that is treated, in Section 11.5. A case study for gas flow control is provided in Section 11.6.

11.1 Overview of Diesel Engine Operation and Control

There are several objectives related to, for example, driving experience and environmental protection, that the engine management system strives to meet. The most important goals and their motives are:

- Fulfill legislated emission limits. The most important emissions from a diesel engine are NO_x, soot or particulates, and also to some extent carbon monoxide, CO, and hydrocarbons, HC.
- Fulfill the torque request from the driver (and powertrain). The torque generated by the engine is the force that drives the vehicle, and torque control is thus fundamental for the driver to achieve vehicle propulsion.
- Maintain low sound levels from the engine. Traditionally, diesel engines have had a characteristic loud combustion sound that has been called "diesel knock." The source of this sound is the steep pressure rise during the first early phase of combustion.
- Achieve low fuel consumption.

Modeling and Control of Engines and Drivelines, First Edition. Lars Eriksson and Lars Nielsen.
© 2014 John Wiley & Sons, Ltd. Published 2014 by John Wiley & Sons, Ltd.
Companion Website: www.wiley.com/go/powertrain

There are several means to fulfill these goals, but they also interact and compromises arise; it can, for example, cost fuel to reduce emissions. We will start with the emissions from the diesel engine, as these have been a main driver for development of engine control systems.

11.1.1 Diesel Engine Emission Trade-Off

The emissions from a diesel engine were discussed in Section 6.4.3, and the main ones are nitrogen oxides NO_x and particulates (PM). When other conditions are constant, there is a compromise between these emissions, mostly depending on the start of combustion. This trade-off is illustrated for heavy duty diesel engines in Figure 11.1, where an early start of combustion increases local temperatures, which results in more NO_x. A late start of combustion, on the other hand, leads to less time for complete combustion and increases the particulates. A late start of combustion also leads to losses in expansion work and thus efficiency, highest efficiency often means high NO_x up to a certain point when the combustion comes too early. The whole curve describing the trade-off can be translated towards the origin by means of higher injection pressure, EGR, or exhaust gas after treatment.

Figure 11.1 also illustrates that the legislated limits on emissions have evolved over the years and drive the development and introduction of new technologies. Nitrogen oxides, NO_x and particulates (PM) have been dominating, but as emission regulations become stricter also CO and HC also need to be taken care of in the process from engine design to controller

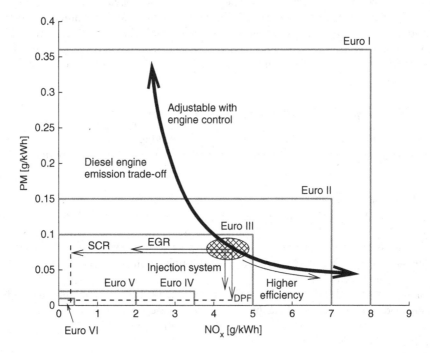

Figure 11.1 Legislated NO_x and PM limits for heavy duty diesel engines (note that the units are in g/kWh). The trade-off between NO_x and PM is also shown as a thick line. Further, different technologies and control measures can be used to achieve emission reduction. The ellipse shows the target region for engine operation in Euro III

calibration. To give an introduction to the technologies used for reaching the limits, it is beneficial to study the evolution from an earlier standpoint to today. For a heavy duty engine, standing at the Euro III emission limits, there were two paths to Euro IV (and Euro V). One path was to use engine measures, like increased injection pressures for PM and cooled EGR for NO_x reduction. The other path was to use after-treatment measures with diesel particulate filters (DPF) for PM and catalysts for NO_x-reduction, either selective catalytic reduction (SCR) or NO_x-storage. Euro V limits could, for some engines, be reached without after-treatment for NO_x by using high amounts of cooled EGR. Euro VI, however, will see the combined usage of both after-treatment systems and engine measures to reduce the engine-out emissions. With a flexible engine system, its operation can be optimized, for example to ensure that the after-treatment system, with DPF and catalyst(s), is heated and receives the right composition of gases for best emission reduction.

11.1.2 Diesel Engine Configuration and Basics

An example of a diesel engine is depicted in Figure 11.2, and it is seen that there are quite a few similarities with the turbocharged gasoline engine, as depicted in, for example, Figure 4.7. The fundamentals of diesel combustion were discussed in Chapter 6 and we recall some properties and differences between gasoline (SI) and diesel (CI) engines. An SI engine operates at $\lambda = 1$ and load control is performed with the throttle and air flow (and fuel since $\lambda = 1$). In a CI engine, fueling is used for load control and it operates with excess air, $\lambda > 1$, typically well above (with lower limits around 1.2–1.4), to avoid soot formation. With a limit of $\lambda > 1.3$, a naturally aspirated CI engine would have 30% less torque than an SI engine of same size. Therefore turbocharging is used as a part of load control on all modern automotive CI engines to increase the specific power. In Figure 11.2, there is also an intake throttle in the CI engine,

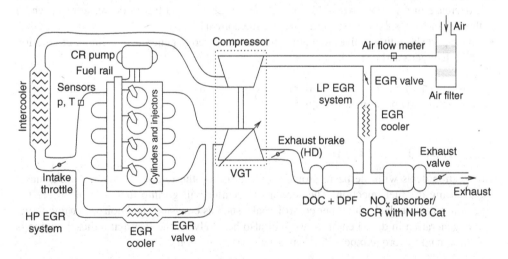

Figure 11.2 A schematic of a diesel engine showing a variety of systems. Fuel injection is done with a common rail system, with pump, rail, and injectors. There is a turbocharger with a variable geometry turbine VGT. There are two cooled EGR systems, high pressure (HP) and low pressure (LP). There are additional gas control valves: intake throttle, exhaust valve, and exhaust brake (used in Heavy Duty (HD) applications)

which has been introduced recently. In SI engines, the throttle is used for load control, while in CI engines the aim is to have it as open as possible, to give best fuel economy, and reduce the area only when it is needed for emission abatement in some operating modes.

Engine Actuators and Options

The CI engine can, as shown in Figure 11.2, be equipped with several actuators and after-treatment systems. The main actuators on the engine are:

- Fuel injector. The basic system takes two control signals, the amount of fuel, m_f in mg/stroke, and the timing, α_{inj} in degrees BTDC. Multiple injections are also used for reducing pollutant and sound emissions.
- Fuel injection pressure. This is a supporting system for the fuel injection where the rail pressure, that influences the injected amount and spray, is controlled using the high pressure pump and a pressure sensor in the fuel rail.
- Turbo control. To control the boosting, a turbine with either wastegate or variable geometry is used. The variable geometry turbine (VGT) has been popular on diesel engines, and one example was depicted in Figure 8.24.
- EGR valve(s). Control valves are used for controlling the amount of EGR delivered to the cylinders. These can be used together with intake and exhaust throttles to produce sufficient pressure drops for EGR flows.
- Exhaust throttle. Used together with EGR valve to produce sufficient pressure drops so that the desired EGR rates can be attained.
- Intake throttle. This has two applications: (i) Used with the HP EGR valve to produce a pressure drop so that the desired EGR rates can be reached. (ii) Used to increase the exhaust temperature during low load operation. The mechanism utilized is shown in (5.19), where a decrease in pressure gives a lower cylinder air charge and thus a decrease in λ, which thereby increases the gas temperature after the combustion.
- Intake port shrouding. Valves that guide the flow are used to generate in cylinder gas motion through either swirl or tumble.

Exhaust after-treatment is also important, and such systems also contain actuators and sensors; these were already been mentioned in Section 6.5.2 and Figure 6.18.

11.2   Basic Torque Control

This section deals with the basic problem of controlling the torque delivered from a diesel engine using the fuel injection. The sections thereafter will go into fuel control, where the injection is used to deal with other control goals, such as emissions and noise. Turning to the torque generation in diesel engines, we will also here rely on the fact that a diesel engine is well described by torque model that was developed in Section 7.9.

A p-V Diagram of a Diesel Engine

An example of a measured p-V diagram from a diesel engine is shown in Figure 11.3a. This p-V diagram can be compared to the ideal cycles presented in Figure 5.5 in Chapter 5, where it

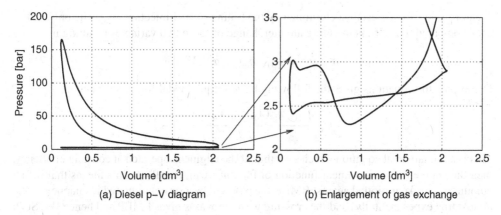

(a) Diesel p–V diagram (b) Enlargement of gas exchange

Figure 11.3 (a) A p-V diagram measured on a diesel engine. Note that it does not conform to an ideal Diesel cycle, and specifically that combustion does not evolve under constant pressure. (b) An enlargement of the lower portion of the p-V diagram showing the pumping work. It is clear that the pumping work is negligible compared to indicated gross power

was pointed out that the ideal cycles are only models of the engine operation. Compare this also to the relation between a real gasoline engine and an Otto cycle that was illustrated in Figure 5.15. It is clear that the p-V diagram from a diesel engine, Figure 11.3, does not fully agree with an ideal Diesel cycle as shown in Figure 5.5. One specific point is that the combustion in a real diesel engine does not evolve under constant pressure. However, basic features such as compression, combustion, and expansion are there.

Torque Model

Recall (7.55) from Section 7.9 that describes the torque produced by an engine. The engine torque, M_e, is modeled using three different components: the gross indicated work per cycle, the pumping work from the difference in intake and exhaust manifold pressures, and friction work consumed by the engine components and auxiliary devices. The equation is

$$M_e = \frac{W_e}{n_r \, 2 \, \pi} = \frac{W_{i,g} - W_{i,p} - W_{fr}}{n_r \, 2 \, \pi} \tag{11.1}$$

As a first approximation, the pumping work can be neglected, since there is basically no throttling in a diesel engine. The small pumping work is also shown in Figure 11.3a. Hence,

$$W_{i,p} = 0 \tag{11.2}$$

is a good approximation for most cases. However, it must be noted that control measures, such as closing of an intake or exhaust throttle or variable geometry turbine, can cause pressure drops and these might be necessary to include in the model.

The indicated gross work is described by (7.56) that couples to the energy that comes from the fuel, $m_f \, q_{LHV}$. The efficiency $\tilde{\eta}_{ig}$ depends on the operating conditions and the fuel, and is modeled as a function of these. A similar equation holds for a diesel engine. The only big difference is the timing of combustion, which for a gasoline engine is determined by the spark event, θ_{ign}, but for a diesel is determined by the timing of fuel injection, α_{inj}. The only difference compared to (7.56) is therefore an interchange of these two variables resulting in

$$W_{i,g} = m_f \, q_{LHV} \, \tilde{\eta}_{ig}(\lambda_c, \alpha_{inj}, V_d, N) \tag{11.3}$$

Combining the equations gives the following basic torque model for a diesel engine

$$M_e = m_f \frac{q_{LHV} \, \tilde{\eta}_{ig}(\lambda_c, \alpha_{inj}, V_d, N)}{n_r \, 2 \, \pi} - \frac{W_{fr}}{n_r \, 2 \, \pi} \tag{11.4}$$

This is a fundamental equation which says that, if the engine is operated at constant efficiency, then the engine torque is a linear function of fuel, m_f, with an offset. This means that if fuel amount per stroke is plotted against BMEP, the plot will be a straight line (which happens to be almost true experimentally) and the crossing with the x-axis gives FMEP and hence W_f. Such an affine function (linear with offset) is often used to also specify component performance and is generally called the *Willans line*. The fact that these lines, that is (11.4) are in good agreement with experimental data, is the basis for the common statement that "engine torque for a diesel engine is proportional to fuel amount."

11.2.1 Feedforward Fuel Control

Feedforward fuel control for the diesel engine is achieved with model (11.4) that can be inverted and solved for the amount of fuel to be injected. This was previously outlined in Section 10.2.2. The resulting control action is

$$m_{f,r} = \frac{M_{e,ref} \, n_r \, 2 \, \pi + W_{fr}}{q_{LHV} \, \tilde{\eta}_{ig}(\lambda_c, \alpha_{inj}, V_d, N)}$$

A *smoke limiter* is needed to ensure that there is sufficient air in the combustion chamber to avoid smoke generation. Therefore, the injected fuel is compared to the estimated cylinder air mass flow $\dot{m}_{ac}$, using the limit on the air fuel ratio λ_{min}

$$m_f = \min(m_{f,r}, \frac{\dot{m}_{ac}}{\lambda_{min} \, (A/F)_s \, n_r \, N})$$

which gives the basic amount of fuel mass that should be injected.

11.3 Additional Torque Controllers

Besides pure fuel control, there are also additional torque controllers that are active, for example the boost pressure is connected to the fueling and the long-term torque. Furthermore, for heavy duty diesels, there are also controller modes for engine braking, and these are described below.

Long-Term Torque and Boost Pressure Control

Long-term torque is controlled with the boost pressure system, where the volumetric efficiency is used to connect the torque demand to set points in boost pressure. The procedure is equivalent to that of the SI engine, described in Section 10.2.2, with the addition of one factor, namely that EGR displaces the air. In particular, the volumetric efficiency, that describes the total flow to the engine, needs to be augmented with the EGR fraction which will give an increase in the required boost pressure. One also needs to remember that the set point λ_r must be included, as it is through λ in (10.4).

Engine Braking in Heavy Duty Applications

In heavy duty applications, there are often additional devices for braking, so that the service brakes are spared from wear and their life is prolonged, leading to reduce maintenance cost. During vehicle braking the engine enters fuel cutoff and various control measures are taken. Three examples are: (i) An *exhaust brake* on the engine exhaust side (see Figure 11.2). A valve is closed on the exhaust side that builds up back-pressure and increases the pumping work. It is most effective if the engine speed is high, so that the high flows can build up high back-pressures. (ii) A *compression brake* that controls the valves to the cylinder, this is discussed in Section 12.1. (iii) A *retarder* on the gearbox, this is mentioned here to give the complete picture but it belongs to the driveline torque actuation, see Section 15.1.3.

11.4   Fuel Control

Fueling has a very strong impact on all engine outputs, such as torque, emissions, and gas flow dynamics. Fueling interacts with the gas dynamics through the cylinder flows, and in particular its output into the exhaust system. Its coupling to the rest of the gas dynamics is through the turbo and EGR systems. The essential fuel dynamics of a diesel engine (evaporation, ignition delay, combustion, etc.) is in the cylinder. Therefore, the view adopted here is that the fuel control is devoted to the processes in the cylinder, while the gas dynamics describes and accounts for the interaction. In fuel control, the first step is fuel preparation that is included in the fuel injection system, and then the next is the fuel injection itself that is controlled by the injector.

Fuel Injection System

Diesel fuel preparation comprises two fuel systems, corresponding to the two stages in the fuel preparation system. A low pressure fuel supply system and a high pressure fuel system that follows after. The low pressure system basically prepares the diesel fuel, for example by filtration and water separation, and delivers it to the high pressure system. There are some control issues in the low pressure system, such as fuel pre-heating in some situations, but the main control issues are in the high pressure system.

In a diesel engine, the fuel is injected directly into the cylinder under high pressure. The requirements are such that this means very small and precise amounts must be delivered within a thousandth of a second, and this is the task of the fuel injection system. There are older systems like, for instance, the distributor pump and unit injector systems, but the major technique today is the common rail system.

Common Rail Injection System

A common rail system is an accumulator system. Diesel fuel is kept in a common high pressure accumulator, the common rail, and the amount and timing of fuel delivery is determined by the opening and closing times of the injector. For this type of system, techniques with

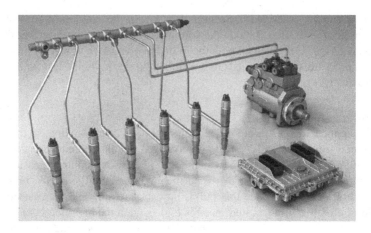

Figure 11.4   Picture of a common rail system, showing the injectors, pressure rail, high pressure pump, and EMS. Reproduced with permission from Robert Bosch GmbH

multiple injections have evolved, and this will be further discussed in Section 11.3. A common rail system with fuel pump, fuel rail, injectors, and EMS for a 6-cylinder engine is shown in Figure 11.4.

The fundamental function of fuel injection is determined by the opening and closing of the internal fuel circuit. Under simplifying assumptions, the amount of fuel is controlled by the duration of the valve opening time. There are, however, some additional steps that need to be taken in order to actuate the fuel.

- Compensation for temperature, since diesel density varies with temperature. One reason for temperature variations is warming up of the recirculated diesel.
- Compensation for diesel compressibility. At the pressure that current systems operate, diesel cannot be regarded as an incompressible fluid.
- Pressure wave compensation. As in Figure 7.17, there is a dependency on injection pressure and also on wave phenomena in the rail, where pressure waves in the rail from the current and preceding injections influence the injected mass.
- Control of injector. There are two injector technologies in common rail systems: piezo and solenoid injectors. Here we will only cover the solenoid and refer to Dietsche (2011) for the piezo. The injector current is measured and it is controlled during the injection, using the voltage to the injector. A high current is used initially to give a fast opening, while the current is reduced when the injector has opened. See the main injection in Figure 11.7 for an example.

11.4.1   Control signal – Multiple Fuel Injections

The fuel injection system has one actuator, the injector. Nevertheless, it is important to realize that the control signal is multi-dimensional since it is possible to use multiple fuel injections, as depicted in Figure 11.5. Further, depending on how advanced the injection system is, one may control the whole injection profile including number of injections, timing, duration, and shape of each injection. This is very important since these injection characteristics determine the

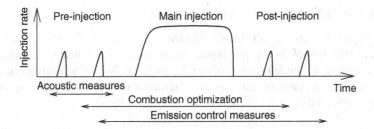

Figure 11.5 A principle sketch of multiple fuel injections in a diesel engine (not just limited to five). The key control authority in diesel fuel control is the injection profile: number of injections, timing, duration, and shape

evaporation process of the diesel fuel in the cylinder, and then of course also the combustion process. Note also that already the control of just the timing and duration of five injections requires control of ten parameters, which emphasizes the multi-dimensionality.

Formation of Sound

Sound levels from a diesel engine can partly be controlled by fuel management. The phenomenological basis for this influence is that *diesel knock* is caused by the abrupt pressure rise at the start of combustion. In particular, there is a strong correlation between sound and the pressure derivative, dp/dt, which in turn basically is proportional to the rate of heat release. Recall Figure 6.6, where the basic heat release curve for a diesel engine was depicted: it shows a premixed combustion phase and then a diffusion combustion phase. The rate of heat release depends on the initial temperature at the start of injection, and a principle behavior is sketched in Figure 11.6. The figure is a sketch of the data in Heywood (1988), and shows the mechanism that is the basis for sound management by fuel pre-injections. A higher temperature promotes evaporation and leads to a shorter ignition delay, which in its turn reduces the rapid premixed diesel combustion that is the source of the high pressure derivatives and thus sound.

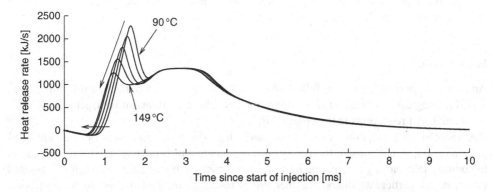

Figure 11.6 Heat release rate as a function of initial temperature. The arrows indicate that higher temperatures give an earlier start of combustion and lower peaks during the premixed combustion. A higher temperature can thus decrease the diesel engine sound since it is related to the pressure rise which depends on the heat release rate

11.4.2   Control Strategies for Fuel Injection

The injection profile influences all control objectives: torque, fuel efficiency, sound, and emissions. The classification of the separate injections and their use is indicated in Figure 11.5. The terminology advanced/retarded refers to the crank angle position, where advanced means earlier and retarded means later, which is the same terminology used in the term spark advance. The principle influence of each injection will now be presented.

Main Injection

The *main injection* is the middle one in Figure 11.5. It contains the majority of the fuel, and its timing, α_{inj}, is selected for optimal efficiency. The objective is to place the combustion relative to TDC to obtain the most indicated gross work, $W_{i,g}$. However, the maximum pressure or pressure rise can also be limiting factors. Further, as mentioned concerning the trade-off between the PM and NO_x and emissions, a delay of this injection results in lower temperatures and less NO_x but at the cost of reduced efficiency and higher fuel consumption.

It is also advantageous with a high injection pressure to reduce soot. The reason being that a high pressure gives better fuel vaporization and air mixing. This decreases the amount of incompletely burned hydrocarbon chains that are the precursors to soot formation.

Pre-Injections

The pre-injections pre-condition the combustion chamber by heating it. This gives a shorter ignition delay of the main injection and the combustion is then smoother with a pressure rise during the combustion of the main injection, see Figures 11.6 and 11.7.

The *advanced pre-injection* is the first depicted in Figure 11.5, and it is mainly used for sound reduction. As described above, the sound is related to the steep pressure rise during diesel engine combustion, and an early pre-injection has the effect that the pressure rise starts earlier and the pressure then evolves much more smoothly resulting in much less noise.

A *retarded pre-injection* closer to the main injection has two objectives. The first objective is to keep the pressure development smooth enough not to result in noise, and the second objective is to reduce the emission formation during combustion.

Post-Injections

An *advanced post-injection* can follow shortly after the main injection while the combustion is still in progress. Its function is to reduce soot by reburning it, and it is possible to achieve substantial soot reductions in the order of 70%.

A *retarded post-injection* can follow later than the main injection or the advanced post-injection, if that is used. Its use depends on the after-treatment system used. One use is to control the exhaust gas temperature in order to increase it substantially when it is needed to regenerate particulate filters. Another use is based on the fact that if the post-injection is sufficiently retarded then the diesel is not burned, it is only vaporized by the warm exhaust gases. This means that the fuel follows the exhaust out in the after-treatment system, and the diesel can then act as a reducing agent for NO_x in an accumulator type catalytic converter.

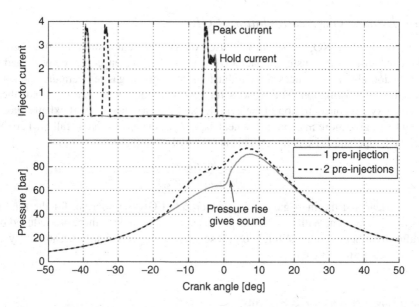

Figure 11.7 Top: Injector current, showing multiple injections and that there is a peak and hold when the injector opening time is long. Bottom: Cylinder pressure, showing that pre-injections heat the gas and change the pressure development. There is a long ignition delay from the second pre-injection at $-35°$ to SOC at $-15°$, when it is cold. The ignition delay of the main injection is reduced with two pre-injections

Post-injections can also be used to keep the temperature up, and this has a possible use in keeping the chemi-kinetic reactions, like the Zeldovich mechanism, going so that NO_x is continually reduced throughout the expansion. This is used to prevent freezing of the chemical reactions. However, the main thermal effect is the increased temperature for exhaust treatment systems.

11.5 Control of Gas Flows

The gas flows in a diesel engine are crucial since they determine air charge and amount of exhaust gas recirculation, and these are fundamental for combustion and thus emissions. Gas flow management is the handling of flows of air and burned gases and how they interact with the turbocharging system. It can be seen as combined control of turbo and exhaust gas recirculation (EGR) and the purposes are the following.

- Turbo or air control has as its purpose to deliver enough air to the cylinders, so that the air to fuel ratio λ is high enough to avoid soot production.
- EGR control has the purpose of delivering enough burned gases to the cylinders so that production of NO_x is avoided.

The fundamentals of gas flows were treated in Chapter 7 and turbocharging in Chapter 8, both contributing to an example of a diesel engine model with EGR and VGT in Section 8.10. Turbocharging has already received a thorough treatment, so our attention is now turned to the EGR system and later to its interaction with the turbocharger.

11.5.1 Exhaust Gas Recirculation (EGR)

One way to control the NO_x emissions is to use exhaust gas recirculation (EGR), where a portion of the exhaust gas is transferred back to the intake. The presence of already burned gas in the combustion chamber reduces the combustion and burned gas temperatures and thereby reducing the NO_x formation. The reasons for lowered temperature are two-fold: the EGR reduces the oxygen concentration in the cylinder and the presence of inert exhaust gases, N_2, H_2O, and CO_2, give more mass and thus heat capacity. This means that locally, where NO_x is formed, the temperature rise is lower. This is obtained from (5.19)

$$\Delta T = \frac{m_f \ q_{LHV}}{(m_f + m_a + m_{egr}) \ c_v} \tag{11.5}$$

where it is clear that the inert gases, m_{egr}, lower ΔT. EGR is important for the total diesel engine emission system. With less NO_x is produced in the cylinders, there is a reduced burden on the after-treatment system. While most diesel engines rely on an after-treatment system to reduce NO_x, they still utilize EGR as well.

High and Low Pressure EGR Systems

A complex system with two loops for the EGR was shown in Figure 11.2. The loop nearest the engine is called high pressure (HP) or short route EGR, while the outer loop is called low pressure (LP) or long route EGR. Of these, the most common EGR configuration is HP EGR. To generate EGR flow, there is a need to have a pressure difference. The exhaust gases have a relatively high pressure, but the intake manifold pressure might be high too (due to boosting). With very little pressure differential between the exhaust and intake manifolds, EGR gases will not flow at the required rate. For example, in well matched engine and turbo systems the exhaust can, at high loads, have back pressures before the turbine that are lower than the pressures after the compressor and intercooler. The pressure differences needed to generate EGR flows can be generated by: (i) increased exhaust gas back pressure by closing the vanes in a VGT, (ii) decreased intake manifold pressure with an intake throttle.

LP EGR is an alternative approach that utilizes exhaust gases that have passed through the turbocharger, DOC, and DPF. It works well when soot is filtered from the EGR gases by the DPF, making the EGR clean and reliable. While the exhaust gases are also at a lower pressure, they can be sent through an EGR cooler to the compressor inlet. This allows a good EGR flow rate, that can be further enhanced with the aid of a downstream exhaust throttle valve.

Properties of LP and HP EGR Systems

The HP EGR has an advantage in that it has faster dynamics than the LP EGR. The latter has a slower response due to the longer gas transport and more filling dynamics. The LP EGR system has two advantages over the HP EGR: the temperatures can more easily be reduced and, from a controls perspective, it is less coupled to the turbine and boosting dynamics. Another division is into hot, partially cooled, and cold EGR, depending on how much the EGR is cooled. *Hot EGR* is internal EGR where the residual gas in the cylinder is controlled with variable inlet and exhaust valves, see Section 12.1. *Cold EGR* is when there is an EGR cooler doing an effective job. *Partially cooled EGR* is when the EGR is extracted but not actively or fully cooled, for example if there is no EGR cooler or if the cooler is bypassed. It is often easier to cool the LP EGR as it has lost more temperature to heat transfer at the extraction point.

Different designs are therefore combined and used to achieve certain goals with HP and LP systems. Depending on the application, the implemented systems can thus be very diverse. For example, the passenger car engine described in Hadler et al. (2008), is similar to the system shown in Figure 11.2, except that it does not have an HP EGR cooler. There, the HP EGR is used at low loads and in cold weather to achieve higher cylinder temperatures which enhance combustion stability. The cooled LP EGR is used at high loads to reduce temperature and thus further lower the formation of NO_x.

LP EGR control is performed with the LP EGR valve, and if more EGR is needed the exhaust valve is also closed. With a combined system, the LP EGR can be used as a base and then fast transients can be handled by the HP EGR system (Hadler et al. 2008). The remainder of this section will focus on HP EGR, since it is both challenging to control and a frequently used configuration.

11.5.2   EGR and Variable Geometry Turbine (VGT)

Control of HP EGR and variable geometry turbine (VGT) is a much studied subject, mainly because it is challenging to control. The reason for the challenge is that it is multivariable and highly nonlinear, with sign reversals as well as non-minimum phase behaviors, as shown in Kolmanovsky et al. (1997). Measurements from an engine with EGR and VGT, configured as in Figure 11.8, give an introduction to the properties. Control inputs are the opening in the VGT, u_{vgt}, and the amount of opening of the EGR valve, u_{egr}. The important gas flows for a diesel engine, as indicated in Figure 1.8, are the air flow $\dot{m}_{air} = \dot{m}_c$ and the exhaust gas recirculation flow $\dot{m}_{egr}$. These are to be controlled to desired levels and their dynamics are important for all overall control objectives discussed in Section 11.1.

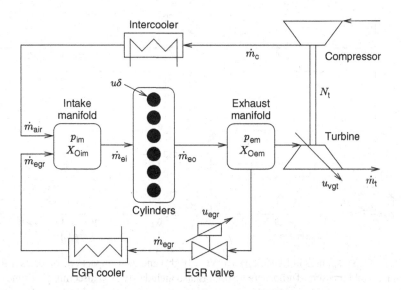

Figure 11.8   A diesel engine with gas flow control using VGT and EGR. The control signals are: fuel injection u_δ, VGT opening u_{vgt}, and EGR valve opening u_{egr}. Measured signals are compressor flow $\dot{m}_c$, intake manifold pressure p_{im}, and turbocharger speed N_t.

For control, the amount of air and recirculated exhaust gas need to be formulated in measured variables or variables available from virtual sensors. Two measured variables are p_{im} and $\dot{m}_{air}$, these are also referred to *manifold air pressure* (MAP) and *mass air flow* (MAF) in the literature. The flow $\dot{m}_{egr}$ is not measured but can be calculated, since we know that p_{im} via the volumetric efficiency η_{vol} gives the total flow into the cylinder $\dot{m}_{cyl}$. In stationary operation, the EGR flow is then the difference between the total flow and air flow, $\dot{m}_{egr} = \dot{m}_{cyl} - \dot{m}_{air}$. Based on this, a first control objective can be formulated: control p_{im} and $\dot{m}_{air}$ using u_{vgt} and u_{egr}. This is a complicated nonlinear multi-variable control problem and some specific characteristics will be given for the different channels. Measurement data for the four channels is shown in Figure 11.9 and these channels will be discussed below.

The Channel u_{egr} to p_{im}

Decreasing u_{egr} corresponds to closing the EGR valve, and gives an immediate decrease in the EGR flow, $\dot{m}_{egr}$. Less EGR flow, reduces the intake pressure through this fast filling and emptying dynamics. However, less EGR flow means that less exhaust gas is recirculated and there is thus more exhaust gas to drive the turbo. With less EGR the exhaust temperature will also increase. Both effects cause the turbo to spin up and give a larger increase in the pressure

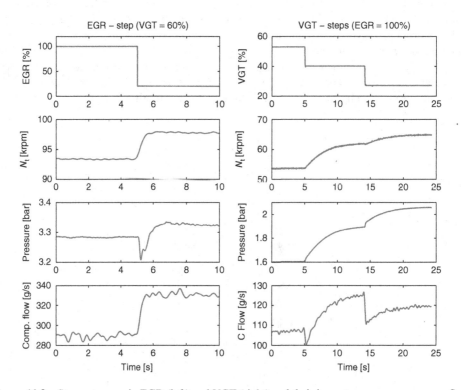

Figure 11.9 Step responses in EGR (left) and VGT (right) and their impact on compressor mass flow and intake manifold pressure. Turbocharger speed is also included to help discerning the time constants of the dynamic responses. Engine speed and fueling are maintained constant. Measured data comes from the engine modeled in Section 8.10

p_{im} than the initial decrease. This effect is slower though due to the slower dynamics of the turbocharger speed, which means that u_{egr} to p_{im} has a typical non-minimum phase behavior.

The Channel u_{egr} to W_{air}

As above, decreasing u_{egr} corresponds to closing the EGR valve and gives an immediate decrease in the EGR flow, $\dot{m}_{egr}$. Less EGR flow reduces the intake (boost) pressure which gives an increased air flow, $\dot{m}_{air}$. As boost pressure builds the flow is maintained. This follows the fast filling and emptying dynamics and the channel can be approximately modeled by a static function (with negative sign). Looking at input–output pairing for control, this indicates an opportunity for using the EGR valve to control the air flow. A special case is that, at low engine speeds, u_{egr} loses control authority over $\dot{m}_{air}$, in which case it may be better to use u_{vgt} for air control instead.

The Channel u_{vgt} to p_{im}

Decreasing u_{vgt} corresponds to closing the VGT, such that the exhaust gas flow is more restricted. If the EGR is open, this forces more EGR into the intake and the filling and emptying dynamics gives an initial small but rapid increase in the pressure, see, for example, the initial pressure increase at $t = 14$ s in the right plot in Figure 11.9. With closed VGT the exhaust pressure builds up and the turbo speed increases, the compressor will therefore increase the mass flow and the pressure p_{im}, which is the cause for the slower dynamics.

The Channel u_{vgt} to W_{air}

As above, decreasing u_{vgt} corresponds to closing the VGT such that exhaust gas flow through it is more restricted. If the EGR valve is sufficiently open when this happens, then the recirculation, $\dot{m}_{egr}$, will increase (since the flow through the turbo is restricted). In that case $\dot{m}_{air}$ will decrease as the intake pressure increases. The closing of the VGT also caused the turbo to spin up, which increases the air flow with slower dynamics. In the first step, at 5 s, there is a non-minimum phase behavior in the u_{vgt} to $\dot{m}_{air}$ channel. At 14 s the same dynamics effects are seen but stationary levels differ. In that case, the VGT is very closed and there is much EGR that gives a higher pressure, but the increase in turbo speed can therefore not produce the same air flow. Hence, the static gain of the u_{vgt} to $\dot{m}_{air}$ channel changes sign and the non-minimum phase behavior changes to an overshoot.

On the other hand, if the EGR valve is closed when the exhaust gas flow through the turbo becomes restricted, then the recirculation cannot increase and instead the effect will be that mass flow only increases with the turbo speed. In that case, $\dot{m}_{air}$ will increase (and so will p_{im}). There is thus a change of sign in the u_{vgt} to $\dot{m}_{air}$ channel, depending on the settings of u_{egr} and u_{vgt}.

Linking Physics and Channel Properties

The fundamental physical explanation of these system properties is that the system consists of two dynamic effects that interact: fast pressure dynamics in the manifolds and slow turbocharger dynamics. These two dynamic effects often work against each other and change in size, which results in the system properties above. For example, when the fast dynamic effect is small and the slow dynamic effect is large, the result is non-minimum phase behavior, see $\dot{m}_c$ for the VGT and the EGR step. However, when the fast dynamic effect is large and the slow dynamic effect is small, the result is an overshoot and a sign reversal. The precise condition for the sign reversal is due to a complex interaction between flows, temperatures, and pressures in the entire engine.

Model of a Diesel Engine with EGR VGT

When developing and testing control designs, it is beneficial to have a model for evaluating their properties and performance. The model summarized in Section 8.10 was developed to capture these phenomena and its structure follows that shown in Figure 11.8. The engine model is a mean value model and the gas path system has three main states: intake manifold pressure p_{im}, exhaust manifold pressure p_{em}, and turbocharger speed N_t. There are also two states, X_{Oim} and X_{Oem}, to separate and track the burned and fresh gases in the system. The inputs to the model are fueling rate u_δ, EGR-valve position u_{egr}, VGT actuator position u_{vgt}, and engine speed N. This model will be used for system analysis and controller tuning of EGR VGT controller of the next chapter, and it is also available for download as mentioned in Section 8.10.

EGR and VGT Control

As can be understood from the phenomena discussed above, EGR VGT control is a challenging control problem and several approaches have been published. The approaches differ in what performance and feedback variables are used, as well as what control design method is used, and to cover all of them is beyond the scope of this book. Therefore, one specific controller is designed, tuned, and evaluated as a case study and references to related approaches will be used to broaden the discussion in the upcoming section.

11.6   Case Study: EGR and VGT Control and Tuning

A realistic control problem in heavy duty diesel engines will be used to illustrate the characteristics of coordinated EGR and VGT control; it is based on Wahlström et al. (2010). The overall control goals are to fulfill the legislated emission levels and safe operation of the engine and the turbocharger. These goals are achieved through regulation of the following performance variables: normalized A/F ratio λ, intake manifold EGR-fraction x_{egr}, as well as turbocharger speed. The characteristics of this type of control problem, will be discussed in this section, showing how different controller tunings influence the behavior when the engine is tested on a demanding part of the European Transient Cycle.

Selecting the Performance and Feedback Variables

The selection of performance and feedback variables is an important step for emission control and, as mentioned in the beginning of Chapter 10, important variables such as NO_x and PM emissions and torque are not available. Instead, a controller has to rely on other feedback signals. Figure 11.10 shows the situation for the EGR VGT controller. With NO_x and PM unavailable, we need to look for other performance variables. EGR-fraction x_{egr} has a strong influence on NO_x and it is therefore natural. Exhaust gases, present in the intake from EGR, also contain oxygen and it can therefore be beneficial to use the oxygen/fuel ratio instead of

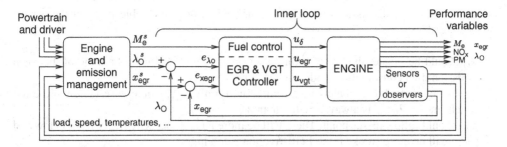

Figure 11.10 A cascade control structure, with an inner loop where EGR and VGT actuators are controlled, using the main performance variables EGR fraction x_{egr} and oxygen/fuel ratio λ_O. This sketch is a simplified illustration of the main idea that will be completed in Section 11.6.3 to also include fuel control and turbo protection

the traditional air/fuel ratio. The main motive is that it is the oxygen content that is crucial for smoke generation, and the idea is to use the oxygen content of the cylinder instead of air mass flow, see, for example, Nakayama et al. (2003). The normalized oxygen/fuel ratio λ_O is defined in (7.36).

EGR-fraction x_{egr} and oxygen/fuel ratio λ_O are natural to select as performance variables since they have a strong coupling to the emissions. Here they will also be used as feedback variables and the emission limits are thus formulated as set-points in EGR-fraction and λ_O. Load control is also necessary, since the driver's demand must be actuated, and this is achieved through basic fuel control using feedforward. Further, turbocharger speed needs to be monitored and controlled, since aggressive transients can cause damage through overspeeding, but this loop is not explicitly shown in Figure 11.10.

Remark on Performance and Feedback Variables

The choice of feedback variables defines the overall controller structure, and this is one selection of performance variables, but λ and EGR fraction have been used in Rajamani (2005), Wahlström and Eriksson (2013). To widen the perspective, we note that this is not the only selection as there are many other possible choices, similar alternatives for performance variables are the burned gas fraction, X_B, defined in (7.18), which is also strongly coupled to NO_x. Air/fuel ratio and burned gas ratio are frequent choices for performance variables (Jankovic et al. 2000; Rajamani 2005; Stefanopoulou et al. 2000; van Nieuwstadt et al. 2000). Concerning feedback variables, several have also been proposed and used. The most common choice in the literature are compressor air mass flow and intake manifold pressure (Nieuwstadt et al. 1998; Rückert et al. 2001; Stefanopoulou et al. 2000; van Nieuwstadt et al. 2000). Other choices are intake manifold pressure and EGR-fraction (van Nieuwstadt et al. 2000), exhaust manifold pressure and compressor air mass flow (Jankovic et al. 2000), intake manifold pressure and EGR flow (Rückert et al. 2004), intake manifold pressure and cylinder air mass flow (Ammann et al. 2003).

11.6.1   Control Objectives

The goal is to follow a driving cycle while maintaining low emissions, low fuel consumption, and suitable turbocharger speeds. Primary variables to be controlled are engine torque M_e, normalized oxygen/fuel ratio λ_O, intake manifold EGR-fraction x_{egr}, and turbocharger speed N_t. From the discussion above we get the following control objectives for the performance variables.

1. λ_O should be greater than a soft limit, a set-point λ_O^s, which enables a trade-off between emission, fuel consumption, and response time.
2. λ_O is not allowed to go below a hard minimum limit $\lambda_{O,min}$, otherwise there will be too much smoke. $\lambda_{O,min}$ is always smaller than λ_O^s.
3. x_{egr} should follow its set-point. There will be more NO_x if the EGR-fraction is too low and there will be more smoke if the EGR-fraction is too high.
4. The engine torque, M_e, should follow the set-point from the driver's demand.
5. The turbocharger speed, N_t, is not allowed to exceed a maximum limit, preventing turbocharger damage.
6. The pumping losses, M_p, should be minimized in order to decrease the fuel consumption.

In the next step, a control structure will be developed that achieves all these control objectives, when the set-points for EGR-fraction and engine torque are reachable.

Remarks Concerning the Choice of Variables

The implications of having EGR-fraction x_{egr} and oxygen/fuel ratio λ_O as both performance and feedback variables are discussed in detail in Wahlström et al. (2010) and only summarized here. In diesel engines, a large λ_O is allowed, and there is thus an extra degree of freedom when λ_O is greater than its set-point. This can be used to optimize the fuel consumption, and a simple mechanism for minimizing the pumping work is implemented. Further, x_{egr} and λ_O are strongly connected to the emissions and give a natural separation within the engine management system. The performance variables are handled in a fast inner loop, whereas trade-offs between, for example, emissions and response time for different operating conditions are made in an outer loop. The idea with two loops is depicted in Figure 11.10. It fits well into the industry's engineering process, where the inner control loops are first tuned for performance. Then, the total system is calibrated to meet the emission limits by adjusting set-points for different operating conditions, different hardware configurations, and different legislative requirements, depending on the measured emissions during the emission calibration process.

Often, neither x_{egr} nor λ_O are measured and thus have to be estimated using observers, see Rajamani (2005). If there is no λ sensor, oxygen estimation needs particular attention, since it is only detectable (an unobservable mode that is stable), see Diop et al. (1999). Here, an observer is used in the experiments, but if a λ-sensor is available then it can be incorporated in the controller.

11.6.2   System Properties that Guide the Control Design

An analysis of the behavior and characteristics of the system gives valuable insight into the control problem and is important for successful design of the control structure (see, for example, Kolmanovsky et al. (1997)). Figure 11.9 illustrated some system properties for the sensor outputs, but since the performance variables are different from these, they also need to be analyzed. An extensive system analysis over the engine operating region is given in Wahlström (2009), and provides input to how the objectives and system properties can be handled.

Steps in VGT Position and EGR-Valve

Model responses to steps in u_{vgt} and u_{egr}, in Figure 11.11, show that λ_O has non-minimum phase behaviors, overshoots, and sign reversals. These properties are also analyzed with respect to where the engine operates in a driving cycle, and the important results are summarized in the following paragraphs.

Both the non-minimum phase behavior and the sign reversal in the channel $u_{vgt} \rightarrow \lambda_O$ occur in operating points where the engine frequently operates. Therefore, these two properties must be considered in the control design. For the other channel, $u_{egr} \rightarrow \lambda_O$, both the non-minimum phase behavior and the sign reversal only occur in operating points where λ_O, pumping loss M_p, and turbocharger speed N_t are high. Consequently, there are significant drawbacks when operating in these operating points. Therefore, the control structure should be designed so that these operating points are avoided (these properties will be returned to in Section 11.5.3).

The channel $u_{egr} \rightarrow x_{egr}$ has a positive DC-gain. The channel $u_{vgt} \rightarrow x_{egr}$ has a negative DC-gain, except for a sign reversal that occurs in a small operating region with low torque, low to medium engine speed, half to fully open EGR-valve, and half to fully open VGT. Further, the relative gain array is also analyzed and shows that the best input–output pairing for SISO controllers is $u_{egr} \rightarrow \lambda_O$ and $u_{vgt} \rightarrow x_{egr}$ in the regions where the engine frequently operates.

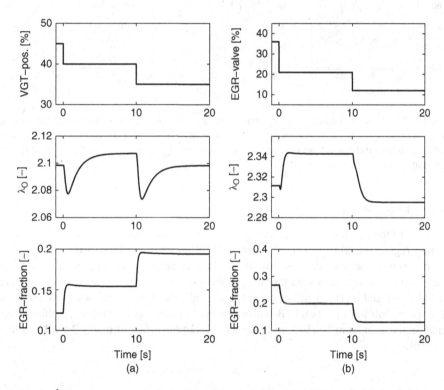

Figure 11.11 Responses to steps in (a) VGT position and (b) EGR valve showing non-minimum phase behaviors and sign reversals in λ_O. Operating point for the VGT steps: $u_\delta = 145$ mg/cycle, $N = 1500$ rpm, and $u_{egr} = 50\%$. Operating point for the EGR steps: $u_\delta = 230$ mg/cycle, $N = 2000$ rpm, and $u_{vgt} = 30\%$

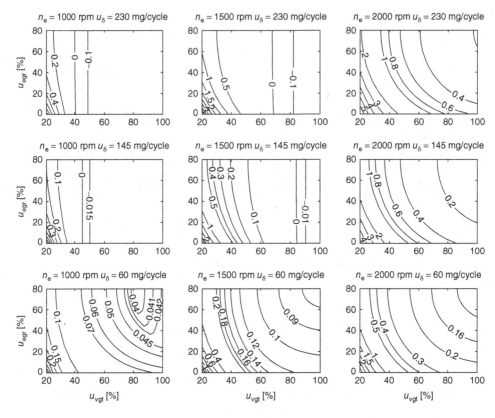

Figure 11.12 Contour plots of $p_{em} - p_{im}$ [bar] in steady-state at three different N and three different u_δ, showing that $p_{em} - p_{im}$ decreases with increasing EGR-valve and VGT opening, except in the left bottom plot where there is a sign reversal in the gain from u_{vgt} to $p_{em} - p_{im}$. (The control input u_{egr} saturates at 80%.)

Pumping Losses in Steady State

A mapping of the pumping losses in steady state is shown in Figure 11.12, covering the entire operating region (at 20 different u_{vgt} points, 20 different u_{egr} points, 3 different N points, and 3 different u_δ points). It gives insight into how to achieve pumping work minimization in the control structure. It is seen that the pumping losses, $p_{em} - p_{im}$, decrease with increasing EGR-valve and VGT openings, except in a small operating region with low torque, low engine speed, half to fully open EGR-valve, and half to fully open VGT, where there is a sign reversal in the gain from VGT to pumping losses, but the effect of this is 2.5 mbar which is very small.

11.6.3 Control Structure

The control design objective is to coordinate u_δ, u_{egr}, and u_{vgt} in order to achieve the control objectives in Section 11.6.1. The diesel engine is a nonlinear and coupled system and one could consider using a multi-variable nonlinear controller. However, based on the system analysis in the previous section, it is possible to build a controller structure using min/max-selectors and SISO controllers for EGR and VGT control, and to use feedforward for fuel control. As will be shown, this can be done systematically by mapping each loop to the control objectives via the system analysis. The resulting structure of loops is the main result together with the rationale for it, but within the structure different SISO controllers could be used. However, throughout the presentation, PID controllers will be used. The foremost reasons are that all control objectives will be shown to be met and that PID controllers are widely accepted by the industry.

The solution is presented step by step in the following sections, but a MATLAB/SIMULINK schematic of the full control structure is shown in Figures 11.13, where all signals and the fuel controller are shown.

Signals, Set-Points, and a Limit

The signals needed for the controller are assumed to be either measured or estimated using observers. The measured signals are engine speed (N), intake and exhaust manifold pressure (p_{im}, p_{em}), and turbocharger speed (N_t). The observed signals are the mass flow into the engine $\dot{m}_{ei}$, oxygen mass fraction X_{Oim}, λ_O, and x_{egr}. All these signals can be seen in the block "Signals" in Figure 11.13. The set-points and the limit in the controller vary with operation

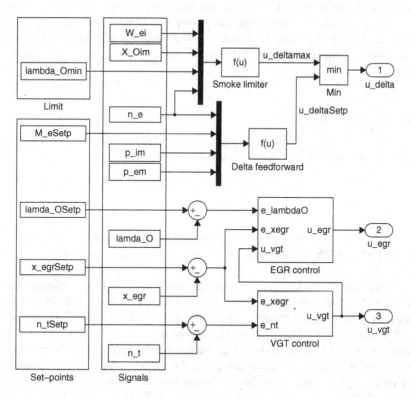

Figure 11.13 The control structure, as MATLAB/SIMULINK block diagram, showing a limit, set-points, measured and observed signals, fuel control with smoke limiter, together with the controllers for EGR and VGT

conditions during driving. These signals are provided by an engine and emission management system, as depicted in Figure 11.10. The limit and the set-points are obtained from measurements and tuned to achieve stable combustion and the legislated emissions requirements. They are then represented as look-up tables being functions of operating conditions.

Main Feedback Loops

The starting point for the design is the structure in Figure 11.10. Based on an analysis of the system properties in Section 11.6.2, the following two main feedback loops are chosen

$$u_{egr} = -\text{PID}(e_{\lambda_O}) \tag{11.6}$$

$$u_{vgt} = -\text{PID}(e_{xegr}) \tag{11.7}$$

where $e_{\lambda_O} = \lambda_O^s - \lambda_O$ and $e_{xegr} = x_{egr}^s - x_{egr}$. These two main feedback loops are selected to handle items 1 and 3 of the control objectives in Section 11.6.1. In the first loop λ_O is controlled to a set-point λ_O^s with the control signal u_{egr}, and in the second loop intake manifold EGR-fraction, x_{egr}, is controlled to its set-point, x_{egr}^s, with the control signal u_{vgt}. The PID controllers have a minus sign since the system analysis showed that the corresponding channels have negative DC-gains in almost the entire operating region.

From the analysis of system properties in Section 11.6.2 it is known that the channel from u_{vgt} to λ_O has a sign reversal and a non-minimum phase behavior. These are avoided in the structure (11.6) because u_{egr} is used to control λ_O. The relative gain array analysis also supports the input–output pairing for the main control loops. A deeper discussion is found in Wahlström et al (2010).

Additional Feedback Loops

In order to achieve the control objectives 3 and 5 in Section 11.6.1, the following two additional feedback loops are added to the main control loops (11.6)–(11.7):

$$u_{egr} = \min(-\text{PID}_1(e_{\lambda_O}), \text{PID}_2(e_{xegr})) \tag{11.8}$$

$$u_{vgt} = \max(-\text{PID}_3(e_{xegr}), -\text{PID}_4(e_{Nt})) \tag{11.9}$$

where $e_{Nt} = N_t^s - N_t$. Note that there is no minus sign for PID_2 since the corresponding channel has positive DC-gain. The additional loops are motivated by the following:

- In operating points with low engine torque there is too much EGR, although the VGT is fully open. To achieve the desired x_{egr}, objective number 3, a lower EGR-fraction x_{egr} is obtained by closing the EGR-valve u_{egr}, using $\text{PID}_2(e_{xegr})$ in (11.8), and a min selector is used. A side effect is a higher λ_O, but this is allowed.
- To avoid overspeeding of the turbo, objective number 5, the VGT is opened with the turbo speed controller N_t in (11.9). In this case, N_t is controlled with u_{vgt} to a set-point N_t^s which has a value slightly lower than the maximum limit N_t^{max} in order to avoid overshoots exceeding N_t^{max}.

Minimizing Pumping Work

The control structure (11.8)–(11.9) is not guaranteed to minimize the pumping work. The reason is that there are many combinations of the EGR valve area and VGT turbine pressure that can give the same flow, and consequently there are many u_{egr} and u_{vgt} that can give the same x_{egr} in cases when $\lambda_O > \lambda_O^s$. Thus, in some cases when $\lambda_O > \lambda_O^s$, both u_{egr} and u_{vgt} are governed by e_{xegr}. In stationary conditions, when $PID_2(e_{xegr})$ and $PID_3(e_{xegr})$ in (11.8)–(11.9) have converged, the controller fulfills the control objectives but the EGR-valve and VGT are not guaranteed to minimize the pumping work.

To achieve control objective 6, that is to minimize the pumping work, the two following control modes are added

$$u_{egr}(t_i) = \begin{cases} \min\left(-PID_1(e_{\lambda_O}), PID_2(e_{xegr})\right), & \text{if } u_{vgt}(t_{i-1}) = 100 \\ -PID_1(e_{\lambda_O}), & \text{else} \end{cases} \qquad (11.10)$$

$$u_{vgt}(t_i) = \begin{cases} 100, & \text{if } (u_{vgt}(t_{i-1}) = 100)\,\&\,(e_{xegr} < 0.01) \\ \max\left(-PID_3(e_{xegr}), -PID_4(e_{Nt})\right), & \text{else} \end{cases}$$

$$\qquad (11.11)$$

In case 1 in (11.11) the VGT is locked to fully open (the value 100) until $e_{xegr} > 0.01$ in order to avoid oscillations between case 1 and 2 in (11.10).

In this structure, u_{egr} is calculated using a minimum selector only when $u_{vgt} = 100$, compared to (11.8). This subtle difference results in minimized pumping work in stationary points by striving to open the actuators as much as possible. Looking at the pumping work minimization in more detail, the important controller action is coupled to λ_O, and in particular the operating conditions where there is a degree of freedom, that is when $\lambda_O > \lambda_O^s$. From the physics we know that opening a valve reduces the pressure differences over the corresponding restriction. Therefore, minimizing pumping work, control objective 6, is achieved through the mechanism explained above that opens the EGR-valve and VGT. These properties are also confirmed in Figure 11.12, which shows that the lowest pumping work is achieved when the EGR-valve and VGT are opened as much as possible while keeping the control objectives.

Remarks on the Control Approach
The approach described above is a complete structure that uses several PID controllers to achieve the gas flow control. It is only one example of an approach, but was selected as it illustrates both how different system properties can be handled with different loops and that a complex system can require several modes to get full functionality. However, as it is only an example, it is also important to highlight that there are other approaches that have been proposed in the literature. Robust control is investigated and used in Amstutz and del Re (1995), Jung et al. (2005), Nieuwstadt et al. (1998), while a nonlinear multi-variable Lyapunov based design is developed in Jankovic et al. (2000). Rückert et al. (2001) describe decoupling control, while several approaches with varying selection of feedback variables are compared in van Nieuwstadt et al. (2000). Model predictive control MPC is also popular since it can handle constraints and multi-variable systems, examples are Alberer and del Re (2010), Ferreau et al. (2007), Garca-Nieto et al. (2008), Ortner and del Re (2007), Rückert et al. (2004), Stewart et al. (2010), and Wahlström and Eriksson (2013) to just mention a few.

Feedforward Fuel Control

Engine torque control, control objective 4, is achieved by feedforward from the set-point M_e^s by utilizing the torque model and calculating the set-point value for u_δ according to

$$u_\delta^s = c_1\,M_e^s + c_2(p_{em} - p_{im}) + c_3\,N^2 + c_4\,N + c_5$$

which is obtained by solving u_δ from (8.77)–(8.79). This feedforward control is implemented in the block "Delta feedforward" in Figure 11.13.

Aggressive transients can cause λ_O to go below its hard limit $\lambda_{O,min}$, resulting in exhaust smoke. The PID controller in the main loop (11.6) is not designed to handle this problem. To handle control objective 2, a smoke limiter is used which calculates the maximum value of u_δ. The calculation is based on engine speed N, mass flow into the engine $\dot{m}_{ei}$ (marked W_ei in the figure), oxygen mass fraction X_{Oim}, and lower limit of oxygen/fuel ratio $\lambda_{O,min}$

$$u_\delta^{max} = \frac{\dot{m}_{ei}\,X_{Oim}\,120}{\lambda_{O,min}\,(O/F)_s\,10^{-6}\,n_{cyl}\,N}$$

which is implemented in the block "Smoke limiter" at the top of Figure 11.13.

Combining these two, the final fuel control command is given by

$$u_\delta = \min(u_\delta^{max}, u_\delta^s) \qquad (11.12)$$

which concludes the description and the motivation of the control structure in Figure 11.13.

11.6.4 PID Parameterization, Implementation, and Tuning

Each PID controller has the following parameterization

$$\text{PID}_j(e) = K_j\left(e + \frac{1}{T_{ij}}\int e\,dt + T_{dj}\frac{de}{dt}\right) \qquad (11.13)$$

where the index j is the number of the different PID controllers in (11.10)–(11.11). All the PID controllers have integral action and since these are mode switching, care must be taken to avoid integrator windup and ensure bump-less transfer in mode switches. The PID controllers are implemented in incremental form, which leads to anti-windup and bump-less transfer between the different control modes (Aström and Hägglund 1995).

Concerning the derivative part, it has been found that the loop from VGT-position to turbocharger speed ($\text{PID}_4(e_{Nt})$) benefits from a derivative part in order to predict high turbocharger speeds. This is due to the large time constant in the corresponding open-loop channel. The channel $u_{egr} \to \lambda_O$ also has a large time constant, but there is a lower demand on the bandwidth for $\text{PID}_1(e_{\lambda_O})$ compared to $\text{PID}_4(e_{Nt})$.

In the proposed structure there are four PID controllers that need tuning. This can be cumbersome work and therefore an efficient and systematic method for tuning the parameters K_j, T_{ij}, and T_{dj} in (11.13) has been developed. The systematic analysis of the control problem was used to couple the control objectives to the controller structure. This coupling to objectives gives the foundation for systematic tuning, be it manual or automatic. There are conflicting goals, as it is not possible to get both good transient response and good EGR tracking at the same time, and trade-offs have to be made.

Manual Tuning

The tuning parameters K_j, T_{ij}, and T_{dj} in (11.13) are obtained by first initializing them (without the derivative part) using the Åström–Hägglund step-response method for pole-placement (Åström and Hägglund 1995). Then the parameters are fine tuned using the methods described below in order to achieve the control objectives. After the initialization, the following tuning order is applied:

$$1\ u_{egr}\ \text{to}\ \lambda\ \text{loop}, \quad 2\ u_{egr}\ \text{to}\ x_{egr}\ \text{loop}, \quad 3\ u_{vgt}\ \text{to}\ x_{egr}\ \text{loop}, \quad 4\ u_{vgt}\ \text{to}\ N_t\ \text{loop}$$

This order follows the causality of the system, that is the order that the signals in Figure 11.13 start changing in a load transient

$$M_{eSetp} \rightarrow u_\delta \rightarrow \lambda \rightarrow u_{egr} \rightarrow x_{egr} \rightarrow u_{vgt} \rightarrow N_t$$

which is obtained by taking the smallest time constants of the different loops into consideration.

Manual Fine Tuning Method

The final tuning is obtained by adjusting the parameters iteratively until the control objectives are achieved. The gain K_j is adjusted in order to change the speed of the controller, and T_{dj} is adjusted in order to improve the performance of the closed-loop system. If it is possible, a derivative part is avoided. The tuning strategy for T_{ij} is to increase it until oscillations appear in the control signal, then decrease it until the oscillations disappear. This provides a systematic approach for the tuning of the PI(D) parameters, but the tuning can be time consuming and therefore it can be beneficial to use automatic controller tuning, and this is covered in the next section.

Automatic Controller Tuning

The control objectives in Section 11.6.1 and the system properties in Section 11.6.2 are mapped to a quadratic performance measure, where each term reflects either control objectives or actuator stress. As a result, the following nonlinear least squares problem is formulated

$$\begin{aligned} \min\ &V(\theta) \\ s.t.\ &\theta > 0 \end{aligned} \tag{11.14}$$

where the parameter vector θ and cost function $V(\theta)$ are

$$\theta = [K_1, T_{i1}, K_2, T_{i2}, K_3, T_{i3}, K_4, T_{i4}, T_{d4}]^T \tag{11.15}$$

$$V(\theta) = \sum_{i=1}^{N} \gamma_{Me} \left(\frac{e_{Me}(t_i, \theta)}{M_{eNorm}} \right)^2 + \gamma_{egr} \left(\frac{e_{xegr}(t_i, \theta)}{x_{egrNorm}} \right)^2 + \gamma_{Nt} \left(\frac{\max(N_t(t_i, \theta) - N_t^{max}, 0)}{n_{tNorm}} \right)^2$$

$$+ \left(\frac{u_{egr}(t_i, \theta) - u_{egr}(t_{i-1}, \theta)}{u_{egrNorm}} \right)^2 + \left(\frac{u_{vgt}(t_i, \theta) - u_{vgt}(t_{i-1}, \theta)}{u_{vgtNorm}} \right)^2$$

$$\tag{11.16}$$

Here t_i is the time at sample number i. All terms in (11.16) are normalized to get the same order of magnitude for the five terms, and this means that the weighting factors have an order of magnitude as $\gamma_{Me} \approx 1$ and $\gamma_{egr} \approx 1$.

These terms have been derived by analyzing the control objectives and system properties, and the connections and motives for them are given in the following paragraphs. Objectives 2 and 6 are fulfilled directly as they are built into the structure in terms of the smoke limiter and the pumping work minimization presented in Section 11.6.3.

Term 1

This term is the most intricate one and it is coupled to objectives 1 and 4, and they are in their turn related to each other through the system properties. They are related since a good transient response, especially during tip-in maneuvers, is connected to availability of oxygen, and thus a fast λ_O-controller will give good transient response. A further motivation for choosing to minimize engine torque deficiency, $e_{Me} = M_e^s - M_e$, comes from the fact that negative values of $e_{\lambda_O} = \lambda_O^s - \lambda_O$ are allowed, and it is positive e_{λ_O} values that have to be decreased. Now, note that torque deficiency occurs when the smoke limiter in Section 11.6.3 restricts the amount of fuel injected, thai is, when $\lambda_O = \lambda_{O,min}$ (see Figure 11.14 between 309 s and 313 s). Since $\lambda_{O,min} < \lambda_O^s$, a positive e_{λ_O} exists when torque deficiency occurs. One could also consider using e_{λ_O} directly, but such a choice is not sufficiently sensitive during transients where there is a need for air. Due to the smoke limiter, e_{λ_O} will be limited to the difference $\lambda_O^s - \lambda_{O,min}$ when the smoke limiter is active, and this does not reflect the actual demand for air and λ_O during transients. Thus the torque deficiency is selected as performance measure.

Terms 2 and 3

Term 2 is directly coupled to objective 3 and strives to minimize the EGR error ($e_{xegr} = x_{egr}^s - x_{egr}$). Term 3 is a direct consequence of objective 5 and avoids the turbocharger speed exceeding its maximum limit. A high penalty is used, $\gamma_{Nt} \approx 10^3$, to capture that this is a safety critical control loop.

Terms 4 and 5

Terms 4 and 5 are coupled to the general issue of avoiding actuator stress, for example oscillatory behavior in the EGR valve or in the VGT control signals. The terms have equal weight since the control signals are of the same magnitude.

In summary, all control objectives are considered and handled in the tuning. Furthermore, the difficulty of tuning of the individual controllers, related to the trade-off between transient response (λ_O) and EGR errors, is efficiently handled by the two weighting factors γ_{Me} and γ_{egr}. This will be further illustrated in Section 11.6.5.

Solving Eq. (11.14)

A methodology for solving the optimization problem has been developed, and the details are described in Wahlström et al. (2008). The important constituents are: a transient selection method and a solver for the optimization problem. Transient selection is made to reduce the computational time and the method identifies representative and aggressive transients where different control modes are excited. As a result, computational time is reduced by a factor of 10 when using only the selected transients instead of a full ETC cycle. The numerical solver has three steps. Firstly, the tuning parameters are initialized using the Åström–Hägglund step-response method for pole-placement (Åström Hägglund 1995). Secondly, a heuristic globalization method is used to scan a large region around the initial values. Thirdly, a standard nonlinear local least-squares search is used. The heuristics in the second step is important to avoid the local solver ending up in an unsatisfactory local minimum.

11.6.5 Evaluation on European Transient Cycle

The control tuning method is illustrated and applied, and a simulation study is performed on the European Transient Cycle (ETC). The cycle consists of three parts representing different driving conditions: urban (0–600 s), rural (600–1200 s), and highway (1200–1800 s) driving.

The closed loop system, consisting of the model in Section 8.10 and the proposed control structure in Section 11.6.3 (depicted in Figure 11.13), is simulated over the ETC cycle. The set points for λ_O and x_{egr} are authentic recordings that have been provided by the industry. A remark is that an observer is not used in the simulations. Instead a low pass filter, with the time constant 0.02 s, is used to model the observer dynamics for all variables assumed to come from an observer. This is done in the block "Signals" in Figure 11.13. The different signals in the cost function (11.16) are calculated by simulating the complete system and sampling the signals with a frequency of 100 Hz.

Further, a tuning rule for avoiding oscillations in the control signals u_{egr} and u_{vgt} is to decrease the sum $\gamma_{Me} + \gamma_{egr}$ until the oscillations in the control signals disappear.

Balancing Control Objectives

The weighting factors γ_{Me}, γ_{egr}, and γ_{Nt} in the cost function (11.16) are tuning parameters. When tuning these, trade-offs are made between torque deficiency, EGR error, pumping losses, and turbo over-speed.

A tuning strategy for the relation between γ_{Me} and γ_{egr} is to increase γ_{Me} when a controller tuner wants to decrease the torque deficiency, and increase γ_{egr} when a controller tuner wants to decrease the EGR error and the pumping losses. It is important that the sum $\gamma_{Me} + \gamma_{egr}$ is constant in order to avoid influence of the third and fourth term in the cost function when tuning the first and the second term. In the following section $\gamma_{Me} + \gamma_{egr} = 2$. A tuning strategy for avoiding turbo over-speeding is to increase γ_{Nt} until the fifth term becomes equal to zero.

Illustration of Performance Trade-Offs

The trade-offs between torque deficiency, EGR error, and pumping losses are illustrated in Figures 11.14–11.15, where the control system is simulated on an aggressive transient from the ETC cycle with two sets of weighting factors. The first set is $\gamma_{Me} = 1$ and $\gamma_{egr} = 1$ and the second set is $\gamma_{Me} = 3/2$ and $\gamma_{egr} = 1/2$. The latter set of weighting factors punishes the torque deficiency more than the first. Figure 11.15 also shows the control modes for the EGR valve

$$mode_{egr} = \begin{cases} 1 & \text{, if PID}_1(e_{\lambda_O}) \text{ active} \\ 2 & \text{, if PID}_2(e_{xegr}) \text{ active} \end{cases} \tag{11.17}$$

and the VGT position

$$mode_{vgt} = \begin{cases} 1 & \text{, if } u_{vgt} = 100 \\ 2 & \text{, if PID}_3(e_{xegr}) \text{ active} \\ 3 & \text{, if PID}_4(e_{Nt}) \text{ active} \end{cases} \tag{11.18}$$

The setting $\gamma_{Me} = 3/2$ and $\gamma_{egr} = 1/2$ gives less torque deficiency but more EGR error and more pumping losses compared to $\gamma_{Me} = 1$ and $\gamma_{egr} = 1$, which is seen in Figures 11.14–11.15 in the following way. Between 305 and 308 s the engine torque is low, which leads to a high λ_O,

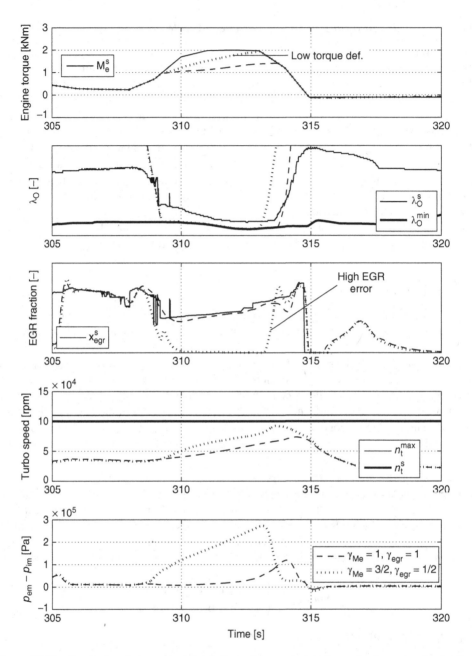

Figure 11.14 Comparison between two simulations of the control system using two sets of weighting factors. The first set is $\gamma_{Me} = 1$ and $\gamma_{egr} = 1$ and the second set is $\gamma_{Me} = 3/2$ and $\gamma_{egr} = 1/2$. The latter set of weighting factors gives less torque deficiency but more EGR error and more pumping losses compared to the first set of weighting factors

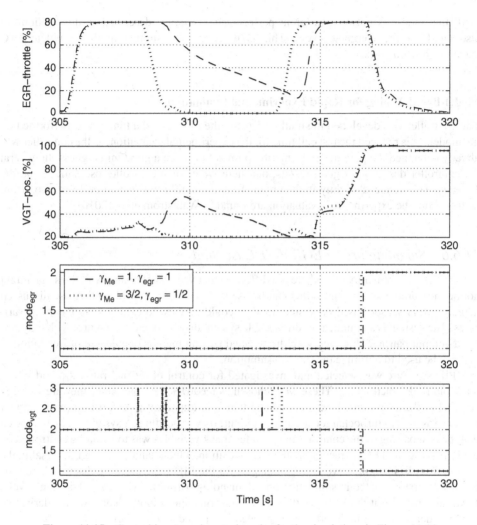

Figure 11.15   Control inputs and control modes for the simulations in Figure 11.14

an open EGR-valve, and that the VGT position controls the EGR-fraction so that the EGR error is low. Thereafter, an increase in engine torque at 308 s leads to a decrease in λ_O and therefore a closing of the EGR-valve. This closing is faster if γ_{Me}/γ_{egr} is increased from 1 to 3, which leads to a lower EGR-fraction (i.e., more EGR error), a more closed VGT position, a faster increase in turbocharger speed, and consequently a lower torque deficiency. Note that the torque deficiency and the EGR error cannot be low at the same time during the aggressive transient. Note also that there are more pumping losses at $\gamma_{Me} = 3/2$ and $\gamma_{egr} = 1/2$ due to the fact that the EGR-valve and the VGT position are more closed. Consequently, in dynamic conditions trade-offs are made between torque deficiency and pumping loss. However, it is important to note that the pumping loss is still minimized in stationary points by the control structure in both cases in Figures 11.14–11.15 compared to the other control structure in Section 11.6.3 that gives higher pumping losses.

All the trade-offs between different performance variables described in this section are also valid for the complete cycle. This is illustrated by simulating the complete ETC cycle (Wahlström 2006).

Model-Based Tuning for Rapid Experimental Evaluation

The controller was developed in simulation using the model, and a tuning was performed on the model. This reduced the test cell time for the experimental evaluation; as the controller was already debugged the parameters from the optimization were a good initial guess. In general the controller design gave good results, but there were some controller oscillations and the K_p gains were reduced slightly in the engine tests to achieve the desired performance. More details about the experimental evaluation are found in Wahlström et al. (2010).

11.6.6 Summing up the EGR VGT Case Study

It is important to remember that there is a difference between performance variables that might not be measurable and variables that can be used for feedback control. In the case discussed here, a virtual sensor based on system models could be used to estimate the feedback variables. The system is a nonlinear multi-variable system that is not easy to control with only one input–output pairing. However, with the selected pairing there is one degree of freedom in λ that can be used to optimize the fuel consumption.

A PID structure was selected and investigated for control of air/fuel ratio, λ_O, and intake manifold EGR-fraction x_{egr}. These were chosen as feedback variables since they are strongly coupled to the emissions. It was also noted that it is important to control torque for driver response as well as turbocharger speed to avoid turbo overspeeding. A system analysis showed that the best pairing of the control inputs and feedback variables was to let λ_O be controlled by the EGR-valve and EGR-fraction by the VGT-position, which handles the sign reversal in the system from VGT to λ_O. Two tuning strategies for the PID structure were discussed, and the result was first evaluated in simulation on a demanding part of the European Transient Cycle. It was also noted that the VGT-position to turbocharger speed loop benefits from a derivative part in order to predict high turbocharger speeds. This is due to the large time constant in the corresponding open-loop transfer function.

In summary, many issues arise that need to be handled in engine control, and the interplay between controller goals, performance variables, and nonlinear system dynamics has been demonstrated in this case study of the complex EGR VGT system.

11.7 Diesel After Treatment Control

As mentioned above and in Section 6.5.2, the emissions of main concern in a diesel engine are NO_x and soot or particulates. A diesel can therefore be fitted with different systems for treatment of the exhaust gas after it has left the cylinder, see Figure 11.2. Due to legal restrictions on emissions, there is an intense development of different exhaust treatment systems that

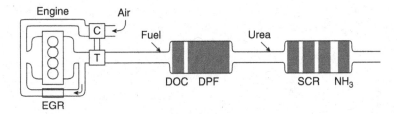

Figure 11.16 Examples of components in a diesel engine gas treatment in a heavy duty engine, see, for example, Charlton et al. (2010) for details. The components are: Compressor (C), Variable Geometry Turbine (T), Exhaust gas recirculation (EGR), Diesel oxidation catalyst (DOC), Diesel particulate filter (DPF), three blocks with Selective catalytic reduction (SCR) catalyst, and last a fourth block with an Ammonia slip catalyst (NH_3). There is control of fuel injection for the DOC and Urea injection for the SCR

will best suit the application; some of the candidates to select from are one or more from the following list:

- Diesel oxidation-type catalytic converter.
- Particulate filter.
- NO_x accumulator-type catalytic converter.
- System for SCR (Selective Catalytic Reduction), Figure 11.16.
- Combination systems (Future four-way systems).

For all these systems, the goal is to strive towards the origin in Figure 11.1, and to get into the boxes defined for upcoming legislation there are several control problems that need to be be solved. To mention just one; control of exhaust gas temperature is needed for SCR (at least 250°C) and for regeneration of particulate filters. In low loads when there is little exhaust enthalpy, this can be achieved by introducing the intake system throttle, as depicted in Figure 11.2, which decreases λ, which in turn increases the engine-out temperature.

Fusion of SI and CI Engines

As a concluding remark on a general trend, we note that technologies that were primarily used on either SI or CI engines are now spreading to each other as demands for performance increase. For example, the SI engine has adopted direct injection from CI engines and the CI engine has adopted intake throttles from SI engines, and both rely on the availability of the control system to handle the complexity and optimize their performance.

Part Four

Driveline – Modeling and Control

12

Engine–Some Advanced Concepts

Engines are computerized machines, and this fact–together with needs and demands from customers and legislative requirements from society–has created vigorous development activities. Two trends of special interest in the context of engine and control system development are:

- New mechanical designs that make the engine hardware more flexible by removing design trade-offs. These designs are made possible by, and rely on, the existence of an engine management system that controls the engine to the optimal points on-line.
- New developments for engine state estimation. The developments in new sensor technologies, and continuous increase in computing and network technology, open up possibilities for extracting information about the engine state.

In fact, many designs that are surfacing are not new, even if sometimes presented so, but the novelty is instead that they can now, with proper **control**, be realized to increase functionality and performance. This chapter illustrates these trends by examples that go beyond the basics of engine control, and the treatment is therefore more brief but with pointers to support material. The first example is variable valve actuation (VVA) that gives flexibility to the engine. It is possible to adapt and tune the performance but it also gives more dimensions to cover and optimize in the control system. The second example, variable compression, also exemplifies an advanced engine where a classical engine design trade-off has been removed. The third example is ion-sense, which is an in-cylinder measurement that gives a signal that is rich in information but that is complicated to interpret. It exemplifies the development of new sensors and advanced signal processing and interpretation techniques, which gives an opportunity for improved performance through feedback control. In particular, extracting more information from available sensors through improved signal interpretation and sensor fusion gives added functionality without any component costs.

12.1   Variable Valve Actuation

The valvetrain, with its inlet and exhaust valve openings and profiles, controls the gas exchange and thus influences the combustion and torque generation of an engine. Valve profiles and timing, in a fixed cam engine, are selected as a trade-off between the performance at high and low

Modeling and Control of Engines and Drivelines, First Edition. Lars Eriksson and Lars Nielsen.
© 2014 John Wiley & Sons, Ltd. Published 2014 by John Wiley & Sons, Ltd.
Companion Website: www.wiley.com/go/powertrain

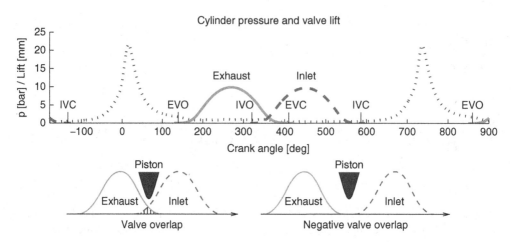

Figure 12.1   Top: Cylinder pressure (dotted) together with exhaust (solid) and inlet (dashed) valve lifts. The events EVO, EVC, IVO, and IVC are also marked. Bottom: Valve profiles, illustrating valve overlap and negative valve overlap, and the fact that the piston motion can restrict the possible valve motions

speeds as well as high and low loads. The important events in the cycle were introduced and discussed previously in Section 5.1.1, and Figure 12.1 shows an example with a cylinder pressure trace and valve lift profiles together with the valve events. With variable valve actuation (VVA), the timings and/or lift profiles are made variable which removes the trade-offs. Instead, the Engine Managements System (EMS) gets the task of selecting the best valve strategy for the current operating point to optimize fuel economy, emissions, and maximum torque and power.

Valve events influence work production and pumping work consumption, residual gas, and gas motion. Some understanding of the mechanisms is gained from the valve events that influence several characteristics in conjunction with the valve overlap, shown in Figure 12.1. The events and main mechanisms are:

- Exhaust Valve Opening (EVO) influences the expansion work, blow-down, and puming losses. A too early opening reduces the pressure and thus gives a loss in expansion work, while too late gives higher pressure that lasts into the gas exchange and gives higher pumping work. An early opening is favorable at high loads and high speeds.
- Exhaust Valve Closing (EVC) also influences the pumping work production, during the switch from open to closed valve. It also influences the amount of residual gases trapped in the cylinder, where it interacts with IVO through the valve overlap, that will be discussed in Section 12.1.2.
- Inlet Valve Opening (IVO) also influences the pumping work and the residual gases through the interaction with the EVC in the valve overlap, as mentioned above.
- Inlet Valve Closing (IVC) influences pumping work, the amount of gas trapped in the cylinders, and the effective compression ratio.

The situation is complex, so we here simplify matters when we say that the most influential valve event for controlling part load pumping work and engine efficiency is the inlet valve closing (in an SI engine). Exhaust valve closing influences the residual gas fraction and gas exchange, and thus NO_x emissions through the residual gas. A deeper review and analysis of valve events and their effects is given in Hong et al.(2004).

12.1.1   Valve Profiles

Examples of individual valve profiles are shown in Figure 12.2, illustrating how the individual profiles can be phased, scaled, or reshaped. A lot of hardware solutions have been proposed, but here the focus is on the valve profiles and only a few comments will be made about the hardware to achieve them. The individual profiles, in Figure 12.2, can be applied to one or both of the inlet and exhaust valves. For example, the top left profile, *cam phasing*, can be used on both intake and exhaust systems and some examples are: intake only (Stivender 1978), dual equal (Stefanopoulou et al. 1998), dual independent (Magner et al. 2005). Cam phasing has become common and can be implemented by rotating the camshaft in relation to the crank and cam chain, with an hydraulic actuator, see for example, Stone (1999).

The second example, *valve lift*, is mostly used on the intake side. Reducing only the lift has a negative effect on pumping work, as it generates higher flow losses over the valves. Its application is therefore to generate higher in-cylinder gas motion to improve the turbulence, mixing, and combustion. There are also possibilities to change the valve open time and also to combine lift and open time. An intermediate solution is to use *cam profile switching* where there is the possibility of switching between two (lift) profiles. The switching can also be combined with cam phasing and an example is given in Sellnau and Burkhard (2006).

So far, only systems that can be implemented with cam profiles have been mentioned, but there are also *fully flexible valve actuation* systems. These are based on mechanic, electrohydraulic (Dittrich et al. 2010), or electromagnetic actuators (Wang et al. 2002) that can control the opening and closing profiles more freely. With fully variable valve actuation it is also possible to switch between four stroke and two stroke operation. Modeling and control of the actuators themselves is also an interesting task but it is outside our scope here.

This concludes the brief introduction and discussion of valve profiles and their generation, and our attention is turned to what these systems can achieve and how.

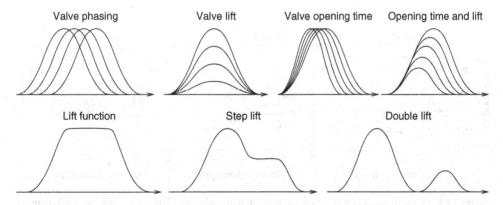

Figure 12.2   Examples of individual valve lift profiles. Upper row: Translation or scaling of the valve profiles. Bottom row: Changes to the appearance of the valve profile

12.1.2 Effects of Variable Valve Actuation

When looking at the engine and its output in terms of work and emissions there are three goals with VVA systems:

- Miminizing the gas exchange losses for high engine efficiency.
- Controlling the gas composition, that is residual gas fractions.
- Preparing the gases for combustion.
    - Generating in-cylinder flows, turbulence or tumble and swirl.
    - Having the right effective compression ratio.

Beside these there are additional features that can be utilized, like additional flows with valve overlap, cylinder deactivation, or combustion control in advanced engines, for example homogeneous charge compression ignition (HCCI). Concerning the potential and applicability, there are several papers that cover the topics of VVA systems and the interested reader is referred the following early survey papers: Asmus (1982), Gray (1988), Ahmad and Theobald (1989), and Asmus (1991).

Pumping Work Reduction for Improved Engine Efficiency

As mentioned earlier, throttling down the intake manifold pressure gives an undesirable loss of efficiency in SI engines, and this is one of the effects that the VVA systems can mitigate. The early work in Stivender (1968) demonstrated increased part load efficiency with inlet valve throttling, even though the focus in the original work was on combustion stability in lean operation. Coming to valve actuation, Figure 12.3 shows an example of how load control can be achieved with throttle free control by varying the inlet valve closing points. The figure also illustrates that there are two options, either early IVC or late IVC. With *early IVC* (Tuttle 1982) the valve closes before the piston is at BDC and locks the gas in the cylinder. When the piston goes to BDC and returns, the gas in the cylinder expands and re-compresses. With *late IVC* (Tuttle 1980), the inlet valve is open over the intake stroke and a part of the compression

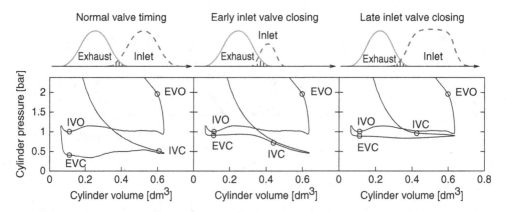

Figure 12.3 Valve lift and p-V diagram for normal, early inlet, and late inlet valve closing, where the latter two exemplify how the pumping work can be reduced

stroke, which causes the gas in the cylinder to re-enter the intake manifold and thus reduces the amount of air (and fuel) inducted into the pressure. Both early and late IVC reduce the effective compression ratio. This type of operation, with a compression stroke that is shorter than the expansion stroke, is referred to as *Miller cycle* operation. A benefit of the operation is that the gas exchange area (and pumping losses) from IVC to BDC is close to zero. A side effect of a reduced effective compression ratio is that cylinder temperatures will be lower, which has negative effects on the combustion rate and cold start but positive effects leading to less NO_x and heat transfer.

If one looks at the high pressure part of the cycle, the IVC influences the effective compression ratio, as discussed above, while the EVO influences the expansion and thus the blow-down and exhaust. In particular, variable EVO timing enables the control system to select the best trade-off between lost expansion work or additional pumping work in the exhaust stroke, depending on the operating condition.

Valve Overlap

Valve overlap has an influence on the pumping work production through its influence on the pressure. However, it foremost influences the gas flows and trapping of residuals and/or possibly scavenging. Residuals can be controlled with the exhaust valve closing. With *early EVC* there is more exhaust gas left in the cylinder, when the valve closes, which gives higher residual gas fractions. With *negative valve overlap* there can also be a re-compression of the exhaust gases as the valve is closed while the piston continues up to TDC. With a late IVO, the work consumed by compression can be recovered during the gas expansion until the inlet valve opens.

With small or no valve overlap and a late EVC that occurs in the intake stroke, some of the exhausted gases are re-inducted into the cylinder, which increases the residual gas fraction. With a *positive valve overlap* and *late EVC* there are two cases: (i) higher pressure in exhaust than intake. (ii) Higher pressure in intake than exhaust. In the first case, residual gases will expand into the intake after the inlet valve opens and exhaust gases can also flow back into the cylinder from the exhaust. When the exhaust valve is closed, the gases that have entered the intake manifold are inducted into the cylinder together with the fresh air–fuel mixture during the intake stroke.

The second case, with higher intake pressure than exhaust, can occur in turbocharged engines. The valve overlap can lead to fresh gases going into and passing the cylinder all the way out to the exhaust. The fresh gases can be seen as blowing the cylinder clean of residual gases, which is called *scavenging*.

With scavenging it is possible to improve the low speed torque in turbocharged gasoline engines through the use of scavenging, twin scroll, and direct injection, see Leduc et al. (2003). In addition it has also been shown that scavenging can improve the turbocharger response by adding more flow and power through the turbine and thus shorten the turbo lag.

This is also a delicate control problem where the traditional feedback λ-control cannot be utilized during the scavenging phase. Instead, models have to be used for the induction and blow through so that the gases trapped in the cylinder have $\lambda \geq 1$, otherwise the fuel rich mixture, with unburned CO and HC, would mix with the air and oxidize with an exothermic reaction in the exhaust or catalyst. If this isn't properly controlled the additional heating can easily damage the catalyst.

12.1.3   Other Valve Enabled Functions

When enabling variable valve events, there are opportunities for implementing other functions. Before proceeding, it is also worth commenting that there are also other simpler solutions that influence the valve flows, like guide vanes in the inlets that guide the flow, or even port and valve deactivation that are used to generate turbulence, or cylinder gas motion like tumble or swirl for better combustion quality. Below are two examples of controllers that use the cylinder valve actuation.

Cylinder Deactivation

Valve profile switching was discussed above and a special case is to switch off one or more cylinders in the engine. This is called *cylinder deactivation*, see Figure 12.4, and improves the part load efficiency of an engine through the following mechanisms. Consider an engine running at a part load operating point with a given torque. When cylinders are deactivated their valves are closed and they have no pumping work. Their work production is also removed, so that the other cylinders will need to operate at a higher load to generate the same total engine torque. This means that the intake manifold pressure is increased and the efficiency of the working cylinders and total engine is increased. There is still friction work from the deactivated cylinders.

In the EMS, control is about both strategies and detailed torque control. *Strategies* concern when to activate or deactivate the cylinders, and naturally depends highly on the torque request but also on strategies for avoiding too many switches. While the *torque control* uses the throttle, ignition timing, and fuel injection to produce a smooth engine torque during activation and deactivation of cylinders. The MVEM models for cylinder air-flow and torque in Sections 7.4 and 7.9 can be extended for this purpose. Activated and deactivated cylinders, are modeled separately, paying attention to the fact that deactivated cylinders have neither volumetric efficiency nor gross or pumping work.

Compression Braking

In larger diesel engines, for example heavy duty vehicles, there is often an exhaust brake that increases the back pressure and pumping work, see also Section 11.1.2. This allows the vehicle to be braked without using the brakes at the wheels, which reduces the maintenance and cost for the owner. A replacement is a compression braking system. In such cases there is no fuel injection and the valves are controlled to be closed during compression and open when the

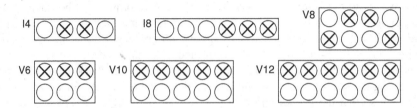

Figure 12.4   Deactivation strategies for engines with different number of cylinders that ensure balanced engine and torque generation

piston reaches TDC, the engine then consumes the compression work and this can be used to brake the vehicle.

12.1.4 VVA and Its Implications for Model Based Control

When extending the mean value engine models to VVA operation, there are foremost two models that have to be adjusted. First is the torque model in Section 7.9, where the basic structure still holds but the components need to be analyzed. The gross indicated work can be influenced slightly by the valve events, as mentioned above. The main effect is on the gas exchange and pumping work models that need to be revisited and analyzed so that the pumping loop describes the valve events. There can be some changes to the friction from the piston rubbing due to higher or lower friction or from the valve mechanism.

The other major concern is the gas exchange with residual gases and volumetric efficiency, in Section 7.4.1, that needs a component for the valve actuation. In Magner et al. (2005) a model structure is proposed for a dual independent system. Cylinder flow there is an affine function in pressure with slope c_s and offset c_o that depend on the engine speed and valve phasing. It can also be formulated in terms of volumetric efficiency, yielding

$$\eta_{\mathrm{vol}} = p_{\mathrm{im}} \, c_s(N, IVO, EVO) + c_o(N, IVO, EVO)$$

More studies on control-oriented modeling of the flow can be found in, for example, Berr et al. (2006), and Leroy et al. (2008).

The important implication for engine control is that it changes the breathing characteristics and entrapment of burned gases (Jankovic and Magner 2002). The added flexibility and improved performance come at the price of a significantly increased engineering effort in the control and calibration. In short, the control challenges relate to the following three related issues: more degrees of freedom, finding the optimum while calibrating the multi-dimensional system, and finding the right trade-off between different properties. Maps can be used to represent the behavior, but there is a need to reduce the dimensions of the data to save both EMS memory and calibration requirement (Magner et al. 2005). Estimators and models, based on cylinder pressure measurements, for the gas exchange describing fresh and residual gases have also been proposed and investigated (Mladek and Onder 2000; Öberg and Eriksson 2006; Worm 2005).

For these systems with a high degree of freedom, many design and optimization methods for finding the optimal control timings have been proposed, and one direction is to utilize gas dynamics simulation models to support the time consuming calibration process, see, for example. Sellnau and Burkhard (2006), and Wu et al. (2007). Several control schemes have been investigated for VVA engines, and the interested reader is referred to Chauvin et al. (2008), Colin et al. (2007), Stefanopoulou and Kolmanovsky (1999).

12.1.5 A Remark on Air and Fuel Control Strategies

A simplified view of engine air and fuel control is that the air is controlled from the driver's demand by the throttle, and the amount of fuel is a supporting loop (earlier supplied by the carburetor) that controls the air/fuel ratio. An alternative is to let the driver control the fuel and use the air loop for control of the (A/F) ratio. This was discussed in Stivender (1978) where the (A/F) control of a variable valve actuation system was studied. In particular, it was shown that it would be a good solution for lean burn engine operation, where it achieves good driver response and good attenuation of sensor noise and other engine control disturbances. However, in today's integrated EMS it is difficult to discern which loop is leading the other; regardless, it is interesting to reflect on the principles.

12.2   Variable Compression

Variable compression ratio engines take two steps towards improved engine efficiency. One is the removal of the design trade-off between compression ratio (efficiency) and engine knocking, so that the EMS can select the best possible compression ratio for the current driving situation. The other is that it offers the possibility for significant downsizing and supercharging for improved fuel economy.

The basic idea is simple: use a high compression ratio during low loads for high efficiency and, as the load increases, the compression ratio is decreased to match the knocking. Figure 12.5 shows a schematic description of a variable compression engine where the cylinder is hinged to the crank case and can be tilted by a hydraulic crank mechanism. Different tilt angles result in different clearance volumes, V_c, which give different compression ratios $r_c = \frac{V_d + V_c}{V_c}$.

Background on the Design Compromises

A key contribution of Nicholas August Otto was his finding that compressing the air and fuel mixture prior to combustion gives an increased efficiency. In Section 5.2.2 the ideal Otto cycle efficiency was derived, yielding $\eta = 1 - 1/r_c^{\gamma-1}$, where r_c is the compression ratio. Current SI engines have a compression ratio around 10, due to autoignition (knock) of the air and fuel mixture. Increased heat transfer, friction losses, mechanical stress, and emission of unburned hydrocarbons also play a role in the useful compression ratios. However, as discussed in Section 6.2.2, autoignition occurs when the air and fuel mixture has been exposed to high temperatures for a certain time. The link between temperature and compression ratio is easy to see by considering an isentropic compression process: $T_2 = T_1 r_c^{\gamma-1}$, relating the initial

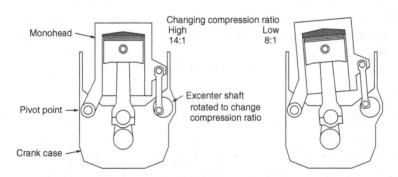

Figure 12.5   Variable compression–a schematic description of the operating principle. The cylinder can be tilted so that the clearance volume changes, and thereby the compression ratio changes

temperature T_1 to the temperature at the end of the compression T_2. Increasing the compression ratio gives a higher temperature at the end of compression.

Another compromise in engine design, which affects the customer, is the size of the engine. A small engine gives good fuel economy but poor driving performance, and a big engine gives a good performance but increases the fuel consumption. Downsizing and turbocharging for improved fuel economy was discussed in Section 4.4 as a solution for improving the system efficiency. With turbocharging or supercharging the compression ratio needs to be reduced to accommodate the higher pressures and temperatures yielding a peak efficiency penalty. A downsized and supercharged engine with variable compression thus addresses both of these two compromises.

12.2.1   Example – The SAAB Variable Compression Engine

Several mechanical designs have been proposed for varying the compression ratio, see, for example, Schwaderlapp et al. (2002) and Nilsson (2007) for surveys. One example is the SAAB Variable Compression (SVC) engine (Drangel et al. 2002). It is a 5-cylinder 1.6-l engine with performance of 225 hp/305 Nm, designed to effectively replace a 3.0-l naturally aspirated engine. A sketch of the construction is shown in Figure 12.6.

Of vital importance for the possible level of downsizing is not only the maximum achieved power and torque output, but also the torque at 1000 rpm and below if the engine is intended to replace a larger engine in a medium or large size car. To meet these objectives, the SVC engine uses a mechanical supercharger (a screw compressor) instead of a turbocharger. The high IMEP necessary for the SVC 1.6-l engine to equal the 3.0-l naturally aspirated engine's performance forces the compression ratio down to 8:1 to avoid knocking and high IMEP variations that cause uneven engine running at high loads. At part load operation, where most of the driving occurs, a fixed compression ratio of 8:1 would have a considerable

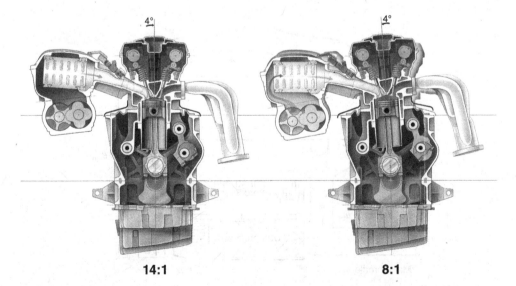

14:1   **8:1**

Figure 12.6 Sketch of the SAAB implementation of a variable compression engine in the states of compression ratio 14 and 8 respectively

negative impact on overall fuel consumption. The SVC engine enables a compression ratio of 14:1 to be used during part load operation, thereby achieving good efficiency.

12.2.2 Additional Controls

The SVC engine has variable compression through the variable size of the clearance volume. This alone adds a new control input that affects almost every other aspect of engine control. The engine is also equipped with a compressor for supercharging, with compressor bypass and compressor-clutch control as additional control inputs, see Figure 12.7. Due to these additional control possibilities, compared to a standard engine, there is of course an added potential to utilize on-line control to optimize engine combustion, engine behavior, and engine performance. We will here just mention two interesting control design issues to exemplify the additional controls. One clear trend is the increased use of model-based control, since the traditional technique with engine maps becomes very complex when there are several control variables.

Control Balancing Compression and Spark Timing

One interesting example of a control issue is that both compression and spark timing influence knock and efficiency. (Review Sections 6.2.2, 7.9.1, and 10.6 for spark timing control and its influence on efficiency, knock, and emissions.) Using spark timing only means that spark advance has to be retarded until knock disappears. Having the possibility of also using compression ratio control opens up the possibility of coordinating compression ratio control and spark timing control, so that the extra freedom can be utilized to optimize behavior and performance. An example of the relation between ignition timing, compression ratio, and work is shown in Figure 12.8.

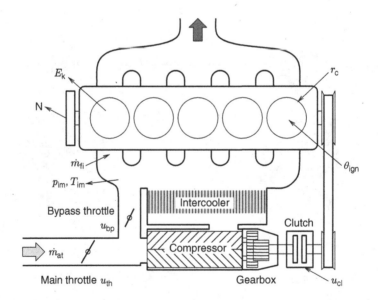

Figure 12.7 Sketch of the SVC engine and its supercharger, with screw compressor, gearbox, and controlled clutch. Sensors and actuators for the control loops are also shown

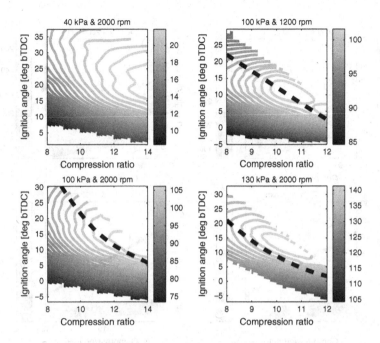

Figure 12.8 Contour plots of engine torque, for the same air charge and fueling, showing the relation between compression ratio, ignition timing, and work output. The fuel has RON=95 and the knock limit is seen as the dashed line. The desirable region for operation is below the knock line

The optimal selection of the compression ratio and ignition timing is studied in Nilsson et al. (2008). The approach is to build a MVEM torque model, with the same structure as (7.55), and use the model to search for the optimum within the knock limit. The indicated gross work captures the connection between the compression ratio and ignition and is expressed by

$$W_{\mathrm{i,g}}(m_{\mathrm{f}}, r_{\mathrm{c}}, \Delta\theta_{\mathrm{ign}}) = m_{\mathrm{f}}\, q_{\mathrm{LHV}}\, \eta_{\mathrm{ig,ch}}\, \left(1 - \frac{1}{r_{\mathrm{c}}^{\gamma-1}}\right) \eta_{\mathrm{ign}}(\Delta\theta_{\mathrm{ign}}). \tag{12.1}$$

$$\eta_{\mathrm{ign}}(\Delta\theta_{\mathrm{ign}}) = 1 - c_2\, (\Delta\theta_{\mathrm{ign}})^2 - c_3\, (\Delta\theta_{\mathrm{ign}})^3 \tag{12.2}$$

$$\Delta\theta_{\mathrm{ign}} = f(\theta_{\mathrm{ign}}, p_{\mathrm{im}}, N) = \theta_{\mathrm{ign}} - \theta_{\mathrm{ign,opt}}(p_{\mathrm{im}}, N) \tag{12.3}$$

$$\theta_{\mathrm{ign,opt}}(p_{\mathrm{im}}, N) = a_0 + a_1 \frac{1}{p_{\mathrm{im}}} + a_2\, N \tag{12.4}$$

There is also a pumping work model that includes the engine speed, and a friction work model that includes the work done by the mechanical compressor. All models with parameters are given in Nilsson et al. (2008). The model is parametrized using data with RON=99 up to the knocking limit; in knocking conditions there is a loss in efficiency, due to heat transfer, which is not modeled since that region must be avoided anyway. A validation of the model on RON=95 data is shown in Figure 12.9. The maximum deviation between the optimum predicted by the model and validation data gives a loss in efficiency that is only 0–0.4 percentage units lower. This illustrates that the MVEM torque model can provide a basis for selecting the optimum control combinations.

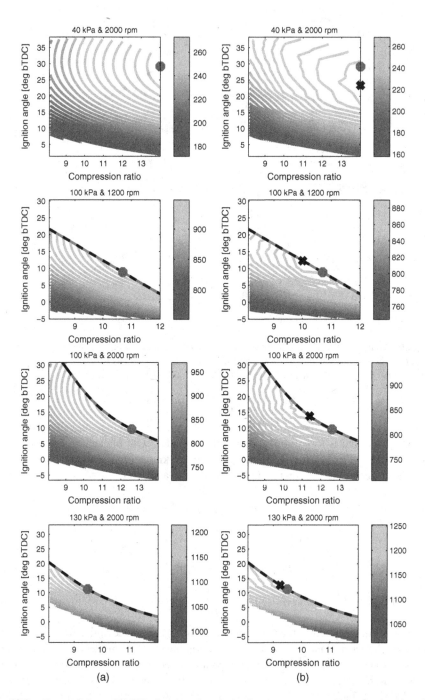

Figure 12.9 Contour plots of IMEP, showing the relation between compression ratio, ignition timing, and efficiency. (a) Torque model based on RON=99 data but with knocking limit set by RON=95 data. (b) Validation data for RON=95. The optimum predicted by the model is marked with ∗ and in validation data with ×

Driveability Control

Another example of an interesting control topic is the control of the supercharger. Two additional control inputs, the extra bypass throttle and the supercharger clutch, should be handled. A schematic of the system can be seen in Figure 12.7.

The supercharger is disconnected during part load operation to avoid parasitic losses. The connection and disconnection of the supercharger is operated by an electro-magnetic clutch. Due to the moment of inertia from the supercharger when it is engaged, a control strategy was developed to make the supercharger switching transparent to the driver. Depending on driving conditions, different engagement times are used to engage the supercharger. At constant speed, driving without the supercharger, when a small increase in load requires the supercharger to be engaged, a smooth connection of the supercharger is necessary to avoid a jerk in the movement of the car, and thus an engagement time of up to 650 ms is used. If, on the other hand, the driver suddenly requires full torque, a quick engagement of the supercharger is necessary to obtain a dynamic vehicle behavior, and thus an engagement time of less than 100 ms is used. Above 3500 rpm, the supercharger is engaged all the time.

Control of Compressor Boost Pressure

Boost pressure control is achieved with the mechanical supercharger and the throttles, as shown in Figure 12.7. The system with two throttles gives an overactuated control problem with one extra degree of freedom, similar to that of turbo control discussed in Section 10.9.2. Control of the mechanical compressor bypass and main throttle for best fuel economy was investigated in Lindell (2009), and it showed that the most fuel efficient strategy was to open the main throttle fully and then perform boosting pressure control by closing the bypass throttle.

12.3   Signal Interpretation and Feedback Control

The availability of algorithms and software has now reached a level where it is possible to extract a great deal of information using single sensors and a combination of sensors. This provides valuable information for the EMS that can be used to improve the control performance. With more feedback information the engine can be adjusted or re-tuned during operation for best performance by changing parameters in the engine control system. Ion-sense is one example that illustrates these possibilities.

12.3.1   Ion-sense

In an ideal combustion reaction, hydrocarbon molecules react with oxygen and generate only carbon dioxide and water, recall the reaction in (4.1). However, in the combustion chamber there are also other reactions and molecules present, that include ions, which go through several steps before they are completed; some examples are (Shimasaki et al. 1993)

$$CH + O \rightarrow CHO^+ + e^-$$

$$CHO^+ + H_2O \rightarrow H_3O^+ + CO$$

$$CH + C_2H_2 \rightarrow C_3H_3^+ + e^-.$$

These electrons and ions, as well as several others, are generated by the chemical reactions in the flame front. Additional ions are also created when the burned gas temperature increases due to the increased cylinder pressure.

In *ion-sensing*, a measurement voltage is applied to the spark plug, when it is not used for firing, and the resulting current through the spark plug is measured. See Figure 12.10 for a measurement technique and the bottom plot of Figure 12.11 for examples of the resulting current. The sensed current is called the ion current, and depends on the electrons and ions that are created in the combustion chamber, and on their relative concentration and recombination, which in turn depends on pressures, temperatures, and the components in the gas, to mention some of the more important factors. The processes that cause the ion current are complex and also vary from engine cycle to engine cycle. Figure 12.11 shows ten consecutive cycles of the cylinder pressure and the ion current operating at constant speed and load. The signal thus is very rich in information but also complex to analyze and, as can be seen, the cycle-by-cycle variations are significant.

Measurement Details

To detect the ions, a DC bias is applied to the spark plug, generating an electrical field. The electrical field makes the electrons and ions in the spark plug gap move, which generates a current. A diagram of the spark plug and the essential components in the measurement circuit is shown in Figure 12.10. Note that the ignition current goes in the opposite direction to the ion current and charges the capacitor at each ignition, the Zener-diode controls the measurement voltage. Measuring the current at the low-voltage side of the ignition coil, avoids the high-voltage pulses associated with the ignition.

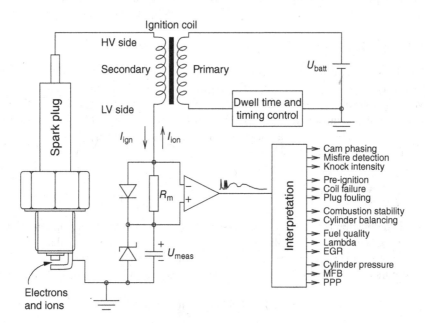

Figure 12.10 Measurement circuit for the ion current. The spark plug-gap is used as a probe, detecting primarily electrons but also ions. The measurement is performed on the low voltage side of the ignition coil. With advanced signal processing the spark plug can be a sensor for several parameters. Knock intensity and misfire are already implemented in production cars. Peak pressure position estimation will be used in a closed loop control example later in this section

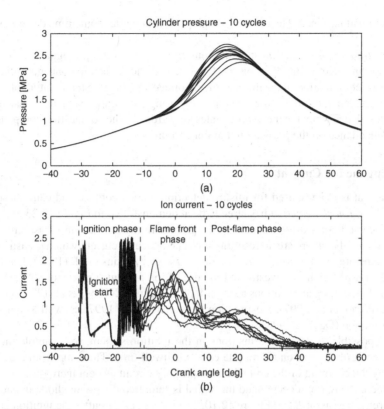

Figure 12.11 (a) cylinder pressure. (b) ion current. Ten consecutive cycles at stationary engine operation, showing that cycle-to-cycle variations are always present in the combustion. The three phases, ignition, flame front, and post-flame, are also shown for the ion current

The ion current is an interesting engine signal to study since it is a direct measure of the combustion, that contains a lot of information. Some of the parameters that affect the ion current are: temperature, air/fuel ratio, time since combustion, exhaust gas recycling (EGR), fuel composition, engine load, and several others. Several challenges remain in its interpretation.

Ion Current Terminology

The ion current typically has three phases: a phase related to ignition, a phase related to ions from the flame development and propagation, and a phase related to pressure and temperature development. In Figure 12.11, the three phases of the ion current are displayed. Each phase has varying characteristics and they also mix together in complicated ways. In the *ignition phase*, the ion current is large with reversed polarity. Due to the high current in the ignition, the measured signal shown in the figure is limited. What can be seen in Figure 12.11 is that the ringing phenomenon in the coil after the ignition is a result of the inductance and the small capacitance that inevitably exists in the circuit.

In the *flame-front phase*, the high level of ions associated with the chemical reactions in the flame produces one or more characteristic peaks. The ions generated by the flame have different recombination rates. Some ions recombine very quickly to more stable molecules, while others have longer residual times. The result is a high peak which after some time decays as the ions recombine.

In the *post-flame phase* the most stable ions remain, generating a signal that follows the cylinder pressure due to its effect on the temperature and molecule concentration. Ions are created by the combination of the measurement voltage and the high temperature of the burned gases, since the temperature follows the pressure during the compression and expansion of the burned gases, that is when the flame propagates outwards and the combustion completes. There is thus a dependence on the pressure in the ion current.

Interpreting the Ion Current

The ion current is already used for misfire detection, knock control, and cam phase sensing (Figure 12.10). *Knock detection* has already been demonstrated in Figure 10.23, where it was seen that the oscillation from a knock is also present in the ion current measurement. When there is a *misfire*, then there are no resulting ions and hence no current which is easily detected. These systems are already used in production cars; see, Auzins et al. (1995), Lee and Pyko (1995). Several papers have investigated the use of ion current for detection of other engine parameters. Some early applications are: spark plug fouling (Collings et al. 1991), combustion phasing (Eriksson et al. 1997), λ estimation (Reinmann et al. 1997), as well as mass fraction burned estimation (Daniels 1998).

It is also possible to detect malfunctions in the ignition system, for example the spike at $-30°$ is a result of the coil on event, related to the dwell time. The ramp from $-28°$ to $-21°$ is caused by the charging of the coil when the primary circuit current increases. The corner at $-21°$ is where the ignition occurs and the signal is saturated due to the diode in parallel with the measurement resistor R_m in Figure 12.10. The ringing comes after the ignition is finished, and then the current is related to combustion reactions as well as temperature and pressure.

Ion Current Modeling

The ion current has been studied by thermodynamical and chemical kinetic modeling, see, for example, Ahmedi et al. (2003), Reinmann et al. (1997), and Saitzkoff et al. (1997, 1996). Concentrating on the pressure-related post-flame phase, a model for the ion current has been presented that is based on the Saha ionization equation (Kittel and Kroemer 1995). Some of the fundamental assumptions in the model are that the gas in the spark plug gap is: fully combusted, in thermodynamic equilibrium, undergoes adiabatic expansion, and that the current is carried in a cylinder extending from the central electrode of the spark plug (Saitzkoff et al. 1996). Given the cylinder pressure, the model for the ion current is

$$\frac{I}{I_m} = \frac{1}{(\frac{p}{p_m})^{\frac{1}{2}-\frac{3}{4}\frac{\gamma-1}{\gamma}}} e^{-\frac{E_i}{2 k T_m}\left[(\frac{p}{p_m})^{-\frac{\gamma-1}{\gamma}}-1\right]}. \tag{12.5}$$

I,	Ion current	I_m,	Ion current maximum
p,	Cylinder pressure	p_m,	Cylinder pressure maximum
T_m,	Maximum temperature	γ,	Specific heat ratio
k,	Boltzmann's constant;	E_i,	Ionization energy

This relation between pressure/temperature and the ion current offers the possibility to extract some information about the pressure and combustion. The extraction of peak pressure position (PPP) is an indicator of how optimal the spark advance is, see Section 10.6, and this will be used as an example of combined signal interpretation feedback control below. Methods for extracting the PPP information started with the double/triple Gaussian method (Eriksson et al. 1996), followed by neural network and peak search (Hellring and Holmberg 2000) as well as other approaches (Moudden et al. 2002). A recent summary of applications is found in Malaczynski et al. (2013).

Virtual Pressure Sensor with Pressure Model and Ion-sense
Another example that illustrates the possibilities of the virtual sensor approach combined with sensor fusion is the approach from Eriksson and Nielsen (2003). A combination of the analytic cylinder pressure model (7.51), summarized in Section 7.8, and the ion current model (12.5), above, is used to generate a virtual sensor for the cylinder pressure. The approach uses normal engine sensors to construct a pressure trace, but since there is an uncertainty in combustion phasing, that piece of information is extracted with the aid of ion-sensing. Information about the pressure is fused together by using the level of the pressure trace from Section 7.8, and the phasing of the combustion from the ion current and (12.5). With the models for pressure and ion current and input data from other sensors the total model produces an estimate of the cylinder pressure trace.

12.3.2 Example – Ion-sense Ignition Feedback Control

We now turn to how the ion current can be used for feedback control. We start by recalling Figure 10.20, showing that an optimal ignition timing positions the peak pressure position (PPP) close to 16° ATDC. Feedback control with this information was demonstrated with cylinder pressure sensors to give an increase of several percent in engine efficiency (Glaser and Powell 1981; Hubbard et al. 1976; Sawamoto et al. 1987). Ion-sense based feedback control of the ignition timing was first demonstrated in Eriksson et al. (1997).

The controller structure for the spark timing is shown in Figure 12.12 and has the following components. The ignition timing manager is the conventional controller discussed in Section 10.6. Here it is augmented so that it gives a reference value for PPP, giving the possibility of having different spark schedules for different operating points, that is meeting goals other

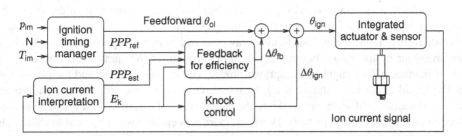

Figure 12.12 The structure of the spark advance control structure, where the spark plug operates as an integrated actuator and sensor. Information is extracted from the raw ion current, and the estimate of the PPP is the input to the spark timing controller; if knock is detected the knock controller takes over. Reference values and feedforward signals are obtained using other sensors, such as engine speed and load

than to maximize the work. For example, in mid-load mid-speed ranges it is desirable to have a spark advance close to MBT, with PPP around 15°, and in high load ranges a more conservative schedule, with late PPP, for reducing engine noise and NO_x emissions. The feedforward shown in Figure 12.12 incorporates information about how changes in reference value and engine transients affect the spark advance. The interpretation algorithm has both knock detection, as described in Section 10.6.1, and PPP estimation from Eriksson and Nielsen (1997) that is used in the feedback controller for positioning the combustion optimally. The feedback controller measures the previous combustion cycle n and updates the spark timing to the coming cycle $n + 1$, and it uses the following integrating controller

$$\Delta\theta_{fb}[n + 1] = \Delta\theta_{fb}[n] - K_I \cdot (PPP_{ref}[n] - PPP_{est}[n]) \qquad (12.6)$$

where $\Delta\theta_{fb}$ is the adjustment to the spark timing, PPP_{ref} the desired peak pressure position, PPP_{est} the PPP estimation from the ion current, and K_I is a gain that has to be tuned.

Closed-loop Controller Parameter

The gain K_I in (12.6) is selected as a balance between attenuation of cycle-to-cycle variations and response speed. The filtering comes at the price of slowing down the feedback loop, but this can be improved with feedforward, shown in Figure 12.12, based on a nominal spark advance table. Very quick response is not an issue since environmental parameters, for example humidity, do not change rapidly. One criterion is that the spark timing should not move more than 1° due to cyclic variations; see Powell (1993). For this engine, the cycle-to-cycle variation for the estimate of the PPP is around 10°.

Another consideration to take into account is how well the PPP estimate correlates with the actual PPP. Moving averages of different lengths have been computed for the measured and the estimated peak pressure positions, see Eriksson et al. (1997), also indicating that $K_I = \frac{1}{10}$ is a good choice for the feedback gain, which is the gain used in the on-line tests.

Experimental Setup

Water is injected into the engine to provide an unknown disturbance to the engine. In Figure 12.13a, the water injection setup is shown together with the engine. As can be seen in the figure, the injection procedure is carried out by hand. The water spray is directed into the induction system towards the throttle plate. The water spray is then drawn, by the lower pressure, into the intake manifold. The sprayer is a color sprayer that has a valve which delivers a liquid spray. This liquid spray is atomized by two opposing holes that blow pressurized air on the spray. Figure 12.13b shows a photo of the sprayer with water spray, and a schematic enlargement of the sprayer nozzle with the liquid spray and the pressurized air. The liquid is not fully atomized by the pressurized air but the droplets are made much smaller. The amount of water sprayed into the engine was not measured, but it had no audible effect on the engine during the tests. However, there was enough water present to change the in-cylinder pressure trace so that the mean peak pressure position moved to a position around four to five degrees later.

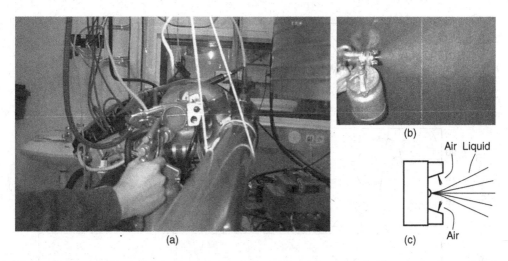

Figure 12.13 (a) The sprayer is directed towards the intake port and throttle plate. At the lower side of the throttle plate, the spray of water can be seen as a pale shade of gray. When the picture was taken, the engine ran at steady state with speed 1500 rpm and load 50 Nm. (b) A picture of the sprayer spraying water. (c) A schematic figure of the sprayer nozzle with the liquid spray, pressurized air, and the atomized liquid drops

Water Injection Experiments

During the experiment, shown in Figure 12.14, the throttle angle and the injection time are held constant. The engine speed is also held constant by the dynamometer. The engine is running at steady state and the (A/F) ratio is tuned to $\lambda = 1$ before the test cycle starts, and then the injection time is locked and held constant during the test cycle.

Figure 12.14 shows a large part of the test cycle. The speed and load condition is 1500 rpm and 55 Nm. Initially in the test cycle, the spark advance controller is running and the controller changes the spark advance, controlling the peak pressure position close to MBT, that is $16-17°$ after TDC. The ion current is used as input to the controller, and the in-cylinder pressure is only used for validation. The signals: PPP, output torque, manifold pressure, and lambda have been filtered off-line with a non-causal zero phase shift filter. A Butterworth filter of order 3 and normalized cut-off frequency of 0.3 is used with `filtfilt` in Matlab.

Around cycle number 100 the feedback controller is turned off and the controller holds the present value close to the optimum. Around cycle 250 the spraying of water is started. Note that the peak pressure position is moved to a position 4° later and that the output torque decreases. Around cycle 400 the feedback controller is turned on again and it controls the peak pressure position back to its optimal value. The controller needs to change the spark advance with around 5° to get back to the optimal position. An identical experiment at a lower load, 35 Nm, in Eriksson and Nielsen (1998) showed that the water injection shifted the optimal ignition 9°. Note that the output torque increases when the feedback controller is switched on, since the spark advance goes back to a point close to optimum. Around cycle 550 the water spraying stops, and the change in spark advance starts to decrease. When the water spraying stops it takes awhile until all the water has passed through the system and states asymptotically return to their initial conditions.

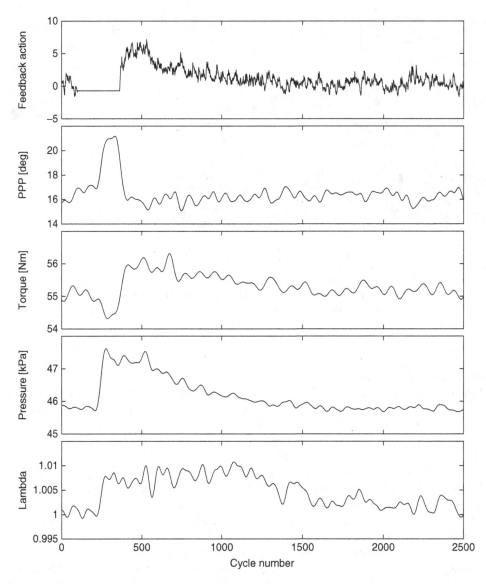

Figure 12.14   The spark advance controller is shut off around cycle 100 and the spark advance is held constant. Water spraying starts around cycle 250 and leads to delayed PPP and decreased output torque. The spark advance controller is switched on around cycle 400, controlling PPP back to MBT and leading to increased output torque. Water spraying stops around cycle 550 and the parameters asymptotically go back to their initial conditions, as the water in the intake manifold evaporates

When the controller is turned off, the spark advance can be viewed as a pre-calibrated schedule with a spark advance close to MBT. The signals that affect the spark advance are then the engine speed and the manifold pressure. A calibrated schedule moves the spark advance in the wrong direction, since a higher intake pressure (Eriksson and Nielsen 1998) is interpreted as a higher load, which normally has later ignition timing.

Analysis of Torque Increase Due to Water Injection

Experiments with water injection and feedback control in Eriksson and Nielsen (1998), showed increase in engine torque with 1.5–3% above the initial level. The increase in torque, just by adding water and controlling the spark advance, may seem surprising at first, but it comes from different sources. Three clues are found Figure 12.14. The first is that the (A/F) ratio increases, which increases the fuel conversion efficiency, and since the amount of fuel is constant this implies an increase in output torque. There is a 1% increase in (A/F) which, according to Figure 5.13, can change the fuel conversion efficiency by 0.4%. The second is that the manifold pressure increases by 2%, which lowers the pumping work (5.13). In Figure 12.14 it can be seen that the manifold pressure does not drop directly when the spraying stops, instead it slowly decreases as the water evaporates. Hence, it is the presence of water in the intake manifold that raises the pressure and not that the sprayer blows air and water on the throttle plate. In particular it is the evaporated water that displaces the air.

The third is that the presence of water also cools the air which, for the same pressure, makes the air density higher. The lowered inlet temperature and the presence of water reduce the temperature in the combustion chamber, which has a favorable influence on the thermodynamic cycle. As a result there is reduced heat transfer (5.32) and dissociation, which contribute to the increased efficiency and output torque. It is important to note that in order to get the increase in output torque with water injection, it is necessary to change the spark advance to gain the benefits. In Figure 12.14 the output torque actually decreases when the water is injected, the increase in efficiency comes when the spark advance controller is switched on.

Water injection for temperature reduction has even more benefits at full load, where it reduces both knocking and NO_x. After-market water-injection systems have been used to reduce the knock tendency, so that the spark advance can be maintained closer to optimum at maximum load and thus increase the maximum torque and power of the engine.

Concluding Remarks on the Ion-sense Application

The tools necessary for performing the analysis and providing an understanding of the mechanisms, come from the thermodynamics described in Chapter 5. This illustrates that both insight about control and a thorough understanding of the physics are needed for successful application and interpretation of complex engine systems.

12.3.3   Concluding Remarks and Examples of Signal Processing

This example on ion-sense feedback control was selected to illustrate that added information extraction from available sensors can be a cost effective solution for improving performance. With an ion-sensing capability already available for knock and misfire, only additional signal interpretation in the electronic engine management system (EMS) is needed to develop it further for more advanced functionality.

Crank Angle Speed Variations – Cylinder Balancing

Another example of an available signal is the engine speed signal (rpm signal). So far we have considered the engine speed as an average value over the cycle, but there are variations in engine speed due to the torque pulsations from each cylinder. Signal processing of the rpm signal together with a crankshaft model can be used for misfire diagnosis by, for example, estimating the torque pulses from the angular velocity fluctuations (Chen and Moskwa 1997; Wang et al. 1997). Misfire diagnosis will be discussed in more detail in Section 16.5.4. The speed variations are also used in diesel engine applications where the torque production of the individual cylinders is compared and cylinder balancing is performed.

Multi-sensor applications can also be developed where a basic signal, like engine speed or ion current, is measured and several other sensor signals can be deduced from it, see, for example, Figure 12.10. A variant of the ion-sense example above is to combine it with the pressure model 7.8, to extract combustion phasing information (Andersson and Eriksson 2009). Another approach is to exchange the ion current for a crankshaft model and torque sensor. The combustion phasing information can then be extracted using a crankshaft torque sensor (Larsson and Schagerberg 2004).

λ-sensor Signal Processing

Another sensor that contains cylinder individual information is the planar λ-sensor. It is fast enough to register fluctuations in λ that come from a spread between the cylinders, see for example Figure 7.22. Signal processing can be used to extract λ for individual cylinders (Cavina et al. 2008; Grizzle 1991; Schick et al. 2011), and the additional information can be used to adjust the fueling and reduce the spread among cylinders.

13

Driveline Introduction

The driveline transfers the engine torque to the wheels, and it is thus fundamental for vehicle propulsion. Together with the engine, the driveline constitutes the powertrain, as already defined in Chapter 3, that also gave several examples of topologies and configurations. According to Merriam-Webster, the word powertrain is the oldest, with a first known use in 1943. The first known use of the word "driveline" is from 1949, and the first use of the synonym "drivetrain" is from 1954. An example is Figure 13.1, that shows a powertrain of a rear-driven vehicle including both engine and driveline.

This introductory chapter on drivelines, together with the coming two chapters on modeling and control, will give the basis for model-based understanding and control. First in this chapter, the next section clarifies the nomenclature. Then, Section 13.2 defines the area of driveline control as a certain subarea of powertrain control. To further explain the background, Section 13.3 describes unwanted vehicle behavior that results from inappropriate driveline control, and by that clarifies the control tasks at hand. The approach taken in the following chapters on modeling and control is briefly discussed in Section 13.4.

13.1 Driveline

With the definition above that the driveline is the powertrain excepting the torque provider, then the main parts of a driveline are clutch, transmission, propeller shaft, final drive, driveshafts, and wheels. Figure 13.2 depicts these main parts in a schematic of a driveline of a rear-wheel-driven vehicle like the one in Figure 13.1.

13.2 Motivations for Driveline Modeling and Control

When approaching driveline control, the first questions are:

- What are the principal objectives and variables to control?
- What are the main physical phenomena causes for difficulties?
- How are the difficulties manifested in vehicle behavior if not being controlled?

An introductory discussion of these questions are the topics for the rest of this chapter.

Modeling and Control of Engines and Drivelines, First Edition. Lars Eriksson and Lars Nielsen.
© 2014 John Wiley & Sons, Ltd. Published 2014 by John Wiley & Sons, Ltd.
Companion Website: www.wiley.com/go/powertrain

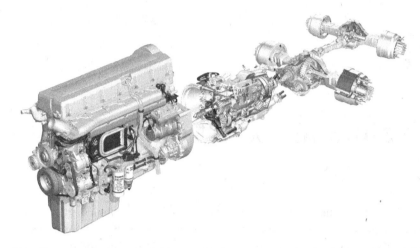

Figure 13.1 Example of a rear-wheel-driven vehicular powertrain showing both engine and driveline. Reproduced with permission from Volvo Trucks

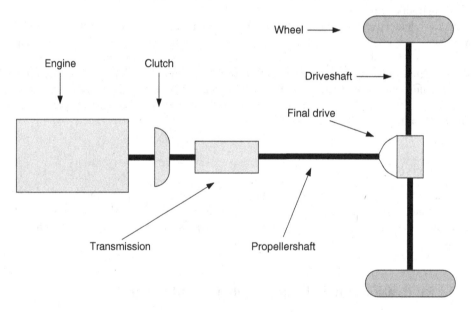

Figure 13.2 Schematic of a rear-wheel-driven vehicular driveline, for example, like the one in Figure 13.1

13.2.1   *Principal Objectives and Variables*

There are two types of variables that are of special interest in driveline control:

- (rotational) velocities
- torques.

Note that there are a number of velocities (engine, transmission, wheel) that are not the same, and there are also several different torques affecting the driveline (engine, transmission, wheel). The principal objectives are to control one or combinations of these variables so that they behave well. There are many specific objectives depending on application, and as will be seen later one example is to control the velocity of the wheel by actively controlling the engine speed to counteract oscillations. Another example of a control objective, related to gear shifting, is to control the torque inside the transmission to zero.

13.2.2   Driveline Control vs. Longitudinal Vehicle Propulsion Control

Driveline control could possibly refer to any issue regarding longitudinal propulsion. However, with the notation adopted above, that the driveline is the powertrain excepting the torque provider, it is common to let driveline control refer to different phenomena on a time scale of seconds or faster. In particular, the focus is on phenomena where the velocities or torques (engine, transmission, wheel) along the driveline are not the same. Powertrain control, propulsion control, or synonymously longitudinal vehicle control, on the other hand, take care of vehicle acceleration, temperature, and many other slowly varying phenomena. These distinctions are not strict, but are adopted in this book.

It should be noted that the same actuator, such as the engine, at the same time can be used for both driveline control and longitudinal vehicle control, where the time-average mean-value engine torque determines vehicle acceleration, and superimposed engine torque oscillations around the mean-value can be used to damp oscillations in the driveline.

13.2.3   Physical Background

The physical reason that the different parts of a vehicle driveline (clutch, transmission, shafts, and wheels) can have different velocities or torques relative each other is that they are elastic. This means that mechanical resonances may occur. The handling of such resonances is basic for functionality and driveability, but is also important for reducing mechanical stress and noise.

The torsional forces and the torsional energies in the driveline can be considerable. For example, the data presented in the top plot of Figure 14.4 show that the driveshaft has a torsion of more than 20°, and the release of that energy in an undesirable way is of course a major problem. New development of even more high-powered engines means that even more energy can be injected in to the driveline, which emphasizes even more the need for driveline control. Already, today, full torque on the lowest gear of a loaded heavy truck can break some shafts in the driveline.

13.2.4   Application-driven Background

Driving feel, together with wear and tear, of course benefit from smooth vehicle motion, that is smooth driveline behavior. Further, smooth driveline behavior is also important in several aspects relating to optimal cruise control, improved emissions, and handling of transients. However, smoothness is not the only requirement—often a change in operating condition should be as fast as possible. This is especially so in case of mode shifts in the driveline. One example is a fast gear shift, where an internal torque should be controlled to zero as quickly as possible without inducing oscillations. In modern drivelines, like in vehicles with advanced multi-mode engines or in hybridized vehicles, there are even more mode shifts that all need good control, preferably so good that the driver does not even notice, for example, a change in energy source.

In relation to the increasingly demanding applications, another trend worth mentioning here is that both sensors and actuators related to driveline control are under strong development enabling both faster and more precise response.

13.3   Behavior without Appropriate Control

As mentioned above, the torsional energy in a driveline can be substantial, and there are a number of problems if it is not controlled appropriately. Three well-known examples that every vehicle design has to cope with are vehicle shuffle or vehicle surge, shunt-and-shuffle, and oscillations after engagement of neutral gear. The topic of the next three subsections is to present these phenomena together with real vehicle measurements as illustrations.

13.3.1   Vehicle Shuffle, Vehicle Surge

The terms "vehicle shuffle" or "vehicle surge" refer to when the whole vehicle rocks back and forth due to driveline oscillations. In addition to the speed oscillation, typically the pitch of the vehicle will vary considerably for strong vehicle shuffle. The phenomenon is illustrated in

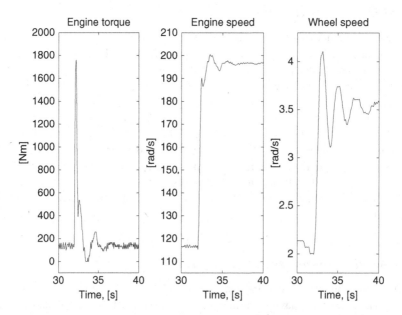

Figure 13.3   Measured speed response of a step in accelerator position at $t = 32$ s. A P-controller is used to control the engine speed to 2000 RPM. The engine speed is well damped, but the resonances in the driveline are seen to give oscillating wheel speed, resulting in vehicle shuffle

Figure 13.3 where a P-controller has been used to control engine speed in a heavy-duty truck from Scania. The figure shows how the measured engine speed and wheel speed respond to a step input in accelerator position. It is seen how the engine speed is well behaved with no oscillations. With a stiff driveline this would be equivalent to also having well damped wheel speed, but as seen the wheel speed oscillates resulting in vehicle shuffle.

When performing the test drives to collect the above and similar data, the truck was provoked. At times the vehicle shuffle was so strong that the driver cabin rocked back and forth in a way such that it was truly difficult to hold on to the measurement laptop computer without dropping it.

13.3.2   Traversing Backlash–Shunt and Shuffle

A special case of shuffle may appear when a tip-in starts from negative torque, traverses through the backlash of the driveline, and continues to accelerate with positive torque. Traversing the backlash is called shunt, and may end in an abrupt impact creating both undesired noise and jerk, which is the time derivative of acceleration. The vehicle may then start to shuffle. The combined phenomenon is called shunt and shuffle. A measurement on a Volvo personal car is seen in Figure 13.4 (Karlsson, 2001).

13.3.3   Oscillations After Gear Disengagement

When driving at steady pace, a steady torque corresponding to the driving resistance is transmitted in the driveline. The torsion in the driveshaft is then proportional to this torque. If the

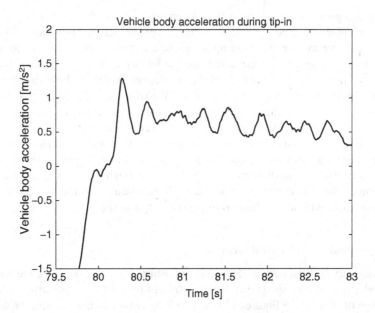

Figure 13.4   Measured vehicle speed illustrating the shunt and shuffle phenomenon. The shunt traversing the backlash occurs at 80 s, and it ends with a fairly abrupt impact at 80.2 s. Then the vehicle starts to shuffle

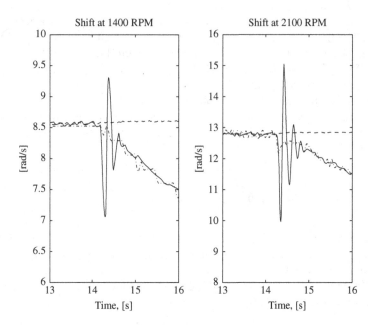

Figure 13.5 Engagement of neutral gear commanded at 14 s, with stationary driveline at 1400 RPM and 2100 RPM on a flat road with gear 1. The engine speed (dashed) and wheel speed (dash-dotted) are scaled to transmission speed (solid) with the conversion ratio of the driveline. After a short delay time, the neutral gear is engaged, causing the driveline speeds to oscillate. The amplitude of the oscillating transmission speed is higher the higher the stationary speed is

neutral gear is engaged in this situation, then the energy stored in the driveshaft is released, and depending on how this is done oscillations occur. The phenomenon is illustrated in Figure 13.5, presenting measurements on a Scania truck, where a torque control phase is not used when the neutral gear is engaged. The measurements are for two constant speeds and the principle behavior is the same, but the amplitude of the oscillations increases the higher the stationary speed is. It is seen that both the engine speed and the wheel speed behave well. This is quite natural, since in neutral gear the engine has no load and the vehicle is free-rolling with a speed that decreases in relation to the driving resistance. Thus, in this situation the effect on vehicle motion of the released torsional energy is negligible. Instead, the transmission will oscillate as is clearly seen in the measurements. The consequence is that the next gear will not be possible to engage until the oscillations have died down enough. This indicates that there is a need for torque control in connection with mode shifts, like a gear shift, in order to reach zero transmission torque without oscillations in the transmission speed.

Complex Oscillations in Neutral Gear

In the previous trials there was no relative speed difference, since the driveline was in a stationary mode. If a relative speed difference is present prior to the gear-shift, there will be a different type of oscillation. Figures 13.6 and 13.7 describe two trials where the neutral gear is engaged with an oscillating driveline without torque control. The oscillations are a result of an engine torque pulse at 11.7 s.

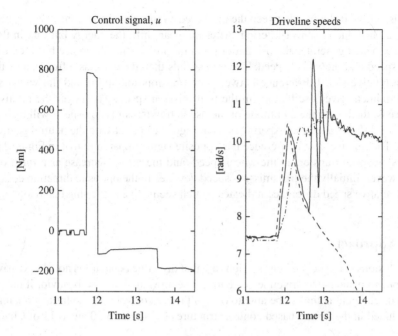

Figure 13.6 Engaged neutral gear without torque control at 12.5 s in a trial with oscillating driveline as a result of a provoking engine torque pulse at 11.7 s in the left figure. Engine speed (dashed) and wheel speed (dash-dotted) are scaled to transmission speed (solid) in the right figure. After the gear shift, the transmission speed oscillates

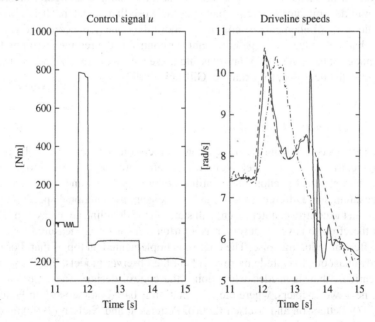

Figure 13.7 Same field trial as in Figure 13.6, but with engaged neutral gear at 13.2 s. Engine speed (dashed) and wheel speed (dash-dotted) are scaled to transmission speed (solid) in the right figure. After the gear shift, the transmission speed oscillates

There is a small difference between the measured engine speed and transmission speed prior to the gear shift due to sensor filtering. After the gear shift, the energy built up in the shafts is released, which generates the oscillations and minimizes the difference between the transmission speed and the wheel speed. The two speeds then decrease as a function of the load. Hence, a relative speed difference between the transmission speed and the wheel speed at the shift moment gives oscillations in the transmission speed. The larger the relative speed difference is, the higher the amplitude of the oscillating transmission speed will be.

Figure 13.7 shows a similar experiment as in Figure 13.6, but with the neutral gear engaged at 13.2 s. The relative speed difference has opposite sign compared to that in Figure 13.6. The transmission speed transfers to the wheel speed, and these two decrease as a function of the load. However, initially the transmission speed deviates in the opposite direction compared to what the relative speed difference indicates, which seems like surprising behavior.

13.4 Approach

The conclusions from the previous sections are that driveline behavior is not always trivial, and that the phenomena due to driveline behavior cause unwanted vehicle behavior if not properly controlled. The goal is thus to be able to design proper controls. The solutions obtained have to be included in the torque-based control structures in Figures 3.10 and 3.11 of Chapter 3.

13.4.1 Timescales

It is clear from the previous content in this chapter, see, for example, all figures in Section 13.3, that a common denominator of the problems is their oscillatory nature. It is also clear that the time scale of the fundamental oscillation is more than one per second and less than ten. This means that the frequency range important for control is the regime including the lowest resonance modes of the driveline. Vibrations and noise contribute to a higher frequency range which is not treated here, see, for example, Gillespie (1992).

13.4.2 Modeling and Control

As described above, driveline modeling and control focus on situations where the velocities or torques (engine, transmission, wheel) along the driveline are not the same. Models used must capture the main physical phenomena regarding torsional effects, and must be able to fit to relevant measurements. Different sensors, usually sensors for rotational speed, may be used, as returned to in the control chapter, but not all states in the driveline are measured. The typical structure is therefore to have an observer as a virtual sensor for unmeasured states, but also having the effect of reducing noise. The control is implemented on top of that. For best effect it may be advantageous to include more details in the observer models, like sensor dynamics and nonlinearities, than is needed when doing the control design. These questions are the topic of the next two chapters, where the presentation is based on the work in Pettersson and Nielsen (1997), Pettersson and Nielsen (2000), Pettersson and Nielsen (2003), and Kiencke and Nielsen (2005).

14

Driveline Modeling

The driveline is fundamental to a vehicle, and depending on purpose and application it can be modeled in different detail. Due to its fundamental character, a general modeling methodology will be presented, and it will be shown how to obtain a model suitable for propulsion studies or a model suitable for design of driveline control. The phenomena to handle are those described in the previous chapter, that is the frequency regime including the lowest resonance modes of the driveline. As mentioned before, vibrations and noise contribute to a higher frequency range not treated here, but the same modeling methodology can be used.

Section 14.1 covers the general modeling methodology for obtaining the fundamental equations describing a driveline. A basic model for a rigid driveline is obtained in Section 14.2, where experimental data from a truck is used for illustrative purposes. The same experiments are used in the modeling in Sections 14.3 and 14.4, where the aim is to find the most important physical effects explaining the oscillations in the measured engine speed, transmission speed, and wheel speed. Section 14.5 includes the behavior of a closed clutch and other nonlinearities like backlash, and Section 14.6 treats the modeling when the driveline is separated in to two parts, which is the case when in neutral gear or when the clutch is disengaged. Then, Section 14.7 models a slipping clutch, and Section 14.8 models a torque converter.

14.1 General Modeling Methodology

A driveline of a rear-wheel-driven vehicle, as in Figure 13.2, consists of such components as clutch, transmission, propeller shaft, final drive, drive shafts, and wheels, that each need a model. The generalized Newton's second law is used to derive the component models, and the result is rotating inertias connected by damped shaft flexibilities. The interfaces between sub-models are torques and rotational velocities, which fits well with their use in a torque-based control structure as introduced in Section 3.3. In addition to the component models, the input torque to the driveline, for example the engine output torque, is needed together with the loads, for example the driving resistance effectively acting on the wheels. The latter are fetched from the modeling of driving resistance in Chapter 2.

Modeling and Control of Engines and Drivelines, First Edition. Lars Eriksson and Lars Nielsen.
© 2014 John Wiley & Sons, Ltd. Published 2014 by John Wiley & Sons, Ltd.
Companion Website: www.wiley.com/go/powertrain

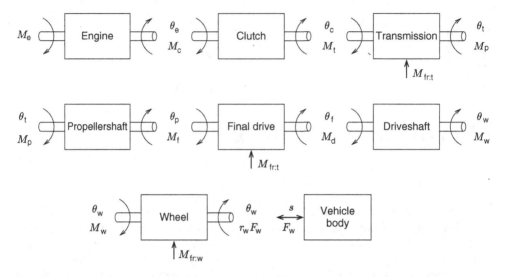

Figure 14.1 Subsystems of a vehicular driveline with their respective angle and torque labels. Also included are the surrounding systems, that is engine and vehicle body, that interact with the driveline. The variable s is the position variable as defined in (2.19)

14.1.1 Graphical Scheme of a Driveline

As a first step when modeling a driveline like the one in in Figure 13.2, it is helpful to use a graphical scheme to define each component and its interface variables. Figure 14.1 shows such a scheme that defines the labels, the inputs, and the outputs of each subsystem of the driveline considered. The next step is to derive the equations.

14.1.2 General Driveline Equations

For each subsystem, the equations are derived systematically. For notation see Figure 14.1.

Engine: The engine is modeled as in Section 7.9 by the model (7.60). Instead of using angular speed ω, we will use angle as the basic variable so that $\dot{\theta} = \omega$, and the reason is that the models will include torsion, that us angular difference. With J_e as the moment of inertia of the engine and θ_e as the angle of the flywheel the dynamics is

$$J_e \ddot{\theta}_e = M_e - M_c \tag{14.1}$$

where, as in (7.60), M_e is the output torque of the engine, and the external load is the load from the clutch M_c.

Clutch: A friction clutch found in vehicles equipped with a manual transmission consists of a clutch disk connecting the flywheel of the engine and the transmission's input shaft. When the clutch is engaged, and no internal friction is assumed, $M_c = M_t$ is obtained. The transmitted torque is a function of the angular difference $(\theta_e - \theta_c)$ and the angular velocity difference $(\dot{\theta}_e - \dot{\theta}_c)$ over the clutch

$$M_c = M_t = f_c(\theta_e - \theta_c, \dot{\theta}_e - \dot{\theta}_c) \tag{14.2}$$

Transmission: A transmission has a set of gears, each with a conversion ratio i_t. This gives the following relation between the input and output torque of the transmission

$$M_p = f_t(M_t, M_{fr:t}, \theta_c - \theta_t i_t, \dot{\theta}_c - \dot{\theta}_p i_t, i_t) \tag{14.3}$$

where the internal friction torque of the transmission is labeled $M_{fr:t}$. The reason for considering the angle difference $\theta_c - \theta_t$ in (14.3) is the possibility of having torsional effects in the transmission.

Propeller shaft: The propeller shaft connects the transmission's output shaft with the final drive. No friction is assumed ($\Rightarrow M_p = M_f$), giving the following model of the torque input to the final drive

$$M_p = M_f = f_p(\theta_t - \theta_p, \dot{\theta}_p - \dot{\theta}_p) \tag{14.4}$$

Final drive: The final drive is characterized by a conversion ratio i_f in the same way as for the transmission. The following relation for the input and output torque holds

$$M_d = f_f(M_f, M_{fr:f}, \theta_p - \theta_f i_f, \dot{\theta}_p - \dot{\theta}_f i_f, i_f) \tag{14.5}$$

where the internal friction torque of the final drive is labeled $M_{fr:f}$.

Driveshafts: The driveshafts connect the wheels to the final drive. Here it is assumed that the wheel speed is the same for the two wheels. Therefore, the driveshafts are modeled as one shaft. When the vehicle is turning and the speed differs between the wheels, both driveshafts have to be modeled. No friction ($\Rightarrow M_w = M_d$) gives the model equation

$$M_w = M_d = f_d(\theta_f - \theta_w, \dot{\theta}_f - \dot{\theta}_w) \tag{14.6}$$

Wheel: The final component, the wheel, drives the vehicle, so to formulate the equation it is necessary to consider the whole vehicle. Recall Chapter 2 where Figure 2.1 showed the forces acting on a vehicle with mass m and speed v. Recall also that the governing equation (2.1) was

$$m\dot{v} = F_w - F_{DR} \tag{14.7}$$

where the braking force F_b is omitted as usual when studying propulsion and powertrain behavior. The wheel force, F_w, can now be solved from (14.7), and the resulting torque, due to F_w is equal to $F_w r_w$, where r_w is the wheel radius. Newton's second law gives

$$J_w \ddot{\theta}_w = M_w - M_{fr:w} - F_w r_w \tag{14.8}$$

where J_w is the mass moment of inertia of the wheel, M_w is given by (14.6), and $M_{fr:w}$ is the friction torque.

Combining (14.7) and (14.8) together with $v = r_w \dot{\theta}_w$ gives

$$(J_w + mr_w^2)\ddot{\theta}_w = M_w - M_{fr:w} - r_w F_{DR} \tag{14.9}$$

The dynamical influence from the tire has been neglected in the equation describing the wheel.

A complete model of the driveline with the clutch engaged is described by (14.1) to (14.8). So far the functions f_c, f_t, f_p, f_f, f_d, and the friction torques $M_{fr:t}$, $M_{fr:f}$, and $M_{fr:w}$ are unknown. In the following, assumptions will be made about these, so far unspecified, model parts, resulting in a series of driveline models with different complexities. Also the driving resistance, F_{DR}, needs to be defined where the models in Sections 2.2 and 2.3 are to be used.

14.2   A Basic Complete Model – A Rigid Driveline

For many studies in propulsion it is quite sufficient to neglect the internal states of a driveline, and thus treat it as a rigid driveline. The clutch and the shafts are then assumed to be stiff, and the transmission and the final drive are assumed to multiply the torque by the conversion ratio, without losses. Such a basic model will now be developed. Assumptions about the fundamental equations in the general treatment above are made in order to obtain a model with a lumped inertia, and labels are still according to Figure 14.1. For driving resistance, the standard model (2.28) is used.

Engine: The engine is modeled as in (14.1)

$$J_e \ddot{\theta}_e = M_e - M_c \tag{14.10}$$

Clutch: The clutch is assumed to be stiff, which gives the following equations for the torque and the angle

$$M_c = M_t, \quad \theta_e = \theta_c \tag{14.11}$$

Transmission: The transmission with conversion ratio i_t is described by one rotating inertia J_t. The friction torque is assumed to be described by a viscous damping coefficient b_t. The model of the transmission, corresponding to (14.3), is

$$\theta_c = \theta_t i_t$$

$$J_t \ddot{\theta}_t = M_t i_t - b_t \dot{\theta}_p - M_p$$

By using (14.11), the model can be rewritten as

$$J_t \ddot{\theta}_e = M_c i_t^2 - b_t \dot{\theta}_e - M_p i_t$$

The equation above illustrates the general methodology that will be returned to in Subsection 14.3.2. For now, a simplified version is used by neglecting the inertia and the damping losses. This is done by using $J_t = 0$ and $b_t = 0$, which results in the following transmission model

$$\theta_c = \theta_t i_t \tag{14.12}$$

$$M_t i_t = M_p \tag{14.13}$$

Propeller shaft: The propeller shaft is also assumed to be stiff, which gives the following equations for the torque and the angle

$$M_p = M_f, \quad \theta_t = \theta_p \tag{14.14}$$

Final drive: In the same way as for the transmission, the final drive can in general be modeled by one rotating inertia J_f and a friction torque that is assumed to be described by a viscous damping coefficient b_f. However, also here, for the basic model, the inertia and the damping are neglected by $J_f = 0$ and $b_f = 0$.

The model of the final drive, corresponding to (14.5), is then

$$\theta_p = \theta_f i_f \tag{14.15}$$

$$M_f i_f = M_d \tag{14.16}$$

Driveshaft: The driveshaft is assumed to be stiff, which gives the following equations for the torque and the angle

$$M_w = M_d, \quad \theta_f = \theta_w \tag{14.17}$$

Wheel: The force and torque balances on the wheel includes the dynamics of the vehicle and is modeled as before in (14.9). Including the standard model, (2.28), for driving resistance, F_{DR}, and neglecting the wheel friction, $M_{fr:w} = 0$, leads to

$$(J_w + mr_w^2)\ddot{\theta}_w = M_w - \frac{1}{2}c_w A_a \rho_a r_w^3 \dot{\theta}_w^2 \tag{14.18}$$
$$-r_w mg \,(f_0 + f_S r_w \dot{\theta}_w) - r_w mg \sin(\alpha)$$

Note that mr_w^2 appears in front of $\ddot{\theta}_w$, effectively as an added inertia–a fact that is returned to after the Basic Driveline Model below.

14.2.1   Combining the Equations

The equations from engine to wheel, (14.10)–(14.18), now constitute a complete chain. Start the elimination of intermediate variables by using (14.11)–(14.17) to obtain the relationships for angles and torques for a stiff driveline ($\theta_e = \theta_c = i_t\theta_t = i_t\theta_p = i_t i_f\theta_f = i_t i_f\theta_w$, and in the same way for the torques). This results in

$$i_t i_f M_c = M_w, \quad \theta_e = i_t i_f\theta_w \tag{14.19}$$

The complete description is now reduced to this equation, (14.19), in combination with (14.10) and (14.18) which are

$$J_e\ddot{\theta}_e = M_e - M_c \tag{14.20}$$

$$(J_w + mr_w^2)\ddot{\theta}_w = M_w - \frac{1}{2}c_w A_a \rho_a r_w^3 \dot{\theta}_w^2 \tag{14.21}$$
$$- r_w mg \,(f_0 + f_S r_w \dot{\theta}_w) - r_w mg \sin(\alpha)$$

There is now an arbitrary choice whether to write the final model in terms of engine speed or wheel speed. Here the wheel speed is chosen. This means that the three variables M_c, M_w, θ_e should be eliminated from the four equations (14.19)–(14.21). The result is one single equation constituting the complete model, named the Basic Driveline Model.

Model 14.1 The Basic Driveline Model

$$(J_w + mr_w^2 + i_t^2 i_f^2 J_e)\ddot{\theta}_w = i_t i_f M_e \tag{14.22}$$
$$- mg f_S r_w^2 \dot{\theta}_w - \frac{1}{2}c_w A_a \rho_a r_w^3 \dot{\theta}_w^2$$
$$- r_w mg(f_0 + \sin(\alpha))$$

The structure of the model (14.22) shows how the driving resistance enters the model

$$(J_w + mr_w^2 + i_t^2 i_f^2 J_e)\ddot{\theta}_w = i_t i_f M_e - r_w F_{DR} \tag{14.23}$$

Alternative Driving Resistance Models

It is easy to change the model for driving resistance just by changing F_{DR} to any of the alternatives given in Sections 2.2 and 2.3. As pointed out there, the standard driving resistance used above includes terms that are constant, first, and second power of velocity v, and dependent on slope α. Often this model structure will be sufficient to get a good fit to experimental data, even though the interpretation of the coefficients will be slightly different. For low gears and speeds, the influence from the air drag is low, and by neglecting $\frac{1}{2}c_w A_a \rho_a r_w^3 \dot{\theta}_w^2$ in (14.22), the model is affine in the state $\dot{\theta}_w$, but nonlinear in the parameters.

14.2.2   Reflected Mass and Inertias

Another important observation is made from (14.22). Changing the variable from wheel angle θ_w to engine angle θ_e gives the dynamics in the form

$$(\frac{J_w}{i_t^2 i_f^2} + \frac{mr_w^2}{i_t^2 i_f^2} + J_e)\ddot{\theta}_e = M_e - \frac{r_w F_{DR}}{i_t i_f} \tag{14.24}$$

Comparing this equation with (14.1), or equivalently the model (7.60) in Section 7.9, shows that they are of exactly the same form. However, the inertia that the engine effectively drives is

$$J_{eff} = \frac{J_w}{i_t^2 i_f^2} + \frac{mr_w^2}{i_t^2 i_f^2} + J_e \tag{14.25}$$

which includes both vehicle mass, m, and wheel inertia, J_w, scaled by the gear ratio squared. These terms are called reflected mass and inertia. Also, the load from the driving resistance is scaled by the gear ratio as seen in (14.24).

14.3   Driveline Surge

The unwanted vehicle behavior known as vehicle shuffle or vehicle surge was introduced in Section 13.3.1, and the underlying reason for these phenomena is driveline oscillations, also called driveline surge. Having obtained a basic model for a rigid driveline in the previous section, it is now time to approach a description of surge, that is internal torsions. First, some experiments are presented to motivate where to put in modeling effort, and then a model is derived and validated.

14.3.1   Experiments for Driveline Modeling

The main part of the experiments used for modeling considers low gears. The reason for this is that the lower the gear is, the higher the torque transferred in the drive shaft is. This means that the shaft torsion is higher for lower gears, and therefore also the problems with oscillations. Furthermore, the amplitudes of the resonances in the wheel speed are higher for lower gears, since the load and vehicle mass appear reduced by the high conversion ratio.

A Scania heavy-duty truck, a Scania 144L 6x2 truck, was used for experiments. It is equipped with a 14-liter V8 turbocharged diesel engine with maximum power of 530 Hp and maximum torque of 2300 Nm. The engine is connected to a manual range-splitter transmission GRS900R via a clutch. The transmission has 14 gears and a hydraulic retarder. It is also equipped with the automated manual transmission (AMT) system OptiCruise. The weight of the truck is $m = 24000$ kg.

Three speed sensors are used to measure the speed of the flywheel of the engine ($\dot{\theta}_e$), the speed of the output shaft of the transmission ($\dot{\theta}_p$), and the speed of the driving wheel ($\dot{\theta}_w$). These rotational velocities are measured by inductive sensors that detect the time when cogs from a rotating cog wheel pass. The transmission speed sensor has fewer cogs than the other two sensors, indicating that the bandwidth of this signal is lower.

The truck is equipped with a set of control units, each connected to a CAN-bus. These CAN nodes are the engine control node, the transmission node, and the ABS brake system node. Each node measures a number of variables and transmits them via the bus.

Test Results

A number of test roads at Scania, with different known slopes, were used for testing. The variables in Table 14.1 were logged during tests that excite driveline resonances. Figure 14.2 shows a test with the 144L truck where step inputs in accelerator position excite driveline oscillations. In Figure 14.2 it is seen that the main flexibility of the driveline is located between the output shaft of the transmission and the wheel, since the largest difference in speed is between the measured transmission speed and wheel speed. This leads to the conclusion that the main flexibility in the driveline is located in the driveshaft. The physical reason for this is, on one hand, the mechanical design, but also that the drive shaft is subject to the relatively largest torsion, which is mainly due to the high torque difference that results from the amplification of the engine torque by the conversion ratio of the transmission (i_t) and the final drive (i_f). This number ($i_t i_f$) can be as high as 60 for the lowest gear.

14.3.2   Model with Driveshaft Flexibility

To model a driveline with driveshaft flexibility it is natural to model the driveline with a lumped engine and transmission inertia connected to the wheel inertia by driveshaft flexibility. This is schematically illustrated in Figure 14.3. The derivation follows the methodology in Section 14.1 so assumptions about the fundamental equations are made in order to obtain a model with a lumped engine and transmission inertia and driveshaft flexibility. Labels are according to Figure 14.1. The clutch and the propeller shafts are assumed to be stiff, and the drive shaft is described as damped torsional flexibility. The transmission and the final drive are assumed to multiply the torque by the conversion ratio, without losses.

Table 14.1   Measured variables transmitted on the CAN-bus

	Measured variables		
Variable	*Node*	*Resolution*	*Rate*
Engine speed, $\dot{\theta}_e$	Engine	0.013 rad/s	20 ms
Engine torque, M_{m_e}	Engine	1% of max torque	20 ms
Wheel speed, $\dot{\theta}_w$	ABS	0.033 rad/s	50 ms
Transmission speed, $\dot{\theta}_p$	Transmission	0.013 rad/s	50 ms

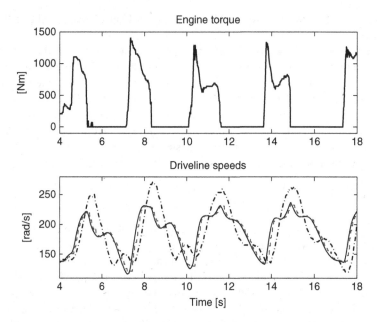

Figure 14.2   Logged data on the CAN-bus during step inputs in accelerator position with the 144L truck. The transmission speed (dashed line) and the wheel speed (dash-dotted line) are scaled to engine speed in the solid line. The main flexibility of the driveline is located between the output shaft of the transmission and the wheel, since the largest difference in speed is between the measured transmission speed and wheel speed

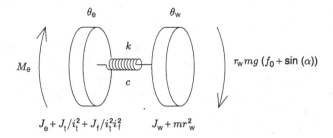

Figure 14.3   The Driveshaft Model consists of a lumped engine and transmission inertia connected to the wheel inertia by damped torsional flexibility, here with (2.24) as road load

Clutch: The clutch is assumed to be stiff, which gives the following equations for the torque and the angle

$$M_c = M_t, \quad \theta_e = \theta_c \tag{14.26}$$

Transmission: The transmission is described by one rotating inertia J_t. The friction torque is assumed to be described by a viscous damping coefficient b_t. The model of the transmission, corresponding to (14.3), is

$$\theta_c = \theta_t i_t \tag{14.27}$$

$$J_t \ddot{\theta}_t = M_t i_t - b_t \dot{\theta}_p - M_p \tag{14.28}$$

By using (14.26) and (14.27), the model can be rewritten as

$$J_t \ddot{\theta}_e = M_c i_t^2 - b_t \dot{\theta}_e - M_p i_t \tag{14.29}$$

Propeller shaft: The propeller shaft is also assumed to be stiff, which gives the following equations for the torque and the angle

$$M_p = M_f, \quad \theta_t = \theta_p \tag{14.30}$$

Final drive: In the same way as for the transmission, the final drive is modeled by one rotating inertia J_f. The friction torque is assumed to be described by a viscous damping coefficient b_f. The model of the final drive, corresponding to (14.5), is

$$\theta_p = \theta_f i_f \tag{14.31}$$

$$J_f \ddot{\theta}_f = M_f i_f - b_f \dot{\theta}_f - M_d \tag{14.32}$$

(14.32) can be rewritten with (14.30) and (14.31) which gives

$$J_f \ddot{\theta}_t = M_p i_f^2 - b_f \dot{\theta}_p - M_d i_f \tag{14.33}$$

Converting (14.33) to a function of engine speed is done by using (14.26) and (14.27) resulting in

$$J_f \ddot{\theta}_e = M_p i_f^2 i_t - b_f \dot{\theta}_e - M_d i_f i_t \tag{14.34}$$

By replacing M_p in (14.34) with M_p in (14.29), a model for the lumped transmission, propeller shaft, and final drive is obtained

$$(J_t i_f^2 + J_f)\ddot{\theta}_e = M_c i_t^2 i_f^2 - b_t \dot{\theta}_e i_f^2 - b_f \dot{\theta}_e - M_d i_f i_t \tag{14.35}$$

Drive shaft: The drive shaft is modeled as damped torsional flexibility, having stiffness k, and internal damping c. Hence, (14.6) becomes

$$M_w = M_d = k(\theta_f - \theta_w) + c(\dot{\theta}_f - \dot{\theta}_w) = k(\theta_e/i_t i_f - \theta_w) \tag{14.36}$$
$$+ c(\dot{\theta}_e/i_t i_f - \dot{\theta}_w)$$

where (14.26), (14.27), (14.30), and (14.31) are used. By replacing M_d in (14.35) with (14.36), the equation describing the transmission, the propeller shaft, the final drive, and the drive shaft, becomes

$$(J_t i_f^2 + J_f)\ddot{\theta}_e = M_c i_t^2 i_f^2 - b_t \dot{\theta}_e i_f^2 - b_f \dot{\theta}_e \tag{14.37}$$
$$- k(\theta_e - \theta_w i_t i_f) - c(\dot{\theta}_e - \dot{\theta}_w i_t i_f)$$

Wheel: If (14.9) is combined with (14.36), the following equation for the wheel is obtained:

$$(J_w + mr_w^2)\ddot{\theta}_w = k(\theta_e/i_t i_f - \theta_w) + c(\dot{\theta}_e/i_t i_f - \dot{\theta}_w) \tag{14.38}$$
$$- b_w \dot{\theta}_w - \frac{1}{2} c_w A_a \rho_a r_w^3 \dot{\theta}_w^2 - mgf_s r_w^2 \dot{\theta}_w - r_w mg(f_0 + \sin(\alpha))$$

where the friction torque is described as viscous damping, with label b_w.

The complete model, named the Driveshaft Model, is obtained by inserting M_c from (14.37) into (14.1), together with (14.38), which gives the following equations.

Model 14.2 The Driveshaft Model

$$(J_e + J_t/i_t^2 + J_f/i_t^2 i_f^2)\ddot{\theta}_e = M_e - (b_t/i_t^2 + b_f/i_t^2 i_f^2)\dot{\theta}_e \qquad (14.39)$$

$$- k(\theta_e/i_t i_f - \theta_w)/i_t i_f$$

$$- c(\dot{\theta}_e/i_t i_f - \dot{\theta}_w)/i_t i_f$$

$$(J_w + mr_w^2)\ddot{\theta}_w = k(\theta_e/i_t i_f - \theta_w) + c(\dot{\theta}_e/i_t i_f - \dot{\theta}_w) \qquad (14.40)$$

$$- (b_w + mgf_S r_w^2)\dot{\theta}_w - \frac{1}{2}c_w A_a \rho_a r_w^3 \dot{\theta}_w^2$$

$$- r_w mg(f_0 + \sin(\alpha))$$

The driveshaft torsion, the engine speed, and the wheel speed are used as states according to

$$x_1 = \theta_e/i_t i_f - \theta_w, \quad x_2 = \dot{\theta}_e, \quad x_3 = \dot{\theta}_w \qquad (14.41)$$

More details of state-space descriptions are given in Section 15.2.1. Further, the comments that were made for the Basic Driveline Model are the same for the Driveshaft Model. It is straightforward to modify the model to another driving resistance, and also here vehicle mass and inertias are reflected depending on the choice of variable. For low gears, the influence from the air drag is low, and by neglecting $\frac{1}{2}c_w A_a \rho_a r_w^3 \dot{\theta}_w^2$ in (14.40), the model is affine in the states, but nonlinear in the parameters.

Validation of the Driveshaft Model

It turns out that the Driveshaft Model is important and useful in several applications. It is, therefore, the place to look at its parameters and its experimental agreement. A data set containing engine torque, engine speed, and wheel speed measurements is used to estimate the parameters and the initial conditions of the Driveshaft Model. The load, l, is the wheel radius, r_w, multiplied by the driving resistance model (2.24) that consists of a constant load and a varying slope. The use of this load model is reasonable since the speeds are low and the goal is to look at the internal dynamics. The parameters to obtain from data sheets or to estimate are

$$i = i_t i_f, \quad l = r_w mg(f_0 + \sin(\alpha))$$

$$J_1 = J_e + J_t/i_t^2 + J_f/i_t^2 i_f^2, \quad J_2 = J_w + mr_w^2 \qquad (14.42)$$

$$b_1 = b_t/i_t^2 + b_f/i_t^2 i_f^2, \quad b_2 = b_w$$

together with the stiffness, k, and the internal damping, c, of the drive shaft. The estimated initial conditions of the states are labeled x_{10}, x_{20}, and x_{30}, according to (14.41).

Estimation remarks. Following established principles in parameter estimation, the data sets are divided into two parts. The parameters are estimated on the estimation data. The results are then evaluated on the validation data, and these are the results shown in this chapter. When estimating the parameters of the Driveshaft Model, there may be problems when identifying the viscous friction components b_1 and b_2. The sensitivity in the model due to variations in the friction parameters is low, and the same model fit can be obtained for a range of friction parameters. However, the sum $b_1 i^2 + b_2$ is constant during these tests and thus easier to estimate. Possible problems with estimation of viscous parameters will be further discussed in connection with additional dynamics, see Section 14.4.1.

Results. Figure 14.4 shows an example of how the model fits the measured data. The measured driveline speeds are shown together with the model output, x_1, x_2, and x_3. According to the model, the clutch is stiff, and therefore the transmission speed should be equal to the engine speed scaled with the conversion ratio of the transmission (i_t). In the figure, this signal is shown together with the measured transmission speed. The plots are typical examples that show that a major part of the driveline dynamics is captured with a linear mass-spring model with the drive shafts as the main flexibility. The conclusions from Figure 14.4 are

- The main contribution to driveline dynamics from driving torque to engine speed and wheel speed is the drive shaft, explaining the first main resonance of the driveline.
- The true driveshaft torsion (x_1) is unknown, but the value estimated by the model has physically reasonable values. These values will be further validated in Chapter 15.
- The model output transmission speed (x_2/i_t) fits the measured transmission speed data reasonably well, but there is still dynamics between model outputs and measurements in the form of a systematic lag.

14.4   Additional Driveline Dynamics

In addition to the important driveshaft flexibility, there are of course other sources for dynamic behavior in the driveline. We will approach some of these by taking a closer look at the validation data presented in Figure 14.4. As mentioned, there is good agreement between model output and experimental data for $\dot{\theta}_e$, and $\dot{\theta}_w$, but there is a slight deviation between measured and estimated transmission speed.

14.4.1   Influence on Parameter Estimation

Before going into further detail on driveline modeling, it is reasonable to ask if it has any value besides the satisfaction of understanding, and of course the fact that modern drivelines are becoming more and more advanced, which increases model requirements in new applications with high performance demands. The Driveshaft Model fits data very well, and it will be shown in Chapter 15 that it is quite sufficient for many advanced applications. So what is the problem with additional dynamics? The consequences show up when estimating the parameters in the model.

Fitting a model like the Driveshaft Model directly to data may give strange values when there are additional dynamics. The phase shift and smoothing between engine speed and transmission speed is minor, but even such minor unmodeled dynamics can cause errors or even non-physical values in parameters. The reason is that algorithms for parameter estimation try to create a curve fit even though the model structure is not correct. Our experience is that the effective damping of the driveshaft flexibility, the parameter c in (14.36), may be particularly

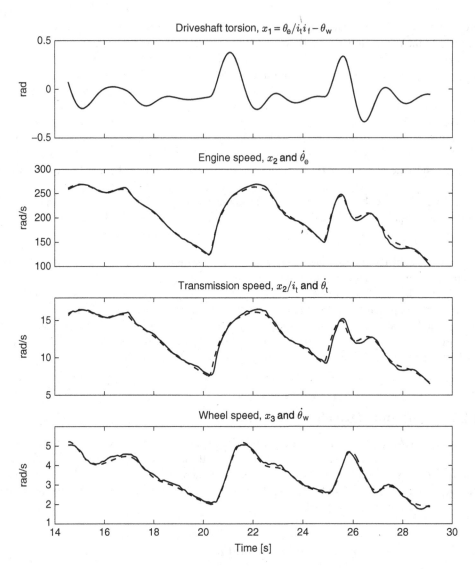

Figure 14.4 The parameters of the Driveshaft Model estimated on data with step inputs in accelerator position using gear 1. The top figure shows the estimated driveshaft torsion, and the bottom figures show the model outputs (x_2, x_3) in dashed lines, together with the measured driveline speeds in a solid line. The plots are typical examples of the fact that a major part of the dynamics is captured by a linear model with driveshaft flexibility

sensitive. With unmodeled dynamics, this parameter can assume unrealistic or even negative values to get a good fit, and therefore more modeling is beneficial.

14.4.2   Character of Deviation in Validation Data

With the Driveshaft Model, stiff dynamics between the engine and the transmission is assumed, and hence the only difference between the model outputs of engine speed and transmission

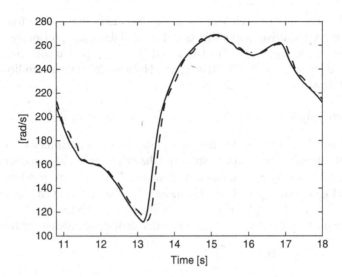

Figure 14.5 Measured engine speed (solid line) and transmission speed (dashed line). The transmission speed is multiplied by the conversion ratio of the transmission, i_t

speed is the gain i_t (conversion ratio of the transmission). However, a comparison between the measured engine speed and transmission speed shows that there is not only a gain difference according to Figure 14.5. Instead, the deviation between engine speed and transmission speed has the character of a phase shift and some smoothing (signal levels and shapes agree). This indicates that there is some additional dynamics between engine speed, $\dot{\theta}_e$, and transmission speed, $\dot{\theta}_p$. Two natural candidates are additional mass-spring dynamics in the driveline, or sensor dynamics, so these will be treated below.

The clutch or backlash are two other general candidates when there are model deviations, but they are ruled out for this particular situation. The motivation for this is that the high stiffness of the clutch flexibility (given from material data) cannot result in a phase shift of the magnitude shown in Figure 14.5. Neither can backlash in the transmission explain the difference, because then the engine and transmission speeds would be equal when the backlash is at its endpoints. However, there are situations where the clutch or backlash are important, and this will be returned to in Section 14.5.

14.4.3 Influence from Propeller-shaft Flexibility

The Driveshaft Model assumes a stiff driveline from the engine to the final drive, and a natural candidate for additional dynamics is the flexibility in the propeller shaft. The propeller shaft and the drive shaft are separated by the final drive, which has a small inertia compared to other inertias, for example, the engine inertia. A model can be derived where the propeller shaft and the driveshafts are modeled as damped torsional flexibilities. As in the derivation of the Driveshaft Model, the transmission and the final drive multiply the torque by the conversion ratio. It turns out, for the validation data in modeling example used, that the model fit is not significantly improved by including propeller shaft dynamics. Instead, the most valuable insight is how it influences parameter estimation for the Driveshaft Model.

The derivation will only be sketched, but a complete derivation can be found in Kiencke and Nielsen (2005). The drive shaft stiffness and internal damping, previously denoted k, c in (14.36), will now be denoted k_d, c_d. The Driveshaft Model is repeated with the difference that the model for the propeller shaft (14.30) is replaced by a model of the flexibility with stiffness k_p and internal damping c_p

$$M_p = M_f = k_p(\theta_t - \theta_p) + c_p(\dot{\theta}_p - \dot{\theta}_p) = k_p(\theta_e/i_t - \theta_p) + c_p(\dot{\theta}_e/i_t - \dot{\theta}_p) \tag{14.43}$$

where (14.26) and (14.27) are used in the last equality. This formulation means that there are two torsional flexibilities: the propeller shaft and the drive shaft. The propeller shaft and the drive shaft are separated by the final drive with inertia J_f and a corresponding new state θ_p. Compared to (14.39) and Figure 14.3, this means that the first lumped inertia $J_e + J_t/i_t^2 + J_f/i_t^2 i_f^2$ is separated into $J_e + J_t/i_t^2$ and J_f. This is illustrated in Figure 14.6.

The complete model with drive shaft and propeller shaft flexibilities is the following

$$(J_e + J_t/i_t^2)\ddot{\theta}_e = M_e - b_t/i_t^2\dot{\theta}_e \tag{14.44}$$
$$- \frac{1}{i_t}(k_p(\theta_e/i_t - \theta_p) + c_p(\dot{\theta}_e/i_t - \dot{\theta}_p))$$
$$J_f\ddot{\theta}_p = i_f^2(k_p(\theta_e/i_t - \theta_p) + c_p(\dot{\theta}_e/i_t - \dot{\theta}_p)) - b_f\dot{\theta}_p \tag{14.45}$$
$$- i_f(k_d(\theta_p/i_f - \theta_w) + c_d(\dot{\theta}_p/i_f - \dot{\theta}_w))$$
$$(J_w + mr_w^2)\ddot{\theta}_w = k_d(\theta_p/i_f - \theta_w) + c_d(\dot{\theta}_p/i_f - \dot{\theta}_w) \tag{14.46}$$
$$- (b_w + mgf_s r_w^2)\dot{\theta}_w - \frac{1}{2}c_w A_a \rho_a r_w^3 \dot{\theta}_w^2 - r_w mg(f_0 + \sin(\alpha))$$

14.4.4 Parameter Estimation with Springs in Series

The model (14.44) to (14.46) describe the Driveshaft Model extended with the propeller shaft with stiffness k_p and damping c_p. The three inertias in the model are

$$J_1 = J_e + J_t/i_t^2$$
$$J_2 = J_f \tag{14.47}$$
$$J_3 = J_w + mr_w^2$$

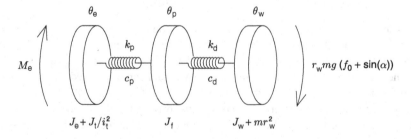

Figure 14.6 Model with flexible propeller shaft and drive shaft

If the magnitude of the three inertias are compared, the inertia of the final drive (J_f) is considerably less than J_1 and J_2 in (14.47). Therefore, the model will act as if there are two damped springs in series. The total stiffness of two undamped springs in series is

$$k = \frac{k_p i_f^2 k_d}{k_p i_f^2 + k_d} \tag{14.48}$$

whereas the total damping of two dampers in series is

$$c = \frac{c_p i_f^2 c_d}{c_p i_f^2 + c_d} \tag{14.49}$$

The damping and the stiffness of the drive shaft in the previous section will thus typically be underestimated due to the flexibility of the propeller shaft. This effect will increase with a lower conversion ratio in the final drive, i_f. The individual stiffness values obtained from parameter estimation are somewhat lower than the values obtained from material data. Note that this is a general conclusion for any additional spring in series.

14.4.5   Sensor Dynamics

Sensor dynamics is another natural candidate for the slight deviation between measured and estimated transmission speed in Figure 14.4 and Figure 14.5. As mentioned before, the bandwidth of the measured transmission speed is lower than the measured engine and wheel speeds, due to fewer cogs in the sensor, see Section 14.3.1. Therefore, of the three sensors, it is assumed that transmission sensor dynamics is the natural candidate, and that the engine speed and the wheel speed sensor dynamics are assumed not to influence the data for the frequencies considered. In summary, after some comparison between sensor filters of different order, the following sensor dynamics is assumed

$$\begin{aligned} f_m &= 1 \\ f_t &= \frac{1}{1 + \gamma s} \\ f_w &= 1 \end{aligned} \tag{14.50}$$

where a first-order filter with an unknown parameter γ models the transmission sensor. Figure 14.7 shows the configuration with the Driveshaft Model and sensor filter f_m, f_t, and f_w. The outputs of the filters are y_m, y_t, and y_w.

Now, the parameters, the initial condition, and the unknown filter constant γ can be estimated such that the model outputs (y_m, y_t, y_w) fit the measured data. The result of this is seen in Figure 14.8 for gear 1. The conclusion is that the main part of the deviation between engine speed and transmission speed is due to sensor dynamics, and the result is:

- If the Driveshaft Model is extended with a first-order sensor filter for the transmission speed, all three velocities ($\dot{\theta}_e, \dot{\theta}_p, \dot{\theta}_w$) are estimated by the model. The model outputs fit the data except for some time intervals where there are deviations between model and measured data. However, these deviations will in the following be related to nonlinearities at low clutch torques.

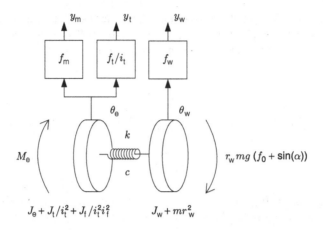

Figure 14.7   The Driveshaft Model with sensor dynamics

14.5   Clutch Influence and Backlash in General

In addition to the candidates for additional dynamics in the previous Section 14.4, there is also the clutch. The situation of clutch closed is the topic of this section, whereas the open clutch is treated in Section 14.6.

14.5.1   Model with Flexible Clutch and Driveshaft

The clutch has so far been assumed to be stiff and the main contribution to low-frequency oscillations is the driveshaft flexibility. However, measured data suggests that there is some additional dynamics between the engine and the transmission, and the candidate that is most flexible is the clutch. Hence, the model will include two torsional flexibilities, the drive shaft, and the clutch. With this model structure, the first and second resonance modes of the driveline are explained, and the reason for their ordering in frequency is the relatively higher stiffness in the clutch, because the relative stiffness of the drive shaft is reduced by the conversion ratio. As in Section 14.4.3, the drive shaft stiffness and internal damping, previously denoted k, c in (14.36), are now denoted k_d, c_d.

A model with a linear clutch flexibility and one torsional flexibility (the drive shaft) is derived by repeating the procedure for the Driveshaft Model with the difference that the model for the clutch is a flexibility with stiffness k_c and internal damping c_c

$$M_c = M_t = k_c(\theta_e - \theta_c) + c_c(\dot{\theta}_e - \dot{\theta}_c) = k_c(\theta_e - \theta_t i_t) + c_c(\dot{\theta}_e - \dot{\theta}_p i_t) \qquad (14.51)$$

where (14.27) is used in the last equality. By inserting this into (14.1) the equation describing the engine inertia is given by

$$J_e \ddot{\theta}_e = M_e - (k_c(\theta_e - \theta_t i_t) + c_c(\dot{\theta}_e - \dot{\theta}_p i_t)) \qquad (14.52)$$

Also, by inserting (14.51) into (14.28), the equation describing the transmission is

$$J_t \ddot{\theta}_t = i_t(k_c(\theta_e - \theta_t i_t) + c_c(\dot{\theta}_e - \dot{\theta}_p i_t)) - b_t \dot{\theta}_p - M_p \qquad (14.53)$$

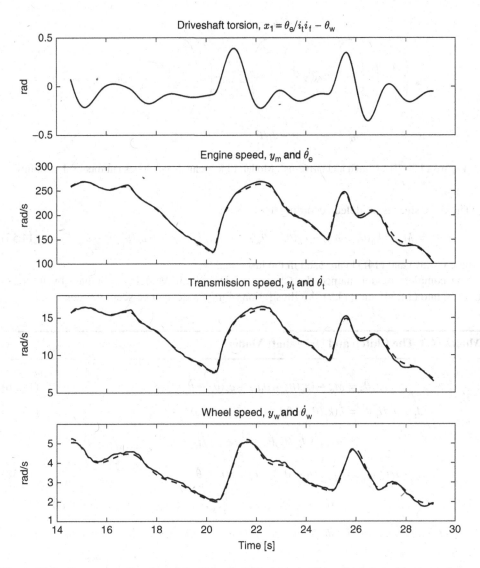

Figure 14.8 Parameter estimation of the Driveshaft Model as in Figure 14.4, but with sensor dynamics included. The top figure shows the estimated driveshaft torsion, and the bottom figures show the model outputs (y_m, y_t, y_w) as dashed lines, together with the measured data as solid lines. The conclusion is that the main part of the deviation between engine speed and transmission speed is due to sensor dynamics

M_p is derived from (14.33) giving,

$$(J_t + J_f/i_f^2)\ddot{\theta}_t = i_t(k_c(\theta_e - \theta_t i_t) + c_c(\dot{\theta}_e - \dot{\theta}_p i_t)) - (b_t + b_f/i_f^2)\dot{\theta}_p - M_d/i_f \qquad (14.54)$$

which is the equation describing the lumped transmission, propeller shaft, and final drive inertia.

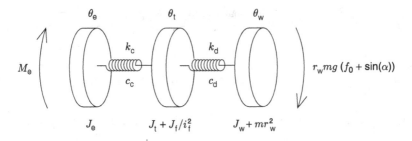

Figure 14.9   The Clutch and Driveshaft Model: Linear clutch and driveshaft torsional flexibility

The drive shaft is modeled according to (14.36) as

$$M_w = M_d = k_d(\theta_f - \theta_w) + c_d(\dot{\theta}_f - \dot{\theta}_w) = k_d(\theta_t/i_f - \theta_w) + c_d(\dot{\theta}_p/i_f - \dot{\theta}_w) \qquad (14.55)$$

where (14.30) and (14.31) are used in the last equality.

The complete model, named the Clutch and Driveshaft Model, is obtained by inserting (14.55) into (14.54) and (14.9). An illustration of the model can be seen in Figure 14.9.

Model 14.3 The Clutch and Driveshaft Model

$$J_e\ddot{\theta}_e = M_e - (k_c(\theta_e - \theta_t i_t) + c_c(\dot{\theta}_e - \dot{\theta}_p i_t)) \qquad (14.56)$$

$$(J_t + J_f/i_f^2)\ddot{\theta}_t = i_t(k_c(\theta_e - \theta_t i_t) + c_c(\dot{\theta}_e - \dot{\theta}_p i_t)) \qquad (14.57)$$

$$- (b_t + b_f/i_f^2)\dot{\theta}_p - \frac{1}{i_f}(k_d(\theta_t/i_f - \theta_w) + c_d(\dot{\theta}_p/i_f - \dot{\theta}_w))$$

$$(J_w + mr_w^2)\ddot{\theta}_w = k_d(\theta_t/i_f - \theta_w) + c_d(\dot{\theta}_p/i_f - \dot{\theta}_w) \qquad (14.58)$$

$$- (b_w + mgf_s r_w)\dot{\theta}_w - \frac{1}{2}c_w A_a \rho_a r_w^3 \dot{\theta}_w^2 - r_w mg(f_0 + \sin(\alpha))$$

The clutch torsion, the driveshaft torsion, and the driveline speeds are used as states according to

$$x_1 = \theta_e - \theta_t i_t, \quad x_2 = \theta_t/i_f - \theta_w, \quad x_3 = \dot{\theta}_e, \quad x_4 = \dot{\theta}_p, \quad x_5 = \dot{\theta}_w \qquad (14.59)$$

For low gears, the influence from the air drag is low, and by neglecting $\frac{1}{2}c_w A_a \rho_a r_w^3 \dot{\theta}_w^2$ in (14.58), the model is affine in the states, but nonlinear in the parameters. The model equipped with the sensor filter in (14.50) gives the sensor outputs (y_m, y_t, y_w).

The parameters and the initial conditions of the Clutch and Driveshaft Model are estimated with the sensor dynamics described above, in the same way as the Driveshaft Model previously. A problem when estimating the parameters of the Clutch and Driveshaft Model is that the bandwidth of the measured signals is not enough to estimate the stiffness k_c in the clutch. Therefore, the value of the stiffness given from material data is used, and the rest of the parameters are estimated.

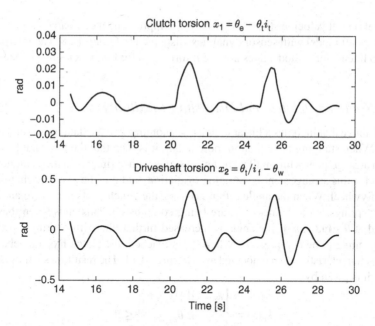

Figure 14.10   Clutch torsion (top figure) and driveshaft torsion (bottom figure) resulting from parameter estimation of the Clutch and Driveshaft Model with sensor filtering, on data with gear 1. The true values of these torsions are not known, but the plots show that the driveshaft torsion has realistic values

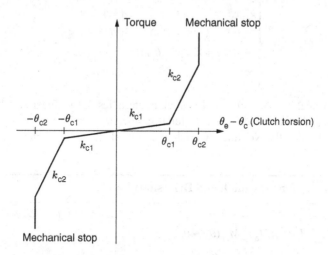

Figure 14.11   Nonlinear clutch characteristics

The resulting clutch torsion (x_1) and the driveshaft torsion (x_2) are shown in Figure 14.10. The true values of these torsions are not known, but the figure shows that the amplitude of the driveshaft torsion has realistic values that agree with material data. However, the clutch torsion does not have realistic values (explained later), which can be seen when comparing with the static nonlinearity in Figure 14.11.

The model output velocities ($\dot{\theta}_e$, $\dot{\theta}_p$, $\dot{\theta}_w$) show no improvement compared to those generated by the Driveshaft model with sensor dynamics, displayed in Figure 14.8. The interpretation of this is that a linear clutch model does not add information for frequencies in the measured data.

14.5.2   Nonlinear Clutch and Driveshaft Flexibility

The model above that includes a linear clutch does not improve the data fit, so something more is needed. When studying the clutch in more detail, it is seen that the torsional flexibility is a result of an arrangement with smaller springs in series with springs with much higher stiffness. The reason for this arrangement is vibration insulation, and it is the same mechanism as for a dual mass flywheel. When the angle difference over the clutch starts from zero and increases, the smaller springs, with stiffness k_{c1}, are being compressed. This ends when they are fully compressed at θ_{c1} radians. If the angle is increased further, the stiffer springs, with stiffness k_{c2}, are beginning to be compressed. When θ_{c2} is reached, the clutch hits a mechanical stop. This clutch characteristic can be modeled as in Figure 14.11. The resulting stiffness $k_c(\theta_e - \theta_c)$ of the clutch is given by

$$k_c(x) = \begin{cases} k_{c1} & \text{if } |x| \le \theta_{c1} \\ k_{c2} & \text{if } \theta_{c1} < |x| \le \theta_{c2} \\ \infty & \text{otherwise} \end{cases} \qquad (14.60)$$

The torque $M_{kc}(\theta_e - \theta_c)$ from the clutch nonlinearity is

$$M_{kc}(x) = \begin{cases} k_{c1}x & \text{if } |x| \le \theta_{c1} \\ k_{c1}\theta_{c1} + k_{c2}(x - \theta_{c1}) & \text{if } \theta_{c1} < x \le \theta_{c2} \\ -k_{c1}\theta_{c1} + k_{c2}(x + \theta_{c1}) & \text{if } -\theta_{c2} < x \le -\theta_{c1} \\ \infty & \text{otherwise} \end{cases} \qquad (14.61)$$

A dual mass flywheel has the same nonlinear characteristics. If the linear clutch in the Clutch and Driveshaft Model is replaced by the clutch nonlinearity according to Figure 14.11, the following model, called the Nonlinear Clutch and Driveshaft Model, is derived.

Model 14.4 The Nonlinear Clutch and Driveshaft Model

$$J_e\ddot{\theta}_e = M_e - M_{kc}(\theta_e - \theta_t i_t) \qquad (14.62)$$
$$- c_c(\dot{\theta}_e - \dot{\theta}_p i_t)$$
$$(J_t + J_f/i_f^2)\ddot{\theta}_t = i_t(M_{kc}(\theta_e - \theta_t i_t) + c_c(\dot{\theta}_e - \dot{\theta}_p i_t)) \qquad (14.63)$$
$$- (b_t + b_f/i_f^2)\dot{\theta}_p - \frac{1}{i_f}(k_d(\theta_t/i_f - \theta_w) + c_d(\dot{\theta}_p/i_f - \dot{\theta}_w))$$
$$(J_w + mr_w^2)\ddot{\theta}_w = k_d(\theta_t/i_f - \theta_w) + c_d(\dot{\theta}_p/i_f - \dot{\theta}_w) \qquad (14.64)$$
$$-(b_w + mgf_s r_w)\dot{\theta}_w - \frac{1}{2}c_w A_a \rho_a r_w^3 \dot{\theta}_w^2 - r_w mg(f_0 + \sin(\alpha))$$

Nonlinear Driveline Model with five states. (The same state-space representation as for the Clutch and Driveshaft Model can be used.) The function $M_{kc}(\cdot)$ is given by (14.61). The model, equipped with the sensor filter in (14.50), gives the sensor outputs (y_m, y_t, y_w).

When estimating the parameters and the initial conditions of the Nonlinear Clutch and Driveshaft Model, the clutch static nonlinearity is fixed with known physical values and the rest of the parameters are estimated, except for the sensor filter which is the same as in the previous model estimations.

The resulting clutch torsion ($x_1 = \theta_e - \theta_t i_t$) and driveshaft torsion ($x_2 = \theta_t/i_f - \theta_w$) are shown in Figure 14.12. The true values of these torsions are not known, as mentioned before. However, the figure shows that both angles have realistic values that agree with other experience. The output velocities ($\dot{\theta}_e$, $\dot{\theta}_p$, $\dot{\theta}_w$) show no improvement compared to those generated by the Driveshaft Model with sensor dynamics, displayed in Figure 14.8. This is natural since the conclusion already from the previous linear clutch model was that there is not information for frequencies in the measured data.

The Driveshaft Model with sensor filtering thus fits the signals well, except for a number of time intervals with deviations. The question is whether this is a result of some nonlinearity. Comparing with Figure 14.5, Figure 14.13 shows the transmission speed plotted together with the model output and the clutch torsion. It is clear from this figure that the deviation between the model and experiments occurs when the clutch angle passes the area with low stiffness in the static nonlinearity (see Figure 14.11).

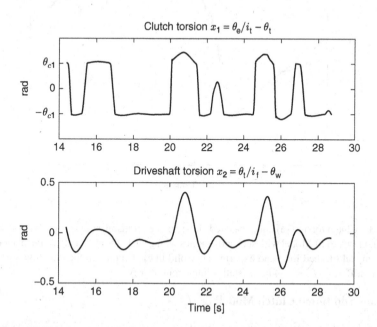

Figure 14.12 Clutch torsion (top figure) and driveshaft torsion (bottom figure) resulting from parameter estimation of the Nonlinear Clutch and Driveshaft Model with sensor filtering, on data with gear 1. The true values of these torsions are not known, but the plots show that they have realistic values

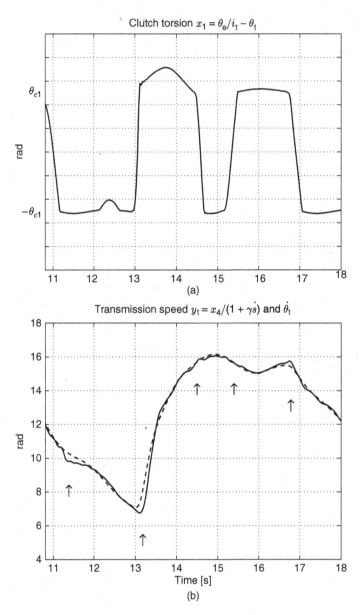

Figure 14.13 Clutch torsion (a) and measured and estimated transmission speeds (a) from the Nonlinear Clutch and Driveshaft Model with sensor dynamics with gear 1. The result is that the main differences between the model (dashed line) and experiments (solid line) occur when the clutch torsion passes the area with low stiffness ($|\theta| < \theta_{c1}$) in the static clutch nonlinearity

Results from Additional Clutch Modeling

- The model including the nonlinear clutch does not improve the overall data fit for frequencies in the measured data.
- The model is able to estimate a clutch torsion with realistic values.

• The estimated clutch torsion shows that when the clutch passes the area with low stiffness in the nonlinearity, the model deviates from the data. The reason is unmodeled dynamics at low clutch torques.

14.5.3   Backlash in General

For a clutch, a small torque is transmitted when traversing zero. Nevertheless, around zero transmitted torque, the clutch acts almost like a backlash due to the specific character of the clutch nonlinearity in Figure 14.11. This is further manifested in the top plot of Figure 14.13.

For a driveline, there is usually some play between cogs, components, and so on, which results in backlash. The principle is illustrated in Figure 14.14. While the input torque changes sign, that is traverses the backlash, there is no transfer of torque, which means that the output torque is zero during the traversal. This principle behavior is shown in Figure 14.15.

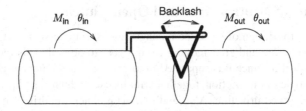

Figure 14.14   Illustration of a backlash with input and output shaft, and a backlash in between

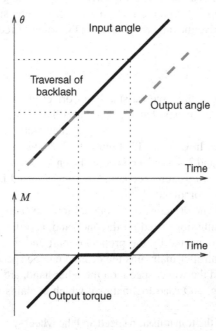

Figure 14.15   Angles and torques during a backlash traversal. While the input traverses the backlash there is no transfer of torque, that is the output torque is zero during the traversal

Importance of Considering Backlash

A backlash will give similar behavior as the clutch behavior in Figure 14.13, which is one area where backlash will add to the understanding of data. Another important example is tip-in control to avoid shunt-and-shuffle, which was introduced in Section 13.3.2. At around 80 s in Figure 13.4 it is seen that the driveline traverses through its backlash. As already mentioned, traversing the backlash is called shunt, and may end in an abrupt impact creating both undesired noise and jerk. The vehicle may then start to shuffle, and the combined phenomenon is called shunt-and-shuffle. To model and handle these situations, an extra state will be needed that describes whether the driveline is in grip or in the zero torque region.

Finally, it should be mentioned that driveline behavior around zero torque adds complexity. One example is when coasting slightly downhill, where gravitation just compensates the other driving resistances such that the driveline acts around zero torque. Then there may be discontinuities and impacts, which complicates even the usual cruise control.

14.6 Modeling of Neutral Gear and Open Clutch

An important basis for design of driveline management is to understand the dynamic behavior of the driveline before and after moving from a gear to neutral. This requires additional modeling of the driveline since it is separated in to two parts when in neutral. Such a decoupled model is the topic of this section, together with its use for analysis of possible oscillation patterns of the decoupled driveline. Also, with an open clutch the driveline is decoupled, at a slightly different point, however. The result is the same model as for neutral gear, but with slightly different parameters.

A decoupled model is used as the basis for a diagnosis system of gear-shift quality. It is also necessary for analysis to cast light on the sometimes, at first sight, surprising oscillations that occur in an uncontrolled driveline, and this indicates the value of feedback control.

14.6.1 Experiments

A number of gear-shifts to neutral with a stationary driveline are performed without using driveline torque control. This means that a speed controller controls the engine speed to the desired level and, when the driveline speeds have reached stationary levels, engagement of neutral gear is commanded. In Chapter 13, Figure 13.5 shows two of these trials where the engine speed is 1400 RPM and 2100 RPM respectively, on a flat road with gear 1. The behavior of the engine speed, the transmission speed, and the wheel speed is shown in the figure. At $t = 14$ s, a shift to neutral is commanded.

After the shift to neutral, the driveline is decoupled into two parts. Then, the engine speed is independent of the transmission speed on the one hand, and the wheel speed on the other hand. The two latter are connected by the propeller shaft and the drive shaft, according to Figure 13.2. The speed controller maintains the desired engine speed after the gear shift too. The transmission speed and the wheel speed, on the other hand, are only affected by the load (rolling resistance, air drag, and road inclination), which explains the decreasing speeds in the figure.

The transient behavior of the transmission speed and the wheel speed differ, however, and the energy built up in the shafts is seen to affect the transmission speed more than the wheel speed, giving an oscillating transmission speed. The higher the speed is, the higher the amplitude of the oscillations obtained. The amplitude value of the oscillations for 1400 RPM is 2.5 rad/s, and 5 rad/s for 2100 RPM.

If the driveline is not in a stationary mode the situation gets more complex. If a relative speed difference is present prior to the gear-shift, there will be a different type of oscillation. Figures 13.6 and 13.7 in Chapter 13 describe two trials where neutral gear is engaged with an oscillating driveline, resulting in quite complex behavior that can be counterintuitive at first sight, or at least possible to miss when designing driveline control if one has not been exposed to it before.

14.6.2 A Decoupled Model

A decoupled model is needed to analyze and explain the different types of oscillations described by Figures 13.5, 13.6, and 13.7. Engaging neutral gear can be described as in Figure 14.16. Before the gear shift, the driveline dynamics is described by the Driveshaft Model. This model assumes a lumped engine and transmission inertia, as described previously. When neutral gear is engaged, the driveline is separated into two parts as indicated in the figure. The two parts move independently of each other, as mentioned before. The engine side of the model consists of the engine, the clutch, and part of the transmission (see also Section 15.5.1). The parameters describing the lumped engine, clutch, and part of transmission are $\bar{J}_1$ and $\bar{b}_1$ according to Figure 14.16. The wheel side of the model consists

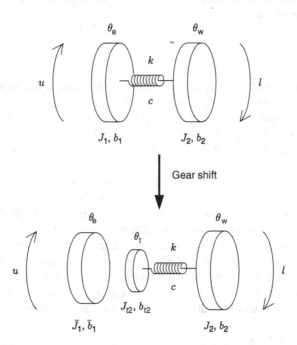

Figure 14.16 Description of how the driveline model changes after engagement of neutral gear. The first model is the Driveshaft Model, which is then separated into two sub-models when the neutral gear is engaged. The left part consists of the engine and one part of the transmission. The right part of the model consists of the rest of the transmission and the drive shaft out to the wheels, termed the Decoupled Model

of the rest of the transmission (characterized by the parameters J_{t2} and b_{t2}, once again see also Section 15.5.1) and the driveshaft flexibility out to the wheels. The model, termed the Decoupled Model, is described by the following equations.

Model 14.5 The Decoupled Model

$$\bar{J}_1 \ddot{\theta}_e = M_e - \bar{b}_1 \dot{\theta}_e \tag{14.65}$$

$$J_{t2} \ddot{\theta}_t = -b_{t2} \dot{\theta}_p - k(\theta_t/i_f - \theta_w)/i_f - c(\dot{\theta}_p/i_f - \dot{\theta}_w)/i_f \tag{14.66}$$

$$J_2 \ddot{\theta}_w = k(\theta_t/i_f - \theta_w) + c(\dot{\theta}_p/i_f - \dot{\theta}_w) - b_2 \dot{\theta}_w - l \tag{14.67}$$

The model equipped with the sensor filter in (14.50) gives the sensor outputs (y_m, y_t, y_w).

All these parameters were estimated in Section 14.3.2, except the unknown parameters J_{t2} and b_{t2}. The model is written in state-space form by using the states x_1 = driveshaft torsion, x_2 = transmission speed, and x_3 = wheel speed.

Note that the decoupled part, (14.66)–(14.67), of the Decoupled Model after the gear shift has the same model structure as the Driveshaft Model, but with the difference that the first inertia is considerably less in the Decoupled Model, since the engine and part of the transmission are decoupled from the model.

Model Validation

The unknown parameters J_{t2} and b_{t2} can be estimated if the dynamics described by the Decoupled Model is excited. This is the case when engaging neutral gear at a transmission torque level different from zero, giving oscillations. One such case is seen in Figure 14.17, where the oscillating transmission speed is seen together with the Decoupled model with estimated parameters J_{t2} and b_{t2}, and initial driveshaft torsion, x_{10}. The rest of the parameters are the same as in the Driveshaft Model, which were estimated in Section 14.3.2. The rest of the initial condition of the states (transmission speed and wheel speed) are the measured values at the time for the gear shift. The model outputs (y_t and y_w with sensor filter) are fitted to the measured transmission speed and wheel speed. The conclusion is that the Decoupled Model captures the main resonance in the oscillating transmission speed.

The different characteristic oscillations seen in the experiments after engaging neutral gear, are explained by the value of the driveshaft torsion and the relative speed difference at the time of engagement. The Decoupled Model can be used to predict the behavior of the driveline speeds if these initial variables are known. The demonstration of problems with an uncontrolled driveline motivates the need for feedback control in order to minimize the oscillations after a gear shift.

14.7 Clutch Modeling

When a clutch is partly open, the friction pads are sliding against each other and a torque is transmitted. This means that the situation is the same as in Figure 14.16, but that the two

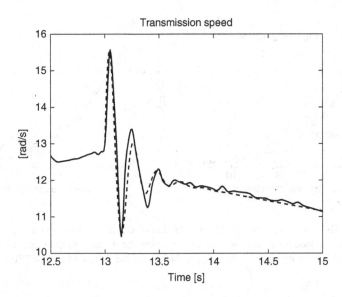

Figure 14.17 Measured oscillations after a gear shift at 13.0 s shown as a solid line. The outputs of the Decoupled Model are fitted to data, shown as a dashed line. The Decoupled model is able to capture the main resonance in the oscillating transmission speed after the gear shift

parts are not separated and a torque, M_c, is transferred. The clutch is then acting as a torque actuator moderating the engine torque, M_e, and the transmitted torque depends on the clutch actuator position, u_c, with $u_c = 0$ being no clutch actuation. When formulating the equations this simply means that $M_c(u_c)$ has to be added with different signs to both (14.65) and (14.66). The result is the Driveline with Slipping Clutch Model as follows

---

Model 14.6 The Driveline with Slipping Clutch Model

$$\bar{J}_1 \ddot{\theta}_e = M_e - \bar{b}_1 \dot{\theta}_e - M_c(u_c) \tag{14.68}$$

$$J_{t2} \ddot{\theta}_t = M_c(u_c) - b_{t2} \dot{\theta}_p - k(\theta_t/i_f - \theta_w)/i_f - c(\dot{\theta}_p/i_f - \dot{\theta}_w)/i_f \tag{14.69}$$

$$J_2 \ddot{\theta}_w = k(\theta_t/i_f - \theta_w) + c(\dot{\theta}_p/i_f - \dot{\theta}_w) - b_2 \dot{\theta}_w - l \tag{14.70}$$

14.7.1 Physical Effects

The final step to get a complete model is to model the transmitted torque, $M_c(u_c)$, as function of actuator position, u_c. The transmitted torque, M_c, in principle depends on several factors such as actuator position, temperatures, rotational speeds, and wear. Furthermore, components are nonlinear, for example spring characteristics, and there may be hysteresis, and so on. Nevertheless, the work in Myklebust and Eriksson (2012b) shows a path for obtaining relatively simple models, in that case for a dry single-plate pull-type normally closed

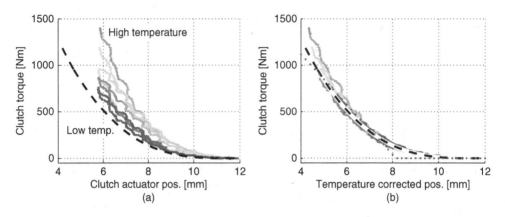

Figure 14.18 The transmitted torque, M_c, as function of clutch actuator position, u_C. (a) the clutch position has been ramped back and forth while measuring the torque, (b) the same data, but with x_{ISP} compensated for temperature. The dashed line is (14.72) and the dotted line is (14.73)

clutch. Original experimental data for such a clutch is seen Figure 14.18a where transferred torque, M_c, is plotted as function of actuator position, u_c, during tests where the temperature varies. It is seen that the transmitted torque varies by up to 700 Nm for a fix actuator position, so the temperature effect is significant. No clear hysteresis is detected and it is therefore not modeled. The sign of the slip speed, $\Delta\dot{\theta} = \Delta\omega$, determines the sign of the torque. Slip speed is recognized as an effect in the literature (Vasca et al. 2011; Velardocchia et al. 1999), whereas others (Dolcini et al. 2010; Myklebust and Eriksson 2012b) report no effect, at least not for slipping where slip speeds are relatively large.

Wear is an effect to be considered, since wear directly affects the thickness of the friction pads, which in turn gives a direct sideways translation of the effect of the actuator position in Figure 14.18. Other than that, wear can be assumed slow enough not to have a significant effect on driveline dynamics, but needs to be estimated and adapted over longer timescales.

Summing up, the main effects are actuator position and temperature leading to

$$M_c = \text{sgn}(\Delta\omega)\, M_c(u_c, T) \tag{14.71}$$

where it is understood that the wear of the friction pads translates the actuator position, u_c, sideways. Usually the factor $\text{sgn}(\Delta\omega)$ is omitted, but it is important in simulation. In an on-line application, the wear would probably need to be adapted continuously, even though it is a slow process.

14.7.2   Clutch Characteristics

To formulate a model for the clutch characteristic, it is useful to define the Incipient Sliding Point, x_{ISP}, also called the kiss point, where the push plate and clutch disc first meet and torque can start to transfer. The shape of the torque transmission curve is mainly due to cushion spring characteristics, and usually a third-order polynomial in the clutch actuator position gives a good approximation of the transmitted torque in a slipping clutch (Myklebust and Eriksson 2012b). Omitting influence of wear, for $u_c \leq x_{ISP}$, it can be expressed as:

$$c_2 \, (x_{ISP}(T) - u_c)^2 + c_3 \, (x_{ISP}(T) - u_c)^3 \tag{14.72}$$

There is no first or zeroth order term in the equation since it is desired to have zero torque and zero derivative at x_{ISP}, and experimental data confirms (14.72) to be a good approximation for a given temperature. Due to thermal expansion of the clutch mechanism, the kiss point is moved away from the actuator position, and this thermal effect is captured in $x_{ISP}(T)$, which is straightforward to compensate for since it is a simple sideways translation of the curve. In Figure 14.18b the same data as in (a) is shown after such a temperature compensation. Also in the figure, a fit of the model (14.72) is seen as a dashed line. All data, at least below 200°C, fit the same curve, that is the same coefficients c_2, c_3, demonstrating that the main thermal effect is a sideways translation of the curve. Sometimes, in driveline studies, a simpler model is used

$$M_c = c_1 \, (x_{ISP} - u_c) \tag{14.73}$$

where the interpretation of x_{ISP} is not physical anymore as is seen for the dotted line in Figure 14.18. In simulations, one may even use direct control, that is $M_c = c_0 u_c$.

14.7.3 Clutch Modes

The clutch is either closed, slipping, or open as is described in Sections 14.5, 14.7, and 14.6, respectively. During operation the clutch shifts between these states. An important physical effect is the torque behavior when going from slipping to closed. From Figure 14.18 it is seen that the torque increases when releasing the clutch, that is moving to the left on the curve (14.72). Then, when it stops slipping, the torque is abruptly reduced to the value for the closed clutch. Appropriate handling of this effect is important in simulation studies (Eriksson 2001).

14.8 Torque Converter

A torque converter is a hydrodynamic fluid coupling that transfers the rotating power from an internal combustion engine (or electric motor) to a rotating driven load, for example the transmission via a fluid. Torque converters are used instead of a mechanical clutch in vehicles with an automatic transmission, and is situated in the same position as the clutch in Figure 14.1.

The two key characteristics of a torque converter are that the input and output speeds of the device are decoupled, and that it has the ability to multiply torque when there is a substantial difference between input and output rotational speed. The latter means that it provides the equivalent of a reduction gear in the driveline. Referring to Figure 14.1, in equations this means that M_c no longer needs to be equal to M_t as was assumed, for example, in (14.51) and in the previous section on a slipping clutch. Some torque converters have a locking mechanism, which rigidly binds the engine to the transmission when the speeds are nearly equal, in which case $M_c = M_t$. This avoids slipping and the associated losses in efficiency.

Torque converter operation and performance depend on the speed ratio between outgoing and incoming speeds $\phi(t) = \frac{\omega_t(t)}{\omega_e(t)}$. The torque on the engine side depends on both $\phi(t)$ and the rotational speed $\omega_e(t)$, and is described with a function $\xi(\phi(t))$. The input (or pump) torque at the converter is then modeled as

$$M_c = M_{tc,e}(t) = \xi(\phi(t))\, \rho_h\, d_p^5\, \omega_e^2(t)$$

where ρ_h is the fluid density and d_p is the pump diameter. The output torque is amplified by a factor ψ that depends on the speed ratio

$$M_t = M_{tc,t}(t) = \psi(\phi(t)) \cdot M_{tc,e}(t)$$

The performance functions $\xi(\phi(t))$ and $\psi(\phi(t))$ are often measured and supplied by the manufacturer, using normalized characteristic curves by a construction where ξ is replaced by a dimensionless ξ_0 in the following manner. For a given device the fluid density and diameter are given, and then $\xi(\phi(t))\, \rho_h\, d_p^5\, \omega_e^2(t)$ can be rewritten as $M_0\, \xi_0(\phi(t))(\omega_e(t)/\omega_0)^2$ by changing the constants. The torque M_0 refers to the input speed ω_0. Then the converter data is represented by the two normalized characteristic curves, as shown in Figure 14.19. A model for the input and output torques in the torque converter is then expressed as follows.

Model 14.7 Torque Converter

$$\phi(t) = \frac{\omega_t(t)}{\omega_e(t)} \tag{14.74}$$

$$M_{tc,e}(t) = M_0\, \xi_0(\phi(t))\left(\frac{\omega_e(t)}{\omega_0}\right)^2 \tag{14.75}$$

$$M_{tc,t}(t) = \psi(\phi(t)) \cdot M_{tc,e}(t) \tag{14.76}$$

where the constants are: M_0 describes the standstill input torque at the input speed, ω_0, which is a nominal speed of the torque converter.

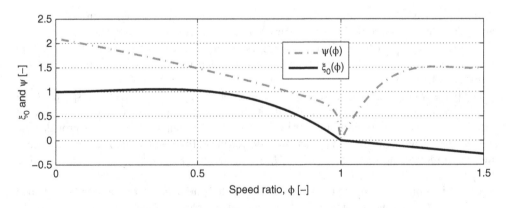

Figure 14.19 Typical characteristic curves of a torque converter: $\xi_0(\phi(t))$ is the input torque characteristic and $\psi(\phi(t))$ is the torque amplification factor

The functions $\xi_0(\phi(t))$ and $\psi(\phi(t))$ specify the characteristics, and typical curves are shown in Figure 14.19. By dividing output power and input power, and using (14.74)–(14.76), torque converter efficiency in traction mode (i.e., $\phi \in [0, 1]$) is obtained as

$$\eta_{tc} = \frac{M_{tc,t}\, \omega_t}{M_{tc,e}\, \omega_e} = \psi(\phi)\, \phi$$

14.9 Concluding Remarks on Modeling

A specific example, using experimental data from the 144 L truck, has been used to guide the presentation of driveline modeling. The measured engine speed, transmission speed, and wheel speed have been explained by deriving a set of models of increasing complexity.

14.9.1 A Set of Models

Figure 14.2 shows that the main difference in speed is between the measured transmission speed and wheel speed. This shows that the main flexibility of the driveline is located between the output shaft of the transmission and the wheel, which is quite common. A first model, the Driveshaft Model, with one torsional flexibility and two inertias is able to fit the measured engine speed and wheel speed in a frequency regime including the first main resonance of the driveline. In order for the model to also fit the measured transmission speed, a first-order sensor model was added to the model, and then all three velocities are fitted accurately enough. Parameter estimation of a model with a nonlinear clutch explains that the difference between the measured data and the model outputs occurs when the clutch transfers zero torque.

The result is a series of models that describe the driveline in increasing detail by, in each extension, adding the effect that seems to be the major cause for the deviation still left. The sequence of models is thus: Driveshaft Model, Clutch and Driveshaft Model, Nonlinear Clutch and Driveshaft Model. In addition to the mechanical models, sensor dynamics was added. The assumption about sensor dynamics in the transmission speed influencing the experiments agrees well with the fact that the engine speed sensor and the wheel speed sensor have considerably higher bandwidth (more cogs) than the transmission speed sensor. For other vehicles, data sheets on relevant sensor dynamics (or simple cog count) should guide whether and where to include sensor modeling. Investigations of other additional dynamics, for example propeller shaft torsion, gave for this example no added explanation, but that may be different for other vehicles.

14.9.2 Model Support

Besides describing experimental data as in Figure 14.2, there is additional support for the models obtained. Supporting the validity of the models is the fact that they give values to the non-measured variables, drive shaft and clutch torsion, that agree with experience from other sources. Furthermore, when estimating the parameters in the Driveshaft Model given by (14.42) the unknown load, l, which varied between the trials, is estimated. The load can be recalculated to estimate road slope, and the calculated values agree well with the known values of the road slopes at Scania. Finally, the estimation of the states describing the torsion of the clutch and the drive shaft shows realistic values. This gives further support to model structure and parameters. Other experimental situations, like when the driveline is separated in two parts, should also be used for the best result, and estimation of parameters in the Decoupled model gives additional support.

14.9.3   Control Design and Validating Simulations

From a controls perspective, the Driveshaft model with some sensor dynamics gives good agreement with experiments within the frequency regime interesting for control Design. It is thus suitable for control design. The additional effects captured by the Nonlinear clutch and driveshaft model make this model suitable for verifying simulation studies in control design. This structure will be used in the next chapter.

15

Driveline Control

Driveline control can now be approached using the models from the previous chapter, Chapter 14, Driveline Modeling. Recall from Chapter 13, that the objective of driveline control is to control

- (all rotational) velocities
- (all internal) torques

despite different driver inputs, different initial states of the driveline, and different disturbances. Recall also from the same chapter that it is essential how much of the torsional energy that is stored in the driveline (and how that energy can be released) when a control action is taken. Consider, for example, a tip-in, that is when the driver presses the accelerator pedal. Handling the driveline in such a case is much simpler if the driveline can be assumed to be in steady state, than if the driveline is already oscillating or if torsional energy is already stored, for example due to previous driver commands or due to external influences like road bumps or impulses from a towed trailer.

Illustrative Problems

Even though there are many important applications for driveline control, two *problems* in automotive behavior have been chosen to illustrate the main concepts. They are

- vehicle surge or vehicle shuffle, see Section 13.3.1
- free driveline oscillations when gear shifting, see Section 13.3.3

where the first requires velocity control and the second requires torque control.

Illustrative Solutions

The *solutions* to the problems above are respectively

- anti-surge control
- torque control for AMT (Automated Manual Transmission) gear shifting

Modeling and Control of Engines and Drivelines, First Edition. Lars Eriksson and Lars Nielsen.
© 2014 John Wiley & Sons, Ltd. Published 2014 by John Wiley & Sons, Ltd.
Companion Website: www.wiley.com/go/powertrain

These two applications will be treated in detail in Section 15.3 regarding anti-surge control and in Sections 15.4–15.6 regarding torque control for gear shifting. Two different approaches to torque control are illustrated, so that Section 15.5 describes a general method and Section 15.6 describes an effective simplification.

15.1 Characteristics of Driveline Control

Driveline control is not an isolated control problem. Control solutions like anti-surge control and torque control for gear shifting have to be integrated and fitted into the structure of torque-based powertrain control that was described in Chapter 3, Powertrain, see Section 3.3 and Figure 3.8. Further, solutions for driveline management and control naturally depend on the sensors and actuators available, and there are several options. When approaching driveline control one has to specifically consider the

- overall control structure to fit into
- sensors and their placement
- torque actuation and its characteristics
- limits depending on operational state.

These topics are elaborated in the following subsections. In addition to these specific control requirements there are general ones, like noise handling, formulation of the control objective, and choice of the control methodology.

15.1.1 Inclusion in Torque-Based Powertrain Control

A first question is how to fit driveline control into torque-based powertrain control. In the lower part of Figure 3.8, the arrows – labeled driveline sensors and actuators – indicate that there is the possibility of including feedback in the scheme for torque propagation. Such driveline control solutions represent short-term interventions, as illustrated in Figure 3.12. The two cases in that figure that represent speed control are repeated in Figure 15.1.

Smoothing Filter or Active Feedback

Figure 15.1a illustrates the case where the anti-surge filter has smoothed the torque request in order not to excite oscillations. That solution for anti-surge is the one illustrated in Figure 3.11 by the block "Anti-surge filter." On the other hand, Figure 15.1b illustrates the situation when feedback control is used instead, and in that case the torque counteracts the oscillation based on feedback. Integrating such behavior in the control scheme requires a modification compared to Figure 3.11, and this is done in Figure 15.2. Then, anti-surge is based on measurements or estimates of driveline states as illustrated in the figure.

Returning to Figure 15.1, it is seen that the responsiveness is better for the feedback solution, which is natural since the main idea behind a smoothing filter is to slow down the response, not to excite oscillations. A feedback solution can handle disturbances, and more importantly, it can handle a variety of initial conditions. As described in Section 13.3.2, shunt and shuffle is a problem at tip-in, and if the driveline is not in steady state, for example due to previous driver

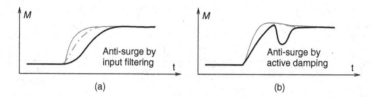

Figure 15.1 Two possibilities for anti-surge management using input filtering (a) or feedback control (b) showing torque demand (thin line), physically possible torque (dash-dotted line), and actually delivered torque (thick line)

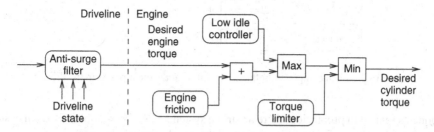

Figure 15.2 Compared to Figure 3.11, that uses a smoothing filter for anti-surge, the scheme here uses feedback, which is made possible by knowledge of the driveline state as depicted by the arrows under the block "Anti-surge filter"

commands or due to external influences like road bumps, these problems are not tractable without using feedback.

15.1.2 Consequence of Sensor Locations

The sensors used are mainly sensors for rotational speeds, and already in Chapter 14, Driveline Modeling, the different characteristics and modeling consequences of using either an engine speed sensor, transmission sensor, or wheel speed sensor were discussed. Recall Figure 14.4 and (14.39)–(14.41). If the driveline was rigid, the choice of sensor would not matter, since the sensor outputs would differ only by a scaling factor. However, the presence of torsional flexibilities implies that sensor choice gives different control problems. The difference can be formulated in control theoretic terms, for example by saying that the poles are the same, but the zeros differ both in number and values. The consequence for feedback control will be given an initial investigation in Sections 15.2.8 and 15.2.6, and will then influence the designs following.

15.1.3 Torque Actuation

Torque is the primary variable to control the dynamic behavior of the driveline, and there are many possibilities for actuating or modifying the torque in the driveline. A number of these options are seen in Figure 15.3, which is Figure 3.2 from Section 3.1. In addition to the

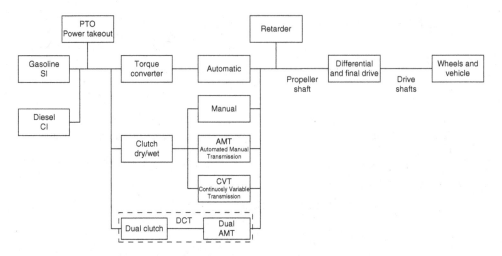

Figure 15.3 Each component influencing torque at different points in the driveline

components related to propulsion, there are more systems like, for example, different systems exerting braking forces. The main torque providers are the primary movers

- combustion engine
- electrical motor

that can provide both positive and negative torque, for example by engine braking. The torque from these providers can be modified by a number of components

- clutch by control of transmitted torque, see Section 14.7
- torque converter, see Section 14.8
- transmission by choice of gear ratio, including control of the continuously variable gear ratio (CVT).

There are also a number of systems that can exert braking torques, often quickly and with large control authority

- retarder
- active differential and other torque vectoring functions
- brakes, including ABS and stability functions like Electronic Stability Control (ESC).

15.1.4 Transmissions

Traditionally, a gear shift is performed by disengaging the clutch, engaging neutral gear, shifting to a new gear, and engaging the clutch again. As indicated in Figure 15.3, there are today more options:

Automatic transmission The classical solution uses a torque converter, Section 14.8, which has a drawback in lower efficiency than manual transmissions.

Automated Manual Transmission (AMT)   With this approach the clutch is replaced by engine control, realizing a virtual clutch. The addition needed to a standard manual transmission is an actuator to move the gear lever. Compared to automatic transmission it has lower cost and higher efficiency.

Continuously Variable Transmission (CVT)   An approach where continuous control of the gear ratio can be used to optimize usage of, for example, the combustion engine.

Dual Clutch Transmission (DCT)   Another approach for Automated Manual Transmission where one clutch is used for the disengaged gear and another for the engaged gear.

From the perspective of driveline control, the configurations above, to different degree, represent mode shifts. For example, for a gear shift with an automated manual transmission, the engine torque is controlled to a state where the transmission transfers zero torque, whereafter the neutral gear is engaged without using the clutch. After the speed synchronization phase, the new gear is engaged, and control is transferred back to the driver. As described in Section 13.3.3, such torque transients can cause oscillations and other unwanted behavior, so the handling of such resonances is basic for functionality and driveability, but is also important for reducing mechanical stress and noise.

Intervention for Gear Shifting

Torque control for gear shifting is thus a short-term intervention modifying the propagated torque in the torque-based control structure. The function is illustrated in Figure 15.4 and a real gear shift is seen in Figure 15.20.

15.1.5   Engine as Torque Actuator

When using the engine as torque actuator, the control signal, u, is the output torque, M_e, from the engine, see Section 7.9. This control signal has the following limitations:

- The engine torque is not smooth, since the combustion in the cylinder results in a pulsating engine torque, see, for example, the cylinder individual pulsations in Figure 7.26.
- The output torque of the engine is not exactly known due to imperfections in the torque model, see, for example, the combustion, pumping, and friction discussed in Sections 7.9.1 to 7.9.3.
- The engine output torque is limited in different modes of operation. For example, the maximum engine torque is restricted as a function of the engine speed, and the torque level is also restricted at low turbo pressures.

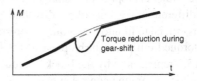

Figure 15.4   A schematic illustration of a gear shift showing torque demand (thin line), physically possible torque (dash-dotted line), and actually delivered torque (thick line). A real gearshift is seen in Figure 15.20

Regarding the first item, the control signal $u = M_e$ can normally, without problems, be treated as a continuous signal for the frequency range considered for control design. A motivation for this is that there are sufficiently many strokes being smoothed out over a transient, for example, an eight-cylinder engine makes 80 strokes/s at an engine speed of 1200 rev/min.

15.1.6   Control Approaches

When approaching driveline control, all aspects in the preceding sections have to be considered. To start with, a model that captures the main physical phenomena, especially torsional dynamics, is available as described in Chapter 14, Driveline Modeling. Then, as pointed out above, different control problems arise depending on the choice of actuators and sensors, which means that some signal in the model is used as control signal, u, and some other signals are measured output signals, y. In the treatment that follows, engine torque will be the control signal, $u = M_e$, and engine speed and wheel speed will be sensor outputs, $y = \dot{\theta}_e$ or $y = \dot{\theta}_w$. These are good examples to describe the general methodology as used in current implementations, but one may note that introduction of novel sensors like, for example, a reliable and economically tractable torque sensor, would change the picture.

Once the model and actuator/sensor configuration is set, a control objective can be formulated. The control problems should preferably be formulated so that it is possible to use established techniques to obtain solutions. Natural candidates would be basic controllers, like PID, but since it was seen in the modeling chapter that at least three states are needed for a good model, it is also natural to look at state estimation with an observer together with feedback from estimated states. The actual formulation of the control objective is then an interplay between model, criterion, and different (nonlinear) limitations, as will be illustrated in Sections 15.3.2–15.3.3, and 15.5.2–15.5.4. One specific circumstance to observe when formulating the control objective is that the driveline is not in steady state, for example during the torque-free phase of a gear shift, since it will lose speed, a fact that influences the formulation in Section 15.5.3.

Important Requirements

Two of the requirements above are particularly important and should be given special attention when evaluating a design. In control terms, these are formulated as robustness to a variety of initial conditions and good behavior in the presence of torque limitations. An example of the first requirement is that a controller for shunt and shuffle at tip-in should be validated for different initial states of the driveline, due to previous driver commands or external influences like road bumps, for example. Other examples are Figures 15.36–15.39 where good behavior is demonstrated despite the different initial state of the driveline.

Handling of torque limits is crucial to fit into the torque-based structure. One example is that a diesel engine may have a torque limit at low turbo pressures to limit smoke exhaust. Such a smoke limiter is in Figure 15.2 represented by the block "Torque limiter." That block enters after the block "Anti-surge filter" where the control signal is computed, and therefore such a block could break the feedback path. It must be verified that the controller works well in such a situation, and a good example of a well-unctioning design is given in Figure 15.18, which is an important figure.

The presentation in the following is mostly based on Pettersson and Nielsen (1997, 2000, 2003), Pettersson et al. (1997), and Kiencke and Nielsen (2005). These texts are research oriented and contain more details, whereas the presentation in the following concentrates on the basic principles, the results, and performance demonstrations in real tests.

15.2   Basics of Driveline Control

Driveline control will now be put on state-space form, which is a standard form enabling use of a vast range of control methods, but the choice of control design methods will not be treated here. Instead we will focus on the formulation of the control problem to capture important vehicle features and obtain feasible and well-behaved solutions. Further, some illustrative observations on driveline control will be made, and some simplified descriptions useful for understanding will be given. The formulations and observations will then be the basis for the presentation of actual designs in Sections 15.3–15.6.

15.2.1   State-Space Formulation of the Driveshaft Model

Depending on the choice of actuators and sensors, some signal in a model is used as the control signal, u, and some other signals are measured output signals, y. As stated above, here engine torque will be the control signal, $u = M_e$, and engine speed and wheel speed will be sensor outputs, $y = \dot{\theta}_e$ or $y = \dot{\theta}_w$. Possible physical state variables in the models of Chapter 14 are torques, angle differences, and angle velocity of any inertia. The angle difference of each torsional flexibility and the angle velocity of each inertia are used as state variables.

The equations depend on the model used for driving resistance. Here the model (2.24) that consists of a constant load and a varying slope is used. The state-space representation is

$$\dot{x} = Ax + Bu + H\,l \tag{15.1}$$

where A, B, H, x, and l for the Driveshaft Model, see Section 14.3.2, are

$$x_1 = \theta_e/i_t i_f - \theta_w$$

$$x_2 = \dot{\theta}_e$$

$$x_3 = \dot{\theta}_w$$

$$l = r_w mg(f_0 + \sin(\alpha)) \tag{15.2}$$

giving

$$A = \begin{pmatrix} 0 & 1/i & -1 \\ -k/iJ_1 & -(b_1 + c/i^2)/J_1 & c/iJ_1 \\ k/J_2 & c/iJ_2 & -(c + b_2)/J_2 \end{pmatrix}, \tag{15.3}$$

$$B = \begin{pmatrix} 0 \\ 1/J_1 \\ 0 \end{pmatrix}, \qquad H = \begin{pmatrix} 0 \\ 0 \\ -1/J_2 \end{pmatrix} \tag{15.4}$$

where, with the same parameterization as in (14.42),

$$i = i_t i_f$$
$$J_1 = J_e + J_t/i_t^2 + J_f/i_t^2 i_f^2$$
$$J_2 = J_w + m r_w^2$$
$$b_1 = b_t/i_t^2 + b_f/i_t^2 i_f^2$$
$$b_2 = b_w \qquad\qquad (15.5)$$

15.2.2   Disturbance Description

The influence from the road is assumed to be described by the slow-varying load l and an additive disturbance v. A second disturbance n is a disturbance acting on the input of the system. This disturbance is considered because the firing pulses in the driving torque can be seen as an additive disturbance acting on the input. The state-space description then becomes

$$\dot{x} = Ax + Bu + Bn + H\,l + Hv \qquad\qquad (15.6)$$

with x, A, B, H, and l defined in (15.2) to (15.5).

15.2.3   Measurement Description

For controller synthesis, it is of fundamental interest which physical variables of the process can be measured. In the case of a vehicular driveline, the normal sensor alternative is an inductive sensor mounted on a cogwheel measuring the angle, as mentioned before. Sensors that measure torque are expensive, and are seldom used in production vehicular applications.

The output of the process is defined as a combination of the states given by the matrix C in

$$y = Cx + e \qquad\qquad (15.7)$$

where e is a measurement disturbance.

Here, only angle velocity sensors are considered and, therefore, the output of the process is one/some of the state variables defining an angle velocity. Especially, the following C-matrices are defined (corresponding to a sensor on $\dot{\theta}_e$ and $\dot{\theta}_w$ for the Driveshaft model).

$$C_m = (0 \quad 1 \quad 0) \qquad\qquad (15.8)$$
$$C_w = (0 \quad 0 \quad 1) \qquad\qquad (15.9)$$

15.2.4   Performance Output

The performance output z is the combination of states that have requirements to behave in a certain way. This combination is described by the matrices M and D in the following way

$$z = Mx + Du \qquad\qquad (15.10)$$

For anti-surge control, the performance variable is vehicle motion which is describer by the wheel speed, that is $z = \dot{\theta}_w$, which means

$$M = (0 \quad 0 \quad 1) \tag{15.11}$$

$$D = 0 \tag{15.12}$$

Another example is in control for gear shifting where, in one approach, the objective is to control an internal torque in the transmission, and then the performance output, z, is this internal torque as described by 15.73.

15.2.5   Control Objective

Once the performance output, z, is determined, it is natural to describe the wanted behavior in terms of a reference signal, r. How well z should follow r can be formulated many ways, and a natural way is to use the criterion

$$\lim_{T \to \infty} \int_0^T (z - r)^2 \tag{15.13}$$

The criterion can be used for performance evaluation, but it can also be used for direct design using control design methods. In automotive applications it turns out that it is crucial to adapt and modify this criterion to get feasible and well-behaved solutions, and this is an important topic in the design sections, Section 15.3 regarding anti-surge control and Section 15.6 regarding torque control for gear shifting.

15.2.6   Controller Structures

Traditional control schemes, like, for example, RQV that will be treated later, use the common PID structure, which for RQV in fact is only P-control, that is

$$u = K(r - y) \tag{15.14}$$

Observer and State Feedback

In modern model-based control the formulation above is utilized. If state-feedback controllers are used, the control signal u is a linear function of the states (if they are all measured) or else the state estimates, $\hat{x}$, which are obtained from an observer, usually called a Kalman filter if it is of same order as the system to be controlled. The control signal is described by

$$u = l_0 r - K_c \hat{x} \tag{15.15}$$

where r is the reference signal with the gain l_0, and K_c is the state-feedback matrix. The equation describing the observer is

$$\dot{\hat{x}} = A\hat{x} + Bu + K_f(y - C\hat{x}) \tag{15.16}$$

where K_f is the observer gain.

The controller structure forms a framework, and there are many methods to find K_c and K_f. One method is Linear Quadratic design with Loop Transfer Recovery (LQG/LTR), see, for

example, Maciejowski (1989). This design method will be used in some of the examples in
the design chapters, but the presentation will not go into the formalism. Instead, the results
will be presented for conceptual understanding using plots.

15.2.7   Notation for Transfer Functions

It is often useful to complement the state-space formulation above with corresponding transfer
functions. The notation used here for the transfer function from signal u to signal y is

$$G_{uy} \qquad\qquad (15.17)$$

15.2.8   Some Characteristic Feedback Properties

The performance output when controlling the driveline to a certain speed is the velocity of the
wheel, defined as

$$z = \dot{\theta}_w = C_w x \qquad\qquad (15.18)$$

When studying the closed-loop control problem, with a sensor on $\dot{\theta}_e$ or $\dot{\theta}_w$, the open-loop
transfer functions are $G_{u\dot{\theta}_e}$ or $G_{u\dot{\theta}_w}$, where the control signal is engine torque, $u = M_e$. This
means that even though the control objective is the same, control of z in 15.18, the consequence
of sensor location is two different control problems. Figure 15.5 shows a root locus with respect
to the gain of a P-controller for two gears using velocity sensor $\dot{\theta}_e$ and $\dot{\theta}_w$ respectively. The
open-loop transfer functions from control signal to engine speed $G_{u\dot{\theta}_e}$ have three poles and
two zeros, as can be seen in Figure 15.5. $G_{u\dot{\theta}_w}$ on the other hand, has the same three poles
but only one zero. Hence, the relative degree, (Kailath, 1980), of $G_{u\dot{\theta}_e}$ is one and for $G_{u\dot{\theta}_w}$ the
relative degree is two. This means that when $\dot{\theta}_w$-feedback is used, and the gain is increased,
two poles must go to infinity which makes the system unstable as seen in Figure 15.5. When
the velocity sensor $\dot{\theta}_e$ is used, the relative degree is one, and the closed-loop system is stable
for all gains, as seen in Figure 15.5. (Remember that $\dot{\theta}_w$ is the performance output and thus
desirable to use.)

The same effect can be seen in step response tests when the P-controller is used. Figure 15.6
demonstrates the problem with resonances that occur with increasing gain for the two cases of
feedback. When the engine-speed sensor is used, the engine speed is well damped when the
gain is increased, but the resonance in the drive shaft makes the wheel speed oscillate. When
using $\dot{\theta}_w$-feedback it is difficult to increase the bandwidth, since the poles moves closer to the
imaginary axis and give a resonant system.

Valuable Insights

The above simple example gives some valuable insights. Using the engine-speed sensor gives
a simple control problem, but care has to be taken in meeting the real control requirements, that
is well-behaved wheel speed, which means smooth vehicle behavior without vehicle shuffle.
On the other hand, using the wheel speed sensor addresses the real control objective, but then
the control design has to consider trade-offs. These observations are more generally valid, as
commented on in the following two subsections.

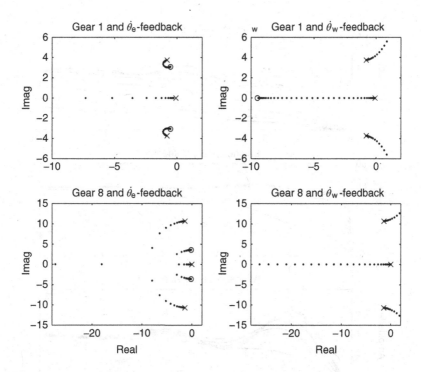

Figure 15.5   Root locus with respect to a P-controller gain, for gear 1 (top figures) and gear 8 (bottom figures), with sensor on $\dot{\theta}_e$ (left figures), or $\dot{\theta}_w$ (right figures). The cross represent the open-loop poles, while the rings represent the open-loop zeros. The system goes unstable when the $\dot{\theta}_w$-gain is increased, but is stable for all $\dot{\theta}_e$-gains

General Structural Properties of Sensor Location

As illustrated in the previous section, different sensor locations result in different control problems with different inherent characteristics. For a general driveline model with n inertias connected by $n - 1$ torsional flexibilities, there are $(2n - 1)$ poles, and the location of the poles is the same for the different sensor locations. The number of zeros depends on which sensor is used, and when using $\dot{\theta}_w$ there are no zeros. When using feedback from $\dot{\theta}_e$ there are $(2n - 2)$ zeros. Thus, the transfer functions $G_{u\dot{\theta}_e}$ and $G_{u\dot{\theta}_w}$, have the same denominators, and a relative degree of 1 and $(2n - 1)$ respectively.

When considering design, a good alternative is to have relative degree one, implying infinite gain margin and high phase margin. In that case, when using $G_{u\dot{\theta}_e}$, one pole has to be moved to infinity, and when using $G_{u\dot{\theta}_w}$, $2n - 1$ poles have to be moved to infinity, in order for the ratio to resemble a first-order system at high frequencies. Further, when the return ratio behaves like a first-order system, the closed-loop transfer function also behaves like a first-order system. This conflicts with the design goal of having a steep roll-off rate for the closed-loop system in order to attenuate measurement noise. Hence, there is a trade-off when using $\dot{\theta}_w$-feedback. When using $\dot{\theta}_e$-feedback, there is no trade-off, since the relative degree of $G_{u\dot{\theta}_e}$ is one.

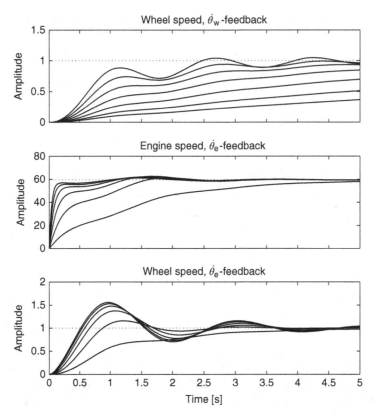

Figure 15.6 Step responses when using a P-controller with different gains on the Driveshaft Model with gear 1. With $\dot{\theta}_w$-feedback (top figure), increased gain results in instability. With $\dot{\theta}_e$-feedback (bottom figures), increased gain results in a well damped engine speed, but an oscillating wheel speed

Excursion: Observations when using LQG/LTR

The characteristic results in Figures 15.5 and 15.6 only depend on the relative degree, and are thus parameter independent. However, this observation may depend on feedback structure, and it is therefore interesting to see how the structural properties show up when using a controller with observer and state feedback. In Pettersson and Nielsen (2003) and Kiencke and Nielsen (2005), the method of LQG/LTR, as mentioned in Section 15.2.6, is used for the same example as was used for P-control shown in Figures 15.5 and 15.6. The results show that the parameter values, that is the actual position of the zeros, are important in the LQG step of the design. Looking at the recovery step, LTR, there are two cases depending on the sensor. For recovery for $\dot{\theta}_w$-feedback there is a trade-off between good attenuation of measurement noise and a good stability margin. A good stability margin requires an increased control signal, which gives a 20 dB/decade roll-off rate for a wider frequency range. On the other hand, for recovery for $\dot{\theta}_e$-feedback there is no trade-off. It is possible to achieve good recovery with reasonable stability margins and control signal, together with a steep roll-off rate. In conclusion, this means that the structural properties, that is the relative degrees, are dominant in determining the LTR step of the design, whereas parameter values dominate in the LQG step.

15.2.9   Insight from Simplified Transfer Functions

The state-space model in Section 15.2.1 is easy to handle numerically both in simulation tools and design tools. Nevertheless, introducing some simplifying assumptions makes the model tractable for hand calculation, and that gives some additional insights. Here this will be done at two levels, where the first is a linearization providing transfer functions. In the second, losses are assumed to be zero which gives simpler pole-zero insights, observations that will be useful in Section 15.6.2 where a useful automotive design is developed.

Somewhat Simplified Linear System

The Driveshaft Model in Section 15.2.1 is used with the main difference being that the driving resistance, see Section 2.3, is assumed to be linear in vehicle speed. Thus comparing with (15.2) the model is

$$l = \gamma x_3 \tag{15.19}$$

This means that the air drag that is normally dependent on the squared velocity is considered to be modeled as linear around the actual operating point. Further, that the constant terms in the driving resistance are included in the stationary point, x_0, u_0. In addition to the linearization, the loss parameters b_1, b_2 are for convenience assumed to be zero.

As before, the input signal is $u = M_e$. We are interested in transfer functions to the engine speed, $\dot{\theta}_e$, the wheel speed, $\dot{\theta}_w$, and the torque in the driveshaft, M_d, as defined in (14.36), namely

$$M_d = k(\theta_e/i_t i_f - \theta_w) + c(\dot{\theta}_e/i_t i_f - \dot{\theta}_w) \tag{15.20}$$

These are collected as outputs in the formulation of y. Thus, let the states be

$$x = \begin{pmatrix} x_1 \\ x_2 \\ x_3 \end{pmatrix} = \begin{pmatrix} \frac{1}{i}\dot{\theta}_e - \dot{\theta}_w \\ \dot{\theta}_e \\ \dot{\theta}_w \end{pmatrix} \tag{15.21}$$

and the outputs be

$$y = \begin{pmatrix} M_d \\ \dot{\theta}_e \\ \dot{\theta}_w \end{pmatrix} \tag{15.22}$$

Using (15.19) in (15.2), with notation as in 15.5, then gives the state-space form

$$\dot{x} = -\begin{pmatrix} 0 & \frac{1}{i} & -1 \\ -\frac{\alpha k}{i} & -\frac{\alpha c}{i^2} & \frac{\alpha c}{i} \\ \beta k & \frac{\beta c}{i} & -\beta (c+\gamma) \end{pmatrix} x + \begin{pmatrix} 0 \\ \alpha \\ 0 \end{pmatrix} u \quad \text{where} \quad \begin{cases} \alpha = \frac{1}{J_1} \\ \beta = \frac{1}{J_2} \end{cases} \tag{15.23}$$

$$y = \begin{pmatrix} k & \frac{c}{i} & -c \\ 0 & 1 & 0 \\ 0 & 0 & 1 \end{pmatrix} x \tag{15.24}$$

In this model the losses are modeled by

- c – the damping in the spring model of the driveshaft.
- γ – from the simplified driving resistance (15.19).

Recall that the transformation from linear state-space form to transfer functions is performed by

$$\begin{cases} \dot{x} = Ax + Bu \\ y = Cx \end{cases} \qquad G(s) = C(sI - A)^{-1}B \tag{15.25}$$

This gives the transfer functions

$$\begin{pmatrix} G_{uM_d}(s) \\ G_{u\dot{\theta}_e}(s) \\ G_{u\dot{\theta}_w}(s) \end{pmatrix} = \begin{pmatrix} \dfrac{\alpha\; c(s + \frac{k}{c})\;(s + \beta\;\gamma)}{n(s)} \\[2mm] \dfrac{i^2\;\alpha\;(s^2 + s\;\beta\;(c + \gamma) + k\;\beta)}{n(s)} \\[2mm] \dfrac{i\;\alpha\;\beta\;c(s + \frac{k}{c})}{n(s)} \end{pmatrix} \tag{15.26}$$

$$n(s) = (k + c\;s)\;\alpha\;(s + \beta\;\gamma) + i^2\;s\;(s^2 + k\;\beta + s\;\beta\;(c + \gamma)) \tag{15.27}$$

Zero Losses

Now, assume zero losses in the above formulation, that is $\gamma = 0$ and $c = 0$. A remark is that constant driving resistance gives the same system as $\gamma = 0$ does, since the constant driving loss will be included in the stationary point, x_0, u_0. Now, the numerator polynomial, $n(s)$, is easy to factorize and it is easy to see pole-zero cancellations.

With $\gamma = 0$ and $c = 0$ the transfer functions are

$$\begin{pmatrix} G_{uM_d}(s) \\ G_{u\dot{\theta}_e}(s) \\ G_{u\dot{\theta}_w}(s) \end{pmatrix} = \begin{pmatrix} \dfrac{\alpha\;k}{i^2}\;\dfrac{1}{s^2 + k\;(\frac{\alpha}{i^2} + \beta)} \\[2mm] \alpha\dfrac{s^2 + k\;\beta}{s\;(s^2 + k\;(\frac{\alpha}{i^2} + \beta))} \\[2mm] \dfrac{\alpha\;\beta\;k}{i}\;\dfrac{1}{s\;(s^2 + k\;(\frac{\alpha}{i^2} + \beta))} \end{pmatrix} \tag{15.28}$$

The poles and zeros for this model will be interpreted in the next section, and are as follows

Complex poles in $\qquad \pm j\sqrt{k\;(\beta + \frac{\alpha}{i^2})}$

Zeros for $G_{u\dot{\theta}_e}(s)$ in $\qquad \pm j\sqrt{k\;\beta}$ $\qquad$ (inside the poles since $k\;\beta < k\;(\beta + \frac{\alpha}{i^2})$)

Useful Insights

The treatment above gives insight into the pole-zero configurations in Figure 15.5. Since that figure has the complete model, including all loss coefficients, the locations are not exact. However, it is easy predict how the pole-zero configuration will change, for example when shifting gear, i, by using the formulas above.

Another important observation is the pole-zero cancellation in $G_{uM_d}(s)$, where the cancellation is due to the fact that the damping c in the driveshaft is zero. In real automotive applications the damping is of course not zero, but it is often small compared to the spring constant k. This may result in almost pole-zero cancellations, and a real example is seen in Figure 15.32. For torque control, of, for example, M_d, this means that the system behaves like a second-order system which makes it possible to use controllers like PID. This is a useful fact as shown, for example, in Section 15.6.

15.3 Driveline Speed Control

As stated in the introduction of this chapter, control of all rotational velocities is one of the two main objectives for driveline control. We will now approach anti-surge control as a major example of this type of control. The problem of vehicle surge or vehicle shuffle was introduced in Section 13.3.1, and illustrated in Figure 13.3. The phenomena around driveline oscillations were also the basis in the driveline modeling chapter, where Figure 14.2 was used as a basis for model development. The same truck that was used to obtain these measurements will be used in the design of anti-surge control based on speed control with active damping. The presentation is based on Pettersson and Nielsen (2003).

15.3.1 RQV control

A traditional control method is reviewed with a twofold purpose. One is to further describe the problem to be solved, but a more important aspect is that it has a specific driving feel. Incorporating such behavior, that customers may be used to, is not uncommon in automotive applications, so the presentation here lays the ground for Section 15.9.

For diesel engines, traditional speed control is often referred to as RQV control (Dietsche, 2011), where the driver's accelerator pedal position is interpreted as a desired engine speed. RQV control is essentially a proportional controller calculating the fuel amount as a function of the difference between the desired speed set by the driver and the actual measured engine speed. The reason for this controller structure is the traditionally used mechanical centrifugal governor for diesel pump control. Thus, the RQV controller has no information about the load, and a nonzero load, for example, when going uphill or downhill, gives a stationary error. The RQV controller is described by

$$u = u_0 + K_p(ri - \dot{\theta}_e) \tag{15.29}$$

where $i = i_t i_f$ is the conversion ratio of the driveline, K_p is the controller gain, and r is the reference velocity. The constant u_0 is a function of the speed but not the load, since this is not known. The problem with vehicle shuffle when increasing the controller gain, in order to increase the bandwidth, is demonstrated in the following example.

Example 15.1 Consider the truck modeled in Chapter 14 traveling at a speed of 2 rad/s (3.6 km/h) with gear 1 and a total load of 3000 Nm (≈ 2 % road slope). Let the new desired velocity be $r = 2.3$ rad/s. Figure 15.7 shows the RQV control law (15.29) applied to the Drive-shaft Model with three gains, K_p. In the plots, u_0 from (15.29) is calculated so that the stationary level is the same for the three gains. (Otherwise there would be a gain dependent stationary error.)

When the controller gain is increased, the rise time decreases and the overshoot in the wheel speed increases. Hence, there is a trade-off between short rise time and little overshoot. The engine speed is well damped, but the flexibility of the driveline causes the wheel speed to oscillate with higher amplitude the more the gain is increased.

The same behavior is seen in Figure 15.8, which shows the transfer functions from load and measurement disturbances, v and e, to the performance output, $z = \dot{\theta}_w$, when the RQV controller is used. The value of the resonance peak in the transfer functions increases when the controller gain is increased.

In Figure 15.7, the value u_0 from (15.29) is calculated so that the stationary level is the same for the three gains. Otherwise there would be a gain dependent stationary error. This means that the RQV controller will maintain the speed demanded by the driver, but with a stationary error (velocity lag), which is a function of the controller gain and the load (rolling resistance, air drag, and road inclination). Thus, the RQV control scheme gives a specific character to the driving feeling, for example, when going uphill and downhill, and a question is whether this can be maintained, which will treated in Section 15.9.

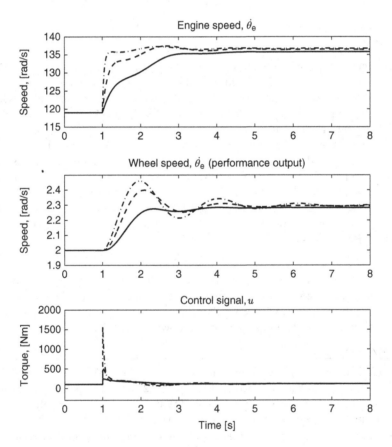

Figure 15.7 Response of step in accelerator position at $t = 1$ s, with RQV control (15.29) controlling Driveshaft Model. Controller gains $K_p=8$, $K_p=25$, and $K_p=85$ are shown in solid, dashed, and dash-dotted lines respectively. Increased gain results in a well damped engine speed and an oscillating wheel speed

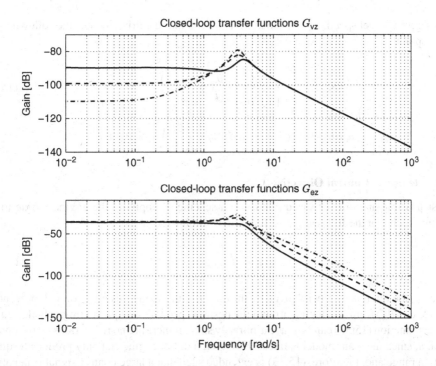

Figure 15.8 Closed-loop transfer functions G_{vz} and G_{ez} when using the RQV control law (15.29) for the controller gains $K_p=8$ (solid line), $K_p=25$ (dashed line), and $K_p=85$ (dash-dotted line). The resonance peaks increase with increasing gain

15.3.2 Formulating the Objective of Anti-Surge Control

The objective of anti-surge control will now be formulated along the lines given in Section 15.2.

Performance variable

Vehicle surge is described by wheel speed, which thus is the performance output, that is $z = \dot{\theta}_w$. This is formalized as in (15.11) and (15.12)

Stationary Point

The operating point, the stationary point to linearize around, $z = r$, is reached if a stationary control signal, u_0, is used. This torque is a function of the reference value, r, and the load, l.
    The stationary point is obtained by solving

$$Ax + Bu + Hl = 0 \tag{15.30}$$

for x and u, where A, B, and H are given by (15.2) to (15.5).

For a given wheel speed, $\dot{\theta}_w$, and load, l, it follows that the driveline has the following stationary point

$$x_0(\dot{\theta}_w, l) = \begin{pmatrix} b_2/k & 1/k \\ i & 0 \\ 1 & 0 \end{pmatrix} \begin{pmatrix} \dot{\theta}_w \\ l \end{pmatrix} = \delta_x \dot{\theta}_w + \delta_1 l \qquad (15.31)$$

$$u_0(\dot{\theta}_w, l) = \left((b_1 i^2 + b_2)/i \;\; 1/i \right) \begin{pmatrix} \dot{\theta}_w \\ l \end{pmatrix} = \lambda_x \dot{\theta}_w + \lambda_1 l \qquad (15.32)$$

First Attempt at Control Objective

A first possible attempt for speed control is a scheme of applying the engine torque to the driveline such that the following cost function is minimized

$$\lim_{T \to \infty} \int_0^T (z - r)^2 \qquad (15.33)$$

where r is the reference velocity given by the driver. Minimizing this criterion will prevent the velocity, $z = \dot{\theta}_w$, from oscillating around the desired speed, r, thus preventing vehicle shuffle. The cost function (15.33) can be made arbitrarily small if there are no restrictions on the control signal, u, since the plant model is linear. However, a diesel engine can only produce torque in a certain range and, therefore, (15.33) is extended such that a large control signal is penalized in the cost function.

Feasible Control Objective

The cost function is modified by using (15.31) and (15.32), such that a control signal that deviates from the stationary value $u_0(r, l)$ adds to the cost function. The extended cost function is given by

$$\lim_{T \to \infty} \int_0^T (z - r)^2 + \eta(u - u_0(r, l))^2 \qquad (15.34)$$

where η is used to control the trade-off between short rise time and control signal amplitude. The result will be an anti-surge speed controller with feasible control signals.

The controller that minimizes (15.34), has no stationary error, since the load, l, is included and thus compensated for. However, it is desirable that the stationary error characteristic for the RQV controller is maintained in the speed controller, as mentioned before. A stationary error comparable with that of the RQV controller can be achieved by using only a part of the load in the criterion (15.34), as will be demonstrated in Section 15.3.3.

15.3.3   Speed Control with Active Damping and RQV Behavior

We are now ready to obtain the anti-surge speed controller. This means computing l_0, K_c, and K_f in (15.15) and (15.16). The problem formulation (15.34) will be treated in two steps. First without RQV behavior, that is using the complete load in the criterion, and then extending to RQV behavior. This first step is done by linearizing the driveline model and rewriting (15.34) in terms of the linearized variables. The derived feedback law is a function of η which is chosen such that high bandwidth together with a feasible control signal is obtained.

The model (15.1)

$$\dot{x} = Ax + Bu + Hl \tag{15.35}$$

is affine since it includes a constant term, l. The model is linearized in the neighborhood of the stationary point (x_0, u_0). The linear model is described by

$$\Delta\dot{x} = A\Delta x + B\Delta u \tag{15.36}$$

where

$$\Delta x = x - x_0$$

$$\Delta u = u - u_0 \tag{15.37}$$

$$x_0 = x_0(x_{30}, l)$$

$$u_0 = u_0(x_{30}, l)$$

where the stationary point (x_0, u_0) is given by (15.31) and (15.32) (x_{30} is the initial value of x_3). Note that the linear model is the same for all stationary points.

Introduce

$$r_1 = Mx_0 - r \tag{15.38}$$

$$r_2 = u_0 - u_0(r, l)$$

The cost function (15.34) is then expressed in terms of Δx and Δu by using 15.11, (15.37), and 15.38

$$\lim_{T\to\infty} \int_0^T (M(x_0 + \Delta x) - r)^2 + \eta(u_0 + \Delta u - u_0(r, l))^2 \tag{15.39}$$

$$= \lim_{T\to\infty} \int_0^T (M\Delta x + r_1)^2 + \eta(\Delta u + r_2)^2 \tag{15.40}$$

The problem is now to find a feedback control law that minimizes this cost function.

An Excursion: Computing the Controller
It is outside the scope of this book to go into explicit control methodology, but one way of minimizing (15.34) using LQG technique, which allows using established techniques and software, will be given here. The example chosen is the truck modeled in Chapter 14 and Example 15, and the resulting controller is the one used for the plots in the rest of this chapter. (One may note that it is useful to be aware of the computational formulation using r_1, r_2.) In order to minimize (15.39) a Riccati equation is used. Then the constants r_1 and r_2 must be expressed in terms of state variables. This can be done by augmenting the plant model (A, B) with models of the constants r_1 and r_2. Since these models will not be controllable, they must be stable in order to solve the Riccati equation (Maciejowski, 1989).

A state-feedback matrix is derived that minimizes (15.34) by solving a Riccati equation. The derived feedback law is a function of η which is chosen such that high bandwidth together with *a feasible control signal is obtained. Therefore the model $\dot{r}_1 = \dot{r}_2 = 0$ is not used because the poles are located on the imaginary axis. Instead, the following models are used*

$$\dot{r}_1 = -\sigma r_1 \tag{15.41}$$

$$\dot{r}_2 = -\sigma r_2 \tag{15.42}$$

which with a low σ indicates that r is a slow-varying constant.

The augmented model is given by

$$A_r = \begin{pmatrix} & & 0 & 0 \\ & A & 0 & 0 \\ & & 0 & 0 \\ 0\ 0\ 0 & -\sigma & 0 \\ 0\ 0\ 0 & 0 & -\sigma \end{pmatrix}, \tag{15.43}$$

$$B_r = \begin{pmatrix} B \\ 0 \\ 0 \end{pmatrix}, \qquad x_r = (\Delta x^T \quad r_1 \quad r_2)^T \tag{15.44}$$

By using these equations, the cost function (15.39) can be written in the form

$$\lim_{T \to \infty} \int_0^T x_r^T Q x_r + R \Delta u^2 + 2 x_r^T N \Delta u \tag{15.45}$$

with

$$Q = (M \quad 1 \quad 0)^T (M \quad 1 \quad 0) + \eta (0 \quad 0 \quad 0 \quad 0 \quad 1)^T (0 \quad 0 \quad 0 \quad 0 \quad 1)$$
$$N = \eta (0 \quad 0 \quad 0 \quad 0 \quad 1)^T \tag{15.46}$$
$$R = \eta$$

The cost function (15.39) is minimized by using

$$\Delta u = -K_c x_r \tag{15.47}$$

with

$$K_c = Q^{-1}(B_r^T P_c + N^T) \tag{15.48}$$

where P_c is the stabilizing solution to the Riccati equation

$$A_r^T P_c + P_c A_r + R - (P_c B_r + N) Q^{-1} (P_c B_r + N)^T = 0 \tag{15.49}$$

The control law (15.47) becomes

$$\Delta u = -K_c x_r = - \begin{pmatrix} K_{c1} & K_{c2} & K_{c3} \end{pmatrix} \Delta x - K_{c4} r_1 - K_{c5} r_2 \tag{15.50}$$

With this controller the phase margin is guaranteed to be at least $60°$ with infinite amplitude margin (Maciejowski, 1989).

Active Damping Obtained

By using (15.37) and (15.38) the control law for the speed controller is written as

$$u = K_0 x_{30} + K_l l + K_r r - \begin{pmatrix} K_{c1} & K_{c2} & K_{c3} \end{pmatrix} x \qquad (15.51)$$

with

$$
\begin{aligned}
K_0 &= \begin{pmatrix} K_{c1} & K_{c2} & K_{c3} \end{pmatrix} \delta_x - K_{c4} M \delta_x + \lambda_x - K_{c5} \lambda_x \\
K_r &= K_{c4} + K_{c5} \lambda_x \\
K_l &= \begin{pmatrix} K_{c1} & K_{c2} & K_{c3} \end{pmatrix} \delta_l - K_{c4} M \delta_l + \lambda_l
\end{aligned}
\qquad (15.52)
$$

where δ_x, δ_l, λ_x, and λ_l are described in (15.31) and (15.32).

When the control law 15.51–15.52 is applied to Example 15 the controller becomes

$$u = 0.230 x_{30} + 4470 r + 0.125 l - \begin{pmatrix} 7620 & 0.0347 & 2.36 \end{pmatrix} x \qquad (15.53)$$

where $\eta = 5 \cdot 10^{-8}$ and $\sigma = 0.0001$ are used. A step-response simulation with the speed controller (15.53) is shown in Figure 15.9.

The rise time of the speed controller is shorter than for the RQV controller. Also the overshoot is less when using speed control. The driving torque is controlled such that the oscillations in the wheel speed are actively damped. This means that the controller applies the engine torque in such a way that the engine inertia works in the opposite direction of the oscillation. Then the engine speed oscillates, but the important wheel speed is well behaved, as seen in Figure 15.9. This means that even in a strong transient vehicle shuffle is avoided.

Extending with RQV Behavior

The RQV control scheme gives a specific character to the driving feeling, for example when going uphill and downhill. This driving character is possible to maintain when extending RQV control with active damping. Traditional RQV control was explained in Section 15.3.1. The RQV controller has no information about the load, l, and therefore a stationary error will be present when the load is different from zero. The speed controller (15.51) is a function of the load, and the stationary error is zero if the load is estimated and compensated for. There is, however, a demand from the driver that the load should give a stationary error, and only when using a cruise controller should the stationary error zero.

The speed controller can be modified such that a load different from zero gives a stationary error. This is done by using βl instead of the complete load l in (15.51). The constant β ranges from $\beta = 0$ which means no compensation for the load, to $\beta = 1$ which means full compensation of the load and no stationary error. The compensated speed control law becomes

$$u = K_0 x_{30} + K_l \beta l + K_r r - \begin{pmatrix} K_{c1} & K_{c2} & K_{c3} \end{pmatrix} x \qquad (15.54)$$

In Figure 15.10, the RQV controller with its stationary error (remember the reference value $r = 2.3$ rad/s) is compared to the compensated speed controller (15.54) applied to Example 15.1 for three values of β. By adjusting β, the speed controller with active damping is extended with a stationary error comparable with that of the RQV controller.

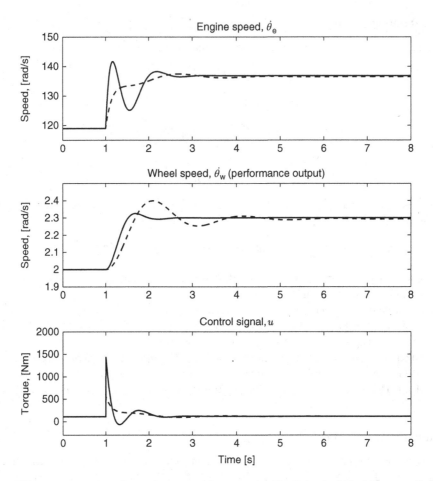

Figure 15.9 Response of step in accelerator position at $t = 1$ s. The Driveshaft Model is controlled with the speed control law (15.53) in solid lines. RQV control (15.29) with $K_p=25$ is seen in dashed lines. With active damping, the engine speed oscillates, resulting in a well damped wheel speed

Observer

The speed controller, (15.53) or (15.54), investigated in the previous section uses feedback from all states ($x_1 = \theta_e/i_t i_f - \theta_w$, $x_2 = \dot{\theta}_e$, and $x_3 = \dot{\theta}_w$). A sensor measuring shaft torsion (x_1) is normally not used, and therefore an observer is needed to estimate the unknown states. The observer for the estimated states $\hat{x}$ is given by

$$\Delta\dot{\hat{x}} = A\Delta\hat{x} + B\Delta u + K_f(\Delta y - C\Delta\hat{x}) \tag{15.55}$$

In the speed control law (15.51), the variables x are replaced by their estimates $\hat{x}$, and control law then becomes

$$u = K_0 x_{30} + K_r r + K_l l - \begin{pmatrix} K_{c1} & K_{c2} & K_{c3} \end{pmatrix}\hat{x} \tag{15.56}$$

with K_0, K_r, and K_l given by (15.52).

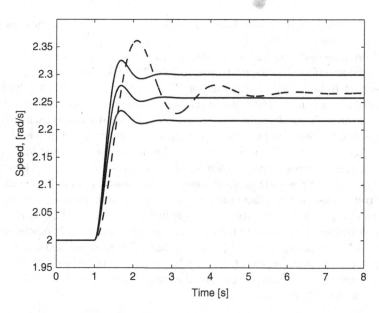

Figure 15.10 Wheel-speed response of step in accelerator position at $t = 1$ s. The Driveshaft Model is controlled with the RQV controller (15.29) in the dashed line, and the speed controller with stationary error (15.54) with $\beta = 0, 0.5, 1$ in the solid lines. The speed controller achieves the same stationary level as the RQV controller by tuning β

15.3.4   Influence from Sensor Location

The characteristics of the observer in the previous section, and thus of the closed loop system, depend on sensor location. Here, either the engine speed or the wheel speed is used as input to the observer, which as before results in different control problems. To illustrate the phenomena, once again the truck in Chapter 14 and Example 15.1 will be used. It is interesting to look at the difference, due to sensor location, in disturbance rejection, both from load disturbance, v, and measurement noise, e.

Excursion:
One way to find the observer gain K_f is to use the Kalman filter with

$$K_f = P_f C^T V^{-1} \tag{15.57}$$

where P_f is found by solving the Riccati equation

$$P_f A^T + A P_f - P_f C^T V^{-1} C P_f + W = 0 \tag{15.58}$$

The observer gain is then calculated using Loop-Transfer Recovery (LTR) (Maciejowski, 1989), using a parameter ρ. There turns out to be a large difference between ρ_e and ρ_w due to the structural difference between the two sensor locations, see Pettersson and Nielsen (2003) for details.

Influence on step response The observer dynamics is canceled in the transfer functions from reference value to performance output ($z = \dot{\theta}_w$) and to control signal (u). Hence, these transfer functions are not affected by sensor location. However, the observer dynamics is included in the transfer functions from disturbances v and e to both z and u, with consequences as in the following items.

Influence from load disturbances Figure 15.11 shows how the performance output and the control signal are affected by the load disturbance v. There is a resonance peak in G_{vz} when using feedback from the engine-speed sensor, which is not present when feedback from the wheel-speed sensor is used. The reason is that when using the wheel-speed sensor, the controller is canceled in the numerator, and when the engine-speed sensor is used, the controller is not canceled. When the zeros in the open-loop system are canceled by the controller, the open-loop zeros become poles in the controller. This means that the closed-loop system will have the open-loop zeros as poles when using the engine-speed sensor. The closed-loop poles become $-0.5187 \pm 3.0753j$, which causes the resonance peak in Figure 15.11.

Influence from Measurement Disturbances The influence from measurement disturbances e is shown in Figure 15.12. The roll-off rates at higher frequencies differ between the two feedback principles. This is due to the fact that the open-loop transfer functions $G_{u\dot{\theta}_e}$ and $G_{u\dot{\theta}_w}$ have different relative degrees. $G_{u\dot{\theta}_w}$ has a relative degree of two, and $G_{u\dot{\theta}_e}$ has a relative degree of one. Therefore, feedback from the wheel has a steeper roll-off rate. The difference in low-frequency level is equal to the conversion ratio of the driveline. Therefore, this effect increases with lower gears.

15.3.5   Load Estimation

The feedback law (15.56) is now completed with the unknown load as

$$ u = K_0 x_{30} + K_r r + K_l \hat{l} - \begin{pmatrix} K_{c1} & K_{c2} & K_{c3} \end{pmatrix} \hat{x} \tag{15.59} $$

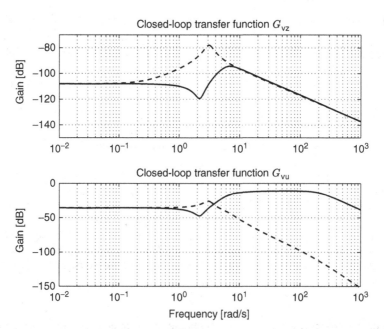

Figure 15.11 Closed-loop transfer functions from load disturbance, v, to performance output, z, and to control signal, u. Feedback from $\dot{\theta}_w$ is shown in solid lines and feedback from $\dot{\theta}_e$ is shown in dashed lines. With $\dot{\theta}_e$-feedback the transfer functions have a resonance peak, resulting from the open-loop zeros

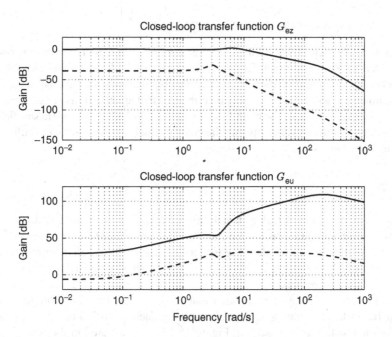

Figure 15.12   Closed-loop transfer functions from measurement noise, e, to performance output, z, and to control signal, u. Feedback from $\dot{\theta}_w$ is shown in solid lines and feedback from $\dot{\theta}_e$ is shown in dashed lines

where $\hat{l}$ is the estimated load. In order to estimate the load, the model used in the observer is augmented with a model of the load. The load is hard to model correctly since it is a function of road slope. However, it can be treated as a slow-varying constant. A reasonable augmented model is

$$x_4 = \hat{l}, \quad \text{with} \quad \dot{x}_4 = 0 \tag{15.60}$$

This gives

$$\dot{\hat{x}} = A_1\hat{x}_1 + B_1u + K_f(y - C_1\hat{x}_1) \tag{15.61}$$

with

$$\hat{x}_l = \begin{pmatrix} \hat{x} & \hat{l} \end{pmatrix}^T, \tag{15.62}$$

$$A_1 = \begin{pmatrix} & & 0 \\ A & & 0 \\ & & -1/J_2 \\ 0\ 0\ 0 & & 0 \end{pmatrix}, \tag{15.63}$$

$$B_1 = \begin{pmatrix} B \\ 0 \end{pmatrix}, \qquad C_l = \begin{pmatrix} C & 0 \end{pmatrix} \tag{15.64}$$

The feedback law is then

$$u = K_0x_{30} + K_r r - \begin{pmatrix} K_{c1} & K_{c2} & K_{c3} & -K_1 \end{pmatrix} \hat{x}_l \tag{15.65}$$

which together with the observer constitutes a **complete anti-surge controller.**

15.3.6   Evaluation of the Anti-Surge Controller

We now have a complete design of an anti-surge controller, and the function now needs to be verified under realistic conditions. To some extent this can be done in simulations, but for the main function experimental verification is needed. Both methods will be used in the following sections. Simulation will be used to study disturbances that are not so easy to generate. Experiments are used to demonstrate the main function of active damping of vehicle shuffle, and also to demonstrate that the controller fits into the control structure with its other ramifications, for example torque limits.

Simulations

An important step in demonstrating feasibility for real implementation is that a controller behaves well when simulated on a more complicated vehicle model than it was designed for. Even more important in a principle study is that such disturbances can be introduced that can hardly be generated in systematic ways in real experiments. One such example is impulse disturbances from a towed trailer.

Already, in Section 14.9.3, it was stated that a control law based on a reduced driveline model can be verified by simulating it with a more complete nonlinear model. This means that the simulation situation is as is seen in Figure 15.13. As illustrated in the figure, the design presented above, based on the Driveshaft Model, is simulated together with the more complicated Nonlinear Clutch and Driveshaft Model as vehicle model. The performance output for the speed controller is the wheel speed, $z = \dot{\theta}_w$, since the wheel speed rather than the engine speed determines vehicle behavior. Figure 15.14 shows the transfer functions from control signal (u) and load (l) to the wheel speed (z) for both the Driveshaft Model and the Clutch and Driveshaft Model. The Clutch and Driveshaft Model adds a second resonance peak originating from the clutch. Furthermore, the high frequency roll-off rate is steeper for the Clutch and Driveshaft Model than for the Driveshaft Model. Note that the transfer function from the load to the performance output is the same for the two models.

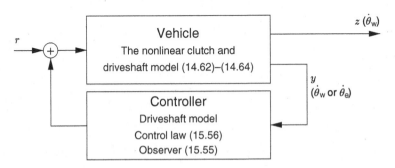

Figure 15.13   Simulation configuration. As a step for demonstrating feasibility for real implementation, the Nonlinear Clutch And Driveshaft Model is simulated with the controller based on the Driveshaft Model

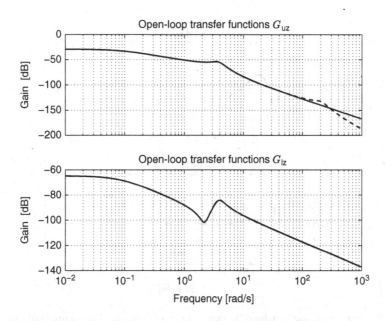

Figure 15.14 Transfer functions from control signal, u, and load, l, to performance output, z. The Driveshaft Model is shown as a solid line and the Clutch and Driveshaft Model is shown as a dashed line. The modeled clutch gives a second resonance peak and a steeper roll-off rate

15.3.7 Demonstrating Rejection of Load Disturbance

Without having to rebuild a vehicle, simulation can be used, for example to study effects from different sensor locations as discussed in Section 15.3.4. The structure in Figure 15.13 is used, which means that the Nonlinear Clutch and Driveshaft Model, given by (14.62) to (14.64), is used as the vehicle model. The steady-state level for the Nonlinear Clutch and Driveshaft Model is calculated by solving the model equations for the equilibrium point when the load and speed are known. The controller used is based on the Driveshaft Model, as was derived in the previous sections. The wheel speed or the engine speed is the input to the observer (15.55), and the control law (15.56) with $\beta = 0$ generates the control signal.

The simulation case presented here is the same as in Example 15.1, and shows the result of a load disturbance. The stationary point is given by

$$\dot{\theta}_w = 2, \quad l = 3000 \ \Rightarrow \ x_0 = \begin{pmatrix} 0.0482 & 119 & 2.00 \end{pmatrix}^T, \quad u_0 = 109 \qquad (15.66)$$

where (15.31) and (15.32) are used, and the desired new speed is $\dot{\theta}_w = 2.3$ rad/s. At steady state, the clutch transfers the torque $u_0 = 109$ Nm. This means that the clutch angle is in the area with higher stiffness ($\theta_{c1} < \theta_c \leq \theta_{c2}$) in the clutch nonlinearity, seen in Figure 14.11. This is a typical driving situation when speed control is used.

At $t = 6$ s, a load impulse disturbance is simulated. The disturbance is generated as a square pulse with 0.1 s width and 1200 Nm height, added to the load according to (15.6). Figure 15.15 shows the result of the simulation, and it is seen that the load impulse disturbance is better attenuated with feedback from the wheel-speed sensor, which is a verification of the behavior that was discussed in Section 15.3.4.

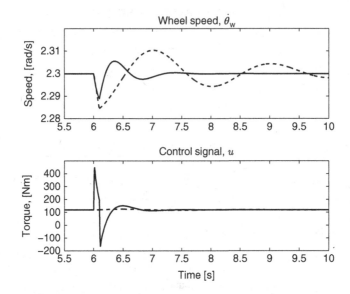

Figure 15.15 Simulated load impulse disturbance at $t = 6$ s. The solid line corresponds to $\dot{\theta}_w$-feedback and feedback from $\dot{\theta}_e$ is seen as the dashed line. The load impulse generates a control signal that damps the impulse disturbance when feedback from the wheel-speed sensor is used, but not with engine-speed feedback

A comment is that the area with low stiffness in the clutch nonlinearity ($\theta_c < \theta_{c1}$) was never entered in this simulation. One should observe that at low clutch torques ($\theta_c < \theta_{c1}$) the clutch nonlinearity can produce limit cycle oscillations. This situation occurs, for example, when the truck is traveling downhill with a load of the same size as the friction in the driveline, resulting in a low clutch torque.

15.3.8 Experimental Verification of Anti-Surge Control

Field trials are used to demonstrate that the method is applicable for real implementations in a truck. The sensors found in Table 14.1 are used. The speed control strategy is implemented by discretizing the feedback law and the observer. The repetition rate of the algorithm is chosen to be the same as the sampling rate of the input variable to the observer. This means that the sampling rate is 50 Hz using feedback from the engine-speed sensor. The controller parameters are tuned for the practical constraints given by the measured signals.

An almost flat test road was used with a minimum of changes from test to test. The focus of the tests was low gears, with low speeds and thus little impact from air drag. Reference values were generated by the computer to generate the same test situation from time to time. Only one direction of the test road was used so that there was no difference in road inclination. Step response tests in engine speed were performed, and the results were compared to the traditionally used RQV controller for speed control. The test presented here is a velocity step response from 2.1 rad/s to 3.6 rad/s (about 1200 RPM to 2000 RPM) with gear 1. In Figure 15.16, the speed controller is compared to traditional RQV control. The engine torque, the engine speed,

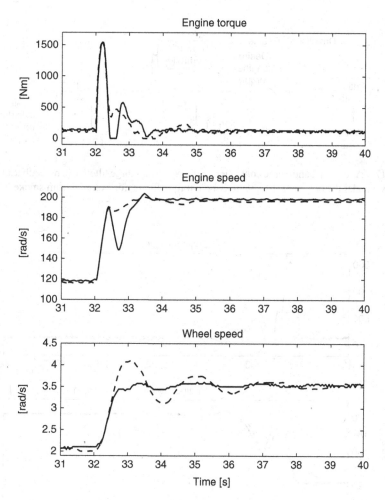

Figure 15.16 Speed step at $t = 32$ s with active damping and engine-speed feedback (solid line) compared to traditional RQV control with K_p=50 (dashed line). Experiments are performed on a flat road. After 32.5 s, the control signals differ depending on the control scheme. With speed control, the engine inertia works in the opposite direction of the oscillations, which are significantly reduced

and the wheel speed are shown. The speed controller uses feedback from the engine speed, and the RQV controller has the gain $K_p = 50$. With this gain the rise-time and the peak torque output are about the same for the two controllers.

With RQV control, the engine speed reaches the desired speed but the wheel speed oscillates, as in the simulations made earlier. Speed control with active damping significantly reduces the oscillations in the wheel speed. This means that the controller applies the engine torque in such a way that the engine inertia works in the opposite direction of the oscillation. This gives an oscillating engine speed, according to Figure 15.16. Hence, it is demonstrated that the assumption about the simplified model structure (Driveshaft Model) is sufficient for control design.

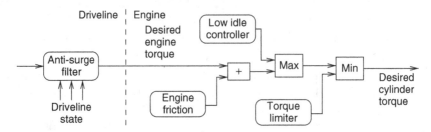

Figure 15.17   A feedback anti-surge controller, in the box "Anti-surge filter" where feedback is depicted by the arrows under the box, must be able to handle torque limits, for example from smoke limitation

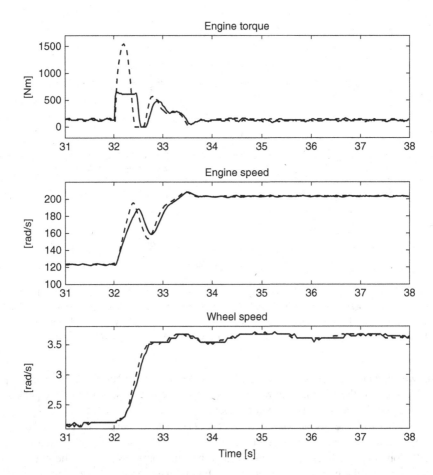

Figure 15.18   Speed control with active damping with and without a smoke limiter. The dashed lines correspond to experiments without a smoke limiter. At $t = 32$ s, a speed step is commanded. When using smoke limiter, a reduced torque level is obtained, and the resulting behavior is showed by the solid lines. As seen, the case with torque limitations is also well handled and active damping is obtained

15.3.9   Experiment Eliminating a Misconception

Most automotive controllers have to work in a complex architecture of other controllers. In the beginning of this chapter, it was noted in Section 15.1.6 that one important requirement is the handling of torque limits. There are several reasons for that, and one reason is that the controller should fit into the torque-based control structure.

Recall Figure 15.2, that is repeated here as Figure 15.17. One important example of torque limitation in diesel engines is the smoke limiter, and Figure 15.18 shows the behavior when the smoke limiter has been active. Looking at the upper plot it is seen that the torque is substantially limited. The important point to be noted here is that the torque limitation is not included in the control design, neither via a model nor any other way, and there is a misconception that such non-linearities that are not considered in the design will cause problems when using the controller in a real setting. The anti-surge speed controller is derived based on the Driveshaft Model, and the behavior is shown in Figure 15.16. This behavior is again shown with dashed lines in Figure 15.18. The torque limitation enters as illustrated in Figure 15.17, and the result of that is shown by solid lines in Figure 15.18. The upper plot shows how the torque is limited and, nevertheless, the lower plot with the wheel speed shows that the situation is well handled and active damping is still obtained. As long as the observer (15.61) is fed with the correct input signal, u, the feedback (15.65) will counteract the oscillations. This clearly demonstrates active damping both with and without a smoke limiter, and it further illustrates the advantages with a feedback solution. Since it is based on the situation at hand, that is on the state of the driveline, it is able to work well under a wide variety of both foreseen and unforeseen circumstances.

15.4   Control of Driveline Torques

Besides controlling the rotational states of the driveline, the other main class of problems is the control of torques along the driveline. There are two main approaches

- **A general approach** with explicit modeling of the torque to be controlled. A control criterion is formulated, and if that is done well a good solution will be obtained using established control theory. This method of explicit torque control is optimal from a theoretical point of view, but more parameters need to be estimated and tracked compared to the other method, called torsion control.
- Torsion control is **an approach based on the physical insight** that if the main torsion of the driveline, is controlled, then the true torque of interest will be close to its desired value. Fewer parameters are needed and the necessary observer is more direct to device (or already available). On top of that a simpler control structure can be used once a good torsion observer is available.

Both these methods will be treated for the case of driveline control for gear shifting.

Mode Shifts

Torque control is crucial for fast mode shifts in the driveline. Typical applications are for mode shifts in hybrid vehicles, for example some switches between primary movers, or for gear shifting. Typical objectives are to make the mode shifts fast and smooth to optimize performance and driving feel. As a consequence, a good solution usually also minimizes noise, wear, and

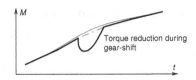

Figure 15.19 A schematic illustration of torque control for a gear shift showing torque demand (thin line), physically possible torque (dash-dotted line), and actually delivered torque (thick line). A real gearshift is seen in Figure 15.20

other unwanted effects. As described at the beginning of this chapter, see Section 15.1, this means that the mode shift is a short-term intervention modifying the propagated torque in the torque-based control structure, as illustrated in Figure 15.4 and repeated in Figure 15.19.

15.4.1   Purpose of Driveline Torque Control for Gear Shifting

The short-term intervention sketched in Figure 15.19 will now be treated in more detail using the real gear shift shown in Figure 15.20. Section 15.1.4 and Figure 15.3 introduced the concept of an automatically controlled manual gear box, that is an automated manual transmission (AMT), and this is a popular concept for heavy trucks. For an AMT engaging and disengaging of gears are enabled by engine torque control, and in Figure 15.20 a typical AMT gear-shifting process is illustrated, in terms of engine torque and speed. Since it is data from a real truck, the torque was not measured, so instead the fuel control signal, u_f, is shown.

When using gear shifting by engine control, there are certain phases. First, control is transferred from the driver to the control unit, entering the *torque control phase*. The engine is controlled to a torque level corresponding to zero transferred torque in the transmission. This is shown in the figure as ramping down the fueling with slope $-C_{ramp}$ reaching the level of zero transmission torque (zt) corresponding to the fueling level $u_{f,zt}$. After the neutral gear is engaged, the *speed synchronization phase* is entered. This is the period shown as t_{adjust}. Then the fueling, $u_{f,shift}$, accelerate the engine to the speed, $\omega_{e,post}$, corresponding to the gear ratio for the new gear. Following that, at time 1, the control turns back to the fueling level, $u_{f,zt}$, for zero torque, whereafter the new gear is engaged. Finally, the torque level is transferred back to the level of the torque demand before the intervention (usually representing the driver demand). In this case, this is done by gradually ramping up the fueling with slope C_{ramp}, and here it concludes with a soft approach with a lower slope, $C_{ramp,sl}$.

15.4.2   Demonstration of Potential Problems in Torque Control

It is desirable to minimize the total time needed for a gear shift not to lose traction. The principle in Figure 15.20, using cautious ramping with slope $-C_{ramp}$, gives a smooth transfer but takes time. With better torque control one could change the torque to the zero torque level much faster. This is the scope of torque control with active damping, and as a background it will now be demonstrated what problems occur if torque is changed abruptly without appropriate control.

Already in Section 13.3.3, Figure 13.5 showed the behavior when the neutral gear is engaged, without a torque control phase, at a constant speed. This means that there is a driving torque

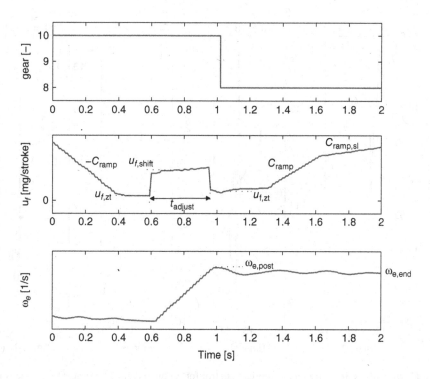

Figure 15.20   A gear shift from the 10th to 8th gear of a 37-ton truck on a 6.5% uphill gradient

transferred in the transmission, which clearly causes the transmission speed to oscillate. The amplitude of the oscillations increases the higher the stationary speed is.

Gear Shifts without Appropriate Control

Figure 15.21 shows the transmission speed when the engine torque is decreased to 46 Nm at 12.0 s. Prior to that, the stationary speed of 2200 RPM was maintained, which requested an engine torque of about 225 Nm. Four trials are performed with this torque profile, with engaged neutral gear at different time delays after the torque step. After 12.4 s there is a small oscillation in the transmission speed, after 13.3 s and 14.8 s there are oscillations with high amplitude, and at 13.8 s there are no oscillations in the transmission speed. This indicates how driveline resonances influence the transmission torque, which is clearly close to zero for the gear shift at 12.4 s and 13.8 s, but different from zero at 13.3 s and 14.8 s. The amplitude of the oscillating transmission torque will be higher if the stationary speed is increased or if the vehicle is accelerating.

The Importance of Driveline State

The previous section and Figure 15.21 clearly show the problems that need to be handled, but one might ask what their main cause is. It could potentially be nonlinearities, noise, or something else. Figures 15.22–15.23 will cast some light on this. These figures are simulations of

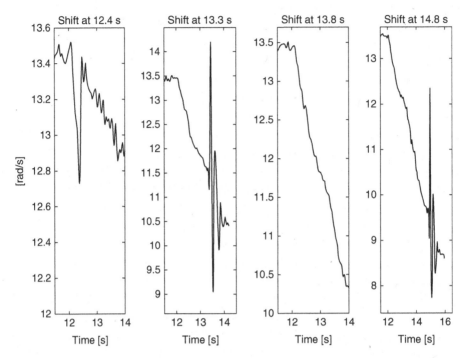

Figure 15.21 Gear shifts without appropriate control for problem demonstration. The engine is at the stationary speed 2200 RPM with gear 1. At 12.0 s there is a decrease in engine torque to 46 Nm in order to reach zero transmission torque. The transmission speed is plotted when the neutral gear is engaged at 12.4 s, 13.3 s, 13.8 s, and 14.8 s (with the same torque profile). The different amplitudes in the oscillations show how the torque transmitted in the transmission oscillates after the torque step. Note that the range of the vertical axes differ between the plots

the model of the decoupled driveline in Section 14.6.2 with parameters that were validated in Figure 14.17. These figures show the influence of initial conditions, that is the state of the driveline, at an intervention. In Figure 15.22 the dashed line shows that there are initial conditions that do not give any oscillations at all, but for slightly different, still realistic, conditions there are substantial oscillations. Figure 15.23 also shows the typical complex behavior that can be experienced in experiments and field trials.

Negative Consequences

A lack of appropriate control leads to the following problems:

- Disturbance to the driver, both in terms of noise and speed impulse.
- Increased wear on transmission.
- Increased time for the speed synchronization phase, since the transmission speed, which is the control goal, is oscillating. The oscillations are difficult to track for the engine and therefore one has to wait until they are sufficiently damped out.

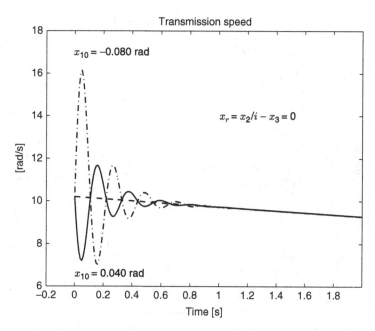

Figure 15.22   Transmission speed oscillations after engagement of neutral gear. The decoupled model is simulated with different initial values of the driveshaft torsion, x_{10}. The relative speed difference, $x_r = x_2/i - x_3$, is equal to zero. The dashed line with no oscillation has the initial value $x_{10} = -0.004$ rad. The higher the driveshaft torsion is, the higher the amplitude of the oscillation will be

15.4.3   Approaches to Driveline Torque Control for Gear Shifting

The specific problem of gear shifting can now be related back to the two approaches at the beginning of Section 15.4. This gives the following two methods:

- For the example of driveline control for gear shifting, the general approach will require the transmission torque to be zero, and it is thus named **Transmission torque control**, presented in Section 15.5.
- For the example of driveline control for gear shifting, torsion control of the Driveshaft Model will lead to **Driveshaft torsion control**, see Section 15.6.

Both these approaches will be treated in the following text. The Transmission torque control will be formulated to illustrate the steps in the general method to obtain a well-formulated problem. In order to implement transmission torque control and validate the results, the unknown parameters describing the transmission torque must be estimated for each gear, which is a non-trivial task. Driveshaft torsion control is easier in this sense and is treated thereafter. It is shown to work well both in simulation and experiments. The Transmission torque control also works well (naturally, since it is more general), so the main conclusion is that both methods work well controlling the torque with active damping, and they both solve the problem of avoiding unwanted oscillations during gear shifts.

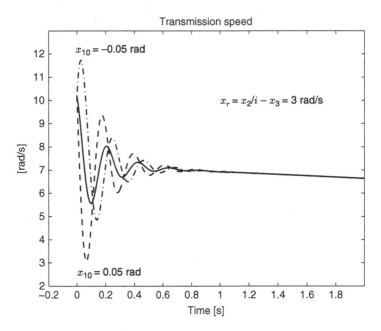

Figure 15.23 Transmission speed oscillations after engagement of neutral gear. The decoupled model is simulated with a relative speed difference, $x_r = x_2/i - x_3$, equal to 3 rad/s, and with three different initial values of the driveshaft torsion, $x_{10} = -0.050$ rad (dash-dotted line), $x_{10} = -0.004$ rad (solid line), and $x_{10} = 0.050$ rad (dashed line)

15.5 Transmission Torque Control

The topic of this chapter is to illustrate the general method for driveline torque control, where the control strategy is based on a model of the torque to be controlled. The example used here is transmitted torque in the transmission, which should be zero in order to engage neutral gear. Thus, a transmission-torque controller is derived that controls the estimated transmission torque to zero while having engine controlled damping of driveline resonances. With this approach, from Pettersson et al. (1997), the specific transmission-torque behavior for each gear is described and compensated for.

A model of the transmission is developed in Section 15.5.1, where the torque transmitted in the transmission is modeled as a function of the states and the control signal of the Driveshaft Model. The controller goal is formulated in mathematical terms as a gear-shift control criterion in Sections 15.5.2–15.5.4, and this is a key point since a great deal of insight is needed for the best result. The control law in Section 15.5.5 minimizes the criterion.

15.5.1 Modeling of Transmission Torque

The performance output, z, for torque control for gear-shifting is the transmission torque transferred between the cogwheels in the transmission. A simplified model of the transmission is depicted in Figure 15.24. The input shaft is connected to bearings with a viscous friction component b_{t1}. A cogwheel is mounted at the end of the input shaft, which is connected to a

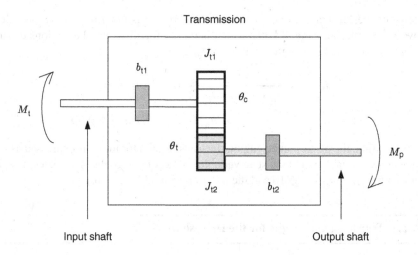

Figure 15.24 Simplified model of the transmission with two cogwheels with conversion ratio i_t. The cogwheels are connected to the input and output shafts respectively, via bearings. The torque transmitted between the cogwheels is the transmission torque, z

cogwheel mounted on the output shaft. The conversion ratio between these is i_t, as mentioned in Chapter 14. The output shaft is also connected to bearings with the viscous friction component b_{t2}. By using Newton's second law, the transmission is then modeled by the following two equations

$$J_{t1}\ddot{\theta}_c = M_t - b_{t1}\dot{\theta}_c - z \tag{15.67}$$

$$J_{t2}\ddot{\theta}_t = i_t z - b_{t2}\dot{\theta}_p - M_p \tag{15.68}$$

The expression for the transmission torque can be derived for any driveline model in Chapter 14, for example the Driveshaft Model or the Clutch and Driveshaft Model. The derivation steps follow the methodology from Section 14.1, and are here illustrated only for the Driveshaft Model.

Transmission Torque for the Driveshaft Model

The Driveshaft Model is defined by (14.39) and (14.40). The model is here extended by the model of the transmission depicted in Figure 15.24, and the expression for the transmission torque is derived. By using the equation describing the engine inertia (14.1)

$$J_e\dot{\theta}_e = M_e - M_c \tag{15.69}$$

together with (14.26)

$$M_c = M_t, \qquad \theta_e = \theta_c \tag{15.70}$$

(15.67) is expressed in terms of engine speed

$$(J_e + J_{t1})\dot{\theta}_e = M_e - b_{t1}\dot{\theta}_e - z \tag{15.71}$$

To describe the performance output on the form 15.10, that is in terms of state variables, the derivative $\dot{\theta}_e$ is replaced using (14.39), which is one of the differential equations describing the Driveshaft Model. This, together with $u = M_e$, gives

$$u - b_{t1}\dot{\theta}_e - z = \frac{J_e + J_{t1}}{J_e + J_t/i_t^2 + J_f/i_t^2 i_f^2}(u - (b_t/i_t^2 + b_f/i_t^2 i_f^2)\dot{\theta}_e \tag{15.72}$$

$$-k(\theta_e/i_t i_f - \theta_w)/i_t i_f - c(\dot{\theta}_e/i_t i_f - \dot{\theta}_w)/i_t i_f)$$

From this equation, the performance output, the transmission torque, z, is expressed as a function of the control signal, u, and the state variables, x, according to the state-space description (15.2) to (15.5). The form of 15.10 and the notation from (15.5) are used.

Model 15.1 Transmission Torque for the Driveshaft Model

$$z = Mx + Du \qquad \text{with}$$

$$M^T = \begin{pmatrix} \frac{(J_e + J_{t1})k}{J_1 i} \\ \frac{J_e + J_{t1}}{J_1}(b_1 + c/i^2) - b_{t1} \\ -\frac{(J_e + J_{t1})c}{J_1 i} \end{pmatrix} \tag{15.73}$$

$$D = 1 - \frac{J_e + J_{t1}}{J_1}$$

Obtaining the Model Parameters

Most of the parameters in (15.73) were estimated in Chapter 14, but there are new unknown parameters $J_e + J_{t1}$ and b_{t1}. These can be obtained by different estimation techniques, but as an alternative one can find expressions based on an approximative analysis as will be sketched now.

In the derivation of the Driveshaft Model in Chapter 14 the performance output, z, is eliminated. If z is eliminated in (15.67) and (15.68) and (15.70) is used, the equation for the transmission is

$$(J_{t1}i_t^2 + J_{t2})\dot{\theta}_e = i_t^2 M_c - i_t M_p - (b_{t1}i_t^2 + b_{t2})\dot{\theta}_e \tag{15.74}$$

By comparing this with the equation describing the transmission in Chapter 14, (14.29)

$$J_t\dot{\theta}_e = i_t^2 M_c - b_t\dot{\theta}_e - i_t M_p \tag{15.75}$$

the following equations relating the parameters are obtained

$$J_t = i_t^2 J_{t1} + J_{t2} \tag{15.76}$$

$$b_t = i_t^2 b_{t1} + b_{t2} \tag{15.77}$$

It is arbitrarily assumed that the gear shift divides the transmission into two equal inertias and viscous friction components, giving

$$J_{t1} = J_{t2} \tag{15.78}$$

$$b_{t1} = b_{t2}$$

(15.76) and (15.77) then reduce to

$$J_{t1} = \frac{J_t}{1 + i_t^2} \tag{15.79}$$

$$b_{t1} = \frac{b_t}{1 + i_t^2} \tag{15.80}$$

The following parameters from the Driveshaft Model were estimated in Chapter 14

$$J_1 = J_e + J_t/i_t^2 + J_f/i_t^2 i_f^2 \tag{15.81}$$

$$b_1 = b_t/i_t^2 + b_f/i_t^2 i_f^2 \tag{15.82}$$

with notations as in (15.5). From (15.79) and (15.81) $J_e + J_{t1}$ is derived as

$$
\begin{aligned}
J_e + J_{t1} &= J_e + \frac{J_t}{1 + i_t^2} = J_e + \frac{i_t^2}{1 + i_t^2}(J_1 - J_e - J_f/i_t^2 i_f^2) \\
&= J_e \frac{1}{1 + i_t^2} + J_1 \frac{i_t^2}{1 + i_t^2} - J_f \frac{1}{i_f^2(1 + i_t^2)}
\end{aligned} \tag{15.83}
$$

A combination of (15.80) and (15.82) gives b_{t1}

$$b_{t1} = \frac{b_t}{1 + i_t^2} = \frac{i_t^2}{1 + i_t^2}(b_1 - b_f/i_t^2 i_f^2) \tag{15.84}$$

For low gears i_t has a large value. This together with the fact that J_f and b_f are considerably less than J_1 and b_1 gives the following approximation about the unknown parameters

$$J_e + J_{t1} \approx J_1 \frac{i_t^2}{1 + i_t^2} \tag{15.85}$$

$$b_{t1} \approx b_1 \frac{i_t^2}{1 + i_t^2} \tag{15.86}$$

There were arbitrary assumptions on the way to obtaining the final approximations, but the formulas should give a right order estimate, and they give a grasp of the principle dependence on i_t. Using such approximations should normally be more cost effective than performing the somewhat tricky experiments to obtain the true parameters.

15.5.2   Transmission-Torque Control Criterion

As indicated in Section 15.2, the control objective is formulated as a criterion, see (15.13). Here this criterion formulation is crucial, as it was when formulating anti-surge control in Section 15.3.2.

The problem of torque control for gear-shifting is to minimize the transmission torque, z, with a control signal, u, that is possible to realize by the diesel engine. Therefore, the criterion consists of two terms. The first term is z^2, which describes the deviation from zero transmission torque. The second term describes the deviation in control signal from the level needed to obtain $z = 0$. Let this level be u_{shift}, which will be speed-dependent as described later. Then the criterion is described by

$$\lim_{T \to \infty} \int_0^T z^2 + \eta(u - u_{\text{shift}})^2 \tag{15.87}$$

The controller that minimizes this cost function will utilize engine controlled damping of driveline resonances (since z^2 is minimized) in order to obtain $z = 0$. At the same time, the control signal is prevented from having large deviations from the level u_{shift}. The trade-off is controlled by tuning the parameter η.

In the following subsections, the influence from the two terms in the criterion (15.87) will be investigated, and then how they can be balanced together for a feasible solution by tuning the parameter η. The truck modeled in Chapter 14 is used in the quantified examples.

Unconstrained Active Damping

The influence from the first term in the criterion (15.87) is investigated by minimizing z^2. The performance output, $z = Mx + Du$, is derived in (15.73), and $z = 0$ is guaranteed by solving $Mx + Du = 0$ for u. The result is

$$u = -D^{-1}Mx \tag{15.88}$$

This control law is called *unconstrained active damping* and the reason for this is illustrated in the following example.

Example 15.2   Consider the 144L truck modeled in Chapter 14 traveling at a speed of 3 rad/s (5.4 km/h) with gear 1 and a total load of 3000 Nm ($\approx 2~\%$ road slope).

Figure 15.25 shows the resulting transmission torque, the control signal, the engine speed, and the wheel speed, when a gear shift is commanded at t=1 s, with the control signal chosen according to (15.88). Unconstrained active, damping is achieved, which obtains $z = 0$ instantaneously. The wheel speed decreases linearly, while the engine speed is oscillating.

Unconstrained active damping (15.88) fulfills the control goal, but generates a control signal that is too large for the engine to generate. It can be noted that despite $z = 0$ being achieved this is not a stationary point, since the speed is decreasing. This means that the vehicle is free-rolling, which can be critical if it lasts too long.

15.5.3   Gear-shift Condition

The influence from the second term in the criterion (15.87) is investigated by minimizing $(u - u_{\text{shift}})^2$, resulting in the control law

$$u = u_{\text{shift}} \tag{15.89}$$

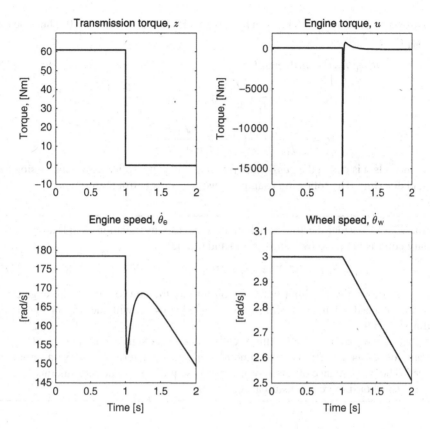

Figure 15.25 Unconstrained active damping of the Driveshaft Model. At $t = 1$ s, a gear shift is commanded and the control law (15.88) calculates the engine torque such that the transmission torque is driven to zero instantaneously. The oscillations in the transmission torque are damped with an unrealizable large control signal. The wheel speed decreases linearly

where the torque level u_{shift} is the control signal needed to obtain zero transmission torque, without using active damping of driveline resonances. Hence, u_{shift} can be derived from a rigid driveline model, by solving for $z = 0$ in the following way.

The differential equation describing the rigid driveline is derived by using the Driveshaft Model in (14.39) and (14.40), and eliminating the torque transmitted by the drive shaft, $k(\theta_e/i - \theta_w) + c(\dot\theta_e/i - \dot\theta_w)$. Then, by using $\dot\theta_e = \dot\theta_w i$ (i.e., rigid driveline), the result is

$$(J_1 i + J_2/i)\dot\theta_w = u - (b_1 i + b_2/i)\dot\theta_w - l/i \tag{15.90}$$

(15.71) expressed in terms of wheel speed is

$$z = u - b_{t1} i\dot\theta_w - (J_e + J_{t1})i\ddot\theta_w \tag{15.91}$$

Combining (15.90) and (15.91) gives, for the rigid driveline,

$$z = (1 - \frac{(J_e + J_{t1})i^2}{J_1 i^2 + J_2})u - (b_{t1} i - \frac{(J_e + J_{t1})i}{J_1 i^2 + J_2}(b_1 i^2 + b_2))\dot\theta_w + \frac{(J_e + J_{t1})i}{J_1 i^2 + J_2}l \tag{15.92}$$

The control signal to force $z = 0$ is given by solving (15.92) for u while $z = 0$. Thus, the torque level u_{shift} is

$$u_{\text{shift}}(\dot{\theta}_w, l) = \mu_x \dot{\theta}_w + \mu_l l \text{ with} \tag{15.93}$$

$$\mu_x = (b_{t1} i - \frac{(J_e + J_{t1})i}{J_1 i^2 + J_2}(b_1 i^2 + b_2))(1 - \frac{(J_e + J_{t1})i^2}{J_1 i^2 + J_2})^{-1}$$

$$\mu_l = -\frac{(J_e + J_{t1})i}{J_1 i^2 + J_2}(1 - \frac{(J_e + J_{t1})i^2}{J_1 i^2 + J_2})^{-1}$$

This control law is called the *gear-shift condition*, since it implies zero transmission torque for a rigid driveline. The following example illustrates the control performance.

Example 15.3   Consider the 144L truck in the same driving situation as in Example 15. The stationary point is obtained by using (15.31) and (15.32).

$$x_{30} = 3, \, l = 3000 \Rightarrow x_0 = (0.0511 \ 178 \ 3.00) \, , \, u_0 = 138 \cdot \tag{15.94}$$

Figure 15.26 shows the resulting transmission torque, the control signal, the engine speed, and the wheel speed when a gear shift is commanded at t=1 s, with the control signal chosen according to (15.93).

This control law achieves $z = 0$ with a realizable control signal, but the oscillations introduced are not damped. Therefore, the time needed to obtain zero transmission torque is not optimized. The performance of this approach is worse if the driveline is oscillating at the time of the gear shift, or if there are disturbances present.

15.5.4   Final Control Criterion

The final control criterion for the transmission-torque controller is obtained by including (15.93) in the cost criterion so that (15.87) becomes

$$\lim_{T \to \infty} \int_0^T z^2 + \eta(u - u_{\text{shift}}(\dot{\theta}_w, l))^2$$

$$= \lim_{T \to \infty} \int_0^T (Mx + Du)^2 + \eta(u - \mu_x \dot{\theta}_w - \mu_l l)^2 \tag{15.95}$$

If the driveline is rigid, there is no difference between the two terms in the cost function (15.95). Furthermore, the condition at which the cost function is zero is not a stationary point, since the speed of the vehicle will decrease because $z = 0$ and $u = u_{\text{shift}}$ implies that the vehicle is free rolling.

15.5.5   Resulting Behavior–Feasible Active Damping

Transmission torque control is now obtained by solving (15.95) for a control law. The derived feedback law is a function of η, which is chosen such that high bandwidth together with a feasible control signal is obtained. Basically, the problem is to tune η so that the response is fast but the unfeasibly high torques in Figure 15.25 are avoided.

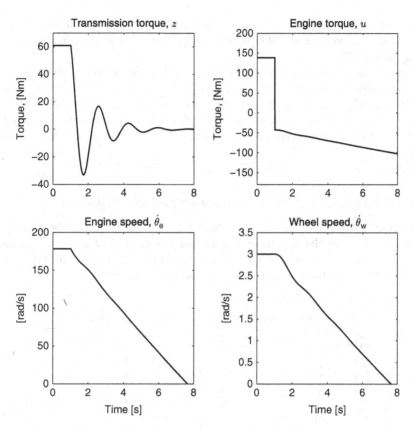

Figure 15.26 The Driveshaft Model controlled with the gear-shift condition (15.93). At $t = 1$ s, a gear shift is commanded. The speed dependent realizable control signal drives the transmission torque to zero. Undamped oscillations in the transmission torque increase the time needed to fulfill the goal of controlling the transmission torque to zero

The result is a controller in the general form

$$u = K_0 x_{30} + K_1 l - \begin{pmatrix} K_{c1} & K_{c2} & K_{c3} \end{pmatrix} x \tag{15.96}$$

Thus, the solution to the gear-shift criterion (15.95) is the transmission-torque controller (15.96), which obtains active damping with a realizable control signal. A detailed derivation using LQG can be found in Pettersson et al. (1997) and Kiencke and Nielsen (2005). The parameter η is tuned to balance the behavior of the unconstrained active damping solution (15.88) and the gear-shift condition (15.93), and a transmission-torque controller with tuned η is presented in the following example.

Example 15.4 Consider the 144L truck in the same driving situation as in Example 15. The transmission-torque controller (15.96) becomes

$$u = 2.37 \cdot 10^{-4} x_{30} - 0.0327\, l - \begin{pmatrix} 4.2123 & 0.0207 & -1.2521 \end{pmatrix} x \tag{15.97}$$

where $\eta = 0.03$ was used.

Figure 15.27 shows the resulting transmission torque, the control signal, the engine speed, and the wheel speed when a gear shift is commanded at t=1 s, with the control signal chosen according to (15.97).

The transmission-torque controller achieves $z = 0$ with a realizable control signal. The oscillations in the driveline are damped, since the controller forces the engine inertia to work in the opposite direction to the oscillations. Therefore, the time needed for the torque control phase and the speed synchronization phase is minimized, since resonances are damped and engagement of neutral gear is commanded at a torque level giving no oscillations in the transmission speed.

15.5.6   Validating Simulations and Sensor Location Influence

As in the case of the speed controller in Section 15.3.6, the feasibility of the gear-shift controller is studied by simulating a more complicated vehicle model than it was designed for. Additionally here, disturbances that are difficult to systematically generate in real experiments are treated in the simulations, especially to load and measurement disturbances.

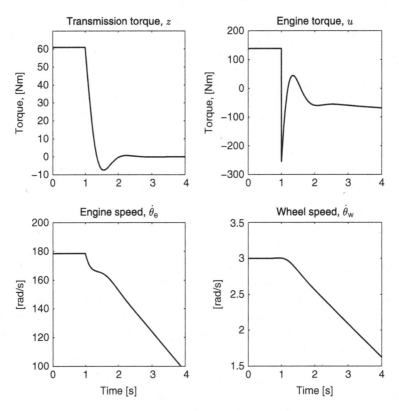

Figure 15.27   The Driveshaft Model controlled with the transmission-torque controller (15.97), solving the gear-shift criterion (15.95). At $t = 1$ s, a gear shift is commanded. A realizable control signal is used such that the transmission torque is driven to zero, while oscillations are actively damped

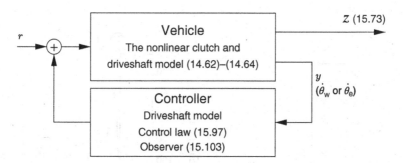

Figure 15.28   Simulation configuration. The Nonlinear Clutch and Driveshaft Model is simulated with the controller based on the Driveshaft Model

Simulation Setup

The control design is simulated with the Nonlinear Clutch and Driveshaft Model, according to Figure 15.28. The effects from different sensor locations are also studied in accordance with the discussion made in Section 15.3.4. The transmission-torque controller used is based on the Driveshaft Model, and was developed in the previous sections. The wheel speed or the engine speed is input to the observer (15.103), and the control law (15.97) generates the control signal.

Scenarios

Three simulations are performed with the driving situation as in Example 15.4, (i.e., with wheel speed $\dot{\theta}_w = 3$ rad/s, and load $l = 3000$ Nm). In the simulations, a gear shift is commanded at $t = 2$ s. The first simulation is without disturbances. In the second simulation, the driveline is oscillating prior to the gear shift. The oscillations are a result of a sinusoid disturbance acting on the control signal. The third gear shift is simulated with a load impulse at $t = 3$ s. The disturbance is generated as a square pulse with 0.1 s width and 1200 Nm height.

Results

Figure 15.29 shows the simulation without any disturbances. This plot should be compared to Figure 15.27 in Example 15.4, where the design is tested on the Driveshaft Model. The result is that the performance does not critically depend on the simplified model structure. The design still works if the extra nonlinear clutch dynamics is added. In the simulation, there are different results depending on which sensor is used. The model errors between the Driveshaft Model and the Nonlinear Clutch and Driveshaft Model are better handled when using the wheel-speed sensor. Neither of the sensor alternatives reaches $z = 0$, which is due to low-frequency model errors related to damping coefficients. In Figure 15.30 the simulation with driveline oscillations prior to the gear shift is shown. The result is that the performance of the controller is not affected by the oscillations. Figure 15.31 shows the simulation with a load disturbance. The disturbance is better damped when using feedback from the wheel-speed sensor, than from the engine-speed sensor, which is a verification of the discussion in Section 15.3.4.

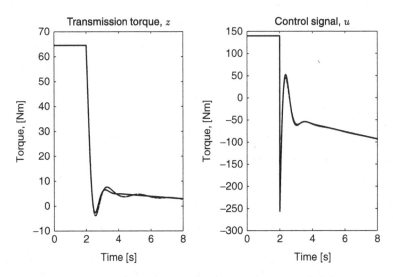

Figure 15.29 Simulation of the Nonlinear Clutch and Driveshaft Model with observer and control law based on the Driveshaft Model. A gear shift is commanded at $t = 2$ s. Feedback from the wheel-speed sensor is shown as solid lines, and feedback from the engine-speed sensor is shown as dashed lines. The design still works when simulated with extra clutch dynamics

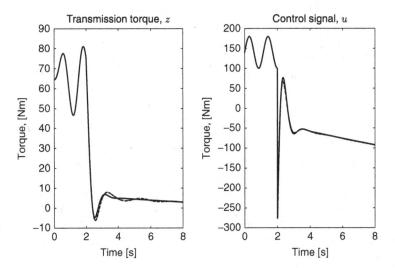

Figure 15.30 Same simulation case as in Figure 15.29, but with driveline oscillations at the start of the transmission-torque controller. Feedback from the wheel-speed sensor is shown as solid lines, and feedback from the engine-speed sensor is shown as dashed lines. The conclusion is that the control law works well despite initial driveline oscillations

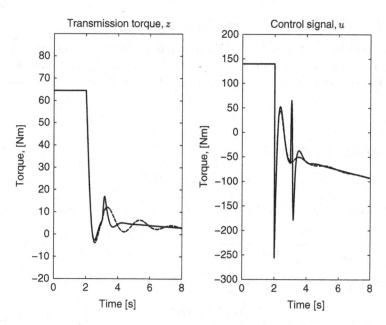

Figure 15.31   Same simulation case as in Figure 15.29, but with a load disturbance at $t = 3$ s. Feedback from the wheel-speed sensor is shown as solid lines, and feedback from the engine-speed sensor is shown as dashed lines. The conclusion is that the load disturbance is better attenuated when using feedback from the wheel-speed sensor

15.6   Driveshaft Torsion Control

Having treated Transmission torque control in the previous section it is now time to treat the other approach listed in Section 15.4, namely Torsion control. In Chapter 14 it was shown using the Driveshaft Model that the oscillations in the transmission speed could be laregly explained by a nonzero driveshaft torsion, x_1. Hence the idea, demonstrated in Pettersson and Nielsen (2000), is that it should be sufficient to control the state x_1 to zero. This is further motivated by the fact that the driveshaft is the main compliance of the driveline. If this torsion is small, it is reasonable to believe that the transmission torque is also small if the higher mode dynamical effects in the transmission are neglected.

The estimation of the driveshaft torsion is more easily performed than estimating the transmission torque. No extra parameters are needed, because the driveshaft torsion is one of the states in the driveline model (15.1)–(15.5). Control of the driveshaft torsion is a more robust approach compared to Transmission torque control, since the different behavior for each gear can be neglected, as the driveshaft is the same for all gears. Another advantage of such a scheme, which utilizes a consistent physical variable, is that extensions to monitoring, supervision, and adaptive control are simpler.

The presentation is structured so that first, assuming the driveshaft torsion (x_1) is known, the controller will be discussed. For the field tests, where there is no sensor for the driveshaft torsion, x_1, an observer is needed. This, together with the experimental setup, gives the basis for the results that are presented thereafter.

15.6.1 Recalling Damping Control with PID

It will turn out to be sufficient to have a control structure consisting of an observer and a PID controller. Therefore, to start with, a small detour is made into basic control. Recall that a second-order system with bandwidth ω_0 and damping ζ_0 is described by the transfer function

$$G(s) = \frac{1}{s^2 + 2\zeta_0\omega_0 s + \omega_0^2} \tag{15.98}$$

A PD-controller is described by

$$G_{PD}(s) = K(1 + T_D s) \tag{15.99}$$

The controller $u = G_{PD}(s)(r - y)$ for the system $y = G(s)$ gives the closed-loop system

$$y = \frac{K(1 + T_D s)}{s^2 + 2\zeta_0\omega_0 s + \omega_0^2 + K(1 + T_D s)}r = \frac{K(1 + T_D s)}{s^2 + (2\zeta_0\omega_0 + KT_D)s + \omega_0^2 + K}r \tag{15.100}$$

For a closed-loop system with bandwidth ω and damping ζ, one has to find K and T_D so that

$$2\zeta_0\omega_0 + KT_D = 2\zeta\omega \tag{15.101}$$

$$\omega_0^2 + K = \omega^2 \tag{15.102}$$

This is easily done by first tuning K in (15.102) to obtain the desired bandwidth. If it is a field test, K can be tuned directly to obtain the desired response time. Then the derivative part is tuned by adjusting T_D in (15.101) until the desired damping is obtained, or if it is a field test, until the step response is well damped, that is well behaved.

15.6.2 Controller Structure

The goal is now to control the driveshaft torsion to zero with damped driveline resonances, and verify that sufficient gear-shift quality is obtained. The first step is to study the transfer function, and an inspiration comes from Section 15.2.9 where it was shown that under simplified assumptions the torque transfer function, $G_{uM_d}(s)$, was especially simple since it became a second-order system.

Analysis

The driveline model (15.1)–(15.5) is a third-order system, and for wheel-speed control it is necessary to have a model-based third-order controller to obtain active damping, as seen in Section 15.3. To investigate if this is also the case for torque control, the open-loop transfer function from control signal, u, to driveshaft torsion, x_1, is investigated. The parameters are those from Example 15.4, that is the usual truck in this chapter. The poles and zeros of the transfer function are shown in Figure 15.32. Here it is seen that the zero and the real pole are close to canceling each other. If they do cancel, the third-order system will act as if it is a second-order system with no zeros. The same result is valid also for higher gears. When controlling a second-order system with no zeros, it is sufficient to have a second-order controller in order to be able to move the poles to any location.

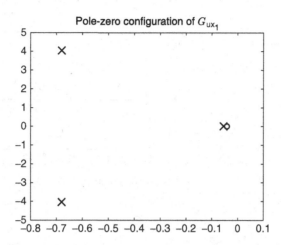

Figure 15.32 Poles and zeros of the transfer function from control signal to driveshaft torsion. The third-order model is reduced to a second-order model by the cancellation of the real pole by the zero. Similar results are found for higher gears

PID-Control

A PID controller controlling the driveshaft torsion is simulated with a gear shift commanded at a stationary speed of 1900 RPM with gear 1. The resulting driveshaft torsion with and without derivative part is shown in Figure 15.33. The controller parameters are obtained by first tuning the proportional parameter such that the negative peak values of the engine torque is possible to generate. (Recall that an engine has to rely on losses like friction or pumping work to generate a negative torque). The integral term is easy and is just adjusted to be slow enough not to interfere with the dynamics. Finally, the derivative parameter is tuned until the driveshaft torsion is well damped, according to Figure 15.33.

Active Damping

Hence, active damping can be obtained with a PID controller structure. This controller structure has natural parameter tuning properties with a derivative part that determines the amount of active damping of driveline resonances.

15.6.3   Observer for Driveshaft Torsion

Since the truck used for experiments uses no torsion sensor, as is the normal case, an observer is used as a virtual driveshaft torsion sensor. A virtual sensor is constructed by using the driveline model (15.1)–(15.5) together with the measured engine speed and wheel speed. This means that the estimated driveshaft torsion is given by (using standard notation)

$$\Delta \dot{\hat{x}} = A\Delta \hat{x} + B\Delta u + K_f(\Delta y - C\Delta \hat{x}) \qquad (15.103)$$

where the observer gain, K_f, has to be tuned.

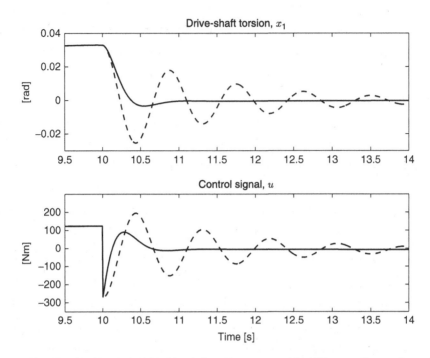

Figure 15.33 Simulated control of the driveshaft torsion to zero with a PID controller starting at 10.0 s. A PI controller (dashed lines) is used that obtains $x_1 = 0$, but with undamped driveline resonances. The solid lines are extension with derivative part in the controller

Excursion: Design of the observer used in the field tests
A Kalman filter was used in the field tests. The linearized model with disturbances is

$$\Delta \dot{x} = A\Delta x + B\Delta u + Gw \tag{15.104}$$

$$\Delta y = C\Delta x + v \tag{15.105}$$

with the output matrix specified by

$$C = \begin{pmatrix} 0 & 1 & 0 \\ 0 & 0 & 1 \end{pmatrix} \tag{15.106}$$

and with w and v corresponding to state disturbances and measurement disturbances respectively.
The observer gain is given by

$$K_f = P_f C^T V^{-1} \tag{15.107}$$

where P_f is found by solving the Riccati equation

$$P_f A^T + AP_f - P_f C^T V^{-1} CP_f + W = 0 \tag{15.108}$$

The covariance matrices W and V corresponding to w and v respectively. The following values are used

$$G = B, \quad W = 10^4 \tag{15.109}$$

$$V = \begin{pmatrix} 1 & 0 \\ 0 & 10^{-4} \end{pmatrix} \tag{15.110}$$

The filter was discretized by Tustin's method (Franklin et al., 1990), and implemented with a sampling time of 20 Hz. An example of how the observer performs on-line, using engine speed and wheel speed inputs, is seen in Figure 15.34. It is seen that the measured signals are closely estimated, which gives support to the estimated driveshaft torsion. The true driveshaft torsion, $x_1 = \theta_e/i_t i_f - \theta_w$, is unknown. The values estimated by the model are validated in a test where neutral gear is engaged at different torsions, while the driveline oscillations are measured, as presented in the Chapter 14.

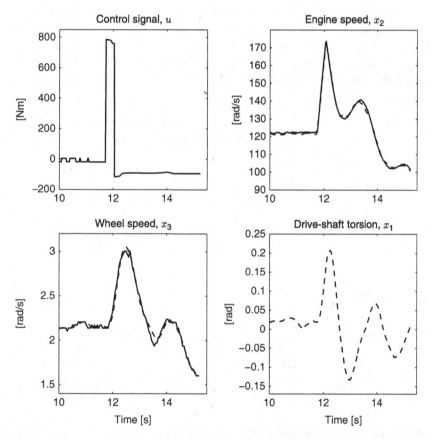

Figure 15.34 Field trial with driveline oscillations excited by an engine torque pulse (at 11.7 s). Measured engine torque, engine speed, and wheel speed are represented by solid lines. The observer estimates the engine speed, the wheel speed, and the driveshaft torsion, on-line (dashed lines)

15.6.4   Field Trials for Controller Validation

For the field trials, the observer, that is the virtual sensor for the driveshaft torsion, and a PID controller with anti-windup were implemented in a truck. The objective is to control the output of the virtual sensor to zero. Figure 15.35 shows the result (note the similarity with Figure 15.33). The dashed lines show the result when only using the PI part, where the proportional part of the controller is tuned to give the desired response time. This alone is not sufficient for damping out the oscillations in the driveline, which is due to the resonant pole-pair seen in Figure 15.32, which cannot be damped by a proportional controller.

The solid lines in Figure 15.35 show the result when the derivative part has been included and tuned. Hence, active damping is obtained in field trials with a PID controller and a virtual sensor measuring the driveshaft torsion.

15.6.5   Validation of Gear Shift Quality

The driveshaft torsion is controlled to zero with damped driveline resonance, which was the goal of the controller. However, it is not yet proved that this actually is sufficient for engaging neutral gear with sufficient quality (short delay and no oscillations). The way to prove this is to use the controller demonstrated in Figure 15.35, and engage neutral gear and measure the oscillations in the transmission speed. This is done in Figures 15.36 and 15.37, where the

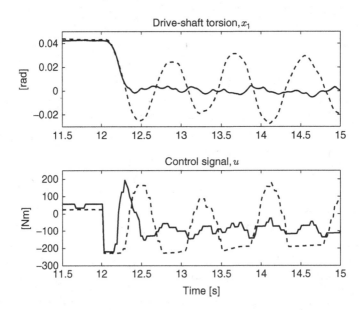

Figure 15.35   Field trials with a PID controller controlling the driveshaft torsion to zero. Control signal and driveshaft torsion when using the gear-shift controller that controls the driveshaft torsion to zero, started at 12.0 s. Prior to that, the engine has the stationary speed 1900 RPM with gear 1 engaged. A PI controller (dashed lines) is used that obtains $x_1 = 0$, but with undamped driveline resonances. The solid lines show the case with a derivative part in the controller. Active damping is thus obtained with a PID controller structure

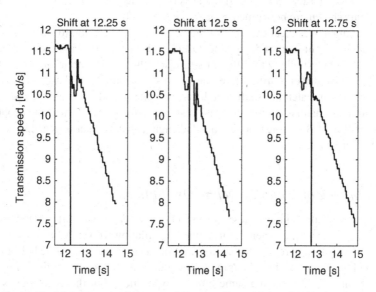

Figure 15.36 Field trials with start of the gear-shift controller at 12.0 s, all with the same PID controller controlling the driveshaft torsion to zero. Engagement of neutral gear is commanded every 0.25 s after the start of the controller, indicated by the vertical lines. The transmission speed is seen when neutral gear is engaged after a delay. The amplitudes of the transmission speed oscillations after the gear shift are less than 1 rad/s with different signs, which is an acceptable level

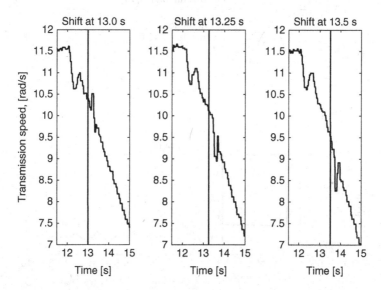

Figure 15.37 Same type of field experiment as in Figure 15.36, but with commanded engagement of neutral gear at 13.0, 13.25, and 13.5 s

controller is started at 12.0 s and gear shifts are commanded every 0.25 s, starting at 12.25 s. The experiments shown are performed at gear 1 at 6.7 km/h.

From these figures, it is clear that controlling the driveshaft torsion to zero is sufficient for obtaining gear shifts with a short delay time. Oscillations in transmission speed are minimized to under 1 rad/s in amplitude with different signs, which is well with in the range for giving no disturbance to the driver. Furthermore, the speed synchronization phase, where the engine speed is controlled to match the shaft speed, can be done quickly, since there are only minor oscillations in the transmission speed. These results are for gear 1, where the problems with oscillations are largest. The time to a commanded engagement of neutral gear can be decreased further for higher gears.

15.6.6   Handling of Initial Driveline Oscillations

One important problem, that it is necessary to handle, is when a gear shift is commanded not at steady state, but, for example, when the driveline is oscillating. To verify that the controller can also handle this situation in real experiments, driveline resonances are excited prior to the gear shift command. This is done by an engine torque pulse at 11.7 s, according to Figure 15.38. Figures 15.38 and 15.39 show the same type of experiment, but the controller is started at different time delays after the engine torque pulse has occurred. For both experiments, the resulting engine torque, calculated by the feedback controller, actively damps the initial driveline oscillations and obtains $x_1 = 0$. The difference in control signal in Figures 15.38 and 15.39 is strong evidence that initial driveline dynamics affects shift performance so much that

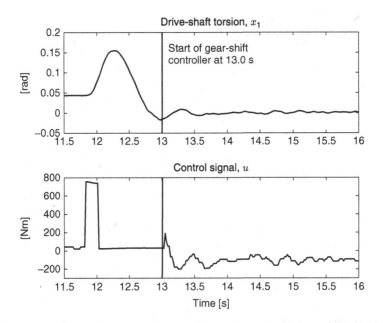

Figure 15.38   Control signal and driveshaft torsion during field trials with the start of the gear-shift controller at 13.0 s. The driveline is oscillating prior to the gear shift due to an engine torque pulse at 11.7 s. The controller controls the driveshaft torsion to zero with damped resonances despite initial driveline oscillations

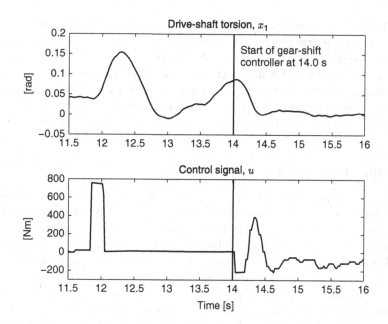

Figure 15.39   Same trial as in Figure 15.38, but with the start of the gear-shift controller at 14.0 s

feedback control is motivated. An open-loop scheme would not be able to handle these initial oscillations, leading to longer time for gear shifts.

15.7   Recapitulation and Concluding Remarks

Having gone through some important applications, we are now in a position to recapitulate and give concluding remarks. In a wider context, driveline control is a topic within powertrain control, which in turn is part of vehicle propulsion control. This means that the boundary conditions for driveline control are on one side the properties of the primary movers, the combustion engine or electrical motor, and on the other side is vehicle longitudinal dynamics, usually formulated as driving resistance as in Chapter 2, Vehicle. In addition to that, solutions to driveline control should fit into the overall powertrain control as described in Chapter 3, Powertrain, where the dominating paradigm now is the torque-based powertrain control, Section 3.3.

Chapter 13, Driveline Introduction, presented examples of unwanted driveline behavior when not having appropriate control, and also gave the physical explanation that it is due to elasticity in the driveline. The energy stored and released in the driveline can be considerable, and as engines are becoming stronger the effects increase. This together, with increased demands on functionality, driveability and reduced wear, increase the importance of driveline control.

15.7.1   General Methodology

Chapters 14 and 15, Driveline Modeling and Driveline Control, give a general methodology for problems in driveline control. The modeling technique in Chapter 14 is general. The control design in Chapter 15 is based on examples but the underlying technique is general. The examples were based on engine torque control, that is using $u = M_e$ as a control signal, but assume, for example, that a clutch control problem was considered instead. Then, the equations would be exactly the same in the modeling phase, but when combining them, for example as in Section 14.3.2, the clutch torque, M_c, should not be eliminated. Thereafter, when following the recipe in Section 15.2 and putting the model in state-space form, the control input should be chosen as $u = M_c$. Thereby the clutch control problem is ready for study and design, and other driveline problems can be handled in the same way.

15.7.2   Valuable Insights

Even though the control problem is put in mathematical form there are a number of insights that give valuable guidance for a best solution. Recall Figures 15.5 and 15.6 showing intrinsic properties, and recall how these properties show up in more advanced settings. One important lesson learned both for the anti-surge controller and the gear shifting controller is that it is not sufficient to study only the response to a command signal. Some important characteristics are not seen in that response (due to cancellations), but show up due to load disturbances, for example, and it is thus important to evaluate a design for such cases and also for different initial conditions of the driveline, especially if it is already oscillating or if substantial energy is stored, for example due to former driver commands or load disturbances like road bumps.

Another insight is that torque control problems sometimes can be treated as second-order problems due to near pole-zero cancellations as introduced in Section 15.2.9, exemplified for a real vehicle in Figure 15.32, and used in the design of the driveshaft torsion controller in Section 15.6.

It is well worth recapitulating the steps leading to the final complete anti-surge controller in 15.65. The basic structure is (15.15) and (15.16), which is extended with RQV for driver feel (15.53), state observer, and load estimation as in Section 15.3.3. It is most instructive to see how the stationary point is obtained in (15.31) and (15.32) and used in the controller coefficients via (15.52) as an addition to the coefficients obtained from the design method used.

15.7.3   Formulation of Control Criterion

To get good behavior, the basic control criterion in 15.13 needs to be tailored to the characteristics of the automotive application. For the speed controller the steps are via 15.34 leading to (15.39) where once again the role of the stationary point should be noted.

For transmission torque control there were even more steps. Already in (15.87) it should be noted that u_{shift} is not representing a stationary point but rather the gear shifting condition (for a rigid driveline) with zero torque, which means that the vehicle will lose speed due to driving resistance. The sequence of Figures 15.25–15.27 together with the reasoning around them lead to the final formulation (15.95).

15.7.4   Validation of Functionality

Both the speed control and the torque control for gear shifting were validated in a test vehicle where the responses to driver command were evaluated. In addition, simulations were used to investigate disturbances and initial conditions that are not so easy to generate in a controlled way in experiments. In these simulations, the vehicle was represented by a more complicated model than the one used for design.

15.7.5   Experimental Verification of Torque Limit Handling

There is a misconception that nonlinearities that are not considered in the design, like, for example, torque limitations, will cause problems when using the controller in a real setting. Therefore, it is important to understand the reasoning and evaluation presented in Figure 15.16, Figure 15.17 and Figure 15.18. The anti-surge speed controller is derived based on a simplified model, the Driveshaft Model.

Figure 15.16 proves the basic function, that is that sensors are sufficient and that the engine control is fast enough to counteract the driveline resonances. Then the test in Figure 15.17 and Figure 15.18, exposes the controller to a situation not included in the design, namely a torque limitation due to the smoke limiter. Nevertheless, the controller works just fine, as seen in Figure 15.18.

15.7.6   Benefits

Active damping of driveline resonances gives a way of optimizing driveline performance, for example response time to speed commands or time needed for the torque control phase leading to a minimized time for a gear shift. Thus, the controller improves performance and driveability since driving response is increased while still reducing vehicle shuffle. Major user advantages are less wear, better comfort, and that more drivers can handle difficult driving situations.

Part Five

Diagnosis and Dependability

16

Diagnosis and Dependability

Previous chapters have presented possibilities and developments based on the fact that vehicles are now computerized machines. As presented, this fact has had an enormous effect on the possibilities for functionality of vehicles, which, together with needs and requirements from customers and from society, have created vigorous development activities. The availability of computing power in vehicles has also strongly influenced another field, namely diagnosis and dependability. Originally the main driving force came from legislation requiring diagnostic supervision of any component or function that when malfunctioning would increase tail-pipe emissions by at least 50%, the well-known On Board Diagnosis (OBD) requirements by the California Air Resource Board (CARB).

Basically, there are observed variables or behaviors for which there is knowledge of what is expected or normal. The task of diagnosis is, from the observations and the knowledge, to generate a fault decision, that is to decide whether there is a fault or not and further to identify the fault. Once a methodology to find faults or malfunctions has been developed, then many new application areas open up, so that the same techniques are used for safety, machine protection, availability, up-time, dependability, functional safety, health monitoring, and maintenance on demand. Examples of consumer value are satisfaction because of dependability, lower costs by maintenance on demand, or increased profit for truck owners through maximum availability.

The two major automotive application areas are

- **Dependability** is a strongly growing field. The objective is to monitor any aspect that could influence safety, reliability, or availability of a vehicle. Thus, the objective is mainly directed toward customer protection and satisfaction.
- **Diagnosis** is the original automotive application. The objective is to point out any faulty component that increases emissions. Thus, the objective is mainly directed toward fulfillment of legislations.

These development trends are intertwined, as they both use similar methodology based on theory for decision support. They are both also important when considering co-design with mechanics and control. Further, as will be seen in the next section, there are several important usages and this is the reason that diagnosis is such an expansive academic research field.

The goal of this chapter is to give insight into these new developments, and to do it in enough depth to show the interplay between the basic physics, the models, and the possibilities for design. After an overview of dependability, there are sections introducing basic concepts,

Modeling and Control of Engines and Drivelines, First Edition. Lars Eriksson and Lars Nielsen.
© 2014 John Wiley & Sons, Ltd. Published 2014 by John Wiley & Sons, Ltd.
Companion Website: www.wiley.com/go/powertrain

methodology, important automotive examples, and legislation where OBD (On-Board Diagnostics) is the major example.

16.1 Dependability

The basic notion of dependability is that the owner of a vehicle should be able to rely on it in all respects including reliable operation, safety, and reasonable maintenance costs. The development of techniques for diagnosis and supervision can thus be used in many ways, and this is one reason for a rather extensive vocabulary in the field. A number of the terms will be presented now, and then in Section 16.1.16 the interrelations will be discussed in connection with the depiction in Figure 16.3.

Some main reasons to incorporate diagnosis techniques are:

- **Safety** In many systems a fault may cause serious personal damage, and this is especially obvious in vehicle dynamics control. For these systems, high reliability of the diagnosis system is required.
- **Environment protection** In automotive emission control systems, a fault may cause increased emissions. It has been concluded that a major part of the total emissions from cars originates from vehicles with malfunctioning emission control systems.
- **Machine protection** A fault can often cause damage to the machine. Therefore it is important that faults are detected as quickly as possible after they have occurred.
- **Availability** For many technical systems it is critical that the systems are running continuously. This is, for example, the case for heavy trucks. The reasons are both economicy as well as safety. A reliable diagnosis system is desirable so that warnings can be obtained before serious breakdown. Important here is a low probability of a false alarm.
- **Up-time** Up-time is similar to availability focusing on the time available for operation.
- **Repairability** Closely connected to availability is repairability. A good diagnosis system will quickly identify the faulty component that should be replaced. In this way, time-consuming fault localization is reduced, which will decrease total repair time.
- **Flexible maintenance** Related to repairability but with the added notion of finding problems before they develop into faults. Compared to traditional scheduled maintenance at regular intervals, flexible maintenance on demand means that the vehicle itself signals when it is time for maintenance. Should lead to lower maintenance cost and more up-time.
- **Health monitoring** Widens the scope even more since it implies that the health and condition of all components are continuously monitored.
- **Supervision of highly automated vehicles** An increasing number of driver support systems or systems for optimal driving are entering modern vehicles, and supervision of these are important. One example would be platooning, see Sections 1.1.4 and 2.2.3, and there are many more such examples.

16.1.1 Functional Safety–Unintended Torque

Functional safety is a wide concept, but when it comes to the vehicle propulsion systems there are usually three main hazards that are pointed out:

- Fire.

- Electric shocks, especially in hybrid and electric cars where high voltages are common.
- Unintended torque, both propulsive and faulty brakes.

Fire and electric shocks are very important, but the analysis and methodology used in these areas are outside the scope of this book. Unintended torque, on the other hand, is at the core of powertrain management and influences the design considerations in many ways. Some more background is therefore given now, and a solution example is given in Section 16.5.2.

Unintended Torque

Media sometimes reports that a car from standstill suddenly leaps forward, hitting a wall, other cars, or in the worst case scenario a person. Such "sudden acceleration incidents" (SAI) have been studied by the National Highway Traffic Safety Administration (NHTSA) and their 1989 Sudden Acceleration Report gives the following definition: Sudden acceleration incidents are defined as unintended, unexpected, high-power accelerations from a stationary position or a very low initial speed accompanied by an apparent loss of braking effectiveness.

Factors causing SAI may include:

- Driver faults, for example, pressing the accelerator when braking is intended. This may depend on pedal design or placement.
- Pedal faults so that the response is wrong or even lacking. This may depend on blockage by floor math or a foreign object, or any other mechanical interference with the operation of either the accelerator or brake pedal.
- Drive by wire faults, for example, a fault in the electronic throttle control or cruise control.
- Stuck throttle.
- Stagnant throttle, where the throttle may temporarily stick in a position, and when the driver increases accelerator position, the throttle can suddenly go wide open.

The first two items are treated by design of car interior, and the two last items are related to the mechanics of the throttle, where there are, for example, solvent sprays available for stagnant throttle.

SAI sometimes require the simultaneous failure of the propulsion system and the brake system, but there are also other cases. In many events in the 1990s (often at car washes) an SAI occurred when shifting to "Drive" or "Reverse" from "Park". Then, the throttle could suddenly move to wide open, and even if the brakes were fully functional, the driver was not able to react quickly enough to prevent an impact. Analysis later showed that a probable cause was a current leakage pathway causing activation of the cruise control servo. This is one reason behind the solutions presented in Section 16.5.2, where hardware duplicated pathways are used with careful separation of electrical ground.

Brake Override System

A so called smart throttle is now becoming common. It is an interlock switch that secures such that the brake pedal always overrides the accelerator. This means that just touching the brake prevents torque being delivered at the wheels, and it prevents, or at least mitigates, the effects of any SAI except driver faults.

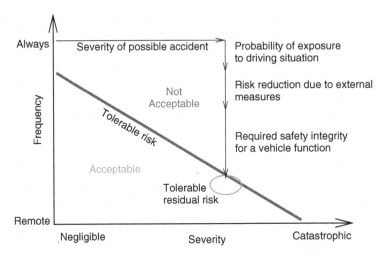

Figure 16.1 A schematic illustration of how risk depends on severity of harm and frequency of harm

16.1.2 Functional Safety Standards

Unintended torque is very important, and it is a good example to show how functional safety is handled. A basis is the standard ISO 26262 where both methods and processes are described, and the goal is to determine the Automotive Safety Integrity Level, ASIL.

Safety analysis consists of hazard analysis and risk assessment. The severity of a hazard, if it occurs, has to be determined on a scale from negligible severity to catastrophic. Further, the probability that the hazard occurs has to be determined, and this can be expressed in frequency on a scale from never to always. Risk is a combination of them, two so that it is a product of probability of occurrence and severity of harm. These concepts are schematically illustrated in Figure 16.1, where the concepts of acceptable risk, tolerable risk, and unacceptable risk are also illustrated.

The goal of safety engineering is to have a process for hazard and risk analysis so that a safety assessment can be made confirming that safety risks are reduced to acceptable levels. The typical steps are

- Find all hazards.
- Determine risks.
- Reduce risks sufficiently.

In the process both systematic and random issues need to be addressed. The main question is of course how to proceed to reduce risks. There are two main means to do so, namely

- Process requirements, usually meaning that fairly strict procedures are to be followed in, for example, software development.
- Technical requirements, which often implies system duplication or diagnosis supervision functions.

For industry, there is the added complexity that the whole process of functional safety needs to be managed and implemented into an efficient process that can be used by several engineering teams working together.

Returning to Figure 16.1, it schematically illustrates the process to arrive at a tolerable residual risk. At the top of the figure there is an arrow "Severity of possible accident" illustrating the severity of a possible accident. The first step in the ASIL analysis is to determine how often a situation can occur "Probability of exposure". The next step, "Risk reduction ...", is an evaluation of the risk reduction, for example based on possible driver actions. Will the driver have the means to act, for example brake, and will there be time to do so, for example depending on speed and other driving conditions. Finally, it is necessary to reduce the residual risk into the region of acceptable risk, which is illustrated by the last arrow "Required safety integrity for a vehicle function." This is an engineering task using solutions like redundancy and supervision.

One example, where functional safety is at the core, is throttle control, described in Section 16.5.2. Functional safety for unintended torque is also a natural transition to the topic in the next section.

16.1.3   Controller Qualification/Conditions/Prerequisites

A controller is an active component, and if for some reason it is given erroneous information it may result in faulty or hazardous behavior. Therefore there are usually quite a few requirements for a controller to be active.

Consider boost   pressure control described in Section 10.9.3. A faulty boost pressure sensor could make the system generate excessive boost pressures which potentially could give excessive torque or destroy the engine. Thus, a diagnostic function is needed for the supervision of the boost sensor, and a prerequisite for running the controller is represented by its output BoostSensor_OK being true in the following pseudo-code. There are also other prerequisites for running the controller. Let's first look at the pseudo-code, where the main line computes the wastegate control u_WG as u_WG = FF(...) + FB(...) combining the feedforward and feedback loops of the controller, and explain the conditions thereafter.

```
if ( BoostSensor_OK & NOT(ExcessiveBoost_DET)
     & Throttle_OK & BoostNeeded )
then
   u_WG = FF(p_b_ref,N_e) + FB(p_b_ref,p_b_sens);
   u_WG = CheckLimits_WG(u_WG);
else
   u_WG = OPEN_WG;
end
I = UpdateIwithAntiWindup_WG(u_WG);
```

The requirement NOT(ExcessiveBoost_DET) is a flag telling us whether an excessive boost pressure has been detected previously, indicating that a fault in the boosting system (i.e., actuator, pneumatic actuator system, wastegate, etc.) has been detected but not isolated. The throttle must be working, thus Throttle_OK. Furthermore, the controller is executed only if boosting is required, implying the condition BoostNeeded. In this example, if the conditions are fulfilled the controller is executed, or otherwise the wastegate is just opened by the line u_WG = OPEN_WG, thus preventing boost build-up that could cause unintended acceleration.

The example further illustrates that the control actions need to be checked against limits, and that if interventions occur the controller states might need to adjusted, for example to avoid windup of integrators, see Example 10.2 in Section 10.2.3. A final comment is that there can be an equal amount of work, or even more, to develop and tune the conditions for running the controller as in the design and calibration of the control computation itself.

In the same way, any controller in the torque-based powertrain control structure, Section 3.3, needs to be examined from a safety perspective (especially unintended torque). The result will be proper supervision of components and a sometimes lengthy list of conditions to be fulfilled for the controller to be active.

16.1.4   Accommodation of Fault Situations

As touched upon in Section 10.10, when a fault or malfunction is detected during operation by an on-board system, then the question arises of what to do.

- **Fault accommodation.**   To reconfigure the system so that the operation can be maintained even in presence of a fault.

Based on the severity of the fault the following actions are possible.

- Stop and/or shut down; call for repair. Used when continued travel would be dangerous or would severely harm the vehicle. A simple example would be no oil in the engine.
- Limp-home. Limp-home is a mode entered if the diagnosis system, after detecting the fault, can exclude the faulty component and use a suboptimal control strategy until the car can be repaired.
- Reconfigure the system so that the operation can be maintained at a basic level.
- Reconfigure the system so that the operation can be maintained at a high level.

Limp-home is part of the legislations around OBD (On-Board Diagnostics), see Section 16.6. The last item is not unusual; if the situation is not safety critical or harmful to the environment, a virtual sensor can be used so that the vehicle can proceed as usual and the fault can be fixed at next maintenance.

Fault Tolerant Control

Fault tolerant control is a general version of the idea in the previous section to reconfigure the system so that the operation can be maintained. The principle of fault tolerant control will be illustrated on the system in Figure 10.5 as seen in Figure 16.2.

16.1.5   Outlook

Even though dependability and diagnosis have already grown into a large field there are many trends indicating continued growth. New infrastructure, Section 1.1.4, for communication and information means that vehicles can be continuously monitored. A vehicle can learn from other vehicles, or can call the manufacturer for advice in situations not anticipated before. For commercial vehicles, typical economic incentives are increased incomes due to high dependability/up-time and lower costs through maintenance on demand, thus both increasing

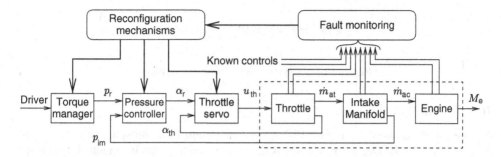

Figure 16.2   A principle sketch of fault tolerant control for the system in Figure 16.2

profit. The concept of buying a vehicle, and then paying for service and spare parts, can be changed into the concept of buying transportation with a price depending on performance, so it is of mutual interest to monitor all aspects of the vehicle in an optimal way.

16.1.6   Connections

As mentioned already in the introduction, the basis that is used both for dependability and diagnosis is observed variables or behaviors for which there are knowledge of what is expected or normal. Methods that from observations and knowledge give decision support are a common basis in the applications, and it thus natural that all these are related. The distinctions between the terms used are not sharp, but see Figure 16.3 for a structure of the most important connections in the field.

16.2   Basic Definitions and Concepts

This section presents definitions and concepts that are central to the area of diagnosis, based on the terminology proposed by the IFAC Technical Committee SAFEPROCESS as a step towards a unified terminology.

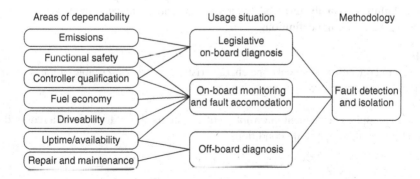

Figure 16.3   An illustration of how the dependability and safety concepts are interlinked

Figure 16.4   A system with actuators and sensors

16.2.1   Fault and Failure

A system can, as shown in Figure 16.4, be separated into three subsystems: actuators, the process, and sensors. Depending on which subsystem a fault occurs in, a fault is classified to be an *actuator fault*, *process fault*, or *sensor fault*. Process faults are sometimes also called *system faults* or *component faults*.

Typical sensor faults are short-cut or cut-off in connectors and wirings, and drifts, that is changes in gain or bias. Also, the time response can degrade, that is the bandwidth is decreased. Examples of process faults are increased friction, changed mass, leaks, and components that get stuck or loose. Examples of faults in an actuator are short-cut or cut-off in connectors and wirings. If the actuator includes an electrical amplifier, there can also be gain and bias faults. Actuators can by themselves be relatively complex systems, containing, for example, DC-motors, controllers, and sensors. Therefore, all examples of sensor and process faults are applicable also to actuators.

In a diagnosis application it may not be sufficient to isolate a faulty (larger) component, for example the DC-motor. Often more detailed knowledge is required about the fault, such as what part of the DC-motor is faulty. Thus, when designing a diagnosis system it is important to have knowledge about what faults can occur or are most common, and also how different faults affect the system. This is because specific diagnostic solutions are often required for each kind of fault.

Faults are often characterized in three basic types describing their time-variant behavior:

- Abrupt, step-faults representing, for example, a component that suddenly breaks.
- Incipient (developing) faults representing, for example, slow degradation of a component or developing calibration errors of a sensor.
- Intermittent fault caused by, for example, loose connectors.

Failure

The word failure is usually reserved for a stronger and more permanent version of faults according to the following definitions.

- **Fault**
  Unpermitted deviation of at least one characteristic property or variable of the system from acceptable/usual/standard/nominal behavior.
- **Failure**
  A fault that implies permanent interruption of a systems ability to perform a required function under specified operating conditions.

Fault Modes

A system, like the one shown in Figure 16.4, is said to be in different modes depending on faults. One mode is normal operation, usually denoted No fault (*NF*). Then there may be faults in the actuators, process, or sensors. A usual notation is to just number these different modes so that a common abstract notation is: No fault (*NF*), fault 1 (F_1), fault 2 (F_2), and so on.

16.2.2   Detection, Isolation, Identification, and Diagnosis

The following definitions are important to describe both the task and the performance of a diagnosis system.

- **Fault detection**
  To determine if faults are present in the system and also the time of detection.
- **Fault isolation**
  Determination of the location of the fault, that is which component has failed.
- **Fault identification**
  Determination of the size and time-variant behavior of a fault.
- **Fault diagnosis**
  For the definition of this term, two common views exist in literature. The first view includes fault detection, isolation, and identification, see for example Gertler (1991). The second view includes only fault isolation and identification, see for example Isermann (1984). Often the word *fault* is omitted so only the word *diagnosis* is used.

The term *fault diagnosis* is here used to denote the whole chain of fault detection, isolation, and identification. This is in accordance with one of the views common in literature. Diagnosis used in this way also serves as a name for the whole area of everything that has to do with diagnosis. If fault detection is excluded from the term *diagnosis*, as in the second view, one has a problem of finding a word describing the whole area. This can be partly solved by introducing the abbreviation FDI (Fault Detection and Isolation), which is common in papers taking the second view of the definition of the term diagnosis. As noted in some papers, FDI does not strictly contain fault identification. To solve this, the abbreviation FDII (Fault Detection, Isolation, and Identification) has also been used.

16.2.3   False Alarm and Missed Detection

The following concepts are central when evaluating a diagnosis system.

- **False alarm**
  The event that an alarm is generated even though no faults are present.
- **Missed alarm**
  The event that an alarm is *not* generated in spite of the fact that a fault has occurred.
- **Missed detection**
  Same as missed alarm.

Naturally, it is desired to have as low a rate as possible for both false alarms and missed detection. In many cases there is a trade-off between the two, so that in the presence of noise a low detection threshold will lead to many false alarms, but a high threshold leads to missed alarms.

These are very important issues in the design of a diagnosis system. If there is a false alarm this could mean that a truck visits a repair shop without needing to, thus reducing availability and up-time. On the other hand, if there is a missed detection this could lead to costly repairs.

16.2.4   Passive or Active (Intrusive)

One classification of diagnosis systems is whether they need to do specific actions to be able to draw conclusions.

- **Active diagnosis**
  When the diagnosis is performed by actively exciting the system so that possible faults are revealed.
- **Intrusive diagnosis**
  Same as active diagnosis.
- **Passive diagnosis**
  When the diagnosis is performed by passively studying the system without affecting its operation.

A passive system will not influence the operation of the vehicle, whereas an active one will, and that may have drawbacks, for example, be disturbing to the driver.

16.2.5   Off-Line or On-Line (On-Board)

A final overall distinction in diagnosis of technical systems is that it can be performed off-line or on-line. In automotive, the typical setting for off-line diagnosis is in a repair shop. A technician may perform own tests or read out error codes stored in the vehicle. When on-line is considered, it is usually called on-board in automotive settings. In the following we mainly deal with automatic diagnosis that is primarily to be performed on-board, even though the same methodology is used also for off-line diagnosis.

16.3   Introducing Methodology

The art of deducing system functionality and component condition of a system is a vast area both as regards engineering applications and theoretical development. As described in Section 16.2.2, the task of diagnosis is to generate a fault decision, that is to decide whether there is a fault or not and also to identify the fault. Thus, the basic problems are what the procedure for generating fault decisions should look like, what parameters or behavior are relevant to study, and how to derive the knowledge of what is expected or normal.

A brief introduction will now be given to methods often used in automotive applications. This will be done by starting with the simplest example and then expanding the ideas via a sequence of examples, where the goal is to find malfunctions in, for example, sensors and actuators. The used input signals are mainly output signals obtained from the sensors, but can also be signals derived from commanded inputs together with models of the system.

16.3.1   A Simple Sensor Fault

Figure 16.5 shows the simplest case where a single sensor observation may be corrupted by a fault, f. Traditionally this type of diagnosis has mainly been performed by limit checking.

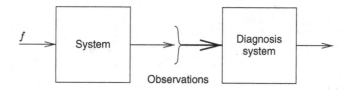

Figure 16.5   A principle illustration of the basic diagnosis situation

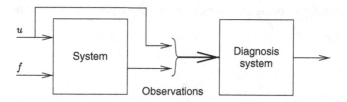

Figure 16.6   Knowledge of the input to the system gives the possibility of predicting expected behavior. A simple example is when there is no reaction to an actuator input

This means that when a sensor signal level leaves its normal operational range, an alarm is generated. The normal range is predefined by using thresholds. An example is when an output voltage from a sensor always should be within an operating range, and when the sensor output goes to battery voltage or ground one can conclude there is an error. Sometimes such a normal range can be dependent on the operating conditions, and then the thresholds for different operating points can be stored in a table. Such use of thresholds as functions of some other variables can be viewed as a kind of model-based diagnosis. In addition to checking signal levels, trends of signals are also often checked against thresholds.

16.3.2   A Simple Actuator Fault

As is true for most components, an actuator usually has some self-diagnosis that checks voltages, currents, and other internal variables. On a system level the result of an actuator fault can be seen as missing reactions when operating the actuator, and as shown in Figure 16.6 a diagnosis system, knowing the actuator signal and the process observations, can make the correct diagnosis.

16.3.3   Triple Sensor Redundancy

Triple sensor redundancy is an important example that illustrates a number of basic techniques in diagnosis. It is a classical technique in safety critical systems where one variable, x, is measured with three independent sensors with possible sensor faults $f = (f_1, f_2, f_3)$ as schematically illustrated in Figure 16.7.

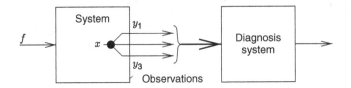

Figure 16.7 Triple sensor redundancy is a classical technique in safety critical systems. One variable, x, is measured with three independent sensors

Denoting the three sensor signals y_1, y_2, y_3 the system equations without faults become

$$y_1 = x$$
$$y_2 = x \qquad (16.1)$$
$$y_3 = x$$

From these three signals, **residuals** can be formed by pairwise comparison as

$$r_1 = y_1 - y_2$$
$$r_2 = y_1 - y_3 \qquad (16.2)$$
$$r_3 = y_2 - y_3$$

In the fault free case all residuals are zero. When a fault occurs in one of the sensors then the residuals react differently.

Let the fault modes, see Section 16.2.1, be denoted F_i and have the interpretation that there is a fault in sensor y_i. Then the reactions of the residuals (16.2) are described by the following **influence structure** showing how faults influence different residuals

	NF	F_1	F_2	F_3
r_1	0	X	X	0
r_2	0	X	0	X
r_3	0	0	X	X

A zero in the influence structure means that there is no reaction in the residual to that fault, but an X means that there is. Assuming a real situation with some sensor noise, the test i typically signals an alarm when the absolute value of residual r_i is larger than a given threshold $|r_i| > J_i$.

As an example, assume that tests 1 and 2 activate alarms. That gives the following conclusions

$$|r_1| > J_1 \Rightarrow F_1 \text{ or } F_2 \qquad (16.3)$$
$$|r_2| > J_2 \Rightarrow F_1 \text{ or } F_3 \qquad (16.4)$$

Under the assumption that only one sensor is faulty, one can conclude that the only possible fault mode is F_1. This means that the fault has been detected and isolated, see Section 16.6.2, which means that F_1 is the diagnosis (and it is the only one). Thus sensor 1 is faulty.

For this case, the internal structure of the diagnosis system can be illustrated as in Figure 16.8

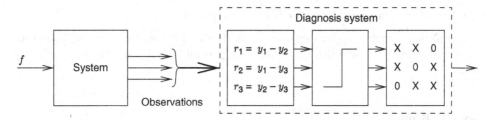

Figure 16.8   As a specialization of Figure 16.5, the internal structure of the diagnosis system for triple sensor redundancy is illustrated

16.3.4   Triple Redundancy Using Virtual Sensors

The previous example is straightforward to generalize to the case of virtual sensors where one variable, x, is measured directly with one sensor, and virtual sensors, for example observers, using models are used to obtain estimates, $\hat{x}_2, \hat{x}_3$.

Let one sensor, y_x, measure x. Assume two other sensors measuring two other variables, z, w, with sensor signals, y_z, y_w, and let there be models, M_2, M_3, relating these sensor signals to x, so that $\hat{x}_2 = M_2(y_z)$ and $\hat{x}_3 = M_3(y_w)$. The equations analogous to 16.1 become

$$y_1 = x = y_x$$
$$y_2 = \hat{x}_2 = M_2(y_z) \tag{16.5}$$
$$y_3 = \hat{x}_3 = M_3(y_w)$$

The residuals are the same as in (16.2), and analogously fault modes and influence structure look the same as in the previous section, Section 16.3.3.

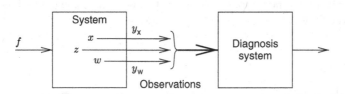

Figure 16.9   Triple redundancy using virtual sensors. One variable, x, is measured directly with one sensor, and virtual sensors, for example observers, using models are used to obtain estimates, $\hat{x}_2, \hat{x}_3$

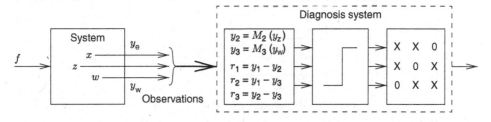

Figure 16.10   The same idea as in Figure 16.8 can be used with virtual sensors replacing some of the real sensors

16.3.5 Redundancy and Model-Based Diagnosis

A common theme in the previous subsections is the use of redundancy. This will be further commented on here, and will also lead to the more general formulation of model-based diagnosis.

Hardware Redundancy

The example seen in Section 16.3.3 is a version of the traditional approach of duplication (or triplication or more) of hardware. This is called hardware redundancy. The method is conceptually simple and robust, but there are at least three drawbacks associated with the use of hardware redundancy: hardware is expensive, it requires space, and adds weight to the system. In addition, extra components increase the complexity of the system which in turn may introduce extra diagnostic requirements.

Analytical Redundancy

Section 16.3.4 illustrates another concept, namely that of analytical redundancy. There is analytical redundancy if there are two or more, but not identical, ways to determine a variable where at least one uses a mathematical process model in analytical form.

A simple example of analytical redundancy is the case where it is possible to both measure an output and, by means of a model, also estimate it. This example is illustrated in Figure 16.11. From the measured and estimated output $y(t)$ and $\hat{y}(t)$, a residual can be formed as

$$r(t) = y(t) - \hat{y}(t)$$

The model used to estimate $\hat{y}(t)$ can be linear or nonlinear. If a fault occurs, it will affect the measured output but not the estimated output. In this way the residual will deviate from zero and the fault is detected.

Model-Based Diagnosis

Model-based diagnosis is a general and useful formulation that includes all the preceding examples. The system to be diagnosed is influenced by control actions (known inputs), disturbances (unknown inputs), and faults. The situation is shown in Figure 16.12, and the task of the diagnosis system is to be performed in all situations independent of the behavior of the inputs or disturbances. Thus, the diagnosis conclusion should only depend on the faults.

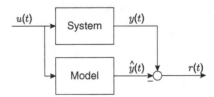

Figure 16.11   A simple example where a residual is formed based on analytical redundancy

The formulation in Figure 16.12 is quite general, but an important case is the one in Sections 16.3.3 and 16.3.4, where the decision is based on residuals and an influence structure. This is depicted in Figure 16.13

Compared to traditional limit checking, model-based diagnosis has great potential to have the following advantages:

- It can provide higher diagnosis performance, for example smaller faults can be detected and the detection time is shorter.
- It can be performed over a large operating range.
- It can be performed passively.
- Isolation of different faults becomes possible.
- Disturbances can be compensated for, which implies that higha diagnosis performance can be obtained despite the presence of disturbances.

Compared to hardware redundancy, model-based diagnosis may be a better solution because of the following reasons:

- It is generally applicable to more kinds of components. Some hardware, like the engine itself, cannot be duplicated.
- No extra hardware is needed, thus avoiding that cost.

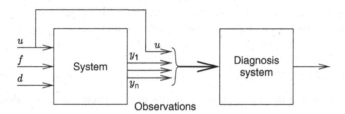

Figure 16.12 Model-based diagnosis is the generalization of Figure 16.10. In addition, the formulation includes the idea that the diagnosis conclusion should only depend on faults and, therefore, should not depend on disturbances or other inputs

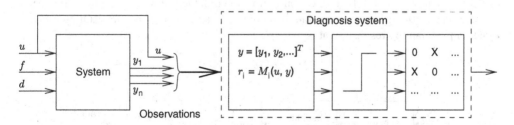

Figure 16.13 A specialization of Figure 16.12 that follows the idea in Figure 16.10, such that the decision is based on residuals and an influence structure

Model-Based Diagnosis is an Active Research Field

Because of its potential, it is natural that model-based diagnosis is an active research field. Looking back, manual diagnosis has been performed as long as there have existed technical systems, but automatic diagnosis started to appear when computers became available. Similar developments were found in other application areas, like chemical plants and aerospace applications. The interest increased after the Harrisburg incident, where the operators of a nuclear plant didn't get the appropriate diagnosis of root causes. The research on diagnosis has since then been intensified with automotives as one main driving application, and today it is still an expansive research area with many unsolved questions.

In principle, the model used can be of any type, from logic-based models to differential equations. Depending on the type of model, different approaches to model-based diagnosis can be used, for example a statistical approach (Basseville and Nikiforov 1993), a discrete event systems approach (Sampath et al. 1995), AI-based approaches (Reiter 1987), and approaches within the framework of control theory, see, for example, Blanke et al. (2006).

16.3.6   Forming a Decision–Residual Evaluation

The purpose of residual evaluation is to generate a *fault decision* by processing the residuals, where a fault decision is the result of all the tasks of fault detection, isolation, and identification. In the presence of disturbances caused by modeling errors and measurement noise, residual evaluation is a non-trivial task.

The residual evaluation scheme used in the examples in Sections 16.3.3 and 16.3.4 is seen in Figure 16.14. In the figure, the residuals are low-pass filtered, and then the absolute values are thresholded. For the generation of a fault decision, the logic scheme generates an alarm when the residual responses matches one of the fault columns in the influence structure.

Thresholding–P(FA) and P(MD)

The thresholding step in the above scheme usually includes some design trade-offs. The test quantity, here the residual, is zero in the fault free case, so a fault is assumed present if $|r(t)| > J$ where J is the threshold. In the ideal case, that is when a perfect model of the process is available and there are no disturbances, the threshold would be $J = 0$, but in a realistic case with model errors and noise it is necessary that $J > 0$.

As pointed out in Section 16.2.3, two important properties of a diagnosis system are the probability of false alarm, P(FA), and the probability of missed detection, P(MD). These concepts are often used in threshold selection as illustrated in the following example. Consider a residual that is affected by additive Gaussian noise $n(t)$, and let $G_f(s)$ be the resulting transfer function from fault to residual, that is

$$r(t) = G_f(s)f(t) + n(t)$$

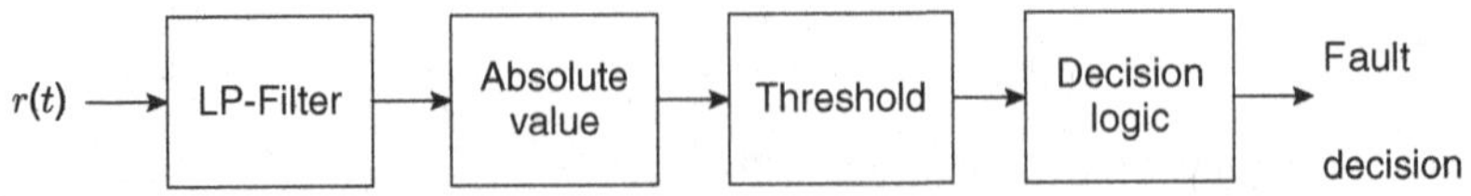

Figure 16.14   The residual evaluation scheme used in the examples in Sections 16.3.3 and 16.3.4

In Figure 16.15, the probability density function of such a residual (after filtering) is shown for the fault free case (solid line) and the case when a fault is present (dashed line). The fault is assumed to be constant, that is $f(t) \equiv c$, and therefore the density function is centered around $G_f(0)c = 4$. The threshold has the value $J = 2$ and is represented by the dotted lines. There are two lines because it is the absolute value of the residual that is thresholded. Integrating the area below the solid line, to the left of the left threshold and to the right of the right threshold, gives the probability of false alarm, P(FA). Similarly the probability of missed detection, P(MD), is calculated by integrating the area below the dashed line between them to the left of the right threshold. It is clear how the values of P(FA) and P(MD) depend on the choice of threshold, J.

The probabilities of false alarm and missed detection, calculated in this way, are obtained for one scalar residual. If the diagnosis system contains several residuals, the calculations become more complicated, and a further aspect to consider is that the probability distribution of the residuals are usually not known, which means that they must be estimated.

α-Risk and β-Risk

Sometimes, in the framework of statistical hypothesis, P(FA) and P(MD) are called α-risk and β-risk (Casella and Berger 2001). They are defined by

α-**risk:** False positive, also called Type I error, with risk size α defined by $\alpha =$P(FA).
β-**risk:** False negative, also called Type II error, where β is defined by $\beta =$P(MD).

Adaptive Thresholding

In the example in the previous section the probability density functions in Figure 16.15 were assumed to be (more or less) constant, which resulted in the use of a fixed threshold, J, representing the optimal trade-off between P(FA) and P(MD). However, in many applications the uncertainties, both regarding model accuracy, $G_f(s)$, and noise, $n(t)$, vary depending on operating conditions. Some examples are that models may be much more precise in stationary operation than during transients, and that for nonlinear systems the transfer gains may vary between operating points. Relating to Figure 16.15, this means that the distributions vary in shape and position.

One important technique in these cases is the use of an adaptive threshold, which is a threshold that varies depending on the situation with the goal to represent a good balance between P(FA) and P(MD) at all instants. Figure 16.16 gives an example from an air system.

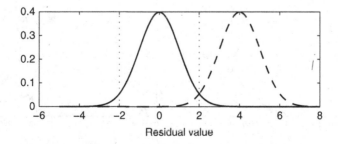

Figure 16.15   An example of probability density functions for a residual in the fault-free case (solid line) and when a constant fault is present (dashed line)

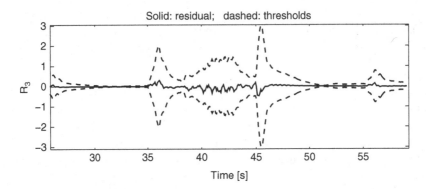

Figure 16.16 An example of an adaptive threshold for a valve in an air system. The residual is shown in the solid line and the thresholds in dashed lines

Advanced Methods

The fault decision should contain information about whether a fault has occurred, and in that case also the following information: when did the fault occur, which is the faulty component, and possibly also the size of the fault. In some cases more advanced methods than thresholding are needed, and it is possible to use statistical tests for abrupt changes. Assume, for the sake of illustration, that the probability density functions in Figure 16.15 have a large overlap, so that, for any given threshold, both P(FA) and P(MD) would be unsatisfactorily large. The resulting signals could then be as in Figure 16.17, where the upper left subplot shows a step in level which is hard to see since the noise level is larger than the step. In the upper right subplot there is another type of change, where the signal changes intensity/variance. Even though it is hard to see, there are methods and algorithms that find these changes. In the lower subplots, the results of such an algorithm, the CUSUM algorithm (Basseville and Nikiforov 1993), show in both cases that the residual is successfully evaluated.

For a comprehensive discussion of these topics, see Basseville and Nikiforov (1993) or Gustafsson (2000). Another approach is *fuzzy thresholding*, see Frank (1994), which is a technique by which the thresholds are made fuzzy and the fault decision is calculated using

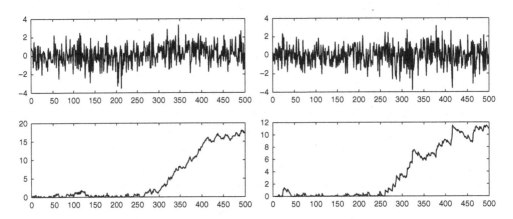

Figure 16.17 In the upper left plot the signal changes level, and in the upper right plot the signal changes intensity/variance. Relating to Figure 16.15, the distributions are so overlapping that it is hard to see where the changes occur. Below, successful detection using CUSUM is shown for both cases

fuzzy rules. In this way the fault decision will contain information about how "probable" it is that the fault has occurred.

16.3.7 Leakage in a Turbo Engine

As an illustration of the methodology, leakage diagnosis in a turbo engine will be used as an example. Consider the engine in Figure 4.7, which is the most frequently used configuration for a turbocharged SI engine. The figure is recaptured in Figure 16.18, but now with the addition of two possible leaks. The **diagnosis objective** is to find leakages before the throttle, but after the intercooler, called a boost leak, or after the throttle, called a manifold leak.

Modeling

The model of the engine is the standard model described in Section 8.9, where the Simulink diagram of the model is seen in Figure 8.26. Using the standard notations from there we have the flow past the mass flow meter, $\dot{m}_{af}$, the throttle flow, $\dot{m}_{at}$, the flow into the cylinder, $\dot{m}_{ac}$, the pressure before the throttle, p_{ic}, and the pressure in the intake manifold, p_{im}.

Leaks are introduced in the model by using a flow restriction model, and here Model 7.10 is used. In the expression (7.10d) the area, A, of the (circular) hole, is chosen by using the diameter 1.5 mm for the manifold leakage and 5 mm for the boost leakage.

Residuals and Influence Structure

The sensors used are engine speed, N, the mass flow meter, $\dot{m}_{af}$, the throttle angle, α, the pressure before the throttle, p_{ic}, and the pressure in the intake manifold, p_{im}. Then, also from Section 8.9, the mass flows in the fault free case schematically relate to the sensors as

$$\dot{m}_{ac} = f_1(N, p_{im})$$
$$\dot{m}_{at} = f_2(\alpha, p_{im}, p_{ic})$$

where other variables, like, for example, manifold temperature T_{im}, have been neglected.

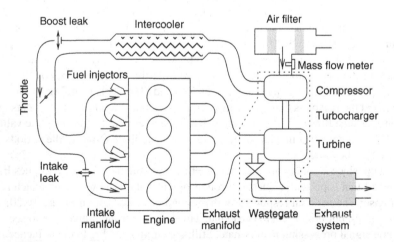

Figure 16.18 A principle illustration of a turbo charged SI-engine with two possible leaks inserted, a boost leak and an intake manifold leak

When the system is working as it should, it is clear that the flow past the mass flow meter, $\dot{m}_{af}$, the throttle flow, $\dot{m}_{at}$, and the flow into the cylinder, $\dot{m}_{ac}$, are the same less transients. Therefore, form the following residuals that should be zero in the fault free case

$$r_1 = \dot{m}_{af} - \dot{m}_{at} = \dot{m}_{af} - f_2(\alpha, p_{im}, p_{ic})$$
$$r_2 = \dot{m}_{af} - \dot{m}_{ac} = \dot{m}_{af} - f_1(N, p_{im})$$
$$r_3 = \dot{m}_{ac} - \dot{m}_{at} = f_1(N, p_{im}) - f_2(\alpha, p_{im}, p_{ic})$$

This leads to the following influence structure showing how the faults influence the residuals

	NF	F_1	F_2
r_1	0	X	0
r_2	0	X	X
r_3	0	0	X

Evaluation and Results

The system above is simulated in three cases, the fault free case and the two cases with either a boost leak, fault F_1, or a manifold leak, fault F_2. In all cases the driving cycle NEDC is used, see Section 2.8 and specifically Figure 2.22. The residuals are computed, and before presentation they are treated according to the first two steps in Figure 16.14, that is they are low pass filtered and the absolute value is taken.

The results are given in Figures 16.19 and 16.20. Both figures include the fault free case, and Figure 16.19 shows the result for the boost leak while Figure 16.19 shows the results for the manifold leak. In both figures the drive cycle NEDC used for evaluation is plotted in the top subplot together with the residual r_1.

First case: Start by examining residual responses in Figure 16.19. In the lower subplot, the two curves are identical for the fault free case and the one with a leak. Both r_1 and r_2 react in the leaky case compared to the fault free case, but not all the time. The reason is that the NEDC has such mild accelerations that boost is not really needed. Nevertheless, the system is preconditioning in some transients leading to boosting, and there are clear responses from around time 1040 s to 1120 s. Note the correlation with NEDC when this happens. This illustrates the fact that there needs to be enough excitation to draw conclusions, and it also relates back to the discussion about adaptive thresholding in Section 16.3.6. Note also that there is a response in the time interval between 780 and 800s. This is during an idle period in the NEDC, and the reason is that the air filter creates a slight vacuum, which means that there is an inward leakage flow (it is thus negative, but remember that it is the absolute value that is presented). This means that in addition to periods with significant boost, idle periods can also be used to detect boost leaks.

Summing up, since r_1 and r_2 are the ones reacting, the influence structure implies F1, that is that it is fault 1 that is present. Hence, the conclusion is that it is a boost leak, which is correct.

Second case: The residual responses in the second case are seen in Figure 16.20. There is no response in r_1, but both r_2 and r_3 respond. This is easy for the human eye to see, but may require some signal processing to automate. Still, signaling r_2 and r_3 to the influence structure implies F2, that is that it is fault 2 that is present. Hence, the conclusion is that it is a manifold leak, which is correct.

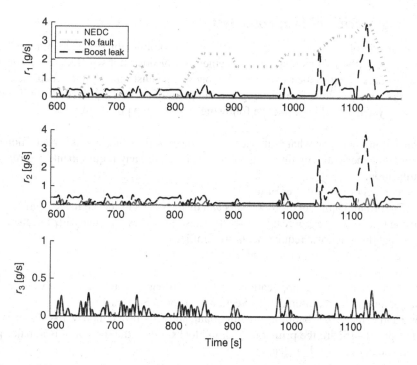

Figure 16.19   The residuals when a boost leak has occurred. NEDC is shown for reference

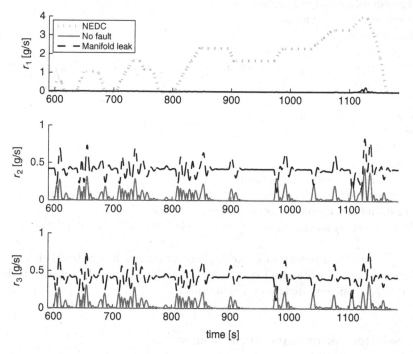

Figure 16.20   The residuals when a manifold leak has occurred. Since r_2 and r_3 react, the influence structure correctly implies that it is a manifold leakage

16.4   Engineering of Diagnosis Systems

The total process of designing a diagnosis system is a combination of many engineering steps coping with usually rather complex requirements. Considerations include developing-time constraints, market requirements, as well as economical constraints, but the focus here will be on the technical side. Thus, from an engineering point of view, the procedure for developing a diagnosis system could look like the following step-by-step procedure:

- Obtain requirements on what faults need to be diagnosed, requirements on detection and/or isolation together with any time constraints on these, and any requirements regarding fault identification.
- Build a model of the fault free case.
- Describe, or build a model of, how faults and disturbances influence the system.
- By using the model, design residuals or other test quantities together with a decision structure so that the diagnosis requirements are fulfilled.
- Test the diagnosis system in simulation and in reality.

Step 1 can, depending on application, be based on legislation, safety analysis, reliability analysis, or any of the requirements in Section 16.1, or in Section 16.6.

Steps 2 and 3, the model building, are important, and established models like the ones in this book, Chapter 1–15, are the primary building blocks. Often the modeling is a major part of the work, and consists of three parts:

- Selection of model and its structure.
- Parameter setting in the model.
- Model validation.

Step 4 is the topic of the present chapter, and looking a bit closer at it gives the following major design steps.

- Choose a model for the fault free case.
- Define fault modes.
- Define fault models.
- Analyze redundancy.
- Design residuals or other test quantities.
- Devise thresholds, adaptive thresholds, or other evaluation schemes depending on need.
- Decide the influence structure or other decision scheme.
- Evaluate sensitivity and performance, for example using P(FA) and P(MD) or other means.

The final step, Step 5, includes the topic of implementation and this can be different depending on whether it is an on-board or off-board application, for example. In conclusion, design of diagnosis systems requires quite some engineering.

16.5   Selected Automotive Applications

In the introduction to this chapter the goal was stated as giving insight into diagnostics, and to do so in enough depth to show the interplay between the basic physics, the models, and the possibilities for design. Having set this goal, it is impossible to cover the field in breadth, so the text has to be a selection of important representatives. The five examples chosen here are: supervision of the three way catalyst, throttle supervision, and diagnosis of the purge evaporative system, of the engine air intake, and of misfire. The examples are chosen because of their importance, but also selected to show some different methodological aspects since there are examples of both active and passive diagnosis, and examples of both hardware redundancy and analytical redundancy. Further, they also represent the major systems in that they are from the air system, the fuel system, or the after-treatment system. The technological principles are treated here, whereas the connection to legislation will be presented in Section 16.6.

16.5.1   Catalyst and Lambda Sensors

Early on, it was realized that malfunction of three way catalysts was a major reason for harmful emissions from SI engines. Typically, misfire in the engine would cause unburned fuel to enter the (hot) catalyst where it would burn and destroy the catalyst. This led to a strong drive to diagnose misfire, see Section 16.5.4, and also to diagnose if the catalyst is working properly.

In OBDII, the diagnosis system must indicate a fault if the efficiency of the catalyst falls below 60%. The technology used today is to use two lambda (oxygen) sensors, one upstream and one downstream of the catalyst. This arrangement was shown already in Figure 7.1, showing the pre-catalyst sensor, λ_{bc}, and the post-catalyst sensor, λ_{ac}. As described in detail in Section 7.7.5 there are different types of lambda sensors, and the diagnostic techniques differ depending on the choice. Here, the principles will be illustrated assuming switching type sensors both pre- and post-catalyst.

A working catalyst stores and releases oxygen as described in Section 6.5.1, and the result is that the catalyst acts as a buffer and almost filters out the oscillations in λ, as seen in the right plot in Figure 7.23. However, the response in the post-catalyst sensor is not completely filtered out even though it may seem so on the resolution in Figure 7.23. The experimental data in Figure 16.21 shows the right subplot in an enlarged scale. There it is seen that the

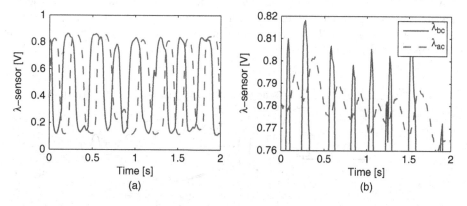

Figure 16.21   Responses in the pre-catalyst and post-catalyst sensors (switch type) for non-working catalyst (a) and a working catalyst (b). When the catalyst is working (b) the oscillations in the incoming gas are almost filtered out due to the oxygen storage capability in the catalyst (note that the scale is changed significantly)

post-catalyst sensor has a response with the same frequency as the pre-catalyst sensor but with some delay. In this figure one can also see typical responses for steps from lean to rich and rich to lean both for the pre-catalyst and post-catalyst sensor.

In the situation described here, the diagnostic tests for the catalyst are using the fact that the variations, due to the limit cycle enforced by the control system, in the upstream lambda sensor should be almost gone in the downstream sensor (recall once again Figure 7.23).

Sensor Diagnosis

Regarding diagnosis of the sensors themselves, the diagnosis system should detect change in the time constant or an offset of the lambda sensors, and this is done by studying the frequency, comparing the two sensors, and studying step responses. OBDII prescribes that the system shall detect a malfunction prior to any failure or deterioration of the oxygen sensor voltage, response rate, amplitude, or other characteristic(s) (including drift or bias corrected for by secondary sensors) that would cause a vehicle's emissions to exceed 1.5 times any of the applicable FTP standards. For response rate, the OBD II system shall detect asymmetric malfunctions (i.e., malfunctions that primarily affect only the lean-to-rich response rate or only the rich-to-lean response rate) and symmetric malfunctions (i.e., malfunctions that affect both the lean-to-rich and rich-to-lean response rates). As an example of a requirement, the system must at a minimum detect a slow rich-to-lean response malfunction during a fuel shut-off event (e.g., deceleration fuel cut event). The overall task is to secure that all sensor responses are normal.

16.5.2   Throttle Supervision

As described in Section 16.1.1, unintended torque is a serious incident. Also mentioned there, the brake override system is a system where the brake pedal always overrides the accelerator. This also disables so called toe-heel operation, that us simultaneous operation of the brake and accelerator to keep engine speed while braking.

To avoid SAI (sudden acceleration incidents), due to drive by wire faults, several measures are taken regarding the throttle. Recall Section 7.13, where the throttle was described. The system is sketched in Figure 7.45, which is repeated here as Figure 16.22. The basic control was described in Section 10.3.

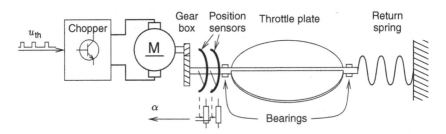

Figure 16.22   Sketch of a throttle from Figure 7.45. For safety reasons the position sensor is duplicated, and two completely separate signal paths are used. Further, the return spring returns the throttle to a predetermined (safe) position if there is no input

Hardware redundancy is used to ensure safe operation. For the position sensor of the throttle, two redundant potentiometers, going in different directions, with separate supply lines, are used. To increase the safety level, two separate channels are used to the ECU, and one channel may be analog and one a digital pulse train. If both channels are analog, then receiving A/D converters are completely separate. In either case, the respective signals enter different CPUs in the ECU. The resulting system is analyzed extensively for possible common cause failures, and this includes all combinations of faulty cables, faulty supply, faulty signal ground, and so on. In recent years, the code handling all this has expanded substantially.

As a remark, one level up there is also supervision of the torque calculations by two different calculations: one in the torque-based structure and one external torque-monitoring calculation. The throttle supervision is thus only one example of all the systems that monitor engine actuators so that the engine does not deliver much more or much less torque than that which the driver requests. If there are severe indications, the car will enter limp home mode. Another remark is that the accelerator pedal also has two sensors and duplicated systems.

16.5.3   Evaporative System Monitoring

The fuel system in personal vehicles must be supervised for leakages, and an on-board diagnostics system sholud monitor the complete evaporative system for vapor leaks to the atmosphere. From a diagnosis point of view it is interesting, since currently no passive method exists, so available methods rely on active diagnosis (intrusive diagnosis).

The fuel system was discussed in Chapter 10. Specifically, it was depicted in Figure 10.19, a figure that is repeated here as Figure 16.23. Already in Section 10.5.3 the main function of the purge system was explained as to take care of fuel vapor from the fuel tank. As explained there, the system contains a carbon canister, which is connected at one end to the fuel tank and at the other end to ambient air. The system has a diagnosis valve which is open during normal operation of the engine, and closed when diagnosis is performed. A purge valve connects the canister to the intake manifold of the engine. The canister is regularly purged of hydrocarbons

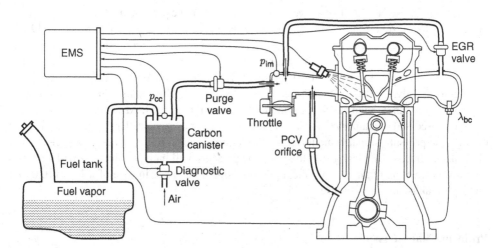

Figure 16.23   Disturbances to the air and fuel metering. Additional flows of air and fuel into the intake manifold

when the purge valve is opened, causing a flow of air through the canister and into the engine where the fuel vapor is combusted. The system is equipped with a pressure sensor that measures the difference in pressure between ambient air and the fuel tank pressure.

Diagnosis Objective

The diagnosis system must be able to detect malfunctioning valves and also a leak in the fuel tank. Quoting the regulation: "The OBD II system shall verify purge flow from the evaporative system and shall monitor the complete evaporative system, excluding the tubing and connections between the purge valve and the intake manifold, for vapor leaks to the atmosphere." It shall detect an evaporative system malfunction when the complete evaporative contains a leak or leaks that cumulatively are greater than or equal to a leak caused by a 0.020" diameter orifice.

Pressure or Vacuum

Two main principles for leakage monitoring exist: vacuum decay and pressure decay principles. With the vacuum decay principle, an underpressure is created in the fuel tank compared to the ambient pressure and the decay of the pressure difference is monitored and analyzed. The pressure decay principle creates an overpressure in the fuel tank and the pressure difference is monitored and analyzed. The two principles have their own set of advantages and disadvantages. A disadvantage with pressure decay methods is that, in case of a leak, the overpressure presses fuel vapor out into the atmosphere. In a vacuum decay method the air flow is into the fuel tank, and thus vacuum decay methods are considered environmentally more safe. In addition, pressure decay methods require an extra component, a pump to pressurize the fuel tank.

System Operation

The technology used is based on active tests, and it will here be presented for the vacuum decay principle. The presentation is entirely based on the paper by Krysander and Frisk (2009). In normal operation of the evaporative emissions control system, the diagnosis valve is open and the purge valve is closed. This means that evaporating fuel will be collected in the carbon canister, which can be purged by opening the purge valve. To initiate a leakage detection sequence, the diagnosis valve is closed and the purge valve is opened. This results in a pressure drop in the tank which can be seen at $t = 0.5$ s and $t = 11$ s in Figure 16.24. After about 2 s, the purge valve is closed and the tank system is, in a fault free case, now sealed. The basic idea is then to monitor the pressure signal behavior in the shaded intervals in Figure 16.24 to detect a possible leakage. In case of a leakage in the tank, the pressure will increase since air will leak into the tank from ambient air, and a typical pressure signal behavior in case of a 1 mm (0.020") diameter leak is shown in Figure 16.25.

Problem Characteristics

A main complication is the effect of evaporating fuel. The effects can be seen in Figure 16.24 where it is clear that even though there is no leakage in the tank, the tank pressure increases.

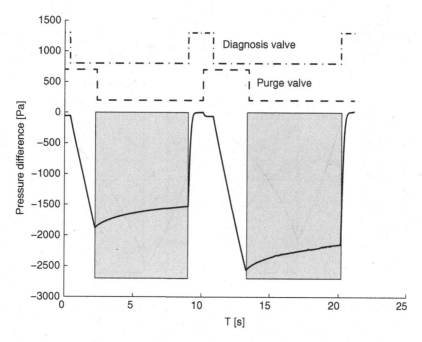

Figure 16.24 Typical cycles for leakage detection for the fault free case. The solid line is the pressure measurement, and the dashed and dashed-dotted lines indicate the position of the purge and diagnosis valve. The gray areas indicate which data are used for detecting leaks. Data are obtained from a production engine fuel system

This means that a leakage with small orifice diameters produces pressure traces similar to the no-leakage case. This similarity increases with decreasing leakage orifice diameter. In the fault free case, the tank pressure increases until it reaches its saturation pressure, and since the saturation pressure is temperature dependent there is a need for the detection algorithm to take this into account in order to be robust towards different temperatures. An additional complication, which can also be seen in both Figure 16.24 and Figure 16.25, is that the pressure sensor is subjected to a slowly time-varying bias. When the diagnosis valve is open and the purge valve is closed, one can expect that the tank pressure equals ambient pressure, that is a sensor reading of 0. However, the pressure reading at $t = 0$ in both Figure 16.24 and 16.25 is distinctly nonzero, and this also needs to be considered when designing the detection algorithm.

Modeling

From the discussion in the previous sections, the diagnostic task is to detect differences in vacuum decay between Figure 16.24 and Figure 16.25. From the experimental data in these figures it is clear that both fuel evaporation in the tank and sensor bias must be considered.

The leakage detection is performed when an underpressure in the tank has been created and both valves are closed. During a leak detection test, it is assumed that the temperature and the volume in the tank are constant. This is reasonable since only about 3 kPa is evacuated, and the leakage test is performed in less than 10 s. In the described situation, the only two ways for the

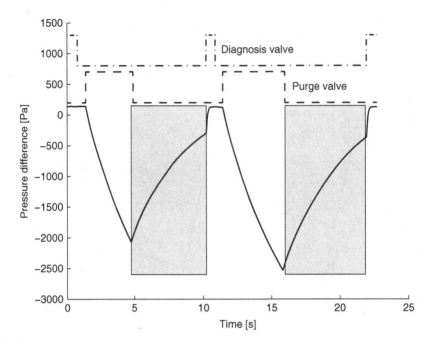

Figure 16.25 Cycles for leakage detection for the case with a 1 mm leak. The solid line is the pressure measurement, and the dashed and dashed-dotted lines indicate the position of the purge and diagnosis valve. The gray areas indicate which data are used for detecting leaks

pressure, p, to increase is by fuel evaporation or a possible leakage. To be able to separate these two processes a physical model of the fuel tank pressure valid in the gray shaded intervals in Figures 16.24–16.25 can be used.

Given a fixed gas volume and temperature in the tank, the ideal gas law implies that the rate of pressure change, $\dot{p}$, is proportional to the sum of fuel evaporation mass flow rate, $\dot{m}_f$, and the leakage mass flow, $\dot{m}_l$ directed into the tank, that is

$$\dot{p} \sim \dot{m}_f + \dot{m}_l \tag{16.6}$$

The total pressure, p, in the tank is according to Dalton's law equal to the sum of the partial pressure of air, p_a, and the partial pressure of fuel vapor, p_f, that is

$$p = p_a + p_f \tag{16.7}$$

A model for the fuel evaporation mass flow rate, $\dot{m}_f$ is that it is proportional to the difference between the saturated fuel pressure, p_f^0, and the fuel vapor partial pressure p_f,

$$\dot{m}_f \sim (p_f^0 - p_f) \tag{16.8}$$

where the saturation pressure is dependent on temperature and fuel composition.

To get a model for the leakage air mass flow, $\dot{m}_l$, into the tank through a hole, incompressible flow is assumed which means that the standard model (7.5) can be used

$$\dot{m}_l \sim \sqrt{p_{amb} - p} \tag{16.9}$$

Combining (16.6), (16.8), and (16.9) gives

$$\dot{p} = k_1(p_f^0 - p_f) + k_2\sqrt{p_{amb} - p} \tag{16.10}$$

where k_1, k_2 are temperature and gas volume dependent proportionality constants. The evaporation constant k_1 is dependent on fuel composition, and the leakage constant k_2 is proportional to the effective leakage area.

Sensor Equation

The sensor measures the overpressure in the tank. With a bias, b, the sensor equation is

$$y = p - p_{amb} + b \tag{16.11}$$

The bias, b, is assumed to be constant during a 10 s leak detection test and this is assumption is consistent with experimental data. This can also be seen in the cycles shown in Figures 16.24–16.25. Assuming also that the ambient pressure p_{amb} is constant, that is $\dot{b} = 0$ and $\dot{p}_{amb} = 0$, then elimination of p and p_f in (16.7), (16.10), and (16.11) results in the first-order model

$$\dot{y} = -k_1 y + k_2\sqrt{b-y} + k_1(p_f^0 + p_a + p_{amb} + b) \tag{16.12}$$

that for the fault free case with no leakage has $k_2 = 0$.

Leakage Monitoring Method

(16.12) is the basis for leakage detection. During a test, it is assumed that sensor bias, b, ambient pressure, p_{amb}, temperature, gas volume, fuel composition, and leakage area are constant parameters. This means that b, p_{amb}, k_1, k_2, and p_f^0 are constants. However, the partial pressure of air, p_a, is constant only if there is no leakage and this needs to be considered in the leakage detection method.

For the slowly varying bias that is assumed constant during the test interval, note that when the diagnosis valve is open and the purge valve is closed, the tank pressure should quickly stabilize around the ambient pressure. This means that the measurement signal, y should be 0 if there is no bias. Thus, the current bias can easily be estimated by taking the mean value over data where the diagnosis valve is open and the purge valve is closed. For example, from the first second in Figure 16.25 it is clear that there exists a bias 150 Pa. The estimated bias can be subtracted from the measurement signal which then can be assumed to be bias free.

Now, based on 16.12 it is straightforward to devise different schemes for leakage detection, for example based on parameter estimation where $k_2 \neq 0$ means a leak, and the size of the leak is proportional to the size of k_2. There are different methods to optimize performance, for example, regarding p_a, see Krysander and Frisk (2009), for one example.

16.5.4 Misfire

Engine misfire detection is an important part of the On-Board Diagnostics (OBDII) legislation to reduce exhaust emissions and avoid damage to the catalytic converters. This is because a misfire means that unburned gasoline reaches the (hot) catalyst where it may ignite, which can result in an overheated and severely damaged catalyst. The diagnosis system must be able to detect a single misfire and also to determine the specific cylinder in which the misfire occurred.

From a perspective of diagnosis methodology, misfire detection is interesting since the modeling is fairly straightforward, but it is advanced decision making that is the key in the more difficult cases.

Crank Speed Variations

The main technology used today is based on signal processing of the rpm signal measuring the engine speed, N. It has been used for many years, but when requirements tighten, still more sophisticated methods may be needed.

The main idea behind analysis of crank speed variations is that a misfire means that there is no torque generated for the misfired cylinder, and thus this will show as a disturbance in the engine crank speed. An example of this is seen in the experimental data in Figure 16.27a. The physical setup to obtain the rpm-signal was presented in Section 9.1.2, see Figure 9.2, where angular velocity measurements are generated by computing the time difference between two teeth when the flywheel rotates, or between two holes in a ring sensed by a Hall effect sensor. Figure 16.26a shows the principle for crankshaft speed measurement by tracking time between cogs, and Figure 16.26b shows an engine with both cogwheel and ring.

Modeling

The basis for modeling is a crankshaft model with torsional flexibilities. The modeling follows the systematic steps presented in Chapter 14, specifically in Section 14.1 and Figure 14.1. However, instead of driveline components, it is the parts of the crankshaft that are considered.

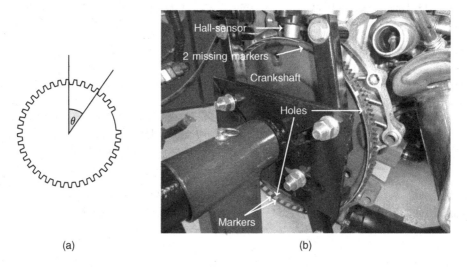

(a)   (b)

Figure 16.26 (a) Cogs at the flywheel are used for tracking engine position as well as for misfire detection. In misfire detection the time between two cogs is detected. See Figure 16.27 for an example of such a measurement. (b) A picture of a flywheel and the Hall effect sensor

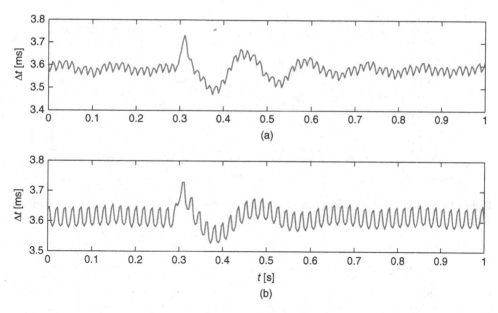

Figure 16.27   In (b) simulated angular velocity measurements at the flywheel are shown during a misfire and in (a) the real measured angular velocity during a misfire

This type of modeling is used in, for example, Kiencke and Nielsen (2005) wherein Section 6.8 gives a detailed presentation of the basic method. In Eriksson et al. (2013) the modeling is taken one step further in that the analytic cylinder pressure model from Section 7.8 is used, see (7.51). Figure 16.27b shows a simulation taken from that paper.

Difficulties

As seen above, the basic method works well in a standard situation. However, there are several different types of disturbances that complicate detection of misfire. These include any phenomenon that could influence the rotational speed, and some examples are

- load and speed
- ignition time
- torsion (road, engine ...)
- flywheel measurement errors
- fuel (different quality)
- auxiliary loads (e.g., AC)
- knock control
- friction.

Cold Start as An Example of a More Difficult Case

One situation that is more difficult than the standard situation is cold start. Figure 16.28 shows the rpm signal during a misfire for a warm engine. The situation with a cold engine

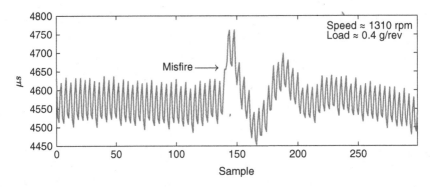

Figure 16.28   Examples of the flywheel angular velocity signal from a 6- cylinder engine. The signals contain misfires which are visible as increased measured time between markers

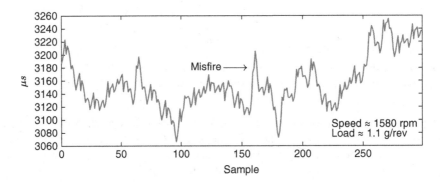

Figure 16.29   An example of the flywheel angular velocity signal from a 6-cylinder engine during a cold start. Compared to signals in Figure 16.28, a misfire is more difficult to detect

and low load is shown in Figure 16.29. It is clear that misfire detection is harder in this latter case.

Advanced Interpretation of Crank Speed Variations for Cold Start

Inspecting Figure 16.29, it is likely that the situation is a case of varying and overlapping distributions as described in Section 16.3.6. It is therefore useful to study distributions for different test quantities, and such a result from Eriksson (2013) is presented in Figure 16.30.

The basis for the analysis is the model from Eriksson et al. (2013), mentioned above. All data are measured on cars from Volvo. The flywheel angular velocity signal is used to detect misfires and the crank counter, load, speed, and the catalytic converter warming flag are used to identify the operating point of the engine, in which cylinder the combustion occurs, and if it is during a cold start or not. All signals are sampled crankshaft angle synchronous and at the same rate. The resolution of the total measurement system is low so it is only possible to get four samples for each combustion. They are obtained at crank count positions named 1, 2, 3, and 4.

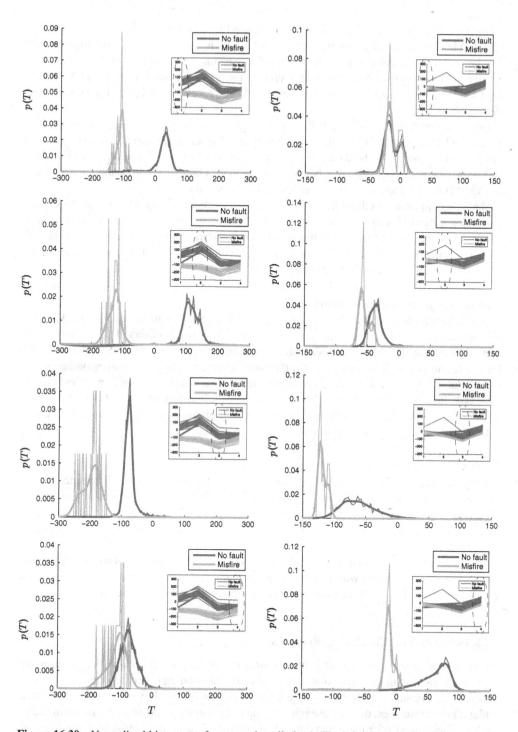

Figure 16.30 Normalized histograms for torque in cylinder 1. The left column is for a warm engine and the right column is for a cold start. The rows represent four different positions during a combustion as indicated by the small inserted box showing torque as a function of crank count

During data collection, misfires have been injected by interrupting the combustion in different cylinders. Using the model, the flywheel signal is used to estimate the torque at the flywheel, and this is done four times for each combustion. For each case of these four crank count positions, a normalized histogram is created for both the case of no fault and the case of misfire. Recall the discussion around Figure 16.15. The results for cylinder 1 are collected in Figure 16.30.

The left column shows the distribution of the torque of cylinder 1 for crank count positions 1, 2, 3, and 4, during normal driving at operating point speed 1500 rpm and load 1.2 g/rev. A normalized histogram of the data and an approximation of each distribution using kernels is shown in each plot. The separation between torque from a fault-free combustion and a misfire is largest at the second sample and smallest at the last sample.

The right column shows distribution of the torque of cylinder 1 for crank count positions 1, 2, 3, and 4, during cold start at operating point speed 1500 rpm and load 1.2 g/rev. A normalized histogram of the data and an approximation of each distribution using kernels is shown in each plot. The separation between torque from fault-free combustion and a misfire is largest at the last sample and smallest at the first sample.

Conclusions from Histogram Study

The conclusions from the study of the histograms in Figure 16.30 are as follows. For normal operation, the difference in torque between misfire or not is observed fairly early in the combustion trace. This corresponds to left column, second row. For a cold start it is different. Using the same detection scheme, that is the same crank count position, corresponding to the right column, second row, would give almost overlapping distributions and an almost impossible task. Instead, one should use a late crank count position, corresponding to the fourth row. Thinking about the physics, it is not surprising that relevant information appears later, because combustion is slower at a cold start than for a warm engine.

Remark

During the development process in Eriksson (2013), the Kullback–Leibler divergence, which is a measure of how separated two distributions are, was used to automate and analyze the ability to detect a misfire given a test quantity and how the misfire detectability performance varies depending on, for example, load and speed. The Kullback–Leibler divergence was also used for parameter optimization to maximize the difference between misfire data and fault free data. Evaluation shows that the resulting misfire detection algorithm is able to have a low probability of false alarms while having a low probability of missed detections. Thus, it is a good example of the use of the more advanced methods as discussed in Section 16.3.6.

V12, Ion-sense, and Concluding Remarks

As mentioned above, a cold start is not the only demanding situation when regulations are tightened regarding misfire detection. One case worth mentioning is very smooth engines, like a large V12. For such an engine, a single misfire may not be clearly visible in the rpm signal under all circumstances, due to the engine inertia and several combustions per revolution. For such engines, in-cylinder measurements are an alternative, and one candidate is ion-sense as discussed in Section 12.3.1.

16.5.5 Air Intake

The final example will combine several of the techniques presented previously. It will also be an outlook, and illustrate how a set of general methods for model-based sequential residual generation and data-driven statistical residual evaluation can be combined. The presentation is based on Svärd et al. (2013), where the aim is an automated design methodology.

The system considered is the diesel engine from Section 8.10, see Figure 8.27, that is studied also in Section 11:4. Here the model is parametrized for a 13-liter 6-cylinder Scania truck diesel engine equipped with EGR, VGT, and an intake throttle. A schematic of the system is repeated in Figure 16.31, where possible faults are also indicated as described below.

Sensors, Actuators, and Faults

The system is equipped with four actuators, $u_{x_{th}}$, $u_{x_{egr}}$, $u_{x_{vgt}}$, u_ρ, and seven sensors, $y_{p_{amb}}$, $y_{T_{amb}}$, $y_{p_{ic}}$, $y_{p_{im}}$, $y_{T_{im}}$, $y_{p_{em}}$, y_{n_e}, see Table 16.1. Faults in all sensors and actuators in Table 16.1 are considered, except in actuator u_ρ and sensor y_{n_e}. The faults, along with their descriptions, can be found in Table 16.2, and the approximate locations of the faults are marked with triangles in Figure 16.31. For the engine considered, all faults found in Table 16.2 are emission critical so they need to be detected and isolated. In addition, if not accommodated in time, the faults in Table 16.2 may lead to decreased safety, increased fuel consumption, decreased driveability, or even engine breakdown.

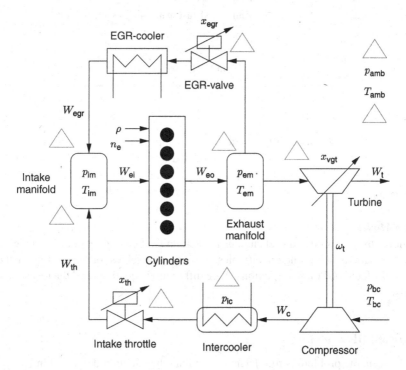

Figure 16.31 Schematic of the automotive diesel engine system. Locations of considered faults are illustrated with triangles

Table 16.1 Sensors and actuators

Sensor	Description	Actuator	Description
y_{n_e}	Engine speed sensor	$u_{x_{th}}$	Throttle position actuator
$y_{p_{amb}}$	Ambient temperature sensor	$u_{x_{egr}}$	EGR-valve position actuator
$y_{T_{amb}}$	Ambient pressure sensor	$u_{x_{vgt}}$	VGT-valve position actuator
$y_{p_{ic}}$	Intercooler pressure sensor	u_{ρ}	Injected fuel actuator
$y_{p_{im}}$	Inlet manifold pressure sensor		
$y_{T_{im}}$	Inlet manifold temperature sensor		
$y_{p_{em}}$	Exhaust manifold pressure sensor		

Table 16.2 List of considered faults

Fault	Description
$\Delta_{y_{p_{amb}}}$	Fault in ambient pressure sensor
$\Delta_{y_{T_{amb}}}$	Fault in ambient temperature sensor
$\Delta_{y_{p_{ic}}}$	Fault in intercooler pressure sensor
$\Delta_{y_{p_{im}}}$	Fault in intake manifold pressure sensor
$\Delta_{y_{T_{im}}}$	Fault in intake manifold temperature sensor
$\Delta_{y_{p_{em}}}$	Fault in exhaust manifold pressure sensor
$\Delta_{u_{x_{th}}}$	Fault in throttle position actuator
$\Delta_{u_{x_{egr}}}$	Fault in EGR-valve position actuator
$\Delta_{u_{x_{vgt}}}$	Fault in VGT-valve position actuator

The faults are modeled as additive signals. For example, fault $\Delta_{y_{p_{im}}}$, representing a fault in the intake manifold pressure sensor $y_{p_{im}}$, is modeled by simply adding $\Delta_{y_{p_{im}}}$ to the equation describing the relation between the sensor value $y_{p_{im}}$ and the actual intake manifold pressure p_{im}, that is, $y_{p_{im}} = p_{im} + \Delta_{y_{p_{im}}}$.

Complete Model
The model of the automotive diesel engine can be found in Section 16.5.6. The model contains in total 46 equations, 43 unknown variables, 11 known variables, of which 4 are actuators, 7 sensors, and 9 faults. Of the 46 equations, 5 are differential equations and the rest are algebraic equations.

Structure of FDI System

As will be seen, the problem is more demanding than the situation described in Figure 16.14, so this situation requires more advanced methods, like the ones illustrated in Figure 16.17. The structure for this is depicted in Figure 16.32.

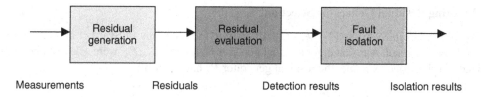

Figure 16.32   Overview of the sub-systems in the FDI system

Similar to the example in Section 16.3.7, measured signals from the actuators and sensors listed in Table 16.1 are used as input to the residual generation block. This block contains a set of residual generators, $R_1, R_2, \ldots, R_n$, and the output from the residual generation block is a set of residual signals, $r_1, r_2, \ldots, r_n$, with $r_i = R_i(y)$. The residual signals are used as input to the residual evaluation block, which contains a set of residual evaluators, $T_1, T_2, \ldots, T_n$. The output from these are test quantities $\lambda_1, \lambda_2, \ldots, \lambda_n$, so that $\lambda_i = T_i(r_i)$. The aim of the residual evaluation is to detect changes in the residual signal behavior in a manner such as that in Figure 16.17, see Figure 16.35 for an example.

Thus, the structure contains three sub-systems: residual generation, residual evaluation, and fault isolation. These will now be treated in order.

Residual Generator Design

In general, it is possible to create a large number of residual generators for complex models. As noted above, the model contains in total 46 equations, so it is natural that there are many possibilities for analytical redundancy, recall Section 16.3.5. A first step in using analytical redundancy is to find overdetermined sets of equations, and this can be done in an exhaustive manner using an algorithm for finding specific overdetermined sets of equations called MSO, see Krysander et al. (2008). Such an MSO set, by definition, contains one more equation than unknown variables, and the algorithm finds 270 MSO sets for the engine model. Given an MSO set, a residual generator can created by removing one equation and then finding a computation sequence for the unknown variables in the remaining just-determined set of equations. The residual is obtained by evaluating the removed redundant equation with the variables solved from the just-determined set. The number of candidate residual generators that can be created from a single MSO set thus equals the number of equations in the MSO set. From the 270 MSO sets, this gives in total 14 242 candidate residual generators for the engine model.

Having the 14 242 candidate residual generators, the next step is to find a computational sequence to realize the residual generator, see for example Svärd and Nyberg (2010) or Sundström et al. (2013). These methods are known as sequential residual generation, and can be carried out slightly differently depending on how differentiation or integration is handled in the computational sequence.

Stability also needs to be considered, especially since the model is non-minimum phase, as pointed out in Section 11:4, which means that the analytical relation between variables will have an unstable zero or pole depending on computation order. Here, for the sake of simplicity, the stability analysis was based on linearization in 20 different equilibrium points parameterized by the injected fuel amount and engine speed. The result was then verified through extensive experimental evaluations.

Resulting Residual Generator Selection

The algorithm in Svärd and Nyberg (2010) and the methodology in Svärd et al. (2013) was used, and the result was that eight residual generators, $R_1, R_2, \ldots, R_8$, were selected and realized. To show one example, the residual generator R_3 has the form

$$\dot{\omega}_t = \frac{P_t \eta_m - P_c}{J_t \omega_t} \tag{16.13a}$$

$$\dot{T}_{em} = \frac{R_e T_{em}}{p_{em} V_{em} c_{ve}} (W_{in} c_{ve}(T_{em,in} - T_{em}) +$$

$$R_e(T_{em,in} W_{in} - T_{em} W_{out})) \tag{16.13b}$$

$$\dot{p}_{em} = \frac{R_e T_{em}}{V_{em}} (W_{eo} - W_{egr} - W_t + \Delta_{W_{em}})$$

$$+ \frac{R_e}{V_{em} c_{ve}} (W_{in} c_{ve}(T_{em,in} - T_{em}) \tag{16.13c}$$

$$+ R_e(T_{em,in} W_{in} - T_{em} W_{out}))$$

$$p_{amb} = y_{p_{amb}} \tag{16.13d}$$

$$p_{bc} = p_{amb} \tag{16.13e}$$

$$x_{vgt} = u_{x_{vgt}} \tag{16.13f}$$

$$\vdots$$

$$T_{em,in} = T_{amb} + (T_e - T_{amb}) \exp\left(-\frac{h_{tot} \pi d_{pipe} l_{pipe} n_{pipe}}{W_{eo} c_{pe}}\right) \tag{16.13g}$$

$$W_{egr} = \frac{(\dot{p}_{im} V_{im} - R_a T_{im} W_{th} + W_{ei} R_a T_{im})}{R_a T_{im}} \tag{16.13h}$$

$$\vdots$$

$$P_c = \frac{W_c c_{pa} T_{bc}}{\eta_c} (\Pi_c^{1-1/\gamma_a} - 1), \tag{16.13i}$$

with the *residual equation*

$$r = y_{p_{em}} - p_{em} \tag{16.14}$$

corresponding to equation e_{43} in Section 16.5.6.

Properties of Selected Residual Generators

Properties for the eight selected residual generators can be found in Table 16.3. The first column in Table 16.3 shows which residual equation the corresponding residual generator uses. It can be noted that a majority of the eight residual generators use either equation e_{41}, or equation e_{43}, as the residual equation, corresponding to $r = y_{p_{im}} - p_{im}$ and $r = y_{p_{em}} - p_{em}$, respectively. Column 2 shows the number of equations contained in the computation sequence on which the corresponding residual generator is based, and the value in parenthesis shows how many

Table 16.3 Properties of the residual generators

	Residual	#Equations	#Inputs
R_1	e_{43}	42 (5)	9
R_2	e_7	43 (5)	10
R_3	e_{43}	43 (4)	10
R_4	e_{41}	44 (4)	10
R_5	e_{41}	44 (4)	10
R_6	e_{43}	44 (4)	10
R_7	e_{43}	41 (3)	10
R_8	e_{41}	43 (5)	10

of those equations are differential equations. Recalling that the model contains in total 46 equations, of which 5 are differential equations, it can be concluded that all residual generators use a substantial part of the complete model. Column 3 in Table 16.3 shows how many of the 11 available signals in Table 16.1 each residual generator uses as input. Clearly, all the residual generators use most of the available signals.

Design of Residual Evaluators

A first step in designing an evaluation scheme for residuals is to find out how they behave when no fault is present. To capture the behavior of the residuals in a variety of operating modes of the diesel engine system, two data sets of different characteristics were used. The first data set is about half an hour long and contains engine test-bed measurements from World Harmonized Transient Cycle (WHTC) test cycle. The second data set is approximately 2 hours long and contains measurements from part of a test drive in the south of Sweden, including both city and highway driving. To reduce the risk of overfit, the data sets were split into an estimation data set and a validation data set, of equal size. The data was sampled at a rate of 100 Hz, and consequently the estimation and validation data sets contain approximately 450 000 samples, each. The eight residual generators were run off-line using the measurements as input to obtain no-fault residual samples. A set of samples from residual r_5 is shown in Figure 16.33. Note the

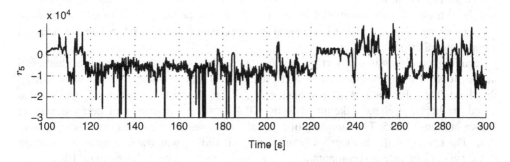

Figure 16.33 A subset of no-fault samples from residual r_5

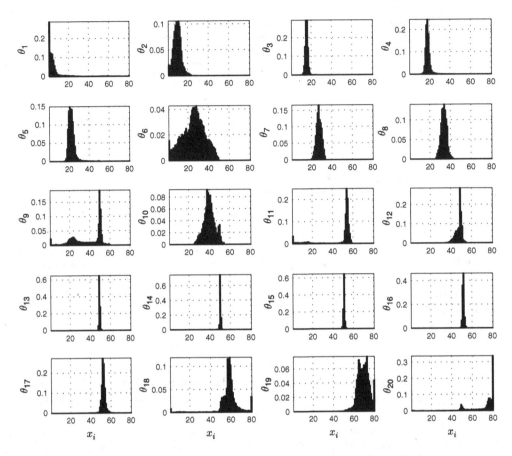

Figure 16.34 The set of 20 estimated no-fault distributions for residual r_5

non-ideal behavior of the residual caused by uncertainties, mainly model errors of time-varying nature and magnitude.

Following the method used to design residual evaluators described in Svärd et al. (2011), the next step is to evaluate the distributions in the fault-free case for different operating points, and as an example Figure 16.34 shows 20 estimated no-fault residual distributions for the residual r_5 obtained as output from residual generator R_5. It turns out that 20 distributions per residual is a good trade-off between model fit and complexity since the gain in model fit obtained when choosing a higher number is marginal in comparison with the corresponding increase in computational effort.

Recalling the discussion in Section 16.3.6 around adaptive thresholding and residual evaluation, it is clear from Figure 16.33 and Figure 16.34 that it is not sufficient with a fixed threshold. Thus, for each of the residuals $r_1, r_2, \ldots, r_8$, a residual evaluator T_i was created based on the Generalized Likelihood Ratio (GLR) test, and examples of performance are shown in Figure 16.35. The sampling of residual values was done by means of a sliding window. The choice of the number of samples in each sliding window is a trade-off between detection performance and computational complexity, and was here chosen to be 1024.

Table 16.4 Fault signature matrix for the eight residual generators. An x in position (i,j) represents the fact that fault mode j may cause residual i to raise an alarm

	$\Delta y_{P_{amb}}$	$\Delta y_{T_{amb}}$	$\Delta y_{P_{ic}}$	$\Delta y_{P_{im}}$	$\Delta y_{T_{im}}$	$\Delta y_{P_{em}}$	$\Delta u_{X_{th}}$	$\Delta u_{X_{egr}}$	$\Delta u_{X_{vgt}}$
T_1		x		x	x	x		x	x
T_2	x	x	x	x		x	x		x
T_3	x	x	x	x	x	x	x		x
T_4				x	x		x		
T_5		x		x	x	x	x		
T_6	x	x	x			x	x	x	x
T_7		x		x	x	x		x	x
T_8	x	x		x	x	x	x	x	x

Influence Structure

The influence structure is based on how the faults enter the equations, but also on residual evaluation tests to see the responsiveness. Table 16.4 shows the *influence structure*, IS, for the eight selected residual generators with respect to the faults in Table 16.2.

The eight selected residual generators are sensitive to several of the 12 faults, as illustrated in Table 16.4. It is a direct consequence of the fact that the engine system contains many physical interconnections such as, for instance, the shaft between the turbine and the compressor, which connects the intake and the exhaust parts of the engine, see Figure 16.31. This leads to a model with coupled equations, in the sense that there are sets of equations containing the same set of unknown variables. This implies that a fault affecting one of these equations influences a number of the other model equations. This, in combination with the relatively small number of sensors, makes fault decoupling non-trivial.

Fault Isolability

In general, given a set of residual generators, a fault Δ_x is said to be *isolable* from a fault Δ_y if the set contains a residual generator that is sensitive to fault Δ_x but not to fault Δ_y, see for example. Thus, this is a property that can be derived directly from the influence matrix. Based on the influence structure, Table 16.4, the resulting isolability matrix is shown in Table 16.5.

For more details about detectability and isolability definitions and specifications, see Krysander and Frisk (2008) and Frisk et al. (2012).

Experimental Evaluation–Fault Detection Performance

To validate the functionality of the diagnosis system it is evaluated experimentally. The first part of the evaluation focuses on the fault detection performance of the individual residual generators and residual evaluators. In the paper of Svärd et al. (2013), this is done systematically using the statistical power of the fault detection tests, for different sizes of the considered faults in Table 16.2. To quantify the power of a test, the power function is used (Casella and Berger 2001). Here we will only illustrate the result in one case. The second part of the evaluation, capturing the detection and isolation performance of the complete FDI system, is presented later.

Table 16.5 Maximum attainable isolability properties for the engine model. An x in position (i,j) indicates that fault mode i cannot be isolated from fault mode j

	$\Delta_{y_{P_{amb}}}$	$\Delta_{y_{T_{amb}}}$	$\Delta_{y_{P_{ic}}}$	$\Delta_{y_{P_{im}}}$	$\Delta_{y_{T_{im}}}$	$\Delta_{y_{P_{em}}}$	$\Delta_{u_{x_{th}}}$	$\Delta_{u_{x_{egr}}}$	$\Delta_{u_{x_{vgt}}}$
$\Delta_{y_{P_{amb}}}$	x	x				x	x		x
$\Delta_{y_{T_{amb}}}$		x				x			x
$\Delta_{y_{P_{ic}}}$			x				x		
$\Delta_{y_{P_{im}}}$				x					
$\Delta_{y_{T_{im}}}$					x	x			
$\Delta_{y_{P_{em}}}$						x			
$\Delta_{u_{x_{th}}}$							x		
$\Delta_{u_{x_{egr}}}$		x			x	x		x	x
$\Delta_{u_{x_{vgt}}}$		x				x			x

In total five data sets are used in the evaluation, where the data are not the same as the data used for finding no-fault distributions. Each data set contains measurements collected during a drive on the Swedish west coast. The data sets contain measurements from approximately 2.5 hours of driving, and include both highway and city driving under different conditions. The fault type used was a gain fault. In the case of, for example, sensor fault $\Delta_{y_{P_{amb}}}$, this means that the sensor signal $y_{P_{amb}}$ fed to the residual generators is $y_{P_{amb}} = \delta \cdot P_{amb}$ where $\delta \neq 1$ indicates a fault. The gain faults were implemented off-line by modification of the corresponding sensor or actuator measurement signals.

Experimental Evaluation–Behaviors of Individual Residuals and Test Quantities

An example of a result will provide some insight into the properties of the residuals and the fault detection tests. Figure 16.35 shows the residuals $r_1, r_2, \dots, r_8$ and test quantities $\lambda_1, \lambda_2, \dots, \lambda_8$ when fault $\Delta_{y_{P_{ic}}}$ of size $\delta = 1.2$ is abruptly injected at time $t = 700$ s. First of all, it is noted that the residuals in Figure 16.35 are all non-zero in both the no-fault and fault cases. In addition, all residuals exhibit non-stationary behaviors. It is clear that a conventional residual evaluation approach by means of, for example, constant thresholding would not be sufficient for these residuals.

Examining Figure 16.35 gives that λ_5 crosses the threshold slightly before 720 s giving a detection time of less than 20 s. Then, at time 760 s, there is a signal from $\lambda_2, \lambda_3, \lambda_4, \lambda_5, \lambda_6$, and comparing with the influence structure, IS, in Table 16.4 this is consistent with a fault in $\Delta_{y_{P_{ic}}}$. Thus the isolation time is around 60 s.

Experimental Evaluation–Performance Measures of Total FDI System

When looking at individual responses as in the previous section there is, of course, always the risk that one response is optimized at the cost of worse behavior of another. Therefore,

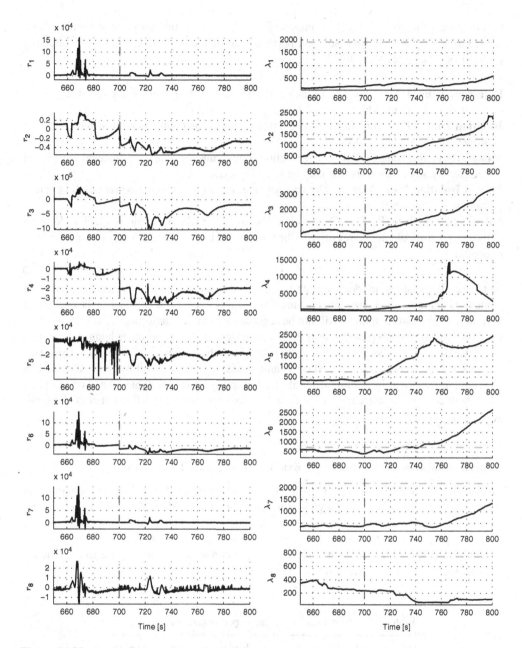

Figure 16.35 Residuals $r_1, r_2, \ldots, r_8$ in the left column, and test quantities $\lambda_1, \lambda_2, \ldots, \lambda_8$ in the right column when fault $\Delta_{y_{p_{ic}}}$ is injected at time $t = 700$ s

it is of great value to evaluate the complete FDI system including all steps in Figure 16.32. To evaluate the detection and isolation performance of the total FDI system the following performance measures are defined.

Detection Time (DT):   Time from fault injection to first detection by any test that may be sensitive to the fault, which is around 20 s in Figure 16.35.

Isolation Time (IT):   Time from fault injection to first correct fault isolation statement, being around 60 s in Figure 16.35.

Missed Detection Rate (MDR):   The fraction of test runs for which the injected fault is not detected by any of the tests that may be sensitive to the fault.

Missed Isolation Rate (MIR):   The fraction of test runs for which a correct fault isolation statement is not obtained.

False Detection Rate (FDR):   The fraction of samples for which the injected fault is detected by a test that should not be sensitive to the fault, or a fault is detected by any test in a no-fault condition.

Note that all measures are defined with respect to the complete FDI system, and not in the context of the individual tests. This means, for instance, that a run in which only one out of several sensitive tests responds, will not be regarded as a missed detection. A situation where only one out of several possible tests responds falsely will, on the other hand, be counted as a false detection. Missed detections and missed isolations are counted on a test run basis, whereas false detections are counted on a sample basis. A correct fault isolation statement is an isolation statement in accordance with the isolability matrix in Table 16.5. For example, when fault $\Delta_{y_{p_{ic}}}$ has occurred, the correct fault isolability statement is that either of the faults $\Delta_{y_{p_{ic}}}$ or $\Delta_{u_{x_{th}}}$ has occurred.

Experimental Evaluation–Setup for Evaluation of Total FDI System

To evaluate the total FDI-system, 12 different data sets were used. The data sets contain measurements from drives with both highway and city parts under different conditions. Each fault specified in Table 16.6 was injected abruptly, one at a time, after a fixed time in each of the 12 data sets, which means that in essence there were in total 12 test runs per fault. The sizes of the faults, as specified in Table 16.6, were chosen in consultation with experienced engineers in order to be realistic for the diesel engine considered.

Table 16.6   Fault sizes used in the experimental evaluation

Fault	Specification	Fault	Specification
$\Delta_{y_{p_{amb}}}$	$y_{p_{amb}} = 0.5 \cdot p_{amb}$	$\Delta_{y_{T_{amb}}}$	$y_{T_{amb}} = 1.3 \cdot T_{amb}$
$\Delta_{y_{p_{ic}}}$	$y_{p_{ic}} = 1.2 \cdot p_{ic}$	$\Delta_{y_{p_{im}}}$	$y_{p_{im}} = 0.9 \cdot p_{im}$
$\Delta_{y_{T_{im}}}$	$y_{T_{im}} = 0.7 \cdot T_{im}$	$\Delta_{y_{p_{em}}}$	$y_{p_{em}} = 0.8 \cdot p_{em}$
$\Delta_{u_{x_{th}}}$	$u_{x_{th}} = 0.3 \cdot u_{x_{th}}$	$\Delta_{u_{x_{egr}}}$	$u_{x_{egr}} = 0.4 \cdot u_{x_{egr}}$
$\Delta_{u_{x_{vgt}}}$	$u_{x_{vgt}} = 0.5 \cdot u_{x_{vgt}}$		

Table 16.7 Summary of detection time (DT), isolation time (IT), missed detection rate (MDR), missed isolation rate (MIR), and false detection rate (FDR), for all considered faults. All time specifications are given in seconds

		$\Delta_{y_{Pamb}}$	$\Delta_{y_{Tamb}}$	$\Delta_{y_{Pic}}$	$\Delta_{y_{Pim}}$	$\Delta_{y_{Tim}}$	$\Delta_{y_{Pem}}$	$\Delta_{u_{Xth}}$	$\Delta_{u_{Xegr}}$	$\Delta_{u_{Xvgt}}$
DT	Mean	48.1	82.9	33.2	41.1	87.0	39.2	66.5	77.8	90.7
	Min	5.0	2.3	18.7	18.7	4.8	11.9	9.4	2.9	6.1
	Max	83.6	35.9	72.5	115.0	290.5	61.3	166.8	116.9	144.3
IT	Mean	168.7	228.6	47.2	148.0	142.7	190.4	246.8	315.7	430.5
	Min	45.5	173.3	28.5	96.6	142.7	57.1	62.0	5.3	129.8
	Max	346.3	283.2	94.0	223.8	142.7	784.7	329.6	545.8	612.8
MDR		0	0	0	0	0	0	0	0	0
MIR		0.42	0.75	0	0.58	0.83	0.25	0.42	0.67	0.67
FDR		0.11	0.082	0.064	0.067	0.053	0.049	0.056	0.063	0.069

Experimental Evaluation–Results for the Total FDI System

Table 16.7 gives the mean, minimum, and maximum, detection time (DT), mean, minimum, and maximum, isolation time (IT), as well as the missed detection rate (MDR), missed isolation rate (MIR), and false detection rate (FDR), for all considered faults. The detection times and isolation times are given in seconds.

The main result is that Table 16.7 shows that all faults can be detected within a reasonable time, meaning that there were no missed detections. Further, isolation performance is satisfactory. The absolute values of the performance measures in Table 16.7 depend mainly on the value of the detection thresholds. The higher the detection thresholds, the lower the rate of false detection. The higher the rate of missed detection, and the longer the detection and isolation times, and vice versa. In addition, as mentioned above, the detection and isolation times are affected by the size of the sliding windows used to collect samples for the residual evaluation.

Concluding Remarks

Air intake diagnosis has been used to show how advanced methods can be used and combined, and even though some explanations were brief, the objective was to give some perspective and outlook. In addition to previously presented material, concepts like sequential residual generators, isolability, methodology for using experimental data, and performance measures, were illustrated. Another message is that today it is possible get a long way with automated methodology to more or less automatically generate the total FDI system. The requirement for this is a model as in Section 16.5.6, once again showing the value of appropriate and validated models as built up in Chapters 7 and 8.

16.5.6   Diesel Engine Model

The total model equations are stated below. The first 37 equations are the diesel engine from Section 8.10, studied also in Section 11:4. The following nine equations, $e_{38}, \ldots ;, e_{46}$, are the considered sensors, actuators, and faults, as described in Tables 16.1, and 16.2, respectively.

$$e_1 : \quad \dot{p}_{ic} = \frac{R_a T_{im}}{V_{ic}} (W_c - W_{th})$$

$$e_2 : \quad \dot{p}_{im} = \frac{R_a T_{im}}{V_{im}} (W_{th} + W_{egr} - W_{ei})$$

$$e_3 : \quad \dot{p}_{em} = \frac{R_e T_{em}}{V_{em}} (W_{eo} - W_{egr} - W_t)$$

$$+ \frac{R_e}{V_{em} c_{ve}} (W_{in} c_{ve}(T_{em,in} - T_{em}) + R_e(T_{em,in} W_{in} - T_{em} W_{out}))$$

$$e_4 : \quad \dot{T}_{em} = \frac{R_e T_{em}}{p_{em} V_{em} c_{ve}} (W_{in} c_{ve}(T_{em,in} - T_{em})$$

$$+ R_e(T_{em,in} W_{in} - T_{em} W_{out}))$$

$$e_5 : \quad W_{in} = \max(W_{eo}, 0) + \max(-W_{egr}, 0) + \max(-W_t, 0)$$

$$e_6 : \quad W_{out} = \max(-W_{eo}, 0) + \max(W_{egr}, 0) + \max(W_t, 0)$$

$$e_7 : \quad W_{th} = \frac{p_{ic} A_{th,max}}{\sqrt{T_{im} R_a}} \Psi_{th}^{\gamma_{th}} (\Pi_{th}) f_{th}(x_{th})$$

$$e_8 : \quad \Pi_{th} = f_{\Pi_{th}}(p_{im}, p_{ic})$$

$$e_9 : \quad W_{ei} = \frac{\eta_{vol} p_{im} n_e V_d}{120 R_a T_{im}}$$

$$e_{10} : \quad \eta_{vol} = c_{vol1} \frac{r_c - \left(\frac{p_{em}}{p_{im}}\right)^{1/\gamma_e}}{r_c - 1} + c_{vol2} W_f^2 + c_{vol3} W_f + c_{vol4}$$

$$e_{11} : \quad W_f = \frac{10^{-6}}{120} \delta n_e n_{cyl}$$

$$e_{12} : \quad W_{eo} = W_f + W_{ei}$$

$$e_{13} : \quad T_e = T_{im} + \frac{q_{HV} f_{T_e W_f}(W_f) f_{T_e n_e}(n_e)}{c_{pe} W_{eo}}$$

$$e_{14} : \quad T_{em,in} = T_{amb} + (T_e - T_{amb}) \exp\left(-\frac{h_{tot} \pi d_{pipe} l_{pipe} n_{pipe}}{W_{eo} c_{pe}}\right)$$

$$e_{15} : \quad W_{egr} = f_{W_{egr}}(p_{im}, p_{em}, T_{em}, x_{egr})$$

$$e_{16} : \quad \dot{\omega}_t = \frac{P_t \eta_m - P_c}{J_t \omega_t}$$

$$e_{17} : \quad P_t \eta_m = \eta_{tm} W_t c_{pe} T_{em} (1 - \Pi_t^{1-1/\gamma_e})$$

$$e_{18} : \quad \eta_{tm} = \eta_{tm,BSR}(BSR) \eta_{tm,\omega_t}(\omega_t) \eta_{tm,x_{vgt}}(x_{vgt})$$

$$e_{19}: \quad BSR = \frac{R_t \omega_t}{\sqrt{2c_{pe}T_{em}(1 - \Pi_t^{1-1/\gamma_e})}}$$

$$e_{20}: \quad \Pi_t = \frac{p_t}{p_{em}}$$

$$e_{21}: \quad W_t = \frac{A_{vgt,max}p_{em}}{\sqrt{T_{em}R_e}} f_{\Pi_t}(\Pi_t) f_{\omega_t}(\omega_{t,corr}) f_{vgt}(x_{vgt})$$

$$e_{22}: \quad \omega_{t,corr} = \frac{\omega_t}{100\sqrt{T_{em}}}$$

$$e_{23}: \quad P_c = \frac{W_c c_{pa} T_{bc}}{\eta_c}(\Pi_c^{1-1/\gamma_a} - 1)$$

$$e_{24}: \quad \Pi_c = \frac{p_{ic}}{p_{bc}}$$

$$e_{25}: \quad \eta_c = \eta_{c,W}(W_{c,corr}, \Pi_c)\eta_{c,\Pi}(\Pi_c)$$

$$e_{26}: \quad W_{c,corr} = \frac{\sqrt{(T_{bc}/T_{ref})}}{\sqrt{(p_{bc}/p_{ref})}} W_c$$

$$e_{27}: \quad W_c = \frac{p_{bc}\pi R_c^3 \omega_t}{R_a T_{bc}}\Phi_c$$

$$e_{28}: \quad \Phi_c = \frac{k_{c1} - k_{c3}\Psi_c}{k_{c2} - \Psi_c}$$

$$e_{29}: \quad k_{c1} = k_{c11}(\min(Ma, Ma_{max}))^2 + k_{c12}\min(Ma, Ma_{max}) + k_{c13}$$

$$e_{30}: \quad k_{c2} = k_{c21}(\min(Ma, Ma_{max}))^2 + k_{c22}\min(Ma, Ma_{max}) + k_{c23}$$

$$e_{31}: \quad k_{c3} = k_{c31}(\min(Ma, Ma_{max}))^2 + k_{c32}\min(Ma, Ma_{max}) + k_{c33}$$

$$e_{32}: \quad Ma = \frac{R_c \omega_t}{\sqrt{\gamma_a R_a T_{bc}}}$$

$$e_{33}: \quad \Psi_c = \frac{2c_{pa}T_{bc}(\Pi_c^{1-1/\gamma_a} - 1)}{R_c^2 \omega_t^2}$$

$$e_{34}: \quad p_{bc} = p_{amb}$$

$$e_{35}: \quad T_{bc} = T_{amb}$$

$$e_{36}: \quad u_\delta = \delta$$

$$e_{37}: \quad y_{n_e} = n_e$$

$$e_{38}: \quad y_{p_{amb}} = p_{amb} + \Delta_{y_{p_{amb}}}$$

$$e_{39}: \quad y_{T_{amb}} = T_{amb} + \Delta_{y_{T_{amb}}}$$

$$e_{40}: \quad y_{p_{ic}} = p_{ic} + \Delta_{y_{p_{ic}}}$$

$$e_{41}: \quad y_{P_{im}} = P_{im} + \Delta_{y_{P_{im}}}$$

$$e_{42}: \quad y_{T_{im}} = T_{im} + \Delta_{y_{T_{im}}}$$

$$e_{43}: \quad y_{P_{em}} = P_{em} + \Delta_{y_{P_{em}}}$$

$$e_{44}: \quad u_{x_{th}} = x_{th} + \Delta_{u_{x_{th}}}$$

$$e_{45}: \quad u_{x_{egr}} = x_{egr} + \Delta_{u_{x_{egr}}}$$

$$e_{46}: \quad u_{x_{vgt}} = x_{vgt} + \Delta_{u_{x_{vgt}}}$$

16.6 History, Legislation, and OBD

Diagnosis in automotives has historically evolved from technological developments leading to possibilities for legislation with the purpose of environmental protection. After a section with a historic perspective, principles in actual legislation are presented.

16.6.1 Diagnosis of Automotive Engines

Diagnosis of automotive engines has a long history. Since the first automotive engines in the 19th century, there has been a need for finding faults in engines. For a long time, the diagnosis was performed manually, but diagnostic tools started to appear in the middle of the 20th century. One example is the stroboscope that is used for determining the ignition time. In the 1960s, exhaust measurements became a common way of diagnosing the fuel system. Until the 1980s, all diagnosis were performed manually and off-board. It was around that time that electronics and gradually microprocessors were introduced in to cars. This opened up the possibility of using on-board diagnosis with the objective to make it easier for the mechanics to find faults.

In 1988, the first legislative regulations regarding On-Board Diagnostics, OBD, were introduced by the California Air Resource Board (CARB). In the beginning, these regulations applied only to California, but the federal Environmental Protection Agency (EPA) adopted similar regulations that applied for the whole USA, which forced manufacturers to include more and more on-board diagnosis capability in their cars. The reason that CARB and EPA required OBD was that a fault can imply increased emission of harmful emission components that are dangerous for the environment. As an example, in 1990, EPA estimated that 60% of the total tailpipe hydro-carbon emissions from light-duty vehicles originated from 20% of vehicles with seriously malfunctioning emission control systems. It is important that such faults are detected so that the car can be repaired as quickly as possible.

In 1994, new and more stringent regulations, OBDII, were introduced in California. It took some time for the rest of the world to follow, but, for example, in Europe, the EU announced regulations, the European On-Board Diagnostics (EOBD), that started to apply in the year 2000. Currently, automotive engines is one of the major application areas for diagnosis, which is mainly because of legislative regulations to protect the environment. One should note that automotive diagnosis systems are heavily constrained economically, since even a small cost, for example for an extra sensor, is emphasized because of the large production volumes. Today, software for fulfilling OBD is a major part of engine management systems, and numbers like at least 50% of the code have been reported.

Increasing Customer Value

As already mentioned before, there is today also a strong trend in using diagnostic techniques to increase consumer value, which can be done by increasing dependability and by lowering costs by maintenance on demand. Thus, some other reasons than legislative for incorporating on-board diagnosis are:

- Mechanics can check the stored fault code and immediately replace the faulty component. This implies more efficient and faster repair work.
- If a fault occurs when driving, the diagnosis system can, after detecting the fault, change the operating mode of the engine to accommodate, for example, by *limp home*. This means that no harm is inflicted.
- The engine can be serviced due to the condition of the engine and not due to a service schedule, thus saving service costs.
- The diagnosis system can make the driver aware of faults that can be damaging, so that the car can be taken to a repair shop in time. This is a way of increasing the *reliability*.

The first three items can be summarized as to increase the *availability* of the car.

16.7   Legislation

As described already in Sections 1.1.1 and 2.8, concerns for the environment have resulted in different standards and regulations. Examples of actors mentioned are the international council on clean transportation (ICCT), the Council of Europe, and the California Air Resource Board (CARB). As also described in the sections mentioned above, examples of outcomes have been the Corporate Average Fuel Consumption (CAFE) for manufacturers, the CO_2 declaration for cars in Europe that is used for taxation of vehicles, the Clean Air Act, and standards such as those illustrated in Figures 1.1 and 1.2.

Not Only for New Cars

As also already mentioned, the above emission requirements should not only be satisfied when the car is new. There are, therefore, legal requirements on a diagnosis system that must detect and isolate faults that can increase emissions. This requirement, to monitor that the vehicle is well-behaved over its lifetime, has been the main driver for legislative regulations regarding On-Board Diagnostics, OBD. It was started in 1988 by the California Air Resource Board (CARB), and has grown to be a major requirement for all vehicles. It has forced manufacturers to include more and more on-board diagnosis capability in the cars. The development has continued, OBDII, and also spread, European OBD (EOBD) and Japanese OBD (JOBD).

16.7.1   OBDII

OBDII is the most extensive on-board diagnosis requirement in use so far. OBDII started to apply in 1994, but its requirements have been successively made harder. The regulations are valid for all passenger cars, light-duty trucks, and medium-duty vehicles.

MIL – Malfunction Indicator Light

The main idea is that an instrument panel lamp, called the Malfunction Indicator Light (MIL), must be illuminated in the case of a fault that can make the emissions exceed the emission limits by more than 50%. The MIL should, when illuminated, display the phrase "Check Engine" or "Service Engine Soon." The OBDII also contains standards for the *scantool*, connectors, communication, and protocols that are used to exchange data between the diagnosis system and the mechanics. Further, it is specified that the software and data must be encoded to prevent unauthorized changes of the engine management system.

Driving Cycle

As described in Section 2.8, emission regulations are formulated as maximum values on emissions from a complete vehicle driven in a vehicle dynamometer following a specified driving profile, called a test cycle or driving cycle. Together with the driving cycle there are also specified limits on how good the tracking of the driving cycle must be during the experiment. For OBDII it is cycle FTP 75 that is used, recall Figure 2.21 and see Figure 16.36 below.

The manufacturer must specify the monitor conditions under which the diagnosis system is able to detect a fault. These monitor conditions must be encountered at least once during the first portion, that is Phase I+II, of the FTP 75 (Federal Test Procedure) test cycle. FTP 75 is a standardized test cycle used in the USA and some other countries. It consists of three phases and is defined in vehicle speed as a function of time. Phase I+II is shown in Figure 2.21. The speed data of Phase III equals Phase I, but follows Phase II after a 9–11 min pause.

General Requirements

Upon detection of a fault, the MIL must be illuminated and a fault code stored in the computer no later than the end of the next driving cycle during which the monitoring conditions occurs, where a driving cycle is defined as engine startup, engine shutoff, and any driving between these two events. The information stored in the memory of the computer is the *Diagnostic Trouble Code* (DTC) and *freeze frame data*. Freeze frame data is all the information available on the current state of the engine and the control system. After three consecutive fault free

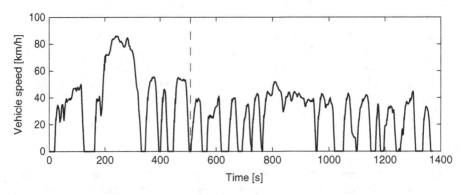

Figure 16.36   Phase I (to the left of the dashed line) and Phase II (to the right of the dashed line) of the FTP 75 test cycle

driving cycles, the MIL should be turned off. Also, the fault code and freeze frame are erased after 40 fault free driving cycles.

Generally, the components that must be diagnosed in OBDII are all actuators and sensors connected to the engine management system. Sensors and actuators must be limit-checked to be in range. Also, the values must be consistent with each other. Additionally, actuators must be checked using active tests.

Specific Monitoring Requirements

In addition to the general specifications, OBDII contains specific requirements and technical solutions for many components of the engine, for example

- Supervision of Throttle and Gas Pedal
- Three Way Catalyst
- Misfire
- Evaporative System Monitoring
- Secondary Air System Monitoring
- Fuel System Monitoring
- Exhaust Gas Sensor Monitoring
- Exhaust Gas Recirculation (EGR) System Monitoring
- Engine Cooling System Monitoring
- Positive Crankcase Ventilation (PCV) System Monitoring
- Air Conditioning (A/C) System Component Monitoring
- Variable Valve Timing and/or Control (VVT) System Monitoring
- Engine Air Intake,

Principle solutions for several of these were presented previously in this chapter. For some of the others, the next section, Section 16.6.4, will give examples of the actual OBDII texts.

16.7.2   Examples of OBDII Legislation Texts

It is instructive to have a look at the actual formulations within OBDII. The two first examples below, MIL Specifications and Secondary Air System Monitoring, are slightly edited to omit extensive details, whereas the two following, Fuel System Monitoring and Exhaust Gas Recirculation (EGR) System Monitoring, are the actual texts.

Note the systematic structure in the latter, that is thorough for all the specific monitoring requirements of components and systems as listed above:

- requirement
- malfunction criteria
- monitoring conditions.

MIL Specifications

(2.1.1) The MIL shall be located on the driver's side instrument panel and be of sufficient illumination and location to be readily visible under all lighting conditions and shall be amber in color when illuminated. The MIL, when illuminated, shall display the phrase "Check Engine" or "Service Engine Soon." The word "Powertrain" may be substituted for "Engine" in the previous phrases. Alternatively, the International Standards Organization (ISO) engine symbol may be substituted for the word "Engine" or for the entire phrase.

In-Use Performance

Description of Software Counters To Track Real World Performance:

Manufacturers are required to track monitor performance by comparing the number of monitoring events (i.e., how often each monitor has run) to the number of driving events (i.e., how often has the vehicle been operated).

$$\text{In-use performance (ratio)} = \frac{\text{Number of Monitoring Events (Numerator)}}{\text{Number of Driving Events (Denominator)}}$$

(3.2.1) Manufacturers shall define monitoring conditions that, in addition to meeting the criteria in section (d)(3.1), ensure that the monitor yields an in-use performance ratio (as defined in section (d)(4)) that meets or exceeds the minimum acceptable in-use monitor performance ratio on in- use vehicles. For the purposes of this regulation, except as provided below in section (d)(3.2.1)(D), the minimum acceptable in-use monitor performance ratio is:

(A) 0.260 for secondary air system monitors and other cold start related monitors utilizing a denominator incremented in accordance with section (d)(4.3.2)(E);

(B) For evaporative system monitors:

(i) 0.260 for monitors designed to detect malfunctions identified in section (e)(4.2.2)(C) (i.e., 0.020 inch leak detection); and

(ii) 0.520 for monitors designed to detect malfunctions identified in section (e)(4.2.2)(A) and (B) (i.e., purge flow and 0.040 inch leak detection);

(C) 0.336 for catalyst, oxygen sensor, EGR, VVT system, and all other monitors specifically required in sections (e) and (f) to meet the monitoring condition requirements of section (d)(3.2);

(3.2.3) Manufacturers may not use the calculated ratio (or any element thereof) or any other indication of monitor frequency as a monitoring condition for any monitor (e.g., using a low ratio to enable more frequent monitoring through diagnostic executive priority or modification of other monitoring conditions, or using a high ratio to enable less frequent monitoring).

Secondary Air System Monitoring

Requirement:

The OBD II system on vehicles equipped with any form of secondary air delivery system shall monitor the proper functioning of the secondary air delivery system including all air switching valve(s). The individual electronic components (e.g., actuators, valves, sensors, etc.) in the secondary air system shall be monitored in accordance with the comprehensive component requirements.

Malfunction Criteria:

For purposes of section (e)(5) (remark: this is the previous paragraph), "air flow" is defined as the air flow delivered by the secondary air system to the exhaust system. For vehicles using secondary air systems with multiple air flow paths/distribution points, the air flow to each bank (i.e., a group of cylinders that share a common exhaust manifold, catalyst, and control sensor) shall be monitored in accordance with the malfunction criteria unless complete blocking of air delivery to one bank does not cause a measurable increase in emissions.

Fuel System Monitoring

(6.1) Requirement:

(6.1.1) The OBD II system shall monitor the fuel delivery system to determine its ability to provide compliance with emission standards.

(6.2) Malfunction Criteria:

(6.2.1) The OBD II system shall detect a malfunction of the fuel delivery system when: (A) The fuel delivery system is unable to maintain a vehicle's emissions at or below 1.5 times any of the applicable FTP standards; or (B) If equipped, the feedback control based on a secondary oxygen or exhaust gas sensor is unable to maintain a vehicle's emissions (except as a result of a malfunction specified in section (e)(6.2.1)(C)) at or below 1.5 times any of the applicable FTP standards; or

(6.2.4) The OBD II system shall detect a malfunction whenever the fuel control system fails to enter closed-loop operation (if employed) within a manufacturer specified time interval.

(6.2.5) Manufacturers may adjust the criteria and/or limit(s) to compensate for changes in altitude, for temporary introduction of large amounts of purge vapor, or for other similar identifiable operating conditions when they occur.

Exhaust Gas Recirculation (EGR) System Monitoring

(8.1) Requirement:

The OBD II system shall monitor the EGR system on vehicles so-equipped for low and high flow rate malfunctions. The individual electronic components (e.g., actuators, valves, sensors, etc.) that are used in the EGR system shall be monitored in accordance with the comprehensive component requirements in section (e)(15).

(8.2) Malfunction Criteria:

(8.2.1) The OBD II system shall detect a malfunction of the EGR system prior to an increase or decrease from the manufacturer's specified EGR flow rate that would cause a vehicle's emissions to exceed 1.5 times any of the applicable FTP standards.

(8.2.2) For vehicles in which no failure or deterioration of the EGR system could result in a vehicle's emissions exceeding 1.5 times any of the applicable standards, the OBD II system shall detect a malfunction when the system has no detectable amount of EGR flow.

(8.3) Monitoring Conditions:

(8.3.1) Manufacturers shall define the monitoring conditions for malfunctions identified in section (e)(8.2) (e.g., flow rate) in accordance with sections (d)(3.1) and (d)(3.2) (i.e., minimum ratio requirements). For purposes of tracking and reporting as required in section (d)(3.2.2), all monitors used to detect malfunctions identified in section (e)(8.2) shall be tracked separately but reported as a single set of values as specified in section (d)(5.2.2).

(8.3.2) Manufacturers may request Executive Officer approval to temporarily disable the EGR system check under specific conditions (e.g., when freezing may affect performance of the system). The Executive Officer shall approve the request upon determining that the manufacturer has submitted data and/or an engineering evaluation which demonstrate that a reliable check cannot be made when these conditions exist. (8.4) MIL Illumination and Fault Code Storage: General requirements for MIL illumination and fault code storage are set forth in section (d)(2).

A

Thermodynamic Data and Heat Transfer Formulas

Thermodynamic data and standard material from thermodynamics, heat transfer, and fluid dynamics are collected here as a support when reading Chapters 4 to 8.

A.1 Thermodynamic Data and Some Constants

Air consists of oxygen, nitrogen, carbon dioxide, argon, water, and some other minor species. Dry air has a molar mass of $M_{air} = 29.05$ g/mol, the relative concentrations of the major species are listed below.

Constituent	Symbol	Molar mass	Volume [%]	Mass [%]
Oxygen	O_2	31.999	20.95	23.14
Nitrogen	N_2	28.013	78.09	75.53
Argon	Ar	39.948	0.93	1.28
Carbon dioxide	CO_2	44.010	0.03	0.05

Thermodynamic data for dry air at atmospheric pressure is listed below.

Temp [K]	ρ [kg/m^3]	c_p [J/kg K]	γ [1]	c_v [J/kg K]	R [J/kg K]	μ [Pa s]
298.15	1.293	1000	1.4	714	286	$1.7 \cdot 10^{-5}$

The ideal gas law is used as state equation and through out the book it is stated in mass basis

$$pV = mRT$$

where R is the ideal gas constant. The universal gas constant $\tilde{R} = 8.3143$ J/mol K, is related to the ideal gas constant through $\tilde{R} = M R$, where M is the molecule weight.

Modeling and Control of Engines and Drivelines, First Edition. Lars Eriksson and Lars Nielsen.
© 2014 John Wiley & Sons, Ltd. Published 2014 by John Wiley & Sons, Ltd.
Companion Website: www.wiley.com/go/powertrain

A.2   Fuel Data

The table below summarizes thermodynamic data for some selected hydrocarbons. (Data for cetane, gasoline, light diesel, heavy diesel, and natural gas, from Heywood (1988).)

Name	Formula	M g/mol	c_{pg} kJ/kg K	T_{ig} °C	q_{HHV} MJ/kg	q_{LHV} MJ/kg	h_{fg} kJ/kg	(A/F) –	RON –	MON –
Pure hydrocarbons										
Methane	CH_4	16.04	2.21	537	55.536	50.048	510	17.2	120	120
Ethane	C_2H_6	30.07	1.75	472	51.902	47.511	489	16.1	115	99
Propane	C_3H_8	44.1	1.62	470	50.322	46.33	432	15.7	112	97
n-Butane	C_4H_{10}	58.12	1.64	365	49.511	45.725	386	15.5	94	90
Isobutane	C_4H_{10}	58.12	1.62	460	49.363	45.577	366	15.5	102	98
n-Pentane	C_5H_{12}	72.15	1.62	284	49.003	45.343	357	15.3	62	63
n-Hexane	C_6H_{14}	86.18	1.62	233	48.674	45.099	335	15.2	25	26
n-Heptane	C_7H_{16}	100.2	1.61	215	48.438	44.925	317	15.2	0	0
n-Octane	C_8H_{18}	114.23	1.61	206	48.254	44.768	301	15.1	20	17
Isooctane	C_8H_{18}	114.23	1.59	418	48.119	44.651	283	15.1	100	100
Cetane	$C_{16}H_{32}$	226.44	1.60	–	47.3	44.0	–	14.8	–	–
Alcohols										
Methanol	CH_3OH	32.04	1.37	385	22.663	19.915	1099	6.5	106	92
Ethanol	C_2H_5OH	46.07	1.42	365	29.668	26.803	836	9	107	89
Commercial fuels										
Gasoline	$C_nH_{1.87n}$	110	1.7	–	47.3	44.0	–	14.6	91–99	82–89
Light diesel	$C_nH_{1.8n}$	170	1.7	–	46.1	43.2	–	14.5	–	–
Heavy diesel	$C_nH_{1.7n}$	200	1.7	–	45.5	42.8	–	14.4	–	–
Natural gas	$C_nH_{3.8n}N_{0.1n}$	18	2	–	50	45	–	14.5	–	–

A.3   Dimensionless Numbers

In fluid dynamics dimensionless numbers are often used to characterize the conditions that are being studied. Below follows some dimensionless numbers that are used in fluid dynamics and in particular in heat transfer.

Re Reynolds number is the ratio between inertia forces and viscous forces.

$$Re = \frac{\rho \upsilon d}{\mu} = / \text{ pipe flow } / = \frac{\rho \frac{W}{\rho A} d}{\mu} = \frac{4 W}{\pi d \mu} \qquad (A.1)$$

Where ρ is the density, μ – viscosity, υ – fluid velocity, and $v = \frac{\mu}{\rho}$ – kinematic viscosity. For flows over a flat plate turbulence occur for Re over $5 \cdot 10^5$. For flows in pipes, turbulence is said to occur for flows with Re over $[2 \cdot 10^3, 5 \cdot 10^3]$.

Pr Prandtl number

$$Pr = \frac{c_p \mu}{\lambda}$$

where λ is the thermal conductivity. For gases the Prandtl number is around 0.7 for all practical cases.

Nu Nusselt number

$$Nu = \frac{h\,l}{\lambda} \tag{A.2}$$

Where h is the heat transfer coefficient, l – characteristic linear dimension, and λ – thermal conductivity of the fluid.

Gr Grasshof number. It is usually used for cases with natural convection

$$Gr = \frac{\beta\,g\,\rho^2\,l^3\,\Delta t}{\mu^2} = \frac{\beta\,g\,l^3\,\Delta t}{v^2}$$

The following example shows the usage of Reynolds number for determining the flow characteristics.

Example A.1 (Gas velocity and Reynolds number)   Consider two ideal flows of air with different velocities in a pipe. The first has a mass flow of $3 \cdot 10^{-3}$ kg/s and the other $2 \cdot 10^{-1}$ kg/s (these flows roughly represent the magnitudes of the smallest and largest gas flows, idle and full power, in a passenger car engine). The diameter of the pipe is 10 cm, the temperature is 20°C and the pressure is 70 kPa. The ideal gas constant is $R = 286$ J/kg K and the viscosity is $\mu = 1.7 \cdot 10^{-5}$ kg/m s.

(a) What are the mean gas velocities?
(b) What are the Reynolds numbers for the two cases?
(c) Is the flow laminar or turbulent?

(a) The mean gas velocities are determined by (7.1), furthermore the density can be determined from ideal gas law $\rho = \frac{1}{v} = \frac{p}{RT}$. This gives

$$U = \frac{\dot{m}}{\rho\,A} = \frac{\dot{m}\,R\,T}{p\frac{\pi\,d^2}{4}}$$

which results in $U_1 = 0.46$ m/s and $U_2 = 31$ m/s. These are below the speed of sound so the flows can be considered incompressible.
(b) The Reynolds number is determined directly from (A.1) which gives $Re = \frac{4\,\dot{m}}{\pi \cdot 0.1 \cdot 1.7 \cdot 10^{-5}}$ and the Reynolds numbers are $Re_1 = 2.2 \cdot 10^3$ and $Re_2 = 150 \cdot 10^3$.
(c) These flows were pipe flows and the highest flow is clearly in the turbulent region (larger than 5000) and the smallest is on the verge of being turbulent. This supports the statement that most flows occurring in the pipes of an engine are turbulent. There are a few situations where the flow is aligned such that it becomes laminar, for example around a laminar air mass flow meter.

A.4   Heat Transfer Basics

Heat transfer in general is a much studied subject and a lot has been written, the interested reader can turn to, for example, Holman (2009) or Schmidt (1993) for good introductions to the subject. In the following sections, basic heat transfer modes, convection, conduction, and radiation are summarized, see also Figure A.1 for some notational conventions. The three modes are also shown in Figure A.1 where the basic equations are also shown. In this section,

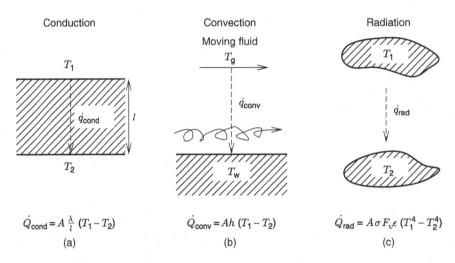

$$\dot{Q}_{cond} = A \tfrac{\lambda}{l} (T_1 - T_2) \qquad\qquad \dot{Q}_{conv} = Ah \, (T_1 - T_2) \qquad\qquad \dot{Q}_{rad} = A \sigma F_v \epsilon \, (T_1^4 - T_2^4)$$

$$(a) \qquad\qquad\qquad\qquad\qquad (b) \qquad\qquad\qquad\qquad\qquad (c)$$

Figure A.1   The three modes of heat transfer that are common in engines. (a) – conduction through a solid, with surface temperatures T_1 and T_2 and thickness l. (b) – convective heat transfer from a moving fluid with bulk temperature T_g to a wall with surface temperature T_w. (c) – radiation from one body with temperature T_1 to another with temperature T_2

the heat transfer rate is denoted $\dot{Q}$ and is measured in W = J/s, the heat flux is denoted $\dot{Q}$ which is the heat transfer rate per unit area $\dot{q} = \dot{Q}/A$ and has the unit W/m^2.

Gas to Wall Heat Transfer

Heat transfer in exhaust pipes is a much investigated topic where several investigations have been made with several extensions to the standard relations for heat transfer in fully developed turbulent flow. From the fluid to the wall there is both radiation and convection but due to high flow velocities and low emissivity of gases only the convection term is of importance. As in most cases with conductive heat transfer, the main difficulty lies in the determination of the heat transfer coefficient, h, where different modeling options have been proposed and investigated in the literature.

Constant Heat Transfer Coefficient for h
The simplest option is to use a constant h which gives the possibility of getting some quantitative values for the heat transfer rate. However, the heat transfer is either overestimated for low flows, giving too low output temperatures, or underestimated for high flows, giving too high temperatures.

Simple Relations for h
Some simple relations for the heat transfer coefficient have been developed in the literature. The following relation for the heat transfer coefficient is given in Inhelder (1996)

$$h_{cv} = \begin{cases} 5.8 + 4\,v & v < 5 \\ 7.12\,v^{0.8} & v \geq 5 \end{cases}$$

For the engines investigated in Eriksson (2003), they give too small values for the exhaust systems. One possibility is to scale the standard relations so that they agree with the heat transfer for the system being modeled.

Nu Re Pr–Relations for h

The majority of the heat transfer relations for convection are expressed as Nusselt–Reynolds–Prandtl relations.

$$Nu = c_0 \, Re^{c_1} \, Pr^{c_2} \left(\frac{\mu_{bulk}}{\mu_{skin}} \right)^{c_3} \tag{A.3}$$

where h is solved from the Nusselt number (defined as $Nu = \frac{h\,l}{\lambda}$). These models offer the most flexibility for capturing the heat transfer and several relations have been proposed for the Nusselt Reynolds numbers for forced convective heat transfer from the exhaust gases to the wall. Summaries of such relations can be found in Chen (1993); Eriksson (2003), Liu et al (1995), Shayler et al. (1999), Wendland (1993), and Zhao and Winterbone (1993). In Table A.1 the heat transfer relations given by the list of authors above are summarized together with additional relations that come from the standard text books. These relations are also plotted in Figure A.2 where the highest number shown is one that is developed for port flows and the lowest comes from relations for fully developed steady flow. The heat transfer coefficient

Table A.1 Heat transfer coefficients for relationships both general from standard text books and those that have been developed especially for exhaust systems. The values in this table refer to the constants shown in (A.3)

Correlation	c_0	c_1	c_2	c_3
CatonHeywood	0.258	0.8	0	0
Shayler1997a	0.18	0.7	0	0
SiederTate	0.027	0.8	$\frac{1}{3}$	0.14
WendlandTakedown	0.081	0.8	$\frac{1}{3}$	0.14
WendlandTailpipe	0.0432	0.8	$\frac{1}{3}$	0.14
Malchowetal	0.0483	0.783	0	0
MeisnerSorenson	0.0774	0.769	0	0
DOHCDownpipe	0.26	0.6	0	0
ValenciaDownpipe	0.83	0.46	0	0
PROMEX	0.027	0.82	0	0
Woods	0.02948	0.8	0	0
Blair	0.02	0.8	0	0
Standard	0.01994	0.8	0	0
std_tu	0.023	0.8	0.3	0
Eriksson	0.48	0.5	0	0
std_lam1	1.86	$\frac{1}{3}$	$\frac{1}{3}$	0.14
Reynolds	0.00175	1	0	0

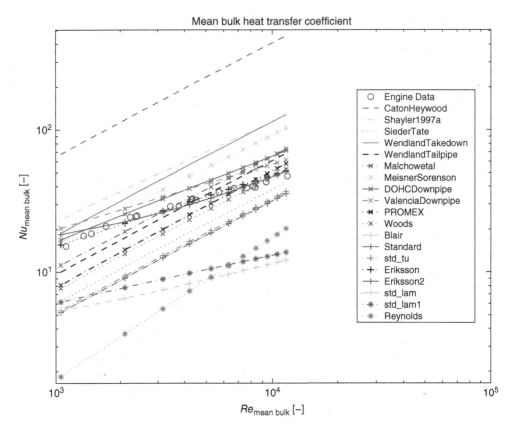

Figure A.2 Average Nusselt and Reynolds number plot for the first section of the exhaust pipe from Engine A. The plot illustrates that there are a variety of correlations and that the results vary by the order of one magnitude

depends on the flow conditions, especially the distance from the exhaust valve. There are a number of investigations of the heat transfer in inlet and exhaust ports, where the pulsating nature of the flow inhibits the boundary layer from fully developing, which leads to a higher heat transfer coefficient than those proposed in the standard heat transfer literature.

Close to the cylinder in the exhaust ports, the instantaneous heat transfer coefficient can reach instantaneous values of $2 \cdot 10^4$ [$\frac{W}{m^2\,K}$] see, for example, Wimmer et al. (2000), and Zapf (1969). For mean value models, it is only the cycle averaged heat transfer coefficient that is of interest and cycle averaged coefficients lie around $4 \cdot 10^2$ [$\frac{W}{m^2\,K}$]. Going further downstream from the engine, the results in Wendland (1993) and Meisner and Sorenson (1986) suggest that the heat transfer coefficients are higher upstream of the catalyst, but the results from Malchow et al. (1979) and Shayler et al. (1995) do not show as high a discrepancy. In Eriksson (2003) experimental data suggested an average heat transfer coefficient that follows the relation (A.3) with an exponent for the Reynolds number that is lower than those in the standard literature for turbulent flows but higher than those for laminar flow. The relation that was determined from the best fit to experimental data was $Nu = 0.48 \cdot Re^{0.5}$.

As a summary comment, these large differences between the Nusselt numbers show that there is a high variability in the exhaust heat transfer coefficient and stress that it is important to validate the heat transfer coefficient used for each application.

Pipe Wall Conduction

The next part of the heat flow path from the exhaust gas to the ambient is the conduction through the wall. The conduction is both perpendicular to the flow direction and also along the pipe walls to, for example, the engine block and turbine that act as heat sinks.

The conductive heat transfer in the radial direction of a pipe section is determined by Fourier's law of conduction and an equivalent conductive heat transfer coefficient can be derived (see for example Holman (2009))

$$
h_{cd} = \frac{2 \pi \lambda L}{A_o \ln \left(\frac{r_o}{r_i} \right)} = \frac{\lambda}{r_o \ln \left(\frac{r_o}{r_i} \right)}
$$

where the outer pipe area A_o has been chosen as the effective area for Newton's law of cooling.

The conduction through the material to the surroundings is so high that it can be neglected when compared to the conductive and radiative heat transfer, and the following example illustrates this. Thermal conductivity of stainless steel is 24.2 W/m K according to Cho et al. (1997). For a pipe with inner radius 21.5 mm and a wall thickness of 1.5 mm, as in Engine A, the heat transfer coefficient becomes 16 000 W/m^2 K. And for a wall with the same inner radius but with thickness 5 mm it becomes 6400 W/m^2 K which is significantly larger than both convection and radiation.

Heat transfer along the pipe is important, this is supported by, for example, the work in Shayler et al. (1997) that describes a model for cycle averaged heat transfer developed by extending the work of Taylor and Toong (1957). They claim that the improvement is foremost from considering an energy balance that takes heat transfer from the exhaust system back to the engine coolant.

Wall to Surroundings Heat Transfer

There are two major modes of heat transfer to the surroundings, convection and radiation, and these are studied below. First a comment on the conduction path sketched in Figure 7.37, which is the conduction to the engine block described above. The conduction is sketched as an external heat transfer mode since it is difficult to discern between the effects of external heat transfer and heat transfer to the engine block when studying lumped mean value models.

External Convection
To get an idea of the magnitude of the convective heat transfer from the pipe to the surroundings, an example that includes both natural and forced convection around a circular pipe is studied.

Natural convection: In many cases approximate heat transfer coefficients are suitable as approximations, and Eastop and McConley (1993) give an example for natural convection around a pipe. Where

$$h_{\mathrm{n}} = \begin{cases} 1.32 \left(\frac{\Delta T}{d} \right)^{1/4} & \text{when} \quad 10^4 < Gr < 10^9 \\ 1.25 (\Delta T)^{1/3} & \text{when} \quad 10^9 < Gr < 10^{12} \end{cases}$$

As an example of the size of the heat transfer coefficient we can use the following numbers, $\Delta T = 300$ K, and $d = 0.05$ m, which gives $h_{\mathrm{n}} = 11.6$ W/K m^2 for the lower Grashof numbers and $h_{\mathrm{n}} = 8.4$ W/K m^2 for the higher Grashof numbers.

Forced convection: In Wendland (1993) external heat transfer coefficients for the exhaust side are given for radiation and convection together in the range of 30 for exhaust manifolds, 30–60 for takedowns, and 30–50 for tailpipe sections. The following correlation was used for the external convection in Zhao and Winterbone (1993)

$$Nu = 0.4 \, Re^{0.6} \, Pr^{0.38}$$

In Chen (1993) a constant value of 20 W/m^2 K was used for external convection. The deviation off different external heat transfer coefficients shown in Shayler et al. (1999) is due to different installations, but also indicates a general uncertainty. For Engine A in the laboratory the air velocity around the pipes varied between 1.5 and 4.4 m/s and the pipe diameter was 25 mm. This corresponds to Reynolds numbers of $Re = 1800$ and $Re = 5400$. The standard correlation in Holman (2009) now yields heat transfer coefficients of 20 and 35 W/K m^2 for these conditions. In laboratory measurements it is possible to measure the velocity at different positions, but surrounding components have a strong influence so the velocity varies with the position. In the engine compartment of a vehicle the air velocity is said to be around 1/3 of the vehicle velocity.

Radiation

From the pipe walls to the environment the heat transfer is through both convection and radiation. In this case the radiation is high compared to convection, so it has to be considered. Here an investigation is made that directly shows the importance of the radiative heat transfer with respect to forced and natural convection around the exhaust pipe. Radiative heat transfer follows the relationship

$$\dot{Q} = F_{\mathrm{v}} \, \epsilon \, \sigma (T_{\mathrm{w}}^4 - T_{\mathrm{a}}^4)$$

Where F_{v} is the gray body view factor, ϵ is the emissivity, and σ the Stefan–Boltzmann constant. Values for the emissivity ϵ are usually in the range of 0.066 for polished steel, 0.14–0.38 for polished iron, 0.4–0.7 for cast iron, and up to 0.8 for sheet steel with a strong oxide layer, Holman (2009). In Shayler et al. (1999) the product of emissivity and the gray body view factor was set to 0.59. Here, a product of 0.6 produced a good fit to the measured engine data.

The radiation equation can be factored to

$$\dot{Q} = F_{\mathrm{v}} \, \epsilon \, \sigma (T_{\mathrm{w}}^3 + T_{\mathrm{w}}^2 T_{\mathrm{a}} + T_{\mathrm{w}} T_{\mathrm{a}}^2 + T_{\mathrm{a}}^3) \, (T_{\mathrm{w}} - T_{\mathrm{a}})$$
$$= F_{\mathrm{v}} \, \epsilon \, \sigma \underbrace{(T_{\mathrm{w}}^2 + T_{\mathrm{a}}^2)(T_{\mathrm{w}} + T_{\mathrm{a}})}_{h_{\mathrm{rad}}(T_{\mathrm{w}}, T_{\mathrm{a}})} \, (T_{\mathrm{w}} - T_{\mathrm{a}}) \tag{A.4}$$

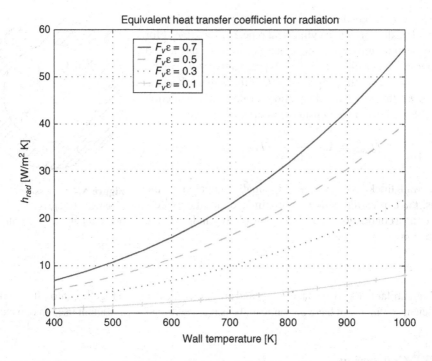

Figure A.3 Wall temperature dependence of the equivalent heat transfer coefficient for radiation plotted for $T_a = 298$ and for different products of $F_v\,\epsilon$

With the lumped nonlinear term h_{rad} this equation gets the same appearance as the expressions for convection and radiation, and this form is convenient for comparison with the convective heat transfer coefficients. The term is plotted in Figure A.3 for ambient temperature $T_a = 298$ and for varying values of the wall temperature and the product $F_v\,\epsilon$. The plot shows that for normal values of $F_v\,\epsilon \approx 0.6$ the heat transfer coefficient varies between 15 and 35 for normal exhaust wall temperatures.

A.4.1 Conduction

Conduction naturally refers to the conduction of heat in the opposite direction of the temperature gradient in a material and it follows

$$\dot{Q}_{\mathrm{cond}} = A \cdot \dot{Q}_{\mathrm{cond}} = -A \cdot \lambda \cdot \frac{\mathrm{d}T}{\mathrm{d}x} \tag{A.5}$$

where λ is the thermal conductivity of the material. When considering heat transfer through a wall with thickness l from the hot side with T_1 to the cold side with T_2 the gradient is exchanged for the finite temperature difference and distance

$$\dot{Q}_{\mathrm{cond}} = A \cdot \dot{Q}_{\mathrm{cond}} = A \cdot \frac{\lambda}{l}(T_1 - T_2) \tag{A.6}$$

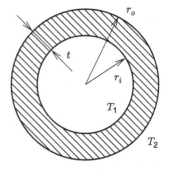

From the general equation another expression can be derived for heat transfer through a cylindrical pipe, see Figure A.4 for a cross-section drawing of the situation. The pipe has length L, inner and outer diameters $d_i = 2\, r_i$ respectively $d_o = 2\, r_o$, and the inner area is thus $A_i = 2\, \pi\, r_i\, L$. If the wall thickness is large, the following general equation for the heat conduction can be derived from (A.5) using cylindrical symmetry

$$\dot{Q}_{cond} = A_i \cdot \frac{\lambda}{r_i \ln \frac{r_o}{r_i}} (T_1 - T_2)$$

Figure A.4 Cross-section of a pipe where there is heat conduction through the pipe wall

If the wall thickness $t = r_o - r_i$ is small compared to the radius, the expression can be approximated and the expression for heat conduction through a thin walled cylindrical pipe then becomes

$$\dot{Q}_{cond} = A_i \cdot \frac{\lambda}{t} (T_1 - T_2)$$

which is similar to (A.6). The factors $\frac{l}{t\lambda}$ and are sometimes lumped together into a heat transfer coefficient so that it can have the same units as the heat transfer coefficient used in convection.

A.4.2   Convection

Heat transfer from a fluid to a solid is called *convection* and is the result of gas movement where heat is transported with the moving gas element. The heat transfer is described by

$$\dot{Q}_{conv} = A \cdot \dot{q}_{conv} = A \cdot h \cdot (T_1 - T_2)$$

where the main difficulty lies in the determination of the heat transfer coefficient h. Depending on the case, the heat transfer coefficient is often determined through empirical relations that have different forms depending on the application and geometry. Most of these correlations express the heat transfer coefficient in the form of the dimensionless *Nusselt number* $Nu = \frac{hL}{\lambda}$, see also Appendix A.3, which is the quotient between convective heat transfer and fluid conductivity. The Nusselt number is determined using a correlation (function) that depends on whether the heat transfer is convective or for mode and geometry.

Convective heat transfer modes are divided into the categories *natural* and *forced* convection. Natural convection refers to the natural gas movement that is induced by, for example, a hot surface, where the induced gas movement transports heat from the surface. For natural convection, the Nusselt number is often determined as a function of the Grasshof and Prandtl numbers, $Nu = f(Gr, Pr)$. Forced convection refers to cases when gases are forced to flow past the hot (or cold) surface, for example inside the pipes of an engine or the wind blowing through a heat exchanger, such as the intercooler. Heat transport to or from the surface occurs with the movement of the gas elements and this movement depends on the geometry and is difficult to describe for the general case. When working with forced convection, the most commonly used expressions for the Nusselt number are based on the Reynolds and Prandtl numbers $Nu = f(Re, Pr)$ and often take the form

$$Nu = c\, Re^m\, Pr^n$$

where the parameters c, m, and n depend on the geometry and flow characteristics. When the Nu is known, the heat transfer coefficient h used in (A.6) can be determined from the definition of Nu (A.2).

A.4.3 Radiation

Heat is also transferred through radiation and the basis for the models comes from the and the black body radiation, for which the emitted power is $\dot{Q}_{rad} = \sigma A T^4$ where σ is Stefan–Boltzmann's constant ($5.669 \cdot 10^{-8}$ W/m^2 K^4). When considering heat exchange by radiation between two bodies, it is also necessary to account for the fact that the bodies are not black bodies and thus emit less heat (this adds an emissivity factor ϵ) and the fact that they do not have the full view of each other (this adds a view factor F_v). The net radiation from body 1 to body 2 is then usually expressed as

$$\dot{Q}_{rad} = A \cdot \dot{q}_{rad} = \sigma F_v \epsilon A (T_1^4 - T_2^4)$$

A.4.4 Resistor Analogy

The resistor analogy is frequently used when analyzing heat transfer effects, and here it will be used to analyze the importance of different heat transfer modes in the exhaust of an engine. In the resistor analogy, the temperature is the driving force and it is represented by the circuit voltage that drives the current which represents the heat flow or heat flux; then the resistor is the reciprocal of the heat transfer coefficient, $R = \frac{1}{h}$.

Resistor Analogy and Radiation

The resistor analogy is useful when analyzing the importance of different heat transfer modes. When the heat transfer is only by convection and conduction it is simple to get an exact solution, but when radiation is taken into account explicitly a fourth-order equation has to be solved. What is readily available with the resistor analogy is the fact that the total heat transfer and the temperatures at different locations can be calculated directly.

Approximate Radiation Solution
If we neglect that the radiation heat transfer coefficient h_{rad} is nonlinear and depends on the temperature and view it as a constant, and also assume that the engine temperature is equal

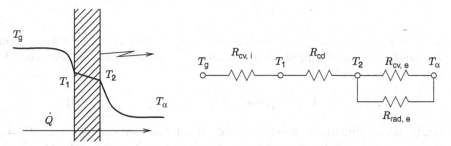

Figure A.5 Illustration of the resistor analogy for heat transfer. In this setup there are four different effects that have to be accounted for: internal convection, wall conduction, external convection, and radiation. The resistance is the reciprocal of the heat transfer coefficient, $R = \frac{1}{h}$

to the ambient temperature, then the resistor analogy gives the total heat flux and the wall temperature directly. The total heat transfer coefficient is

$$\frac{1}{h_{\text{tot}}} = \frac{A_i}{A_o}\frac{1}{h_{\text{cv,i}}} + \frac{1}{h_{\text{cd}}} + \frac{1}{h_{\text{cv,e}} + h_{\text{cd,e}} + h_{\text{rad}}}$$

In this expression, the external area is used as the reference area. In exhaust systems, the change in area from the inner to the outer pipe walls can be neglected. For most exhaust systems the heat transfer coefficient can be neglected. The conductive coefficient h_{cd} is so large, compared to the convective and radiative coefficients, that it can be neglected. Given the ambient temperature and the gas temperature the total heat flux can easily be calculated.

$$\dot{q} = h_{\text{tot}}\,(T_g - T_a)$$

From this the wall temperature (at the inner surface) can easily be calculated as follows

$$T_w = T_g - \frac{\dot{q}}{h_{\text{cv,i}}} \tag{A.7}$$

The outer surface temperature does not differ much from this as the heat transfer coefficient was so large.

The radiative heat transfer can be calculated directly from Eqs. (A.7) and (A.4) by using the technique of fixed point iteration. The procedure is to first assume a value of $h_{\text{rad}}(T_w, T_a)$ and solve (A.7) for T_w, then use to calculate $h_{\text{rad}}(T_w, T_a)$ and start over with (A.7) to get a new temperature and iterate until T_w and h_{rad} have converged. A variant of this technique is utilized in Nasser and Playfoot (1999).

Exact Radiation Solution
The iteration technique described above is simple and it converges within a few steps, but it is possible to get an analytical solution that has a deterministic convergence and this will be done below. This section will derive an explicit solution for the case when the radiation is included exactly.

First let T_w denote the outer surface temperature of the pipe temperature at the outer side of the exhaust pipe. Given the temperature, the heat transfer from the interior of the pipe can be calculated from

$$\dot{Q}_i = A_o\frac{1}{\frac{A_o}{A_i}\frac{1}{h_{\text{cv,i}}} + \frac{1}{h_{\text{cd}}}}(T_g - T_w) = A_o\,h_i\,(T_g - T_w) \tag{A.8}$$

where both heat transfer coefficients have been collected into one coefficient for the pipe internal heat transfer h_i. The heat transfer to the surroundings through radiation, convection, and conduction to the engine is

$$\dot{Q}_e = A_o\{h_{\text{cv,e}}\,(T_w - T_a) + h_{\text{cd,e}}\,(T_w - T_e) + F_v\,\epsilon\,\sigma\,(T_w^4 - T_a^4)\}$$

At steady-state conditions the internal and external heat flows are identical, that is $Q_e = Q_i$. Which gives the following fourth-order equation in the unknown wall temperature

$$T_w^4 + \frac{h_e + \frac{A_i}{A_o}h_i}{F_v\,\epsilon\,\sigma}T_w = \frac{\frac{A_i}{A_o}h_i\,T_g + h_{\text{cv,e}}\,T_a + h_{\text{cd,e}}\,T_e + F_v\,\epsilon\,\sigma\,T_a^4}{F_v\,\epsilon\,\sigma} \tag{A.9}$$

The four solutions to this equation for the outer wall temperature are given in Appendix A.4.5, where the real valued and positive solution is the only one that of interest. From the outer wall temperature, the heat transfer can now easily be calculated using for example, (A.8).

A.4.5 Solution to Fourth-order Equations

Here a solution to the following fourth-order equation is given

$$x^4 + ax = b$$

It is space consuming to write the solution in closed form, so the following intermediate variables are declared and used.

$$c_1 = \frac{4\sqrt[3]{2/3}\,b}{}$$
$$c_2 = \sqrt[3]{9a^2 + \sqrt{81\,a^4 + 768\,b^3}}$$
$$c_3 = -\frac{c_1}{c_2} + \frac{c_2}{\sqrt[3]{18}}$$

And with these variables defined, the four solutions can be written as follows

$$x = \begin{cases} \frac{1}{2}\left(-\sqrt{c_3} + \sqrt{-c_3 + \frac{2a}{\sqrt{c_3}}}\right) \\ \frac{1}{2}\left(-\sqrt{c_3} - \sqrt{-c_3 + \frac{2a}{\sqrt{c_3}}}\right) \\ \frac{1}{2}\left(\sqrt{c_3} + \sqrt{-c_3 - \frac{2a}{\sqrt{c_3}}}\right) \\ \frac{1}{2}\left(\sqrt{c_3} - \sqrt{-c_3 - \frac{2a}{\sqrt{c_3}}}\right) \end{cases}$$

Of these four solutions we are looking for the positive and real valued one which, for the heat transfer situation considered in Section 7.11.2, is the first solution given above.

References

Ahmad T and Theobald MA, 1989. A survey of variable, valve, actuation technology SAE Technical Paper 891674.

Ahmedi A, Franke A, Soyhan HS, Mauss F and Sundén B. 2003. Prediction tool for the ion current in SI combustion. SAE Technical Paper 2003-01-3136.

Ainouz F and Vedholm J. 2009. Mean value model of the gas temperature at the exhaust valve. Master's thesis. Linköping University, SE-581 83 Linköping.

Alberer D and del Re L. 2010. On-line abatement of transient NOx and PM diesel engine emissions by oxygen based optimal control. SAE Technical Paper 2010-01-2201.

Ammann M, Fekete NP, Guzzella L, and Glattfelder AH. 2003. Model-based control of the VGT and EGR in a turbocharged common-rail diesel engine: Theory and passenger car implementation. SAE Technical Paper 2003-01-0357.

Amstutz A and del Re L. 1995. EGO sensor based robust output control of EGR in diesel engines. *IEEE Transactions on Control System Technology* 3(1): 37–48.

Andersson I and Eriksson L. 2009. A parametric model for ionization current in a four stroke SI engine. ASME *Journal of Dynamic Systems, Measurement, and Control* 131(2). 11 pages.

Andersson P. 2005. Air charge estimation in turbocharged spark ignition engines. PhD thesis. Linköpings Universitet.

Andersson P and Eriksson L. 2004. Cylinder air charge estimator in turbocharged SI-engines. SAE Technical Paper 2004-01-1366.

Annand WJD. 1963. Heat transfer in the cylinders of reciprocating internal combustion engines. *Proceedings of the Institution of Mechanical Engineers*. Vol. 177, 36.

Aquino C. 1981. Transient A/F control characteristics of the 6 liter central fuel injection engine. SAE Technical Paper 810494.

ASME. 1997. *PTC 10–1997, Performance Test Code on Compressors and Exhausters*. American Society of Mechanical Engineers, New York.

Asmus T. 1982. Valve events and engine operation. SAE Technical Paper 820749.

Asmus T. 1991. Perspectives on applications of variable valve timing. SAE Technical Paper 910445.

Åström KJ and Hägglund T. 1995. *PID Controllers: Theory, Design and Tuning* 2nd edn. Research Triangle Park, Instrument Society of America.

Åström KJ and Hägglund T. 2006. *Advanced PID Control*. ISA: The Instrumentation, Systems, and Automation Society.

Auckenthaler T, Onder C, and Geering H. 2002. Modelling of a solid-electrolyte oxygen sensor. SAE Technical Paper 2002-01-1293.

Auckenthaler T, Onder C, and Geering H. 2004. Aspects of dynamic three-way-catalytic converter behaviour including oxygen storage. IFAC Symposium Advances in Automotive Control, Salerno, Italy.

Auzins J, Johansson H and Nytomt J. 1995. Ion-gap sense in missfire detection, knock and engine control. SAE Technical Paper 950004.

Backman R. 2011. Open source rapid prototyping environment. SAE Technical Paper 2011-01-0711.

Basseville M and Nikiforov I. 1993. *Detection of Abrupt Changes*. PTR Prentice-Hall, Inc.

Bayraktar H and Durgun O. 2004. Development of an empirical correlation from combustion durations in spark ignition engines. *Energy Conversion and Management* 45(9–10): 1419–1431.

Berggren P and Perkovic A. 1996. Cylinder individual lambda feedback control in an SI Engine. Master's thesis. Linköping University.

Bergman M. 1997. Real-time simulation of model-based lambda-control for SI-engines. Master's thesis. Linköping University.

Bergström J and Brugård J. 1999. Modeling of a turbo charged spark ignited engine. Master's thesis. Linköping University.

Berr FL, Miche M, Colin G, Solliec GL and Lafossas F. 2006. Modelling of a turbocharged SI engine with variable camshaft timing for engine control purposes. SAE Technical Paper 2006-01-3264.

Bigler RS and Cole DJ. 2011. A review of mathematical models of human sensory dynamics relevant to the steering task. Proceedings from 22nd IAVSD Symposium.

Blair GP. 1999. *Design and Simulation of Four-Stroke Engines*. SAE International.

Blanke M, Kinnaert M, Lunze J and Staroswiecki M. 2006. *Diagnosis and Fault-Tolerant Control* 2nd edn. Springer.

Blizard NC and Keck JC. 1974. Experimental and theoretical investigation of turbulent burning model for internal combustion engines. SAE Technical Paper 740191.

Borman GL and Ragland KW. 1998. *Combustion Engineering*. McGraw-Hill.

Brand D. 2005. Control-oriented modeling of NO emissions of SI engines. PhD thesis. ETH Zürich.

Butler KL, Ehsani M and Kamath P. 1999. A Matlab-based modeling and simulation package for electric and hybrid electric vehicle design. *IEEE Transactions on Vehicular Technology* 48(6): 1770–1778.

Canudas de Wit C, Kolmanovsky I, and Sun J. 2001. Adaptive pulse control of electronic throttle. *Proceedings of the American Control Conference* 4: 2872–2877.

Carlsson C. 2012. Modeling and experimental validation of a rankine cycle based exhaust WHR system for heavy duty applications Master's thesis. Linköping University, SE-581 83 Linköping.

Casella G and Berger RL. 2001. *Statistical Inference* 2nd edn. Duxbury Press.

Casey M and Fesich T. 2009. On the efficiency of compressors with diabatic flows. Proceedings of ASME Turbo Expo: GT2009-59015.

Cavina N, Corti E, and Moro D. 2008. Closed-loop individual cylinder air-fuel ratio control via UEGO signal spectral analysis Proceedings of the 17th IFAC World Congress, Seoul, Korea.

Cavina N, Corti E, and Moro D. 2010. Closed-loop individual cylinder air-fuel ratio control via UEGO signal spectral analysis. *Control Engineering Practice* 18(11): 1295–1306.

Chang CF, Fekete N, Amstutz A and Powell J. 1995. Air-fuel ratio control in spark-ignition engines using estimation theory. *IEEE Transactions on Control Systems Technology* 3(1): 22–31.

Chang J, Güralp O, Filipi Z, Assanis D, Kuo TW, Najt P, and Rask R. 2004. New heat transfer correlation for an HCCI engine derived from measurements of instantaneous surface heat flux. SAE Technical Paper 2004-01-2996.

Chapman KS and Shultz J. 2003. Guidelines for: Testing large-bore engine turbochargers. Technical report, The National Gas Machinery Laboratory, Kansas State University, 245 Levee Drive.

Charlton S, Dollmeyer T, and Grana T. 2010. Meeting the US heavy-duty EPA 2010 standards and providing increased value for the customer. SAE Technical Paper 2010-01-1934.

Chauvin J, Corde G, Petit N, and Rouchon P. 2008. Motion planning for experimental airpath control of a diesel homogeneous charge-compression ignition engine. *Control Engineering Practice* 16(9): 1081–1091.

Chen DKS. 1993. A numerical model for thermal problems in exhaust systems. SAE Technical Paper 931070.

Chen SX and Moskwa JJ. 1997. Application of nonlinear sliding mode observers for cylinder pressure reconstruction. *IFAC Control Engineering Practice* 5(8): 1115–1121.

Chevalier A, Müeller M and Hendricks E. 2000. On the validity of mean value engine models during transient operation. SAE Technical Paper 2000-01-1261.

Chiavola O and Giulianelli P. 2001. Modelling and simulation of common rail systems. SAE Technical Paper 2001-01-3183.

Cho YS, Kim CR, Kim DS, Lee J, Lee SW, Myung KH, and Choo JS. 1997. Prediction of exhaust gas temperature at the inlet of an underbody catalytic converter during FTP-75 test. Topics in General and Advanced Emissions SAE Technical Paper 972913.

Colin G, Chamaillard Y, Bloch G, and Corde G. 2007. Neural control of fast nonlinear systems – application to a turbocharged SI engine with VCT. *IEEE Transactions on Neural Networks* 18(4): 1101–1114.

Collings N, Dinsdale S, and Hands T. 1991. Plug fouling investigations on a running engine – an application of a novel multi-purpose diagnostic system based on the spark plug. SAE Technical Paper 912318.

Cormerais M, Hetet J, Chesse P, and Maiboom A. 2006. Heat transfer analysis in a turbocharger compressor: modeling and experiments. SAE Technical Paper 2006-01-0023.

Crane D, Jackson G and Holloway D. 2001. Towards optimization of automotive waste heat recovery using thermoelectric. SAE Technical Paper 2001-01-1021.

Csallner P. 1981. Eine Methode zur Vorausberechnung der änderung des Brennverlaufes von Ottomotoren bei geänderten Betriebsbedingungen. PhD thesis. Technischen Universität München.

Curtis EW, Aquino CF, Trumpy DK, and Davis GC. 1996. A new port and cylinder wall wetting model to predict transient air/fuel excursions in a port fuel injected engine. SAE Technical Paper 961186.

Daniels CF. 1998. The comparison of mass fraction burned obtained from the cylinder pressure signal and spark plug ion signal. SAE Technical Paper 980140.

Dec JE. and Canaan R. 1998. PLIF imaging of NO formation in a DI diesel engine. SAE Technical Paper 980147.

Dec JE. 1997. A conceptual model of DI diesel combustion based on laser-sheet imaging. SAE Technical Paper 970873.

Degobert P. 1995. *Automobiles and Pollution*. Society of Automotive Engineers, Inc.

Deur J, Pavkovic D, Peric N, Jansz M, and Hrovat D. 2004. An electronic throttle control strategy including compensation of friction and limp-home effects. *IEEE Transactions on Industry Applications* 40(3): 821–834.

Dietsche KH (ed.) 2011. *Bosch Automotive Handbook* 8th edn. Robert Bosch GmbH.

Diop S, Moraal PE, Kolmanovsky IV, and van Nieuwstadt, M. 1999. Intake oxygen concentration estimation for DI diesel engines. Proceedings of the 1999 IEEE International Conference on Control Applications.

Dittrich P, Peter F, Huber G, and Kühn M. 2010. Thermodynamic potentials of a fully variable valve actuation system for passenger-car diesel engines. SAE Technical Paper 2010-01-1199.

Dixon S. 1998. *Fluid Mechanics and Thermodynamics of Turbomachinery* 4th edn. Butterworth-Heinemann.

Dolcini PJ, de Wit CC and Bechart H. 2010. *Dry Clutch Control for Automotive Applications Advances in Industrial Control*. Springer.

Dönitz C, Vasile I, Onder C, and Guzzella L. 2009. Realizing a concept for high efficiency and excellent driveability: The downsized and supercharged hybrid pneumatic engine. SAE Technical Paper 2009-01-1326.

Drangel H, Olofsson E, and Reinmann R. 2002. The variable compression (SVC) and the combustion control (SCC)–two ways to improve fuel economy and still comply with world-wide emission requirements. SAE Technical Paper 2002-01-0996.

Eastop TD and McConkey A. 1993. *Applied Thermodynamics for Engineering Technologists*. 5th edn. Wiley.

Eichelberg G. 1939. Some new investigations of old internal combustion engine problems. *Engineering* 149: 463–547.

Ellman A and Piche R. 1999. A two regime orifice flow formula for numerical simulation. *Journal of Dynamic Systems, Measurement, and Control* 121(4): 721–724.

Eriksson D. 2013. Diagnosability analysis and FDI system design for uncertain systems. Licentiate thesis. Linköping University, SE-581 83 Linköping.

Eriksson D, Eriksson L, Frisk E, and Krysander M. 2013. Flywheel angular velocity model for misfire and driveline disturbance simulation. 7th IFAC Symposium on Advances in Automotive Control, Tokyo, Japan.

Eriksson L. 1998. Requirements for and a systematic method for identifying heat-release model parameters. SAE Technical Paper 980626.

Eriksson L. 1999. Spark advance modeling and control. PhD thesis. Linköping University.

Eriksson L. 2001. Simulation of a vehicle in longitudinal motion with clutch engagement and release. 3rd IFAC Workshop "Advances in Automotive Control" Karlsruhe, Germany.

Eriksson L. 2003. Mean value models for exhaust system temperatures. SAE Technical Paper 2002-01-0374.

Eriksson L. 2005. CHEPP – A chemical equilibrium program package for Matlab. SAE Technical Paper 2004-01-1460.

Eriksson L. 2007. Modeling and control of turbocharged SI and DI engines. *Oil & Gas Science and Technology* 62(4): 523–538.

Eriksson L and Andersson I. 2003. An analytic model for cylinder pressure in a four stroke SI engine. *SAE Transactions, Journal of Engines* 111(3). SAE Technical Paper 2002-01-0371.

Eriksson L and Nielsen L. 1997. Ionization current interpretation for ignition control in internal combustion engines. *IFAC Control Engineering Practice* 5(8): 1107–1113.

Eriksson L and Nielsen L. 1998. Increasing the efficiency of SI-engines by spark-advance control. IFAC Workshop "Advances in Automotive Control" (Preprints), pp. 211–216.

Eriksson L and Nielsen L. 2000. Non-linear model-based throttle control electronic engine controls: Controls. SAE Technical Paper 2000-01-0261.

Eriksson L and Nielsen L. 2003. Towards on-board engine calibration with feedback control incorporating combustion models and ion-sense. *Automatisierungstechnik* 51(5): 204–212.

Eriksson L, Frei S, Onder C, and Guzzella L. 2002a. Control and optimization of turbo charged spark ignited engines. IFAC World Congress, Barcelona, Spain.

Eriksson L, Lindell T, Leufven O and Thomasson A. 2012. Scalable component-based modeling for optimizing engines with supercharging, E-boost and turbocompound concepts. SAE Technical Paper 2012-01-0713.

Eriksson L, Nielsen L and Glavenius M. 1997. Closed loop ignition control by ionization current interpretation. SAE Technical Paper 970854.

Eriksson L, Nielsen L, and Nytomt J. 1996. Ignition control by ionization current interpretation. SAE Technical Paper 960045.

Eriksson L, Nielsen L, Brugård J, Bergström J, Pettersson F, and Andersson P. 2001. Modeling and simulation of a turbo charged SI engine. 3rd IFAC Workshop "Advances in Automotive Control", Preprints, pp. 379–387. Elsevier Science.

Eriksson L, Nielsen L, Brugård J, Bergström J, Pettersson F, and Andersson P. 2002b. Modeling and simulation of a turbo charged SI engine. *Annual Reviews in Control* 26(1): 129–137.

Ferreau H, Ortner P, Langthaler P, del Re L, and Diehl M. 2007. Predictive control of a real-world diesel engine using an extended online active set strategy. *Annual Reviews in Control* 31(2): 293–301.

Finn C .1998. *Thermal Physics* 2nd edn. Stanley Thornes Ltd.

Fitzgerald AE, Jr., CK and Umans SD. 2003. *Electric Machinery* 6th edn. McGraw-Hill.

Frank P. 1994. Application of fuzzy logic to process supervision and fault diagnosis. In: *Fault Detection, Supervision and Safety for Technical Processes*. IFAC, Espoo, Finland, pp. 507–514.

Franklin GF, Powell JD and Workman ML. 1990. *Digital Control of Dynamical Systems*. Addison-Wesley.

Frei SA, Guzzella L, Onder CH and Nizzola C. 2006. Improved dynamic performance of turbocharged SI engine power trains using clutch actuation. *Control Engineering Practice* 14(4): 363–373.

Frisk E, Bregon A, Åslund J, Krysander M, Pulido B, and Biswas G. 2012. Diagnosability analysis considering causal interpretations for differential constraints. *IEEE Transactions on Systems, Man, and Cybernetics – Part A: Systems and Humans* 42(5): 1216–1229.

Fröberg A and Nielsen L. 2008. Efficient drive cycle simulation. *IEEE Transactions on Vehicular Technology* 57(2): 1442–1453.

Garcia CE and Morari M. 1982. Internal model control. 1. A unifying review and some new results. *Ind. Eng. Chem. Process. Des. Dev.* 21: 308–323.

Garća-Nieto S, Martínez M, Blasco X. and Sanchis J. 2008. Nonlinear predictive control based on local model networks for air management in diesel engines. *Control Engineering Practice* 16(12): 1399–1413.

Gatowski JA, Balles EN, Chun KM, Nelson FE, Ekchian JA, and Heywood JB. 1984. Heat release analysis of engine pressure data. SAE Technical Paper 841359.

Gerhardt J, Hönninger H, and Bischof H. 1998. A new approach to functional and software structure for engine management systems - Bosch ME7. SAE Technical Paper 980801.

Gertler J. 1991 Analytical redundancy methods in fault detection and isolation; survey and synthesis. In: *IFAC Fault Detection, Supervision and Safety for Technical Processes*. Baden-Baden, Germany, pp. 9–21.

Gillespie TD. 1992. *Fundamentals of Vehicle Dynamics*. SAE International.

Glaser I and Powell J. 1981. Optimal closed-loop spark control of an automotive engine. SAE Technical Paper 810058.

Glover L, Douglas R, McCullough G, Keenan M, Revereault P and McAtee C. 2011. Performance characterization of a range of diesel oxidation catalysts: Effect of Pt:Pd ratio on light off behaviour and nitrogen species formation. SAE Technical Paper 2011-24-0193.

Gravdahl JT. 1998. Modeling and control of surge and rotating stall in compressors. PhD thesis. Norwegian University of Science and Technology Department of Engineering Cybernetics, N-7034 Trondheim, Norway.

Gray C. 1988. A review of variable engine valve timing. SAE Technical Paper 880386.

Greitzer EM. 1981. The stability of pumping systems - the 1980 Freeman lecture. *Journal of Fluids Engineering* 103: 193–242.

Gustafsson F. 2000. *Adaptive Filtering and Change Detection*. John Wiley & Sons.

Guzzella L and Amstutz A. 1998. Control of diesel engines. *IEEE Control Systems* 18(5).

Guzzella L and Amstutz A. 1999. CAE tools for quasi-static modeling and optimization of hybrid powertrains. *IEEE Transactions on Vehicular Technology* 48(6): 1762–1769.

Guzzella L and Onder CH. 2009. *Introduction to Modeling and Control of Internal Combustion Engine Systems*. (1st edn 2004.) 2nd edn. Springer Verlag.

Guzzella L, Wenger U, and Martin R. 2000. IC-engine downsizing and pressure-wave supercharging for fuel economy. SAE Technical Paper 2000-01-1019.

Hadler J, Rudolph F, Dorenkamp R, Stehr H, Hilzendeger J, and Kranzusch S. 2008. Volkswagen's new 2.0l TDI engine for the most stringent emission standards – part 1. MTZ Motortechnische Zeitschrift 69(5), 12–18.

Hansen KE, Jorgensen P, and Larsen PS. 1981. Experimental and theoretical study of surge in a small centrifugal compressor. *Journal of Fluids Engineering* 103: 391–395.

Heintz N, Mews M, Stier G, Beaumont A, and Noble A. 2001. An approach to torque-based engine management systems. SAE Technical Paper 2001-01-0269.

Hellring M and Holmberg U. 2000. An ion current based peak-finding algorithm for pressure peak position estimation. SAE Technical Paper 2000-01-2829.

Hellström E, Åslund J, and Nielsen L. 2010 Horizon length and fuel equivalents for fuel-optimal look-ahead control. 6th IFAC Symposium on Advances in Automotive Control, Munich, Germany.

Hendricks E. 1986. A compact, comprehensive model of large turbocharged, two-stroke diesel engines. SAE Technical Paper 861190.

Hendricks E. 2001. Isothermal vs. adiabatic mean value SI engine models. In: 3rd IFAC Workshop, Advances in Automotive Control, Karlsruhe, Germany, Preprint: pp. 373–378.

Hendricks E and Sorenson SC. 1990. *Mean Value Modelling of Spark Ignition Engines*. SAE Technical Paper 900616.

Hendricks E, Chevalier A, Jensen M, Sorenson SC, Trumphy D and Asik J. 1996. Modelling of the intake manifold filling dynamics. SAE Technical Paper 960037.

Hendricks E, Vesterholm T and Sorenson SC. 1992. Nonlinear, closed loop, SI engine control observers. SAE Technical Paper 920237.

Heywood JB. 1988. *Internal Combustion Engine Fundamentals*. McGraw-Hill series in mechanical engineering. McGraw-Hill.

Heywood JB. 2003. An improved friction model for spark-ignition engines modeling of SI Engines. SAE Technical Paper 2003-01-0725.

Heywood JB, Higgins JM, Watts PA, and Tabaczynski RJ. 1979. Development and use of a cycle simulation to predict SI engine efficiency and NOx emissions. SAE Technical Paper 790291.

Hires S and Overington M. 1981. Transient mixture strength excursions – an investigation of their causes and the development of a constant mixture strength fueling strategy. SAE Technical Paper 810495.

Höckerdal E, Eriksson L, and Frisk E. 2012. Off- and on-line identification of maps applied to the gas path in diesel engines. In: *Lecture Notes in Control and Information Sciences*, eds. D. Alberer, H. Hjalmarsson and L. del Re. Springer Verlag. pp. 241–256. Volume 418.

Höckerdal E, Frisk E, and Eriksson L. 2011. EKF-based adaptation of look-up tables with an air mass-flow sensor application. *Control Engineering Practice* 19(5): 442–453.

Hoffman J, Lee W, Litzinger T, Santavicca D, and Pitz W. 1991. Oxidation of propane at elevated pressures. *Experiments and Modeling. Combustion and Flame* 77: 95–125.

Hohenberg GF. 1979. Advanced approaches for heat transfer calculations. SAE Technical Paper 790825.

Holman JP. 2009. *Heat Transfer* 10th edn. McGraw-Hill.

Hong H, Parvate-Patiland G, and Gordon B. 2004. Review and analysis of variable valve timing strategies–eight ways to approach. *Proceedings of the Institution of Mechanical Engineers, Part D: Journal of Automobile Engineering* 218(10): 1179–1200.

Hu Q, Wu SF, Stottler S, and Raghupathi R. 2001. Modelling of dynamic responses of an automotive fuel rail system, part II: Entire system. *Journal of Sound and Vibration* 245(5): 815–834.

Hubbard M, Dobson P, and Powell J. 1976. Closed loop control of spark advance using a cylinder pressure sensor. *Journal of Dynamic Systems, Measurement and Control* 98(4): 414–420.

Hunter J, Wang H, Litzinger T, and Frenklach M. 1994. The oxidation of methane at elevated pressures Experiments and Modeling. *Combustion and Flame* 97: 201–204.

Inhelder J. 1996. Verbrauchs und Schadstoffoptimiertes Ottomotor-Aufladekonzept. PhD thesis. Swiss Federal Institute of Technology, Zürich, Diss. ETH No. 11948.

Isermann R. 1984. Process fault detection on modeling and estimation methods – a survey. *Automatica* 20(4): 387–404.

Jankovic M and Magner S. 2011. Disturbance attenuation in time-delay systems – a case study on engine air-fuel ratio control. American Control Conference, pp. 3326–3331.

Jankovic M and Magner SW. 2002. Variable cam timing: Consequences to automotive engine control design. 15th IFAC World Congress.

Jankovic M, Jankovic M, and Kolmanovsky I. 2000. Constructive lyapunov control design for turbocharged diesel engines. *IEEE Transactions on Control System Technology* 8(2): 288–299.

Jensen JP, Kristensen AF, Sorenson SC, Houbak N, and Hendricks E. 1991 Mean value modeling of a small turbocharged diesel engine number. SAE Technical Paper 910070.

Jensen PB, Olsen MB, Poulsen J, Hendricks E, Fons M, and Jepsen C. 1997. A new family of nonlinear observers for SI engine air/fuel ratio control. SAE Technical Paper 970615.

Johansson J and Waller M. 2005. Control oriented modeling of the dynamics in a catalytic converter. Master's thesis. Linköpings Universitet, SE-581 83 Linköping.

Johansson R. 2012. Modeling of engine and driveline related disturbances on the wheel speed in passenger cars. Master's thesis. Linköping Universitet, SE-581 83 Linköping.

Jung M, Ford RG, Glover K, Collings N, Christen U, and Watts MJ. 2002. Parameterization and transient validation of a variable geometry turbocharger for mean-value modeling at low and medium speed-load points. SAE Technical Paper 2002-01-2729.

Jung M, Glover K, and Christen U. 2005. Comparison of uncertainty parameterisations for H-infinity robust control of turbocharged diesel engines. *Control Engineering Practice* 13(1): 414–420.

Grizzle, JW and Dobbins KL, and Cook JA. 1991. Individual cylinder air-fuel ratio control with a single ego sensor. *IEEE Transactions on Vehicular Technology* 40(1).

Kailath T. 1980. *Linear Systems*. Prentice Hall.

Kainz J and Smith J. 1999. Individual cylinder fuel control with a switching oxygen sensor. SAE Technical Paper 1999-01-0546.

Kao M and Moskwa JJ. 1995. Turbocharged diesel engine modeling for nonlinear engine control and state estimation. *ASME, Journal of Dynamic Systems Measurement and Control* 117, 20–30.

Karlsson J. 2001. Powertrain modeling and control for driveability in rapid transients with backlash. Technical report. Licentiate Thesis, Chalmers, Göteborg, Sweden.

Keynejad F and Manzie C. 2011. Cold start modelling of spark ignition engines. *Control Engineering Practice* 19(8): 912–925.

Kiencke U and Nielsen L. 2005. *Automotive Control Systems, For Engine, Driveline, and Vehicle* 2nd edn. Springer Verlag.

Kittel C and Kroemer H. 1995. *Thermal Physics* 2nd edn. W. H. Freeman and Company.

Klein M. 2007 Single-zone cylinder pressure modeling and estimation for heat release analysis of SI engines. PhD thesis. Linköpings Universitet.

Klein M and Eriksson L. 2005. A specific heat ratio model for single-zone heat release models. SAE Technical Paper 2004-01-1464.

Kolmanovsky I, Stefanopoulou A, Moraal P, and van Nieuwstadt M. 1997. Issues in modeling and control of intake flow in variable geometry turbocharged engines. Proceedings of 18th IFIP Conference on System Modeling and Optimization, Detroit.

Krieger R and Borman G. 1967. *The Computation of Apparent Heat Release for Internal Combustion Engines*. ASME.

Krysander M. 2000 Air mass flow through a throttle. Master's thesis. Linköping University, SE-581 83 Linköping.

Krysander M and Frisk E. 2008. Sensor placement for fault diagnosis. *IEEE Transactions on Systems, Man and Cybernetics, Part A: Systems and Humans*: 38(6), 1398–1410.

Krysander M and Frisk E. 2009. Leakage detection in a fuel evaporative system. *Control Engineering Practice* 17(11): 1273–1279.

Krysander M, Åslund J and Nyberg M. 2008. An efficient algorithm for finding minimal over-constrained sub-systems for model-based diagnosis. *IEEE Transactions on Systems, Man, and Cybernetics – Part A: Systems and Humans*.

Kummer JT. 1986. Use of noble metals in automobile exhaust catalysts. *Journal of Physical Chemistry* 90(20): 4747–4752.

Larsson S and Schagerberg S. 2004. Si-engine cylinder pressure estimation using torque sensors. SAE Technical Paper 2004-01-1369.

Lavoie GA, Heywood JB and Keck JC. 1970. Experimental and theoretical study of nitric oxide formation in internal combustion engines. *Combustion Science and Technology*.

Leduc P, Dubar B, Ranini A, and Monnier, G. 2003 Downsizing of gasoline engine: an efficient way to reduce CO2 emissions. *Oil & Gas Science and Technology* 58(1): 115–127.

Lee A and Pyko JS. 1995. Engine misfire detection by ionization current monitoring. SAE Technical Paper 950003.

Lee TK and Filipi ZS. 2011 Synthesis of real-world driving cycles using stochastic process and statistical methodology. *International Journal of Vehicle Design* 57(1): 17–36.

Leonhard W. 1996. *Control of Electrical Drives* 2nd edn. Springer-Verlag.

Leroy T, Chauvin J, Berr FL, Duparchy A, and Alix G. 2008. Modeling fresh air charge and residual gas fraction on a dual independant variable valve timing SI engine. SAE Technical Paper 2008-01-0983.

Leufven O. 2013. Model for control of centrifugal compressors. PhD thesis. Linköping University.

Leufven O and Eriksson L. 2011. Surge and choke capable compressor model. IFAC World Congress, Milano, Italy.

Lewis RI. 1996. *Turbomachinery Performance Analysis*. Arnold.

Lindell T. 2009. Model-based air and fuel path control of a VCR engine. Master's thesis Linköping University, SE-581 83 Linköping.

Lindström F, Ångström HE, Kalghatgi G, and Möller CE. 2005. An empirical SI combustion model using laminar burning velocity correlations. SAE Technical Paper 2005-01-2106.

Liu Z, Hoffmanner AL, Skowron JF, and Miller MJ. 1995. Exhaust transient temperature response. SAE Technical Paper 950617.

Locatelli M, Onder CH, and Geering HP. 2003. Exhaust-gas dynamics model for identification purposes. SAE Technical Paper 2003-01-0368.

Locatelli M, Onder CH and Geering HP. 2004. An easily tunable wall-wetting model for PFI engines. SAE Technical Paper 2004-01-1461.

M. Sellnau, T. Kunz JS and Burkhard J. 2006. 2-step variable valve actuation: System optimization and integration on an SI engine. SAE Technical Paper 2006-01-0040.

Maciejowski JM. 1989. *Multivariable Feedback Design*. Addison-Wesley.

Magner S, Jankovic M and Cooper S. 2005. Methods to reduce air-charge characterization data for high degree of freedom engines. SAE Technical Paper 2004-01-0903.

Malaczynski G, Roth G, and Johnson D. 2013. Ion-sense-based real-time combustion sensing for closed loop engine control. SAE Technical Paper 2013-01-0354.

Malchow G, Sorenson S, and Buckius R. 1979. Heat transfer in the straight section of an exhaust port of a spark ignition engine, SAE Technical Paper 790309.

Martin G, Talon V, Higelin P, Charlet A, and Caillol C. 2009. Implementing turbomachinery physics into data mapbased turbocharger models. SAE Technical Paper 2009-01-0310.

Massey B. 1998. *Mechanics of Fluids* 7th edn. Stanley Thornes.

Matekunas FA. 1983. Modes and measures of cyclic combustion variability. SAE Technical Paper 830337.

Matekunas FA. 1984. *Spark Ignition Engines – Combustion Characteristics, Thermodynamics, and the Cylinder Pressure Card*. Central States Section, The Combustion Institute. Minneapolis.

Matekunas FA. 1986. Engine combustion control with ignition timing by pressure ratio management US Patent, A, 4622939.

Meisner S and Sorenson S. 1986. Computer simulation of intake and exhaust manifold flow and heat transfer. SAE Technical Paper 860242.

Merker G, Schwarz C, Stiesch G and Otto F. 2006. *Simulating Combustion: Simulation of Combustion and Pollutant Formation for Engine Development*. Springer.

Miller D. 1990. *Internal Flow Systems*. BHR Group Limited.

Millo F and Vezza D. 2012. Characterization of a new advanced diesel oxidation catalyst with low temperature NOx storage capability for ID diesel. SAE Technical Paper 2012-01-0373.

Mladek M and Onder CH. 2000. A model for the estimation of inducted air mass and the residual gas fraction using cylinder pressure measurements. SAE Technical Paper 2000-01-0958.

Moraal P. 1995. Adaptive compensation of fuel dynamics in an SI engine using a switching EGO sensor, IEEE – 34th Conference on Decision & Control, pp. 661–666 number WM06 2:10.

Moraal P and Kolmanovsky I. 1999. Turbocharger modeling for automotive control applications. SAE Technical Paper 1999-01-0908.

Moraal P, Meyer D, Cook J, and Rychlick E. 2000. Adaptive transient fuel compensation: Implementation and experimental results. SAE Technical Paper 2000-01-0550.

Morris MJ and Dutton JC. 1989. Aerodynamic torque characteristics of butterfly valves in compressible flow. *Transactions of the ASME* 111: 392–400.

Moudden Y, Seghouane AK, and Boubal O. 2002. Extraction of peak pressure position information from the sparkplug ionization signal. *Computing Standard Interfaces* 24(2): 161–170.

Müller M, Hendricks E, and Sorenson SC. 1998. Mean Value Modelling of Turbocharged Spark Ignition Engines. SAE Technical Paper 980784.

Muske K, Jones J, and Franceschi E. 2008. Adaptive analytical model-based control for SI engine air–fuel ratio. *IEEE Transactions on Control Systems Technology* 16(4): 763–768.

Myklebust A and Eriksson L. 2012a. Road slope analysis and filtering for driveline shuffle simulation. E-COSM'12 – IFAC Workshop on Engine and Powertrain Control, Simulation and Modeling, Paris, France.

Myklebust A and Eriksson L. 2012b. Torque model with fast and slow temperature dynamics of a slipping dry clutch. IEEE VPPC 2012 – The 8th IEEE Vehicle Power and Propulsion Conference, Seoul, Korea.

Nakayama S, Fukuma T, Matsunaga A, Miyake T, and Wakimoto T. 2003. A new dynamic combustion control method based on charge oxygen concentration for diesel engines. SAE Technical Paper 2003-01-3181.

Nasser S and Playfoot B. 1999. A turbocharger selection computer model. SAE Technical paper 1999-01-0559.

Nielsen L and Sandberg T. 2003. A new model for rolling resistance of pneumatic tires. *SAE Transactions, Journal of Passenger Cars: Mechanical Systems* 111(6): 1572–1579.

Nieuwstadt M, Moraal P, Kolmanovsky I, Stefanopoulou A, Wood P, and Widdle M. 1998. Decentralized and multivariable designs for EGR–VGT control of a diesel engine IFAC Workshop, Advances in Automotive Control.

Nilsson Y. 2007. Modelling for fuel optimal control of a variable compression engine. PhD thesis. Linköpings Universitet.

Nilsson Y and Eriksson L. 2001. A new formulation of multi-zone combustion engine models 3rd IFAC Workshop "Advances in Automotive Control." Preprints, pp. 379–387. Elsevier Science, Karlsruhe, Germany.

Nilsson Y, Eriksson L, and Gunnarsson M. 2008. Torque modeling for optimising fuel economy in variable compression engines. *International Journal of Modeling, Identification and Control* 3(3).

Öberg P. 2009. A DAE formulation for multi-zone thermodynamic models and its application to CVCP engines. PhD thesis. Linköpings Universitet.

Öberg P and Eriksson L. 2006. Control oriented modeling of the gas exchange process in variable cam timing engines. SAE Technical Paper 2006-01-0660.

Onder CH, Roduner CA, and Geering HP. 1997. Model identification for the A/F path of an SI engine. SAE Technical Paper 970612.

Ortner P and del Re L. 2007. Predictive control of a diesel engine air path. *IEEE Transactions on Control Systems Technology* 15(3): 449–456.

Pacejka HB. 2002. *Tire and Vehicle Dynamics*. SAE.

Patton KJ, Nitschke RG, and Heywood JB. 1989. Development and evaluation of a friction model for spark ignition engines. SAE Technical Paper 890836.

Pavković D, Deur J, Jansz M, and Perić N. 2006. Adaptive control of automotive electronic throttle. *Control Engineering Practice* 14: 121–136.

Pettersson M and Nielsen L. 1997. Driveline modeling and RQV control with active damping of vehicle shuffle. SAE Technical Paper 970536.

Pettersson M and Nielsen L. 2000. Gear shifting by engine control. *IEEE Transactions Control Systems Technology* 8(3): 495–507.

Pettersson M and Nielsen L. 2003. Diesel engine speed control with handling of driveline resonances. *Control Engineering Practice* 11(10): 319–328.

Pettersson M, Nielsen L and Hedström LG. 1997. Transmission-torque control for gear-shifting with engine control. SAE Technical Paper 970864.

Peyton Jones J. 2003. Modeling combined oxygen storage and reversible deactivation dynamics for improved emissions predictions. SAE Technical Paper 2003-01-0999.

Peyton Jones J, Spelina JM, and Fray J. 2013. Likelihood-based control of engine knock. *IEEE Transactions on Control Systems Technology*: doi:10.1109/TCST.2012.2229280.

Powell J. 1993. Engine control using cylinder pressure: Past, present, and future. *Journal of Dynamic System, Measurement, and Control* 115: 343–350.

Rajamani R. 2005. Control of a variable-geometry turbocharged and wastegated diesel engine. *Proceedings of the Institution of Mechanical Engineers, Part D: Journal of Automobile Engineering* 219: 1361–1368.

Rassweiler GM and Withrow L. 1938. Motion pictures of engine flames correlated with pressure cards. SAE Technical Paper 380139.

Reinmann R, Saitzkoff A, and Mauss F. 1997. Local air-fuel ratio measurements using the spark plug as an ionization sensor. SAE Technical Paper 970856.

Reiter R. 1987. A theory of diagnosis from first principles. *Artificial Intelligence* 32(1): 57–95.

Richter JM, Klingmann R, Spiess S, and Wong KF. 2012. Application of catalyzed gasoline particulate filters to GDI vehicles. SAE Technical Paper 2012-01-1244.

Ringler J, Seifert M, Guyotot V and Hübner W. 2009. Rankine cycle for waste heat recovery of IC engines SAE. SAE Technical Paper 2009-01-0174.

Rivera DE, Morari M, and Skogestad S. 1986. Internal model control. 4. PID controller design. *Ind. Eng. Chem. Process. Des. Dev.* 25: 252–265.

Roduner CA, Onder CH, and Geering HP. 1997. Automated design of an air/fuel controller for an SI engine considering the three-way catalytic converter in the H1 approach. *Proceedings of the 5th Mediterranean Conference on Control and Systems*.

Roth JA and Guzzella L. 2010. Modelling engine and exhaust temperatures of a mono-fuelled turbocharged compressed-natural-gas engine during warm-up. *Proceedings of the Institution of Mechanical Engineers, Part D: Journal of Automobile Engineering* 224(1): 99–115.

Rousseau A, Sharer P. and Besnier F. 2004. Feasibility of reusable vehicle modeling: Application to hybrid vehicle. SAE Technical Paper 2004-01-1618.

Rückert J, Richert F, Schloßer A, Abel D, Herrmann O, Pischinger S, and Pfeifer A. 2004. A model based predictive attempt to control boost pressure and EGR–rate in a heavy duty diesel engine. *IFAC Symposium on Advances in Automotive Control*, Salerno, Italy.

Rückert J, Schloßer A, Rake H, Kinoo B, Krüger M, and Pischinger S. 2001. Model based boost pressure and exhaust gas recirculation rate control for a diesel engine with variable turbine geometry. IFAC Workshop: Advances in Automotive Control, Karlsruhe, Germany.

Saitzkoff A, Reinmann R and Mauss F. 1997. In cylinder pressure measurements using the spark plug as an ionization sensor. SAE Technical Paper 970857.

Saitzkoff A, Reinmann R, Berglind T, and Glavmo M. 1996. An ionization equilibrium analysis of the spark plug as an ionization sensor. SAE Technical Paper 960337.

Sampath M, Sengupta R, Lafortune S, Sinnamohideen K, and Teneketzis D. 1995. Diagnosability of discrete-event systems. *IEEE Transactions on Automatic Control* 40(9): 1555–1575.

Sawamoto K, Kawamura Y, Kita T, and Matsushita K. 1987. Individual cylinder knock control by detecting cylinder pressure. SAE Technical Paper 871911.

Scarpati J, Wikström A, Jönsson O, Glav R, Händel P, and Hjalmarsson H. 2007. Prediction of engine noise using parameterized combustion pressure curves. SAE Technical Paper 2007-01-2373.

Scattolini R, Siviero C, Mazzucco M, Ricci S, Poggio R. and Rossi C. 1997. Modeling and identification of an electromechanical internal combustion engine throttle body. *Control Engineering Practice* 5(9): 1253–1259.

Schick W, Onder C, and Guzzella L. 2011. Robustness analysis of a Fourier-based strategy for cylinder-individual air-fuel ratio control. *IMechE Part D - Journal of Automobile Engineering* 225: 1671–1682.

Schmidt FW, Henderson RE, and Wolgemuth CH. 1993. Introduction to Thermal Sciences 2nd edn. John Wiley & Sons.

Schwaderlapp M, Habermann K, and Yapici KI. 2002. Variable compression ratio – a design solution for fuel economy concepts. SAE Technical Paper 2002-01-1103.

Sellnau MC and Matekunas FA. 2000. Cylinder-pressure-based engine control using pressure-ratio-management and low-cost non-intrusive cylinder pressure sensors. SAE Technical Paper 2000-01-0932.

Shaaban S. 2004. Experimental investigation and extended simulation of turbocharger non-adiabatic performance. PhD thesis. Leibniz Universität Hannover.

Shayler P, Chick J and Ma T. 1997. Correlation of engine heat transfer for heat rejection and warm-up modeling. SAE Technical Paper 971851.

Shayler P, Harb C, and Ma T. 1995. Time dependent behaviour of heat transfer coefficients for exhaust systems. *VTMS 2 Conference Proceedings*, pp. 195–205. Paper C496/046/95. IMechE.

Shayler P, Hayden D, and Ma T. 1999. Exhaust system heat transfer and catalytic converter performance. SAE Technical Paper 1999-01-0453.

Shiao Y and Moskwa JJ. 1996. Model-based cylinder-by-cylinder air-fuel ratio control for si engines using sliding observers. IEEE International Conference on Control Applications, pp. 347–354.

Shimasaki Y, Kanehiro M, Baba S, Maruyama S, Hisaki T, and Miyata S. 1993. Spark plug voltage analysis for monitoring combustion in an internal combustion engine. SAE Technical Paper 930461.

Sihling K and Woschni G. 1979. Experimental investigation of the instantaneous heat transfer in the cylinder of a high speed diesel engine. SAE Technical Paper 790833.

Sirakov B and Casey M. 2011. Evaluation of heat transfer effects on turbocharger performance. Proceedings of ASME Turbo Expo GT2011-45887, Vancouver, Canada.

Sokolov A and Glad T. 1999. Identifiability of turbocharged IC engine models. SAE Technical Paper 1999-01-0206.

Soltic P. 2000. Part-load optimized SI engine systems. PhD thesis. Swiss Federal Institute of Technology, Zürich.

Sorenson SC, Hendricks E, Magnusson S, and Bertelsen A. 2005. Compact and accurate turbocharger modelling for engine control. SAE Technical Paper 2005-01-1942.

Spring P, Onder C, and Guzzella L. 2007. Optimized control of a pressure-wave supercharger: A model-based feedforward approach. *IEEE Transactions on Control Systems Technology* 15(3): 457–464.

Stefanopoulou A and Kolmanovsky I. 1999. Analysis and control of transient torque response in engines with internal exhaust gas recirculation. *IEEE Transactions on Control Systems Technology* 7(5): 555–565.

Stefanopoulou A, Cook J, Grizzle J and Freudenberg J. 1998. Control-oriented model of a dual equal variable cam timing spark ignition engine. *ASME Journal of Dynamic Systems Measurement and Control* 120: 257–266.

Stefanopoulou AG, Kolmanovsky I, and Freudenberg JS. 2000. Control of variable geometry turbocharged diesel engines for reduced emissions. *IEEE Transactions on Control System Technology* 8(4): 733–745.

Stewart G, Borrelli F, Pekar J, Germann D, Pachner D, and Kihas D. 2010. Toward a systematic design for turbocharged engine control. In: *Automotive Model Predictive Control: Models, Methods and Applications*, L. del Re, F. Allgöwer, L. Glielmo, C. Guardiola, and I. Kolmanovsky (eds.). Springer Verlag. pp. 59–79.

Stivender DL. 1968. Intake valve throttling (IVT) – a sonic throttling intake valve engine. SAE Technical Paper 680399.

Stivender DL. 1978. Engine air control – basis of a vehicular systems control hierarchy. SAE Technical Paper 780346.

Stobart R and Milner D. 2009. The potential for thermo-electric regeneration of energy in vehicles. SAE Technical Paper 2009-01-1333.

Stöckli M. 1989. Reibleistung von 4-Takt Verbrennugnsmotorenx. Technical report, Internal report of the Laboratory of Internal Combustion Engines, Swiss Federal Institute of Technology, Zürich.

Stone R. 1999. *Introduction to Internal Combustion Engines* 3rd edn. Macmillan.

Sundström C, Frisk E and Nielsen L. 2013. Selecting and utilizing sequential residual generators in FDI applied to hybrid vehicles. *IEEE Transactions on Systems, Man and Cybernetics, Part A: Systems and Humans*.

Svärd C and Nyberg M. 2010. Residual generators for fault diagnosis using computation sequences with mixed causality applied to automotive systems. *IEEE Transactions on Systems, Man and Cybernetics, Part A: Systems and Humans* 40(6): 1310–1328.

Svärd C, Nyberg M, Frisk E, and Krysander M. 2011. Residual evaluation for fault diagnosis by data-driven analysis of non-stationary probability distributions. Proceedings of the 50th IEEE Conference on Decision and Control and European Control Conference (CDC-ECC 2011), pp. 95–102.

Svärd C, Nyberg M, Frisk E, and Krysander M. 2013. Automotive engine FDI by application of an automated model-based and data-driven design methodology. *Control Engineering Practice* 21(4): 455–472.

Taylor C and Toong T. 1957. Heat transfer in internal combustion engines. ASME Paper 57-HT-17.

Taylor CF. 1985. The Internal Combustion Engine in Theory and Practice, Vol 1 & 2. The MIT Press.

Teng H, Regner G, and Cowland C. 2007a. Waste heat recovery of heavy-duty diesel engines by organic rankine cycle part I: Hybrid energy system of diesel and rankine engines. SAE Technical Paper 2007-01-0537.

Teng H, Regner G and Cowland C. 2007b. Waste heat recovery of heavy-duty diesel engines by organic rankine cycle part II: Working fluids for WHR-ORC. SAE Technical Paper 2007-01-0543.

SAE. 1995. SAE J1826 – Turbocharger Gas Stand Test Code 1995 SAE Standard.

SAE. 1995. SAE J922 – Turbocharger Nomenclature and Terminology 1995 SAE Standard.

Thomasson A and Eriksson L. 2011. Model-based throttle control using static compensators and pole placement. *Oil & Gas Science and Technology* 66(4): 717–727.

Thomasson A, Eriksson L, Leufven O, and Andersson P. 2009. Wastegate actuator modeling and model-based boost pressure control. IFAC Workshop on Engine and Powertrain Control, Simulation, and Modeling, Paris, France.

Thomasson A, Eriksson L, Lindell T, Peyton Jones J, Spelina J. and Frey J. 2013. Tuning and experimental evaluation of a likelihood-based engine knock controller. 52nd IEEE Conference on Decision and Control, Florence, Italy.

Turns SR. 2000. *An Introduction to Combustion – Concepts and Applications Mechanical Engineering Series* 2nd edn. McGraw Hill.

Tuttle JH. 1980. Controlling engine load by means of late intake-valve closing. SAE Technical Paper 800794.

Tuttle JH. 1982. Controlling engine load by means of early intake-valve closing. SAE Technical Paper 820408.

van Nieuwstadt M, Kolmanovsky I, Moraal P, Stefanopoulou A, and Jankovic M. 2000. EGR VGT control schemes: Experimental comparison for a high-speed diesel engine. *IEEE Control Systems Magazine*: 63–79.

Vasca F, Iannelli L, Senatore A, and Reale G. 2011. Torque transmissibility assessment for automotive dry-clutch engagement. *IEEE/ASME Transactions on Mechatronics* 16(3): 564–573.

Vašak M, Baotíc M, Morari M, Petrovíc I and Períc N. 2006. Constrained optimal control of an electronic throttle. *International Journal of Control* 79: 465–478.

Velardocchia M, Ercole G, Mattiazzo G, Mauro S, and Amisano F. 1999. Diaphragm spring clutch dynamic characteristic test bench. SAE Technical Paper 1999-01-0737.

Vibe I. 1970. *Brennverlauf und Kreisprocess von Verbennungsmotoren*. VEB Verlag Technik Berlin. German translation of the Russian original.

Vigild CW. 2001. The internal combustion engine – modeling, estimation and control issues. PhD thesis. Technical University of Denmark.

Wahlström J. 2006. Control of EGR and VGT for emission control and pumping work minimization in diesel engines. Technical report. Licentiate thesis, Linköping University.

Wahlström J. 2009. Control of EGR and VGT for emission control and pumping work minimization in diesel engines. PhD thesis. Linköping University.

Wahlström J and Eriksson L. 2011a. Modelling diesel engines with a variable-geometry turbocharger and exhaust gas recirculation by optimization of model parameters for capturing non-linear system dynamics. *Proceedings of the Institution of Mechanical Engineers, Part D, Journal of Automobile Engineering* 225(7): 960–986.

Wahlström J and Eriksson L. 2011b. Nonlinear EGR and VGT control with integral action for diesel engines. *Oil & Gas Science and Technology* 66(4): 573–586.

Wahlström J and Eriksson L. 2013. Output selection and its implications for MPC of EGR and VGT in diesel engines. *IEEE Transactions on Control Systems Technology* 21(3): 932–940.

Wahlström J, Eriksson L, and Nielsen L. 2008. Controller tuning based on transient selection and optimization for a diesel engine with EGR and VGT electronic engine controls. SAE Technical Paper 2008-01-0985.

Wahlström J, Eriksson L, and Nielsen L. 2010. EGR-VGT control and tuning for pumping work minimization and emission control. *IEEE Transactions on Control Systems Technology* 18(4): 993–1003.

Wang Y, Megli T, Haghgooie M, Peterson KS, and Stefanopoulou AG. 2002. Modeling and control of electromechanical valve actuator. SAE Technical Paper 2002-01-1106.

Wang YY, Krishnaswami V, and Rizzoni G. 1997. Event-based estimation of indicated torque for IC engines using sliding-mode observers. *IFAC Control Engineering Practice* 5(8): 1123–1129.

Watson N and Janota M. 1982. *Turbocharging the Internal Combustion Engine.* The Macmillan Press Ltd.

Weber F, Guzzella L, and Onder C. 2002. Modeling of a pressure wave supercharger including external exhaust gas recirculation. *IMechE, Journal of Automobile Engineering* 216(3): 217–235.

Wendland DW. 1993. Automobile exhaust-system steady-state heat transfer. SAE Technical Paper 931085.

Wimmer A, Pivec R, and Sams T. 2000. Heat transfer to the combustion chamber and port walls of IC engines - measurement and prediction. SAE Technical Paper 2000-01-0568.

Wipke KB, Cuddy MR, and Burch SD. 1999. Advisor 2.1: A user-friendly advanced powertrain simulation using a combined backward/forward approach. *IEEE Transactions on Vehicular Technology* 48(6): 1751–1761.

Woermann RJ, Theuerkauf HJ, and Heinrich A. 1999. A real-time model of a common rail diesel engine. SAE Technical Paper 1999-01-0862.

Wong J. 2001. *Theory of Ground Vehicles* 3rd edn. John Wiley & Sons.

Worm J. 2005. The effect of exhaust variable cam phaser transients on equivalence ratio control in an SI 4 stroke engine. SAE Technical Paper 2005-01-0763.

Woschni G. 1967. A universally applicable equation for the instantaneous heat transfer coefficient in the internal combustion engine. SAE Technical Paper 670931.

Wu B, Filipi Z, Prucka R, Kramer D, and Ohl G. 2007. A simulation-based approach for developing optimal calibrations for engines with variable valve actuation. *Oil & Gas Science and Technology* 62(4): 539–553.

Yildiz Y, Annaswamy A, Yanakiev D, and Kolmanovsky I. 2008. Adaptive air fuel ratio control for internal combustion engines. American Control Conference, Seattle, Washington.

Zapf H. 1969. Beitrag zur untersuchung des wärmeüberganges wärend des ladungswecsels im viertakt-dieselmotor. *MTZ Motortechnische Zeitschrift* 30(12): 461–465.

Zeldovich YB. 1946. The oxidation of nitrogen in combustion and explosions. *Acta Physicochimica* 21(4): 577–628.

Zhan R, Eakle ST, and Weber P. 2010. Simultaneous reduction of PM, HC, CO and NOx emissions from a GDI engine, SAE Technical Paper 2010-01-0365.

Zhao H (ed.) 2010a. *Advanced Direct Injection Combustion Engine Technologies and Development – Volume 1: Gasoline and Gas Engines.* Woodhead Publishing Limited.

Zhao H (ed.) 2010b. *Advanced Direct Injection Combustion Engine Technologies and Development – Volume 2: Diesel Engines.* Woodhead Publishing Limited.

Zhao Y and Winterbone D. 1993. A study of warm-up processed and si engine exhaust systems. SAE Technical Paper 931094.

Index

Modeling and Control of Engines and Drivelines, First Edition. Lars Eriksson and Lars Nielsen.
© 2014 John Wiley & Sons, Ltd. Published 2014 by John Wiley & Sons, Ltd.
Companion Website: www.wiley.com/go/powertrain

Printed in the United States
by Bookmasters

Printed in the United States
By Bookmasters